PLANT NUTRITION – MOLECULAR BIOLOGY AND GENETICS

Plant Nutrition – Molecular Biology and Genetics

Proceedings of the Sixth International Symposium on Genetics and Molecular Biology of Plant Nutrition

Edited by

G. Gissel-Nielsen

Risø National Laboratory,
Roskilde, Denmark

and

A. Jensen

Risø National Laboratory,
Roskilde, Denmark

SPRINGER-SCIENCE+BUSINESS MEDIA, B.V.

A C.I.P. Catalogue record for this book is available from the Library of Congress.

ISBN 978-90-481-5225-4 ISBN 978-94-017-2685-6 (eBook)
DOI 10.1007/978-94-017-2685-6

Printed on acid-free paper

Originally published by Kluwer Academic Publishers in 1999

Contents

Genetical background for plant uptake: Breeding

Preface

The sixth International Symposium on Genetics and Molecular Biology of Plant Nutrition was held in Elsinore, Denmark from August 17-21, 1998 and organised by Risø National Laboratory in the year of its 40th anniversary. The 98 participants represented 23 countries and 80 scientific contributions with 43 oral and 37 poster presentations.

The symposium addressed the molecular mechanisms, physiology and genetic regulation of plant nutrition. The Symposium brought together scientists from a range of different disciplines to exchange information and ideas on the molecular biology of mineral nutrition of plants. The symposium emphasised:

- Bridging the gab between molecular biology, applied genetics, plant nutrition and plant breeding.
- The development of methodologies to improve the efficiency and effectiveness of nutrition of plants
- Quality of plant products.

With sessions on: Nitrogen; Phosphorous; Micronutrients; Symbiosis; Membranes; Stress; Heavy Metals and Plant Breeding.

In comparison with the previous conferences in this series more emphasis was placed on use of molecular techniques to clarify physiological mechanisms and processes, gene expression and regulation, as well as genetic marker assisted analysis. Significant progress was reported in exploitation of molecular genetic markers and other biotechnologies in breeding programmes.

I wish to thank the contributors for the high scientific standard of their presentations and for thorough preparation of their manuscripts. The serious attempts to reach the high scientific standard and to meet the suggestions by the referees are highly appreciated. I am grateful to the International Council of Genetics and Molecular Biology of Plant Nutrition and members of the Organising Committee for their generous support and co-operation, from the planning of the scientific programme to holding the symposium.

Members of the International Council: N. El Bassam, B. Berkasem, K. G. Briggs, G. Ferrari, T. Fujiwara, R. R. Duncan, J. Dunlop, R. D. Graham, W. J. Horst, P. G. C. Kuiper, B. C. Loughman, Y. Masaoka, P. J. Randall, D. W. Rains, V. Römheld, P. N. Takkar and R. M. Welch.

Members of the Organizing Committee: H. Giese, K. Hjortsholm, P. B. Holm, A. Jensen, G. Gissel-Nielsen and J. Schjørring.

The Organising Committee expresses its gratitude to Risø National Laboratory, the technical staff at The Plant Biology and Biogeochemistry Department and other people who helped to make the symposium a success. We thank the Danish Veterinary- and Agricultural Research Council, Norsk Hydro A/S, KEMIRA A/S, ENERO and Risø National Laboratory for financial support.

Arne Jensen

Introduction

For many years the scientific disciplines Plant Nutrition and Plant Genetics developed side by side, but with very little interaction. Genetically studies of plants were carried out at different nutrient levels, and nutrition studies were carried out with different plant species or cultivars. In both cases differences in the growth of plants were accepted as a result of interaction between the genetic and the environmental factors. –And that was it.

Very little was achieved concerning improving the plant utilisation of nutrients in the soil by combining the two disciplines. Among the existing lines of crop plants the best species or cultivars for utilisation of a particular nutrient were found, and by using conventional plant breeding methods some improvement was achieved. However, it was to a great extend like shooting in the dark: Sometimes you had a hit, but you did not really know why.

An important break came with the development of biotechnological knowledge and methods. By using DNA-marker techniques it is possible to identify the complex interactions and to localise the genes behind the processes and enzymes responsible for the plant influence on nutrient availability in the soil. This opens up for a much more specific effort to combine the most effective genes in one plant. For that purpose a number of methods can be used including modern techniques such as gene transformation and marker assisted selection.

A great deal of the frontier research to day takes place in the cutting edge between the conventional scientific disciplines. This does not diminish the importance for basic research in these disciplines. On the contrary. To describe the interface you need to know exactly what it interfaces. This counts for the actual subjects, too. Therefore, research in pure plant nutrition as well as in pure plant genetics and plant molecular biology are prerequisites for the main subject of this proceeding: The interaction between these disciplines.

This is the background for this series of meetings of which the Elsinore meeting in 1998 was number 6. This is also the reason why the contributions to the proceedings from these meetings cover the range from pure plant nutrition to pure plant genetics.

Gunnar Gissel-Nielsen

INORGANIC NITROGEN ABSORPTION BY PLANT ROOTS:

Physiology and Molecular Biology

ANTHONY D.M.GLASS[1], YAIR ERNER[2], TAMERA HUNT[1], HERBERT J. KRONZUCKER[1], MAMORU OKAMOTO[1], SUMAN RAWAT[1], SALIM SILIM[1], JAN K. SCHJOERRING[3], M. YAEESH SIDDIQI[1], J. JOHN VIDMAR[1], M.Y. WANG[4] AND DEGEN ZHUO[1]

[1]University of British Columbia, 6270 University Boulevard, Vancouver, British Columbia, V6T 1Z4, Canada; [2]Department of Citriculture, ARO, The Volcani Center, Beth Dagan, Israel; [3]Plant Nutrition Laboratory, Department of Agricultural Sciences, Royal Veterinary and Agricultural University, Copenhagen, Denmark; [4]Ag. Biotech Center, Rutgers University, Newbrunswick, NJ, U.S.A.

1. Abstract

Physiological experiments have established that plant roots possess multiple transport systems for nitrogen absorption. These include at least three active transporter systems for nitrate absorption, and two transporters, one active and one passive, for ammonium absorption. In addition, there exist amino acid transporters, although their importance for nitrogen acquisition in agricultural soils is uncertain. The integration of these transport systems, in the context of satisfying whole plant demand for nitrogen, requires mechanisms for root to shoot signaling as well as signal transduction within the root. Evidence from physiological studies has been interpreted in terms of a nitrogen cycling model, in which reduced nitrogen, in the form of amino acids, returns to the root via the phloem to signal plant demand for nitrogen uptake. Thus far, details of the mechanisms of these putative regulatory processes have not been established.

The recent cloning of genes encoding the low- and high-affinity nitrate transporters (*Nrt1* and *Nrt2*, respectively), as well as the high affinity ammonium transporter

G. Gissel-Nielsen and A. Jensen (eds.), Plant Nutrition – Molecular Biology and Genetics, 1–16.

(*Amt1*), now provides the opportunity to investigate the proposed regulatory mechanisms at the molecular level. Both the uptake of nitrate and ammonium by plant roots and the levels of transcripts for *Nrt1, Nrt2* and *Amt1* respond rapidly to perturbations of nitrogen supply, there being clear evidence for control of gene function at the transcriptional level. In the case of nitrate uptake, both *Nrt1* and *Nrt2* are nitrate-inducible, while down-regulation of *Nrt2* appears to be determined by tissue levels of ammonium and amino acids, notably glutamine. In the case of ammonium uptake, physiological evidence is consistent with effects of cellular levels of ammonium, glutamine, and/or other amino acids, in regulating influx.

2. Introduction

Plant roots are exposed to diverse forms of nitrogen in soil solution (Barber, 1995). In warm aerobic agricultural soils, nitrate (NO_3^-) and ammonium (NH_4^+) are present at millimolar concentrations, and represent the major available forms for plant uptake (Reisenauer, 1966; Barber, 1995). Reisenauer examined 879 agricultural soil samples and reported values for soil solution NO_3^- that varied by orders of magnitude, with a median value around 7 mM. In a compilation of 35 agricultural soil solution analyses, the average NO_3^- concentration was 6.0 mM, compared to 0.8 mM for NH_4^+(Wolt, 1994). This is typical of such soils, and has led to the presumption that NO_3^- is the principal source of nitrogen for plant growth (Schrader and Thomas, 1981). However, several points are worthy of emphasis in this regard. First, soil solution NH_4^+ concentration may represent only ~10 % of that held by adsorption on anionic soil colloids; as soil solution is depleted by plant uptake, this reserve of NH_4^+ serves to buffer soil solution (Barber, 1995). By contrast, there is little adsorption of NO_3^- and thus NO_3^- is readily leached from agricultural soils. Second, a value of 0.8 mM represents a saturating concentration for high-affinity NH_4^+ uptake (Becking, 1956; Ullrich et al., 1984; Wang et al., 1993). Thirdly NH_4^+ fluxes are typically as high as or higher than corresponding NO_3^- fluxes, and NH_4^+ is typically inhibitory to NO_3^- influx, so that the absorption of N from mixtures of NO_3^- and NH_4^+ commonly favours NH_4^+ over NO_3^- (Glass and Siddiqi, 1995). In a commercial tomato crop, we have recently observed NH_4^+ uptake rates that were equivalent to those of NO_3^- (Siddiqi and Glass, unpublished), despite the fact that the concentration of NO_3^- in the feed solution was 10 times that of NH_4^+. Thus, the importance of NH_4^+ as a source of N for the growth of crop species may have been substantially underestimated.

In natural, as in cultivated soils, heterogeneity of nutrient concentrations is the norm. For example, measurements of soil nutrient concentrations, in 362 samples taken from a 10 by 12 m grid in a native sagebrush steppe in Utah, revealed similar values for NO_3^- and NH_4^+, which varied over 3 orders of magnitude (Jackson and Caldwell, 1993). In other habitats, e.g. many mature forests (Stark and Hart, 1997) and cold tundra soils (Kielland, 1994), NO_3^- may be undetectable and NH_4^+ and amino acids are probably more important sources of N. As well as this spatial and habitat heterogeneity, there is considerable seasonal variation associated with microbial and plant activity (Mengel and Kirkby, 1979), while major soil disturbances such as fire, and clear-cut harvesting are followed by dramatic changes in the balance between the different N forms (Likens et al., 1969). Many plant species have adapted to this heterogeneity by becoming specialists, favouring one form of N over another (Lavoie et al., 1992). Thus *Picea glauca*, a late successional forest species, favours NH_4^+ over NO_3^-, both in uptake and assimilation (Kronzucker et al., 1997). At the other end of this continuum, species such as radish and tomato experience nutrient imbalances and growth inhibition when grown with even modest levels of NH_4^+ (Goyal et al., 1982; Magalhaes and Wilcox, 1984). Many early-successional forest species and crop species are opportunist, capable of absorbing and assimilating NO_3^- or NH_4^+ equally well (Bloom, 1988; Min et al., 1998).

In the context of such heterogeneity of supply, it is crucial that N transport systems be regulated so as to respond rapidly to fluctuations of N availability. Furthermore, when plants are simultaneously assimilating different forms of N, there is a need to integrate signals from several potential N pools in order to regulate N uptake, and co-ordinate root N absorption so as to satisfy shoot and whole-plant demand. One current model, (Fig. 1) based on physiological observations (Muller and Touraine, 1992; Lee et al.,1992), suggests that this is achieved through the cycling and recycling of amino acids between shoots and roots and their negative feedback inhibition of nitrogen fluxes (Cooper and Clarkson, 1989; Muller and Touraine, 1992; Marschner et al., 1997). Nevertheless, specific details of the mechanisms involved in the model are lacking. At the same time, studies using nitrate reductase mutants, and methionine sulfoximine (MSX) pretreatments to relieve putative repression of transporter gene expression by glutamine, or other amino acids, have failed to confirm the anticipated up-regulation of N uptake (King et al., 1993; Causin and Barneix, 1993; Feng et al.,1994; Ryan and Walker, 1994). As a consequence, root cytoplasmic NO_3^- and NH_4^+ have been proposed as participants in the regulation of N fluxes.

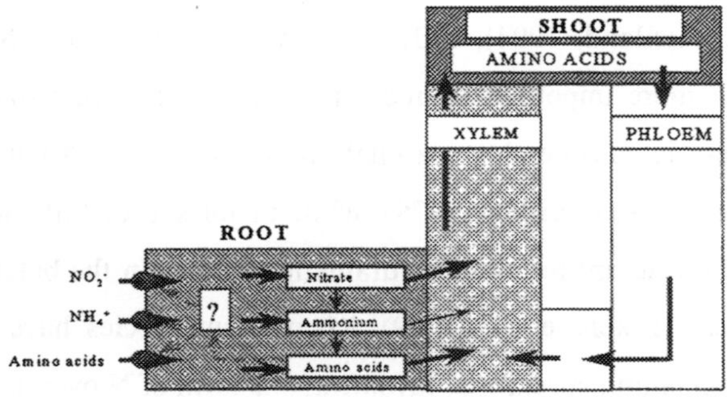

Figure 1. Model for regulation of N fluxes by recycling of amino N (based on Cooper and Clarkson, 1989). Solid lines represent nitrogen fluxes; dotted lines represent feedback processes.

The recently cloned genes encoding high-affinity transport systems for NO_3^- and NH_4^+ now provide an opportunity to address these questions and to test some aspects of the model at the molecular level. The remainder of this paper outlines what is presently understood of the physiological and molecular characterization of transport systems responsible for NO_3^- and NH_4^+ uptake by plant roots, and their regulation so as to satisfy plant demand in the face of the noted heterogeneity of supply.

3. Nitrate Transport

3.1. CONSTITUTIVE HIGH-AFFINITY NITRATE TRANSPORT

Physiological studies, using $^{14}NO_3^-$ to measure net uptake, and $^{13}NO_3^-$ to measure unidirectional influx, have revealed the presence of a constitutive high-affinity, low-capacity transport system (designated CHATS), in barley (Behl et al., 1988; Lee and Drew, 1986; Siddiqi et al., 1990; Aslam et al., 1992) in *Arabidopsis* (Wang and Crawford, 1996) and in white spruce (Kronzucker et al., 1995). This transporter mediates the active transport of NO_3^- into the root cytosol, leading to the induction of a high-capacity, high-affinity NO_3^- transport system (IHATS). Recently, Wang and Crawford (1996) have isolated an *Arabidopsis* mutant (*chl8*), defective in CHATS through chlorate selections. However, no genes encoding this transporter have yet been cloned. Despite its designation as a constitutive transporter, CHATS activity is increased ~ 3 fold following exposure to NO_3^- (Aslam et al., 1992; Kronzucker et al., 1995).

3.2. INDUCIBLE NITRATE TRANSPORT

After 18 h of exposure to NO_3^-, $^{13}NO_3^-$ influx through the IHATS in *Klondike* barley was 30 times that of the pre-induced CHATS (Siddiqi et al., 1990), while in *Steptoe* barley CHATS activity was relatively high before exposure to NO_3^-, and exposure to NO_3^- failed to produce such a large increase of NO_3^- influx (King et al., 1993). Nevertheless, plants typically respond rapidly to variations of N supply (Jackson et al., 1976; Lee, 1982; Siddiqi et al., 1990; Rodgers and Barneix, 1993). The IHATS brings about thermodynamically active transport of nitrate, generally considered to be a proton cotransport system (Ullrich and Novacky, 1981; Glass and Siddiqi, 1995; Miller and Smith, 1996). A characteristic feature of the induction of the IHATS is that it overshoots steady-state requirement for NO_3^- uptake and is then downregulated (Fig.2). The rapidity of induction and subsequent downregulation of IHATS is a function of the concentration of NO_3^- provided in barley (Siddiqi et al., 1989), as well as species and N preference. For example in white spruce full induction required 3 days

(Kronzucker et al., 1995); subsequent down regulation was also slow in this species. In barley $^{13}NO_3^-$ influx was positively- and negatively- correlated with root $[NO_3^-]$ during induction and down-regulation, respectively. However, during NO_3^- loading other N pools may also change and thus the correlation does not establish causal links. In separate experiments, using nitrate reductase mutants and by blocking NO_3^- assimilation through to amino acids with MSX, results consistent with, and in conflict with, a primary role of amino acids in regulating NO_3^- influx have been generated (Breteler and Siegerist, 1984; Ingemarsson et al., 1987; Lee et al., 1992; King et al.,1993). Thus, both amino acids and tissue NO_3^- have been advanced as responsible for regulating NO_3^- influx, and a consensus is lacking.

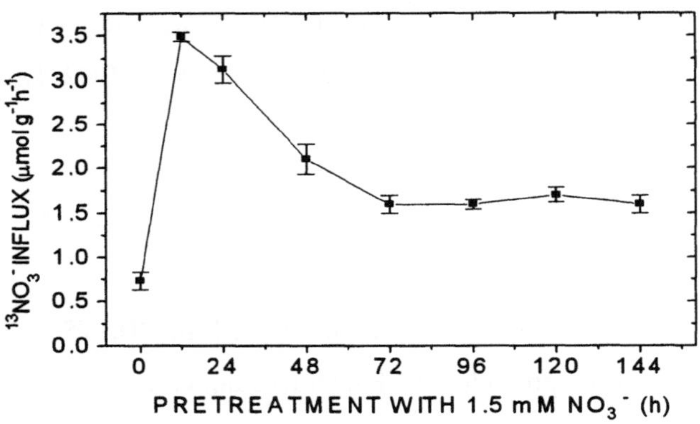

Figure 2. Induction and negative feedback inhibition of high-affinity $^{13}NO_3^-$ influx (measured at 0.1 mM NO_3^-) in *Klondike* barley.

The cloning of a high affinity transporter gene family from barley, *Nicotiana plumbaginofolia* and *Arabidopsis thaliana* (Trueman et al., 1996; Quesada et al., 1997; Zhuo et al., unpublished) now provides a means to explore the regulation of NO_3^- influx at the molecular level. The family of high-affinity NO_3^- transporters (*Nrt2*) are highly conserved from *Aspergillus nidulans* through *Chlamydomonas* to higher plants (Table 1).

	NRT2;2At[a]	NRT2;1Np[b]	BCH1[c]	BCH2[c]	NAR-3[d]	CRNA[e]	CRNB[f]	YNT1[g]
NRT2;1At	87.5	77	72.7	72.4	47	36.8	31.1	33.2
NRT2;2At		74	68.4	67.8	46.3	32	32.3	33.6
NRT2;1Np			71.4	70.2	47.8	32.5	31	32.2
BCH1				92.1	49	33.6	31.8	30.3
BCH2					48.3	32.5	32.8	31.6
NAR-3						31.7	32.7	31.9
CRNA							59.8	39.3
CRNB								40.3

Table 1: Sequence identities of predicted high-affinity nitrate transporters from different organisms: (a) NRT2;1At and NRT2;2At from *A. thaliana*; (Zhuo et al., unpublished) (b) NRT2;1Np from *N. plumbaginifolia* (Quesada et al., 1997), (c) BCH1 and BCH2 from *H. vulgare* (Trueman et al., 1996), (d) NAR-3 from *C. reinhardtii* (Quesada et al., 1994), (e) CRNA and (f) CRNB from *A. nidulans* (Unkles et al., 1991; Zhuo et al.,unpublished), and (g) YNT1 from *H.polymorpha* (Perez et al., 1997).

In *Arabidopsis* transcript abundance was strongly correlated with $^{13}NO_3^-$ influx during induction and subsequent down-regulation, while under steady-state conditions of NO_3^- supply transcript abundance and $^{13}NO_3^-$ influx were again correlated (Zhuo et al., unpublished). Under steady-state conditions of NO_3^- supply between 100 μM to 15 mM, *Nrt2* expression and $^{13}NO_3^-$ influx were inversely correlated with external NO_3^- provision. Metabolic inhibitors were employed to investigate which N pools might be responsible for downregulating *Nrt2* transcript levels. Table 2 shows transcript abundance in plants grown for 3 weeks in 1 mM ammonium citrate and then exposed to various treatments and inhibitors for 3 hours. The results of this experiment support the following conclusions : Treatment # (1): induction of the *Nrt2* gene depends upon exposure to NO_3^-; Treatment #(2): elevated levels of NO_3^- (2 mM KNO_3 compared to 1 mM KNO_3) caused down-regulation of *Nrt2* expression; Treatment #(3): elevated root $[NO_3^-]$, associated with tungstate treatment, is itself capable of reducing *Nrt2* expression; Treatment #(4): elevated root $[NH_4^+]$, associated with MSX treatment, reduced *Nrt2* expression to a greater extent than treatments 1-3; Treatment #(5): elevated [glutamine], associated with azaserine treatment, reduced *Nrt2* expression levels to the level of uninduced plants (Treatment # 6); Treatments 7 and 8, 1 mM NH_4NO_3 with or without MSX, confirm the potent effect of tissue [NH_4^+] in reducing *Nrt2* expression levels.

TREATMENT	***Nrt2;1At***
(1) 1 mM KNO_3	100
(2) 2mM KNO_3	74
(3) 1mM KNO_3 + 0.5 mM tungstate	58
(4) 2mM KNO_3 + 0.25 mM MSX	22
(5) 1mM KNO_3 + 0.5 mM AZA	13
(6) 1mM NH_4 citrate	9
(7) 1mM NH_4NO_3	50
(8) 1mM NH_4NO_3+ 0.25 mM MSX	10
(9) 1mM KNO_3 + 1 mM arginine	18
(10) 1mM KNO_3 + 1 mM asparagine	38
(11) 1mM KNO_3 + 1mM glutamine	77

Table 2. Nrt2;1At mRNA levels in *Arabidopsis* roots in response to 3h treatments with different N sources and various inhibitors of N assimilation. Prior to the 3 h treatments, plants were grown on NH_4^+ as sole source of N. *Nrt2;1At* mRNA levels were quantified by densitometric methods using the Molecular Analyst computer software and are adjusted according to the amounts of RNAs applied, based on ubiquitin mRNA levels. All values are expressed as a percent of the value obtained for plants pretreated with 1 mM KNO_3.

Exogenous application of arginine, asparagine or glutamine reduced *Nrt2* expression in the order arginine > asparagine > glutamine, however, when metabolites are supplied exogenously, it is unclear to what extent differential uptake and/or metabolism influence the results of such treatments. In summary, therefore, the *Arabidopsis* data indicate that *Nrt2* gene expression is induced by NO_3^-, and down-regulated by glutamine, NH_4^+ and, to a limited extent, by NO_3^-.

3.3 LOW AFFINITY NITRATE TRANSPORT

At nitrate concentrations beyond the level of IHATS saturation, a constitutive low-affinity nitrate transport system (LATS) is evident (see Siddiqi et al., 1990 for references). This transport system is also thermodynamically active and thought to be powered by a proton cotransport system (Glass et al., 1992). *Arabidopsis* mutants defective in this transport system were investigated by Doddema and Telkamp (1978), and a gene (*Nrt1*) encoding this transporter was cloned from *Arabidopsis* (Tsay et al., 1993) by selecting for chlorate resistance among T-DNA- tagged lines. Expression of this gene in *Xenopus* oocytes resulted in NO_3^- accumulation as well as the characteristic membrane depolarization associated with exposure to NO_3^-- containing solutions. LATS activity of a mutant line (*chl1-5*) with a major deletion in the *Nrt1* gene was shown to be roughly one third the rate of wild-type transport in plants grown on NH_4NO_3 (Touraine and Glass, 1997). However, no differences were observed when plants were grown on KNO_3. To explain these results, it has been proposed that two transport systems, encoded by *Nrt1* and its homologue, contribute to measured LATS activity (Huang et al., 1996). Furthermor, that differential regulation of these genes by NH_4^+ make the deletion of transport function evident only when NH_4^+ is present in the growth medium. Support for this hypothesis has been provided from studies of *Nrt1* homologues in tomato, from which two *LeNRT1* genes have been identified (Lauter et al., 1996). In barley roots LATS activity was shown to be present in plants grown without NO_3^- (Siddiqi et al.,1990; Glass et al., 1992). Yet, in *Arabidopsis* plants previously grown on ammonium as sole source of N, *Nrt1* expression required exposure to NO_3^- (Tsay et al., 1993). Clearly, more physiological, as well as molecular work is required to resolve these apparent problems.

4. Ammonium absorption

4.1. HIGH-AFFINITY AMMONIUM TRANSPORT

Influx of NH_4^+ at low external [NH_4^+] is thermodynamically active (Ullrich et al., 1984; Wang et al., 1994) and occurs via a saturable high-affinity transport system. In many species NH_4^+ uptake has been shown to be up-regulated in response to N-starvation, and down-regulated when N is resuppplied as NH_4^+ or various amino acids, particularly asparagine and glutamine (Lee and Rudge, 1986; Morgan and Jackson,

1988; Causin and Barneix, 1993; Wang et al; 1993; Kronzucker et al., 1996; Glass et al., 1997). Candidates for the regulation of NH_4^+ uptake include NH_4^+ itself (Morgan and Jackson, 1988; Ryan and Walker, 1994; Feng et al., 1994; Glass et al., 1997) as well as products of NH_4^+ assimilation such as glutamine, asparagine and other amino acids (Lee et al., 1992; Causin and Barneix; 1993; Feng et al., 1994; Glass et al., 1997). Using a *Saccharomyces* double mutant (*mep1 mep2*), defective in NH_4^+ transport, Ninemann et al., (1994) successfully cloned the *Amt1* gene, encoding a high-affinity NH_4^+ transporter from *Arabidopsis* by complementation of the yeast mutant with a plant cDNA. The plant gene has high sequence homology with that of the yeast Mep1 NH_4^+ transporter.

In previous studies of NH_4^+ influx, it has been ably demonstrated that influx is rapidly modulated, either up or down as the NH_4^+ supply is perturbed (Morgan and Jackson, 1988; Wang et al., 1993). The results of a study of NH_4^+ influx in rice during transfer from N-limited conditions to N-sufficiency and vice versa, in the presence or absence of MSX, suggested that NH_4^+ itself was incapable of down-regulating influx when N-starved plants were resupplied with NH_4^+ (Glass et al., 1997). Rather, it appeared that glutamine and possibly other amino acids were responsible for this down-regulation. However, MSX prevented the up-regulation of influx normally observed when N-sufficient rice plants are transferred to low-N conditions, consistent with a role for NH_4^+ itself in controlling the up-regulation of NH_4^+ influx. Clearly, this effect may be operating at the transcriptional level, at a post-transcriptional level or through direct effects on the NH_4^+ transporter itself. (Glass et al., 1997). Unfortunately, it is possible only to speculate regarding the underlying mechanisms responsible for the responses documented in these experiments. However, measurements of root NH_4^+ and glutamine concentrations during down-regulation of NH_4^+ influx in *Arabidopsis thaliana* in the presence and absence of MSX (Tables 3A, 3B) showed that blocking glutamine synthetase activity, by use of MSX, resulted in a 28-fold increase of root NH_4^+. Yet NH_4^+ influx in the presence of MSX, as was the case in our earlier rice experiments (Glass et al.,1997), was substantially higher in the MSX treatment, supporting the conclusion that glutamine, possibly in addition to other amino acids, and not NH_4^+ itself, is responsible for down-regulation of NH_4^+ influx. In agreement with this hypothesis, $^{13}NH_4^+$ influx values were strongly correlated with root glutamine levels (Fig. 3) during the resupply of NH_4^+ to N-deprived plants, while in the MSX-treated plants glutamine concentration declined very slightly by 9 h (Table 3B). This result is consistent with a primary role for glutamine in regulating $^{13}NH_4^+$ influx, but does not preclude direct effects of tissue $^{13}NH_4^+$ on transport activity.

Table 3A. Root NH_4^+ concentrations (μmol g^{-1})

Treatment	Time 0 h	Time (+9 h)
5 mM NH_4NO_3	0.35 ±0.01	1.60 ± 0.048 (5)
5 mM NH_4NO_3 + 1 mM MSX	0.35 ± 0.01	9.70 ± 1.6 (28)

Table 3B. Root glutamine concentrations (μmol g^{-1})

Treatment	Time 0 h	Time (+9 h)
5 mM NH_4NO_3	2.58 ± 0.7	18.30 ± 1.0 (7)
5 mM NH_4NO_3 + 1 mM MSX	2.58 ± 0.7	2.41 ± 0.7 (0.9)

Table 3A: Root ammonium concentrations, and 3B: Root glutamine concentrations at time 0 and 9 h after transfer of low-N plants to 5 mM NH_4NO_3. Means ± standard deviations are based on 4 replicates. Numbers in parentheses represent relative increase/decrease, over zero time values of root [NH_4^+] or [gln].

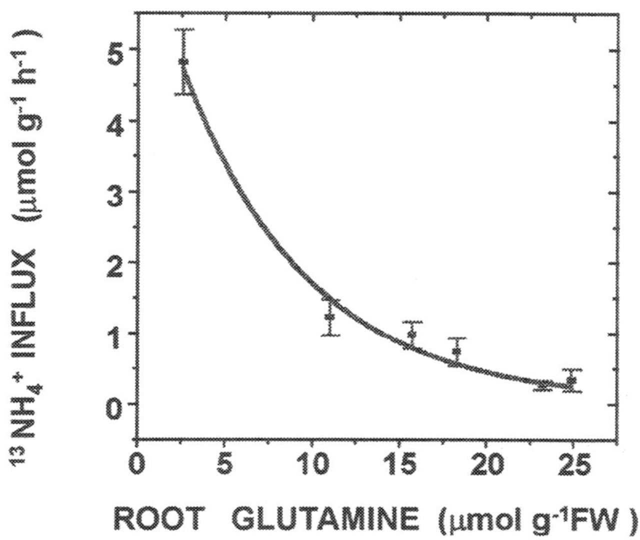

Figure 3. Root glutamine concentrations (μmol g^{-1}FW) and $^{13}NH_4^+$ influx in *Arabidopsis thaliana.*

4.2. LOW-AFFINITY AMMONIUM TRANSPORT

At NH_4^+ concentrations between 0.5 and 1 mM, a low-affinity, non-saturating transport system (LATS) becomes apparent (Ullrich et al.,1984; Wang et al., 1993; Kronzucker et al., 1996). In *Lemna* (Ullrich et al.,1984) and in rice (Wang et al., 1993), this system mediates passive NH_4^+ transport. The $[NH_4^+]$ at which the switch from active to passive transport occurs depends upon the ammonium status of the system. In *Lemna*, this was estimated to be at 67 μM (Ullrich et al., 1984). In roots of rice, the values were 42μM and 655 μM, respectively, depending upon prior NH_4^+ provision (Glass et al., 1997). Because of the absence of saturation of the LATS, as well as the passive nature of the flux, Ullrich et al., suggested that this flux might result from diffusion of NH_4^+ or NH_3. In rice roots, in contrast to the situation in *Lemna*, the LATS strongly depolarized membrane electrical potentials and pH profiles were inconsistent with significant NH_3 permeation (Wang et al.,1993, 1994). An interesting feature of the LATS is the apparent lack of down-regulation of this transport system when plant N status is increased. This is in marked contrast to the situation for the high-affinity transport of NH_4^+. This apparent lack of down-regulation has been observed in rice (Wang, 1993), trembling aspen and lodgepole pine (Min et al., unpublished). This is similar to the apparent lack of down-regulation of the low-affinity transport system for K^+ influx (Glass and Dunlop, 1978; Kochian and Lucas, 1982). In the case of NH_4^+, it may account for the toxic effects of NH_4^+ observed in many plants at elevated NH_4^+ concentrations. In this regard, regulation of the LATS appears to be quite different from that of the HATS.

In summary, plants may absorb nitrogen in the form of NO_3^-, NH_4^+ or amino acids in various combinations, or principally in one form, depending upon species and habitat. For each ion there appear to be multiple transport systems operating at different concentration ranges. Typically, the high-affinity transport systems which operate at low external concentration, mediate active transport and are subject to regulation based upon various internal N pools. Genes encoding the low and high affinity NO_3^- transporters have been cloned from several plants and characterization is underway. Genes encoding the high-affinity NH_4^+ transporter has been characterized from *Arabidopsis*, and some details of its physiological function and regulation have been characterized.

References

Aslam, M., Travis, R.L., Huffaker, R.C. (1992) Comparative kinetics and reciprocal inhibition of nitrate and nitrite uptake in roots of uninduced and induced barley (*Hordeum vulgare* L.) seedlings. Plant Physiol. **99**, 1124-1133.

Barber, S.A. (1995) Soil Nutrient Bioavailability. John Wiley &Sons, New York.

Becking, J.H. (1956) On the mechanism of ammonium uptake by maize roots. Acta. Bot. Neerl. **5**, 2-79.

Behl, R., Tischner, R. and Raschke, K. (1988) Induction of a high capacity nitrate uptake mechanism in barley roots prompted by nitrate uptake through a constitutive low-capacity mechanism. Planta **176**, 235- 240

Bloom, A.J. (1988) Ammonium and nitrate as nitrogen sources for plant growth. ISI Atlas. Animal and Plant Sciences **1**, 55-59.

Breteler, H. and Siegerist, M. (1984) Effect of ammonium on nitrate utilization by roots of dwarf bean. Plant Physiol. **75**, 1099-1103.

Causin, H.F. and Barneix, A.J. (1993) Regulation of ammonium uptake in wheat plants: effect of root ammonium concentration and amino acids. Plant and Soil **151**, 211-218.

Cooper, H.D. and Clarkson, D.T. (1989) Cycling of amino-nitrogen and other nutrients between shoots and roots in cereals - A possible mechanism integrating shoot and root in the regulation of nutrient uptake. J. Exp. bot. **40**, 753-762.

Doddema, H. and Telkamp, G.P. (1979) Uptake of nitrate by mutants of *Arabidopsis thaliana*, disturbed in uptake or reduction of nitrate II. Kinetics, Physiol. Plant. **45**, 332-338.

Feng, J., Volk, R.J. and Jackson, W.A. (1994) Inward and outward transport of ammonioum in roots of maize and sorghum: contrasting effects of methionine sulfoximine. J.Exp. Bot. **45**, 429-439.

Glass, A.D.M. and Dunlop, J. (1978).The influence of potassium content on the kinetics of potassium influx into excised ryegrass and barley roots, Planta, **141**, 117-119.

Glass, A.D.M. and Siddiqi, M.Y (1995) Nitrogen Absorption by Plant Roots In: "Nitrogen Nutrition in Higher Plants" Ed. H.S.Srivastava and R.P.Singh. pp. 21-56, Associated Pub.Co.New Delhi.

Glass, A.D.M., Shaff, J.E. and Kochian, L.V. (1992) Studies of nitrate uptake in barley. IV Electrophysiology. Plant Physiol. **99**, 456-463.

Glass, A.D.M., Erner,Y., Kronzucker, H.J., Schjoerring, J.K., Siddiqi, M.Y.,and Wang, M.Y. (1997). Ammonium fluxes into plant roots: energetics, kinetics and regulation. Z.Pflanzenernahr. Bodenk., **160**, 261-268.

Goyal, S.S., Lorenz, O.A. and Huffaker, R.C. (1982) Inhibitory effects o ammoniacal nitrogen on growth of radish plants.I. Characterization of toxic effects of NH_4^+ on growth and its alleviation by NO_3^-. J.Amer. Soc. Hort.Sci. **107**, 125-129

Huang, N.C., Chiang, C.S., Crawford, N.M. and Tsay Y-F (1996) *CHL1* gene encodes a component of low-affinity nitrate-inducible nitrate transporter in *Arabidopsis* and shows cell-specific expression in roots. The Plant Cell **72**, 705-713.

Ingermarsson, B., Oscarson, P., Af Ugglas, M., and Larrson, C-M. (1987) Nitrogen utilization in *Lemna*. II Stusies of nitrate uptake using $^{13}NO_3^-$. Plant Physiol. **85**, 860-864

Jackson, W.A., Kwik, K.D., Volk, R.J., Butz, R.G. (1976) Nitrate influx and efflux by intact wheat seedlings: Effects of prior nitrate nutrition. Planta **132**, 149-156.

Jackson,R.B. and Caldwell, M.M. (1993) The scale of nutrient hetrogeneity around individual plants and its quantification with geostatistics. Ecology **74**, 612-614.

Kielland, K. (1994) Amino acid absorption by artic plants: implications for plant nutrition and nitrogen cycling. Ecology **75**, 2373-2383

King, B.J. Siddiqi, M.Y., Ruth, T.J., Warner, R.L. and Glass, A.D.M. (1993). Feedback regulation of nitrate influx in barley roots by nitrate, nitrate and ammonium. Plant Physiol. **102**, 1279-1286.

Kochian LV, Lucas WJ (1982) Potassium transport in corn roots.

I. Resolution of kinetics into a saturable and linear component. Plant Physiol. **70**, 1723-1731.

Kronzucker, H.J., Siddiqi, M.Y., Glass, A.D.M. (1995). Kinetics of NO_3^- influx in spruce. Plant Physiol. **109**, 319-326.

Kronzucker, H.J., Siddiqi, M.Y., Glass A.D.M. (1996). Kinetics of NH_4^+ influx in spruce. Plant Physiol. **110**, 773-779.

Kronzucker, H.J., Siddiqi, M.Y. and Glass, A.D.M. (1997) Root discrimination against soil nitrate and the ecology of forest succession. Nature **385**, 59-61.

Lauter, R.-R., *et al.* (1996) Preferential expression of an ammonium transporter and of two putative nitrate transporters in root hairs of tomato, *Proc. Natl. Aca. Sci. USA* **93**, 8139-8144.

Lavoie, N., Vezina, L-P and Margolis, H (1992) Absorption and assimilation of nitrate and ammonium ions by jack pine seedlings. Tree Physiol. **11**, 171-183.

Lee, R.B. and Rudge, K.A. (1986) Effects of nitrogen deficiency on the absorption of nitrate and ammonium by barley plants. Ann. Bot. **57**, 471-486.

Lee, R.B. (1982) Selectivity and kinetics of ion uptake by barley plants following nutrient deficiency. Ann. Bot. **50**, 429-449.

Lee, R.B. and Drew, M.C. (1986) Nitrogen-13 studies of nitrate fluxes in barley roots. II. Effect of plant N-status on the kinetic parameters of nitrate influx. J. Exp. Bot. **185**, 1768-1779.

Lee, R.B., Purves, J.V., Ratcliffe, R.G., Saker, L.R. (1992) Nitrogen assimilation and the control of ammonium and nitrate absorption by maize roots. J. Exp. Bot. **43**, 1385-1396.

Likens, G.E.,Bormann, F.H.,Johnson, N.M.(1969) Nitrification: importance to nutrient losses for a cutover forest ecosystem. Science **163**, 1205-1206.

Magalhaes, J.R. and Wilcox, G.E. (1984) Ammonium toxicity development in tomato plants relative to nitrogen form and light intensity. J. Plant Nutr. **7**, 1477-1496.

Marschner, H., Kirkby, E.A., and Engels,C. (1997) Importance of cycling and recycling of mineral nutrients within plants for growth and development. Bot. Acta. **110**, 65-273.

Mattsson, M., Hausler, R.E., Leegood, R.C., Lea, P.J. and Schoerring, J.K. (1997) Leaf-atmosphere NH_3 exchange in barley mutants with reduced activities of glutamine synthetase. Plant Physiol. **114**, 1307-1312.

Mengel, K and Kirkby, E.A. (1979) Principles of Plant Nutrition . Int. Pot. Inst. Bern, Switzerland.

Miller, A.J. and Smith S J. (1996) Nitrate transport and compartmentation in cereal root cells. J. Exp. Bot. **47**, 43-854.

Min, X., Siddiqi, M.Y., Guy, R.D., Glass, A.D.M. and Kronzucker, H.J. (1998) Induction of nitrate uptake and nitrate reductase activity in trembling aspen and lodgepole pine. Plant Cell and Environment ***in press***

Morgan, M.A., Jackson, W.A. (1988) Inward and outward movement of ammonium in root systems: transient responses during recovery from nitrogen deprivation in presence of ammonium. J. Exp. bot. **39**, 179:191.

Muller, B. and Touraine B. (1992) Inhibition of NO_3^- uptake by various phloem-translocated amino acids in soybean seedlings. J.Exp. Bot. **43**, 617-623.

Ninnemann, O., Jauniaux, J.-C. and Frommer, W.B. (1994) Identification of a high affinity NH_4^+ transporter from plants. EMBO J. **13**, 3463-3471.

Perez, M.D., Gonzalez, C., Avila, J., Brito, N., and Siverio, J.M. (1997). The *YNT1* gene encoding the nitrate transporter in the yeast *Hansenula polymorpha* is clustered with genes YNI1 and YNR1 encoding nitrite and nitrate reductase and its disruption causes inability to grow in nitrate. Biochem J. **321**, 397-403.

Quesada, A., Galvan, A and Fernandez E (1994). Identification of nitrate transporter genes in *Chlamydomonas reinhardttii*. Plant J. **5**, 407-419.

Quesada A, Krapp A, Trueman LJ, Daniel-Vedele F, Fernandez E, Forde BG, Caboche M (1997). PCR-identification of a *Nicotiana plumbaginifolia* cDNA homologous to the high-affinity nitrate transporters of the crnA family. Plant Mol. Biol. **34**, 265-274.

Reisenauer, H.M. (1966). Mineral nutrients in soil solution. In: Environmental Biology. P.L. Altman and D.S. Dittmer, eds. Fed. Amer. Soc. Exp. Biol. Bethesda. pp 507-508.

Rodgers C O. and Barneix A J. (1993). The effect of amino acids and amides on the regulation of nitrate uptake by wheat seedlings. Journal of Plant Nutrition **16**, 337-348.

Ryan, P.R. and Walker, N.A. (1994) The regulation of ammonia uptake in *Chara australis*. J.Exp. Bot. **277**, 1057-1067.

Schrader, L.E. and Thomas, R.J. (1981). Nitrate uptake, reduction and transport in the whole plant. In: Nitrogen and Carbon Metabolism ed by J.D.Bewley, Martinus Nijhoff Dr.W.Junk Pubs. The Hague. Pp 49-93.

Siddiqi, M.Y., Glass, A.D.M., Ruth, T.J., Fernando, M. (1989) Studies of the regulation of nitrate influx by barley seedlings using $^{13}NO_3^-$. Plant Physiol **90**, 806-813.

Siddiqi, M.Y., Glass, A.D.M., Ruth, T.J., Rufty, T.W. (1990). Studies of the uptake of nitrate in barley. I. Kinetics of $^{13}NO_3^-$ influx. Plant Physiol. **93**, 1426-32.

Stark, J.M. and Hart,S.C.(1997) High rates of nitrification and nitrate turnover in undisturbed coniferous forests. Nature **385**, 61-64.

Touraine, B. and Glass, A.D.M. (1997) Nitrate and chlorate fluxes in the chl1-5 mutant of *Arabidopsis thaliana*. Does the *CHL1-5* gene encode a low affinity nitrate transporter. Plant Physiol. **114**, 137-144.

Trueman LJ, Richardson A, and Forde BG (1996) Molecular cloning of higher plant homologues of the high-affinity nitrate transporters of *Chlamydomonas reinhardtii* and *Aspergillus nidulans*. *Gene* **175**, 223-231.

Tsay, Y-F, Schroeder, J.I., Feldmann, K.A., Crawford, N.M. (1993) A herbicide sensitivity gene *CHL1* gene encodes a nitrate-inducible nitrate transporter. Cell **72**, 705-713.

Ullrich, W.R., and Novacky, A. (1981) Nitrate-dependent membrane potential changes and their induction in *Lemna gibba* G1. Plant Sci. Let. **22**, 211-217.

Ullrich, W.R., Larsson, M., Larsson, C-M., Lesch, S., Novacky, A. (1984) Ammonium uptake in *Lemna gibba* G1, related membrane potential change and inhibition of anion uptake. Physiol. Plant. **61**, 369-376.

Unkles, S.E., Hawker, K.L., Grieve, C., Campbell, E.I., Montague, P. and Kinghorn, J.R. (1991) *crnA* encodes a nitrate transporter in *Aspergillus nidulans*. Proc. Natl. Acad. Sci, USA. **88**, 204-208.

Wang, M., Siddiqi, M.Y., Ruth, T.J. and Glass, A.D.M. (1993) Ammonium uptake by rice roots. I. Fluxes and subcellular distribution of $^{13}NH_4^+$. Plant Physiol. **103**, 1249-1258.

Wang, M., Glass, A.D.M., Shaff, J.E. and Kochian, L.V. (1994). Ammonium uptake by rice roots. III. Electrophysiology. Plant Physiol., **104**, 899-906.

Wang RC. and Crawford NM. (1996) Genetic identification of a gene involved in constitutive high-affinity nitrate transport in higher plants. Proc. Natl. Acad. Sci, USA. **93**, 9297-9301.

Wolt,J.D. (1994) Soil Solution Chemistry. John Wiley &Sons, Inc. New York.

Zhuo, D., Okamoto, M., Vidmar, J.J. and Glass, A.D.M. (1998) Expression and Regulation of Two differentially-Expressed *Arabidopsis (Nrt2)* Genes Encoding High-Affinity Nitrate Transporters of the *crnA* Gene Family. The Plant Journal *submitted*.

EFFECT OF METHIONINE SULFOXIMINE ON THE INDUCTION OF NITRATE REDUCTASE AND METABOLIC NITRATE POOL SIZE IN BARLEY ROOTS

M. ASLAM, R.L. TRAVIS, D. W. RAINS
Department of Agronomy and Range Science,
University of California, One Shields Avenue,
Davis, CA 95616-8515 USA

1. Abstract

The effect of methionine sulfoximine (MSO), an inhibitor of glutamine synthetase, on the induction of nitrate reductase activity (NRA) and metabolic NO_3^- pool size in roots of 8-day-old intact barley (*Hordeum vulgare* L.) seedlings was studied. Enzyme activity was induced at different NO_3^- concentrations (0 to 100 mM) in the presence or absence of 0.25 mM MSO and assayed *in vitro.* Metabolic NO_3^- pool concentration was determined using the minus NO_3^- *in vivo* assay. In the absence of MSO, induction of NRA plateaued at 0.1 mM external NO_3^-, even though total NO_3^- concentration in roots increased with increasing external NO_3^-. Treatment with MSO enhanced the induction of NRA above 0.1 mM external NO_3^-. Thus, in roots supplied with MSO at 100 mM NO_3^-, the level of the enzyme activity was 70 to 80% higher than in roots not supplied with MSO, even though total NO_3^- concentration in MSO-treated roots was 20 to 30% lower. Induction of NRA also plateaued after 6 h of exposure to 100 mM NO_3^-; whereas in the presence of MSO, NRA continued to increase up to 24 h. Glutamine, supplied exogenously or accumulated internally by inhibiting its assimilation, had no effect on the induction of NRA. Metabolic NO_3^- concentration in roots treated with MSO at 100 mM NO_3^- was higher than in those not treated with MSO. In contrast, MSO had no effect on metabolic NO_3^- concentration in roots induced with 0.1mM NO_3^-. Metabolic NO_3^- concentration remained constant between the 6 and 24 h induction period. However, MSO treatment almost doubled the metabolic NO_3^- concentration during the same period. The results indicate that the metabolic NO_3^- pool is saturated at 0.1 mM exogenous NO_3^- and the enhancement of NRA by MSO induced with NO_3^- concentrations greater than 0.1 mM may be due to increased availability of NO_3^- in the metabolic pool.

Keywords: Barley, *Hordeum vulgare*, induction, metabolic nitrate, methionine sulfoximine, nitrate reductase, roots.

G. Gissel-Nielsen and A. Jensen (eds.), Plant Nutrition – Molecular Biology and Genetics, 17–20.

2. Introduction

The requirement of NO_3^-, as a substrate inducer, for the induction of nitrate reductase (NR) is well established (Guerrero et al., 1981). NO_3^- within the plant cell exists in two compartments or pools: a relatively small metabolic or active pool and a large non metabolic or storage pool (Ferrari et al., 1973; Aslam et al., 1976; Zhen et al., 1991). The induction of nitrate reductase activity (NRA) is regulated by NO_3^- availability in the metabolic pool (Ferrari et al., 1973; Aslam et al., 1976). In studies on the regulation of NO_3^- assimilation L-methionine sulfoximine (MSO), an inhibitor of glutamine synthetase (GS), is often used. The treatment of plant tissue with MSO leads to the accumulation of NH_4^+ (Lee and Ratcliffe, 1991; Aslam et al., 1997). Lee and Ayling (1993) reported that while most of the NH_4^+ produced after MSO treatment of corn roots accumulated in the vacuolar pool, its concentration in the cytoplasmic pool also increased about ten times compared to that in control roots. However, there is no evidence that the cytosolic NO_3^- pool is also enlarged by MSO. In this study we show that MSO enhanced the induction of NRA by increasing the accumulation of NO_3^- in the metabolic pool.

Abbreviations: GS, glutamine synthetase; MSO, L-methionine sulfoximine; NR(A), nitrate reductase (activity).

3. Materials and methods

3.1. PLANT CULTURE

Barley (*Hordeum vulgare* L var. CM-72) seeds were germinated and grown hydroponically in 0.2 mM $CaSO_4$ in the dark at 25°C for 5 days and then transferred into N-free one-quarter strength Hoagland's solution and placed in a growth chamber under continuous light (400 µmol m^{-2} s^{-1}) at 25° C and 60 to 65 % relative humidity. After 24 h in light the seedlings were induced with different concentrations of NO_3^- (0 to 100 mM) in the absence or presence of MSO.

3.2. *IN VITRO* NRA ASSAY

Roots (1.5 to 2 g) were homogenized with 4 mL of extraction buffer g^{-1} in a chilled mortar and pestle with acid-washed sand. The extraction buffer consisted of 0.05 mM Tris-HCl (pH 8.5), 1 mM DTT, 10 µM flavin adenine dinucleotide, 1 µM Na_2MoO_4, 1 mM EDTA, and 10 µM leupeptin. The homogenates were centrifuged at 30000 *g* for 15 min, and the supernatants were used for the measurement of NRA, NO_3^- and NH_4^+. NRA was assayed using NADH as a source of electrons.

3.3. METABOLIC NO_3^- POOL

Metabolic NO_3^- pool size was determined by following total production of NO_2^- in NO_3^--free assay after it reached a plateau. Excised roots were incubated in 0.1 M potassium phosphate buffer solution (pH 7.5) for 2 h under anaerobic conditions.

3.4. ION DETERMINATION

NO_3^- from the extracts was determined by measuring A_{210} after separation by HPLC on a Partisil-10 SAX anion exchange column. NO_2^- from the enzyme assay reaction mixtures was determined by measuring A_{540} after color development for 15 min with a 1:1 mixture of 1% (w/v) sulfanilamide in 1.5 N HCl and 0.02% (w/v) *n*-naphthylethylenediamine dihydrochloride.

4. Results

4.1. EFFECT OF MSO AT DIFFERENT NO_3^- CONCENTRATIONS

In the absence of MSO induction of NRA plateaued at 0.1 mM external NO_3^-, although NO_3^- accumulation in the roots increased with increasing external NO_3^-. When roots were supplied with MSO, NRA levels increased gradually with the increase of external NO_3^- above 0.1 mM, even though total NO_3^- concentration in MSO-treated roots decreased 20 to 30% as compared to that in control roots.

4.2. TIME-COURSE OF MSO RESPONSE

The induction of NRA increased upto 6 h when roots were exposed to NO_3^-; thereafter no further increase in enzyme activity occurred. Exposure to MSO had little effect on the induction of NRA during the initial 6 h; thereafter NRA continued to increase up to 24 h. NO_3^- accumulation in the roots increased with time of exposure to NO_3^-; however, in the presence of MSO considerably less NO_3^- accumulated under these conditions.

4.3. EFFECT OF MSO ON METABOLIC NO_3^-

NO_3^- concentrations in the metabolic pool of roots induced with 0.1 or 100 mM NO_3^- were similar. Treatment of roots with MSO had no effect on metabolic NO_3^- concentration when induced with 0.1 mM NO_3^-. However, metabolic NO_3^- concentration was increased by MSO when induced with 100 mM NO_3^-. Likewise, in roots induced with 100 mM NO_3^- for 6 h, metabolic NO_3^- concentrations were similar with or without MSO. With longer exposure (24 h), while in the absence of MSO no increase in metabolic NO_3^- concentration occurred, it was nearly doubled in MSO-treated roots.

5. Discussion

The enhanced levels of NRA by MSO at higher external NO_3^- and with longer exposure indicates that the effect of MSO may be to increase accumulation of NO_3^- in the metabolic pool. Thus, the exposure of roots to MSO enhanced the concentration of metabolic NO_3^- at 100 mM NO_3^- supply but not at 0.1 mM NO_3^-. Likewise, in roots of seedlings induced with 100 mM NO_3^-, after 6 h of induction metabolic NO_3^- concentrations were similar with or without MSO. With longer exposure (24 h), while in the absence of MSO no increase in metabolic NO_3^- concentration occurred, it was doubled in MSO-treated roots. Recently Oaks and Sivasankar (1997) reported that treatment of corn roots with MSO also enhanced the induction of NRA. They argued that the stimulation of NRA by MSO was due to the inhibition of glutamine formation in the presence of MSO (Sivasankar et al., 1997). However, in barley roots glutamine had no effect on the induction of NRA at 10 mM NO_3^-, where NO_3^- uptake was not limiting. Our results indicate that stimulation of NRA induction by MSO in barley roots was more likely due to increased availability of NO_3^- in the metabolic pool.

6. Conclusions

Presence of MSO gradually enhanced the induction of NRA in barley roots when supplied with increasing external NO_3^- concentrations above 0.1 mM, even though NO_3^- accumulation in MSO-treated roots decreased. The stimulation of induction of NRA by MSO was not due to inhibition of glutamine formation in the roots but due to increased accumulation of NO_3^- in the metabolic pool.

7. References

Aslam, M., Oaks, A. and Huffaker, R.C. (1976) Effect of light and glucose on the induction of nitrate reductase and the distribution of nitrate in etiolated barley leaves, *Plant Physiol* 58, 588-591.

Aslam, M., Travis, R.L., Rains, D.W, and Huffaker, R.C. (1997) Differential effect of ammonium on the induction of nitrate and nitrite reductase activities in roots of barley (*Hordeum vulgare* L.) seedlings, *Physiol Plant* 101, 612-619.

Ferrari, T.F., Yoder, O.C. and Filner, P. (1973) Anaerobic nitrite production by plant cells and tissues: evidence for two nitrate pools, *Plant Physiol* 51, 423-431.

Guerrero, M.G, Vega, J, and Losada, M. (1981) The assimilatory nitrate-reducing system and its regulation, *Annu Rev Plant Physiol* 32, 169-204.

Lee, R.B. and Ayling, S.M. (1993) The effect of methionine sulphoximine on the absorption of ammonium by maize and barley roots over short periods, *J Exp Bot* 44, 53-63.

Lee, R.B. and Ratcliffe, R.G. (1991) Observations on the subcellular distribution of the ammonium ion in maize root tissue using *in vivo* ^{14}N-nuclear magnetic resonance spectroscopy, *Planta* 183, 359-367.

Oaks, A. and Sivasankar, S. (1997) The specificity of inhibition caused by methionine sulfoximine and azaserine in plant tissues (abstract No 922), *Plant Physiol* 114, S-186.

Sivasankar, S., Rothstein, S. and Oaks, A. (1997) Regulation of the accumulation and reduction of nitrate by nitrogen and carbon metabolites in maize seedlings, *Plant Physiol* 114, 583-589.

Zhen, R-G., Koyro, H-W., Leigh, RA., Tomos, A.D. and Miller, A.J. (1991) Compartmental nitrate concentrations in barley root cells measured with nitrate-selective microelectrodes and by single-cell sap sampling, *Planta* 185, 356-361.

EFFECT OF N-FERTILISER LEVELS AND SOIL pH ON UPTAKE OF TRACE ELEMENTS BY PLANTS

ALLAN BIBAK, STEFAN STÜRUP, PERNILLE GUNDERSEN, AND VAGN GUNDERSEN
Risø National Laboratory, Plant Biology and Biogeochemistry Department, PBK-124
P.O.Box 49, DK-4000 Roskilde, Denmark

1. Abstract

The study aims to determining the influence of N-fertiliser levels and soil pH on uptake of trace elements by plants. Application of three levels of N fertiliser (0, 60, and 120 kg N/ha) tends to split potatoes into three groups with different element profile. The effect of three levels N fertilisation on elemental concentrations of the potatoes is evaluated by use of Discriminant Partial Least Squares Regression (PLS). The contents of Cd, Co, Fe, Mn, Pb, V, and Zn in the cabbages, onion, peas, and sprouts showed negative correlation with pH, whereas the correlation for Mo were positive. Our finding suggest that any kind of acidifying processes due to agricultural practices lead to decrease in soil pH, levels of Mo will tend to decrease in edible crop, whereas those of the other elements will show an overall increase.

2. Introduction

Trace elements availability are closely correlated with fertility practices, pH, and soil type (Bibak, 1994, Bibak et al., 1998a,b, c). The trace element contents of plants is influenced by the level of N fertiliser in the soil (Bibak, 1994, Bibak et al., 1998b). The beneficial action of N-fertiliser may partly due to the acidifying reaction, e.g., of acid forming N fertiliser in the soil and the resulting improvement in the solubility and availability of native and/or applied trace elements and partly due to synergistic effect of NO_3^- on cations uptake by plants.

The purposes of this investigation were: (1) to evaluate the elemental concentrations of crops as affected by application of three levels of N fertilisation, (2) and their correlation with surface soil pH (0-25 cm).

3. Materials and Methods

The strategy for site selection were to cover a wide range of soil types with respect to pH, soil texture and contents of organic matter with use of intensive cultivation, with

G. Gissel-Nielsen and A. Jensen (eds.), Plant Nutrition – Molecular Biology and Genetics, 21–24.

use of fertiliser and pesticides and same sort of the crop, so big differences in site area occur.

The details description on the selection of sampling sites, laboratory modification, special equipment for sample preparation, homogenisation, digestion, laboratory ware, cleaning procedure, instrumental parameter for HR-ICPMS analysis, quantification, quality control procedure, and data processing have been described in (Bibak et al., 1998a,b, c).

Quality control procedures include the analysis of blanks and certified standard reference materials, cabbage (Food Detection Science Institute Ministry of Commerce, Beijing, China), NIST 1567a wheat flour, and Leek in-house material from the National Food Agency of Denmark.

The results were analysed statistically using partial least square regression (PLS) model and simple correlation coefficient (r) and stepwise multiple regression in the Statgraphics statistical software packages. The adjusted regression coefficient (R^2_{adj}) is given in the text.

4. Results and Discussions

The effects of three levels of N-fertilisation on elemental concentrations of potato on two field sites with different cultivation history are evaluated by the use of Discriminant Partial Least Squares regression (PLS) model. The X-matrix used in the PLS model comprise: the concentrations of 54 elements in the 101 potato samples, 47 concentrations of each element at 0-N level, 24 at 1-N level, and 30 at 2-N levels resulting in a 101 x 54 X-matrix. The score and loading weight plot for PLS model of X are shown in Fig. 1. Each N-fertiliser level is given the code of 1 for 0 kg N /ha, 2 for 60 kg N /ha, and 3 for 120 kg N/ha.

The score plot shows that 0-N, 60 kg N/ha and 120 kg N/ha tends to split in 3 groups separated by the 3 lines (Fig. 1). The PLS-component 1 (PC1) and PLS-component 2 (PC2) explain 23% and 10% of the variation in X, respectively. The loading plot in Fig. 1 is used for the selection of variables. The elements with a low influence on the model are removed because they only contribute by noise.

The results of the correlation analysis are shown in Table 1. The contents of Cd, Co, Cu, Fe, Mn, Pb V, and Zn in the crops showed negative correlation with pH, whereas the correlation's for Mo were positive (Table 1). For molybdenum, soil pH is one of the most important single factors affecting the uptake of Mo in plant. The MoO_4^{2-} concentration in soil solution increases with increase in pH (Lindsay, 1972).

Stepwise multiple regression were carried out with the studied crops in which the elements contents were related to soil pH (0-25 cm) as well as contents of clay and organic carbon (0-25 cm). A combination of these factors explained 63, 11, 40 and 18% of total variations in the Co contents of cabbage, onion, pea, and sprouts, respectively (Table 2). In the stepwise multiple regressions the pH was a significant factor negatively correlated with the Co, Cu, Fe, Mn, Pb, and Zn contents of cabbages, the Mn and Zn contents of onions, the Cd, Fe, and Zn contents of peas, and Cd, Co, and V contents of sprouts. It was a significant factor positively correlated with the Mo contents of cabbage, pea, and sprouts.

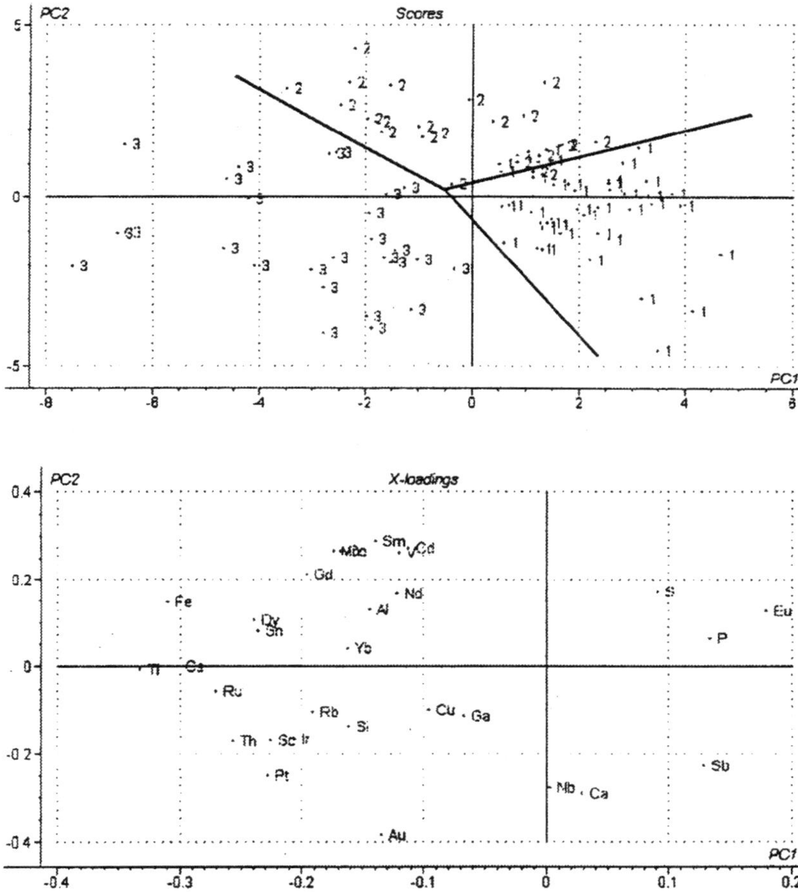

Figure 1. Score and loading plots, the PLS- component 1(PC1) and PLS- component 2 (PC2) explain 23% and 10%, respectively. The lines in the plot indicate the 3 level of N-fertilisation tends to split in the groups.

TABLE 1. Simple correlation coefficients (r) between soil properties and the elemental concentration (μg/kg) in the crops.

Soil properties	Crops	Cd	Co	Cu	Fe	Mn	Mo	Pb	V	Zn
pH	Cabbage	-0.284^{b}	n.s.	-0.440^{c}	n.s.	-0.466^{c}	0.689^{c}	-0.369^{c}	n.s.	-0.2
	Onion	n.s.	n.s.	n.s.	n.s.	n.s.	n.s.	n.s.	n.s.	-0.2
	Pea	n.s.	n.s.	n.s.	-0.637^{c}	n.s.	0.458^{c}	-0.438^{c}	n.s.	-0.7
	Sprouts	-0.332^{c}	-0.316^{b}	n.s.	-0.742^{c}	-0.253^{a}	0.537^{c}	n.s.	-0.325^{b}	n.s.
Clay (%)	Cabbage	0.575^{c}	0.605^{c}	0.338^{c}	0.650^{c}	0.414^{c}	0.523^{c}	n.s.	0.348c	0.54
	Onion	n.s.	0.313^{c}	0.386^{c}	n.s.	n.s.	n.s.	0.418^{c}	n.s.	n.s.
	Pea	0.639^{c}	n.s.	n.s.	n.s.	n.s.	-0.469^{c}	n.s.	n.s.	0.48
	Sprouts	n.s.	n.s.	n.s.	n.s.	0.244^{a}	0.387^{c}	n.s.	n.s.	n.s.
Org.C (%)	Cabbage	n.s.	-0.300^{b}	n.s.	-0.230^{a}	n.s.	n.s.	0.230^{a}	n.s.	n.s.
	Onion	n.s.	-0.314$_{b}$	n.s.	-0.381^{b}	-0.551^{b}	n.s.	n.s.	0.377^{c}	0.53
	Pea	n.s.	-0.682^{c}	n.s.	-0.386^{b}	-0.474^{c}	0.570^{c}	0.295^{a}	n.s.	n.s.
	Sprouts	n.s.	-0.248^{a}	n.s.	n.s.	n.s.	n.s.	n.s.	0.232^{a}	n.s.

a, b, c, significant at $P < 0.05$, 0.01, and 0.001 level, respectively. n.s.= not significant.

5. Conclusion

Application of three levels of N fertiliser (0, 60, and 120 kg N/ha) tends to split potatoes into three groups with different element profile. The contents of Cd, Co, Fe, Mn, Pb, V, and Zn in the cabbages, onion, peas, and sprouts showed negative correlation with pH, whereas the correlation for Mo were positive. Our finding suggest that any kind of acidifying processes due to agricultural practices lead to decrease in soil pH; levels of Mo will tend to decrease in edible crops, whereas those of the other elements will show an overall increase.

6. References

Bibak, A. (1994). Uptake of cobalt and manganese by winter wheat from a sandy loam soil with and without added farmyard manure and fertiliser nitrogen. *Commun. Soil Sci. Plant Anal.* **25**, 2675-2684.

Bibak, A., Behrens, A., Stürup, S., Knudsen, L., and Gundersen, V. (1998a). Concentration of 63 major and trace elements in Danish agricultural crops measured by inductively coupled plasma mass spectrometry.1. Onion (Allium cepa Hysam). *J. Agric. Food Chem.* **46**, 3139-3145.

Bibak, A., Stürup, S., Haahr V., Gundersen P., and Gundersen, V. (1998b). Concentration of 54 major and trace elements in Danish agricultural crops measured by inductively coupled plasma mass spectrometry.4. Potato (*Solanium tu*berosum, Folva). *J. Agric. Food Chem.* (submitted).

Bibak, A., Behrens, A., Stürup, S., Knudsen, L., and Gundersen, V. (1998c). Concentration of 55 major and trace elements in Danish agricultural crops measured by inductively coupled plasma mass spectrometry.2. Pea (*Pisum sativum* Ping Pong). *J. Agric. Food Chem.* **46**, 3146-3149.

Lindsay, W.L. (1972). Inorganic phase equilibria of micronutrients in soils. in: J.J. Mortvedt, P.m. Giordano, and W.L. Lindsay (eds.), *Micronutrients in Agriculture*. Soil Sci. Soc. Am. Madison, Wisconsin. 41-56

NUTRIENT COMPOSITION DIVERSITY□ IN WILD AND COMMON WHEAT GRAINS

David J. Bonfil* and Uzi Kafkafi
Department of Field Crops, Vegetables and Genetics, The Hebrew University of Jerusalem, P.O.Box 12, Rehovot 76100, Israel
* Present address: Department of Field Crops and Natural Resources, Agricultural Research Organization, Gilat Experiment Station, M.P. Negev 2, 85280, Israel

Summary

Wild emmer wheat, *Triticum dicoccoides*, concentrates in the grains the mineral nutrient that is most deficient in the soil of its natural habitat. The nutrient concentration in the grains could affect the baking quality as well as the growing ability of these genotypes in deficient soils. The objective of this research was to determine whether there is nutrient composition diversity in wild and common wheat grains. The nutrient content in the grains was tested in relation to the nitrogen uptake under different fertilization regimes of nine *T. dicoccoides* populations (from nine different habitats), one *T. durum* and two *T. aestivum* wheat cultivars. The plants were sown in soils taken from their natural habitats, and grown in a greenhouse for three successive years. In the first year, no fertilizer was added. Only N was added in the second, and NPK or NPK and S were added in the third year. Grain yields and grain nutrients content were measured. Grain nitrogen yield ranged between 8 and 183 mg/plant and N concentration varied between 18 and 49 mg/g DM. All genotypes exhibited lowest N concentration, and usually the highest nutrient/nitrogen ratio, when no fertilizer was applied. However, without fertilization, grains of genotypes 3, 4, 5 and *T. aestivum* cv. Dariel had the lowest Ca/N ratio, and genotypes 6 and 7 had the lowest Fe/N ratio. The mineral content range (expressed as mg element per g N in the seed) which depended upon soil nutrient level and genotype, was as follows: P 83-250; K 78-310; S 28-81; Mg 34-73; Ca 9-37; Na 0.4-78; Mn 0.7-2; Zn 0.5-14; and Fe 0.5-3.6. The ratio of a nutrient element to unit of N was stable in some genotypes (e.g. Zn/N, genotype 1) and variable in others (Zn/N, genotype 9). Large differences existed in the ratio of S per gram N in the grains of different genotypes, and this was affected by fertilization regime. The Mg/N ratio in the grains of genotype 2 and *T. aestivum* cv. Dariel under NPK fertilization was as high as that without any fertilization; however, in genotype 3 and *T. aestivum* cv. Bethlehem, this ratio was the same as when N or NPKS was added. It was concluded that genetic diversity exists in discrimination and mineral concentration of both wild and common wheat seeds. This diversity can be used in breeding and selection programs for different soil conditions and flour quality.

G. Gissel-Nielsen and A. Jensen (eds.), Plant Nutrition – Molecular Biology and Genetics, 25–33.

Index words: *Triticum dicoccoides, Triticum durum, Triticum aestivum*, nutrition.

1. Introduction

Wild wheat possesses genetic information that is used to improve the quality of wheat flour. A high percentage of grain protein and genetic diversity of wild wheats are attractive for wheat breeders (Ciaffi *et al.*, 1991,1992; Feldman and Sears, 1981; Grama *et al.*, 1987). Many wild cereal populations are grown in different habitats in Israel, and exhibit great genetic diversity (Nevo, 1986; Nevo and Beiles, 1989).

Seed storage protein composition is a genotype trait, but its synthesis is regulated by nutrient content, especially N and S (Randall and Wrigley, 1986; Shewry *et al.*, 1995). Nitrogen, sulphur, and especially S/N ratio, influence seed storage protein both qualitatively and quantitatively more than any other mineral nutrient (Bonfil *et al.*, 1997; Randall and Wrigley, 1986; Wrigley *et al.*, 1980). Phosphorus, K, S and Mg contents were also found to affect the storage proteins of pea (Randall *et al.*, 1979). In several nutrition deficiency circumstances, certain protein components (hordein, gliadin and glutenin bands) from normal wheat and barley grain are either absent or greatly reduced in quantity (Bonfil *et al.*, 1997; Moss *et al.*, 1981; Rahman *et al.*, 1983; Shewry *et al.*, 1983). The content of several minerals (especially P, K, Mg, Ca, Na, Fe, and Mn) in the grains affects both seed storage protein composition and the baking quality of flour (Bequette *et al.*, 1963; Douglas and Dyson, 1985).

The concentration of mineral elements in wheat plants is dependent on genotype, mineral content in the soil, and fertilization (Bequette *et al.*, 1963; Boila *et al.*, 1993; Douglas, 1987). However, wheat carries genes that can control mineral uptake, resulting in K/Na discrimination (Gorham, 1993). Wild emmer wheat, *T. dicoccoides*, concentrates in the grains the mineral nutrient that is most deficient in the soil of its natural habitat (Bonfil and Kafkafi, in preparation). Nitrogen concentration is the main mineral affecting baking quality, but when the ratio of N to other minerals changes, as in the case of S/N, the baking quality is reduced (Randall and Wrigley, 1986). It is necessary, therefore, to know whether the mineral discrimination and concentration in grains is constant, or if it perhaps varies as a result of nutrient deficiency in the soil. The aim of this research was to determine for wild wheat seeds 1) whether there is genetic diversity in mineral concentration and if so, its magnitude, and 2) if each genotype produces grain containing equal mineral ratios when grown on different soils.

2. Materials and Methods

Grain mineral nutrient content in nine *T. dicoccoides* populations (from nine different habitats, Table 1), one *T. durum* (Bareket) and two *T. aestivum* (Bethlehem and Dariel) wheat cultivars was measured and related to nitrogen content under different fertilization regimes. The plants were sown in ten soils taken from natural habitats, and grown in a greenhouse for three successive years. In the first year, no fertilizer was added; only N was added in the second year (85 mg N as NH_4NO_3 kg^{-1} soil); and NPK or NPK and S were added in the third year (120 mg N as NH_4NO_3, 40 mg P and 50 mg K as KH_2PO_4, 25 mg S as $CaSO_4 \cdot H_2O$ kg^{-1} soil). The experimental details and the soil's mineral content have been published elsewhere (Bonfil *et al.*, 1997).

TABLE 1. Wild wheat habitat location characteristics and soil type description.

Habitat No. and location	Coordinates[a] Ln- Lt	Altitude m	Geological epoch	Soil type[b]
1 -Yehudiya	35°41'-32°53'	70	Pliocene	dark basaltic grumusol
2 -Kokhav Yarden	35°32'-32°36'	0	Pliocene	calcitic basaltic protogrumusol
3 -Rosh Pinna	35°32'-32°57'	520	Eocene	red terra rossa
4 -Rihaniya	35°29'-33°03'	620	Eocene	red terra rossa
5 -Dishon	35°31'-33°05'	330	Turonian	reddish-brown terra rossa
6 -Ramot Naftali	35°33'-33°06'	450	Cenomanian	reddish-brown terra rossa
7 -Ramot Naftali	35°33'-33°07'	440	Cenomanian	reddish-brown terra rossa
8 -Agas Valley	35°42'-33°18'	1240	Jurassic	red terra rossa
9 -Barhata Valley	35°43'-33°18'	1335	Jurassic	brown (mediterranean) soils
10-Mount Admon[c]	35°29'-33°02'	780	Senonian	pale rendzina

[a] Ln=longitude in decimals; Lt=latitude, in decimals.
[b] Dan *et al.* (1976).
[c] No natural wild wheat population was found in this habitat.

Grain yields and grain mineral nutrients content were measured. Grains were dried at 60°C and their mineral contents were analyzed. Nitrogen in the grain was determined by the indophenol-blue procedure following Kjeldahl digestion (Jones and Case, 1990; Smith, 1980). P, K, S, Ca, Mg, Na, Fe, Zn, Mn, Cu, Co, Al, B and Mo in the grain were determined by ICP (Inductively Coupled Plasma) spectrometry following digestion in nitric-perchloric acids (Jones and Case, 1990). Two to four replicates of five grains were taken for each digestion.

3. Results

Grain nitrogen yield (Fig. 1A) ranged between 8 and 183 mg/plant, and N concentration varied between 18 and 49 mg/g DM (Fig. 1B). Soil type, fertilization, and plant genotype affected growth, yield per plant, and almost all yield components (Bonfil and Kafkafi, in preparation). All genotypes exhibited the highest N yield in the 3rd year following NPK and S fertilization. Under these growth conditions, the yield advantages of the common varieties over wild wheat were apparent (Fig. 1A). All genotypes exhibited the lowest N concentration in the 1st year, when no fertilizer was applied (Fig. 1B). Most genotypes showed that after N application (2nd and 3rd years) N concentration was kept almost in the same level for each genotype.

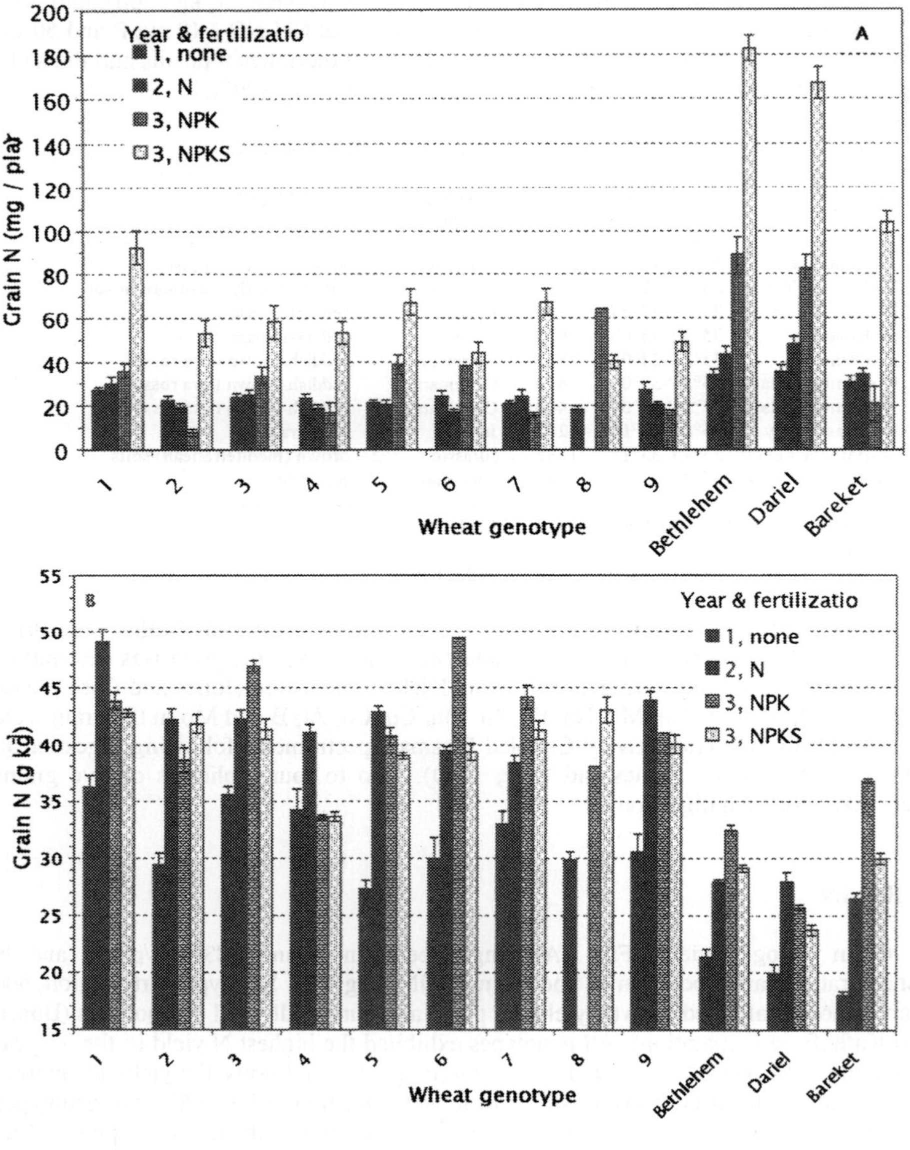

Figure 1. Nitrogen yield (A) and content (B) in grains of 12 wheat genotypes (genotypes 1 to 9 - *T. dicoccoides*; cvs. Bethlehem and Dariel - *T. aestivum*; cv. Bareket - *T. durum*). Plants were grown in natural soils from ten different *T. dicoccoides* habitats over three consecutive years. In the 1st year no fertilization was added, in the 2nd year N was supplied, and in the 3rd year NPK or NPKS was added. Data represent the average of each genotype that was grown in all soils; vertical bars represent the SE.

The amount of the mineral content range (expressed as mg element per g N in the seed) depended upon soil nutrient level and genotype, was as follows: P, 83-250; K, 78-310; S, 28-81; Mg, 34-73; Ca, 9-37; Na, 0.4-78; Mn, 0.7-2; Zn, 0.5-14; and Fe, 0.5-3.6 (Table 2). Usually, genotypes exhibited the highest nutrient/nitrogen ratio when no fertilizer was applied (Table 2, 1st year), presumably a result of nitrogen deficiency in the 1st year. However, even under these growth conditions, grains of wild wheat genotypes 3, 4, 5 and of *T. aestivum* cv. Dariel had the lowest Ca/N ratio. Habitats 3, 4 and 5 soils originated from Eocene and Turonian epochs and differed from the other soils mainly in their Ca/Mg ratio. Habitats 3-5 have a Ca/Mg ratio between 8.7 and 10.4; in habitats 6 and 7 the ratio is 4.0-5.8; in habitats 1 and 2 the ratio is 3.6-4.8; and in habitats 8 and 9 the ratio is 16.3-17.0. All genotypes produced the highest seed yields also on soils from habitats 3-5, suggesting that the optimum Ca/Mg be around 9-10. Under the same growth conditions, genotypes 6 and 7 had the lowest Fe/N ratio. Habitats 6 and 7 soils originated from the dolomite parent rock of the Cenomanian epoch, and have the lowest Ca/Mg ratio of all the terra rossa soils (lime rock).

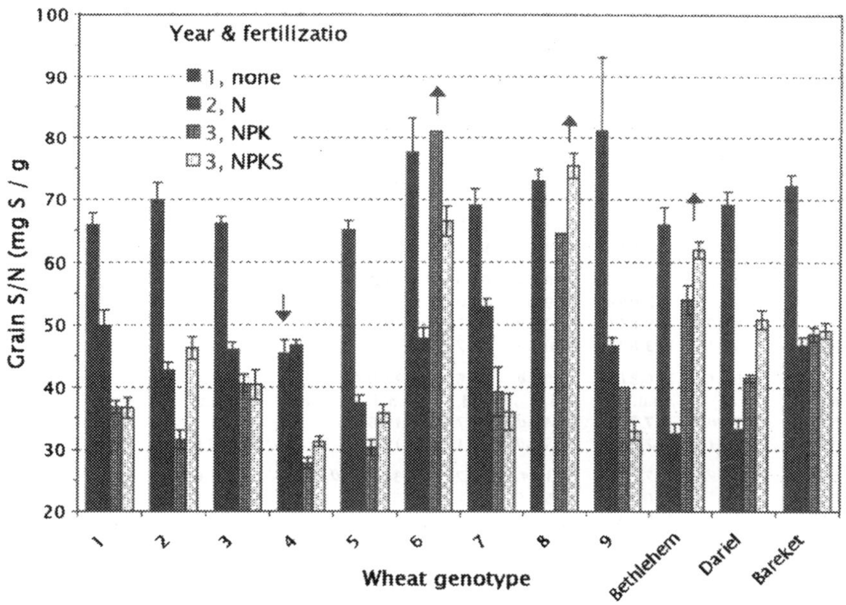

Figure 2. Sulphur-to-nitrogen ratio (mg S / g N) in grains of 12 wheat genotypes (genotypes 1 to 9 - *T. dicoccoides*; cvs. Bethlehem and Dariel - *T. aestivum*; cv. Bareket - *T. durum*). Plants were grown in natural soils from ten different *T. dicoccoides* habitats over three consecutive years. In the 1st year no fertilization was added, in the 2nd year N was supplied, and in the 3rd year NPK or NPKS was added. Data represent the average of each genotype that was grown in all soils; vertical bars represent the SE. The symbols ↓, ↑ represents lowest or highest deviation from the norm data.

TABLE 2. Elements-to-nitrogen ratio (mg element / g N) in grains of 12 wheat genotypes (genotypes 1 to 9 - *T. dicoccoides*; cvs. Bethlehem and Dariel - *T. aestivum*; cv. Bareket - *T. durum*). Plants were grown in natural soils from ten different *T. dicoccoides* habitats over three consecutive years. In the 1st year no fertilization was added, in the 2nd year N was supplied, and in the 3rd year NPK or NPKS was added. Data represent the average of each genotype that was grown in all soils, ±SE (in *italics*).

Genotype /cultivar	Year & fertilization	P/N	K/N	S/N	Mg/N	Ca/N	Na/N	Mn/N	Zn/N	Fe/N
1	1, none	169.8 *9.3*	141.0 *9.4*	65.9 *3.8*	50.5 *3.0*	32.4 *5.4*	59.5 *10.6*	1.7 *0.1*	3.2 *0.2*	1.5 *0.13*
1	2, N	86.5 *6.6*	77.7 *4.1*	49.9 *4.9*	34.5 *1.8*	23.2 *2.5*	21.6 *3.5*	0.9 *0.1*	2.8 *0.3*	0.7 *0.04*
1	3, NPK	120.8 *6.2*	118.2 *5.6*	36.7 *2.1*	50.2 *2.7*	12.1 *0.9*	3.5 *0.1*	1.3 *0.1*	2.6 *0.2*	1.0 *0.12*
1	3, NPKS	105.4 *6.2*	112.0 *5.7*	36.7 *3.4*	42.3 *2.1*	19.9 *7.0*	3.6 *0.4*	1.4 *0.4*	2.0 *0.2*	1.1 *0.26*
2	1, none	214.7 *18.0*	200.3 *14.2*	69.9 *5.6*	70.5 *6.3*	19.8 *6.5*	39.5 *7.4*	1.3 *0.1*	3.7 *1.3*	1.4 *0.15*
2	2, N	104.7 *9.4*	119.5 *5.8*	42.7 *2.5*	48.9 *3.1*	26.6 *1.5*	26.7 *1.7*	0.8 *0.1*	2.1 *0.3*	1.0 *0.11*
2	3, NPK	155.2 *0.2*	154.6 *6.2*	31.6 *3.0*	64.9 *2.8*	16.2 *0.2*	7.4 *0.6*	1.1 *0.0*	4.4 *0.7*	1.6 *0.50*
2	3, NPKS	109.5 *9.0*	140.3 *16.5*	46.2 *3.7*	49.5 *2.6*	18.5 *6.3*	7.3 *3.0*	0.9 *0.2*	2.7 *0.6*	1.4 *0.30*
3	1, none	187.1 *7.4*	174.8 *8.1*	66.0 *2.3*	65.6 *2.7*	14.4 *1.1*	40.5 *5.8*	1.9 *0.1*	4.6 *1.2*	3.5 *0.97*
3	2, N	105.4 *9.2*	125.9 *5.0*	45.9 *2.3*	49.1 *2.8*	32.3 *1.9*	25.4 *1.8*	1.3 *0.1*	3.7 *0.3*	1.1 *0.04*
3	3, NPK	109.5 *10.1*	126.1 *8.8*	40.6 *2.9*	48.3 *5.1*	15.6 *0.7*	6.3 *0.8*	1.2 *0.3*	3.0 *0.2*	1.3 *0.21*
3	3, NPKS	101.5 *9.9*	140.0 *5.2*	40.4 *4.7*	47.4 *4.6*	21.4 *1.7*	8.7 *0.5*	1.1 *0.1*	2.1 *0.2*	1.5 *0.11*
4	1, none	161.9 *22.5*	161.1 *17.8*	45.3 *4.5*	54.5 *7.6*	9.3 *1.2*	40.2 *7.4*	1.3 *0.1*	1.8 *0.2*	1.3 *0.21*
4	2, N	97.4 *5.3*	123.5 *3.2*	46.6 *1.8*	39.4 *1.7*	28.1 *2.7*	28.7 *4.1*	0.9 *0.1*	4.3 *0.6*	1.4 *0.08*
4	3, NPK	138.4 *3.7*	204.4 *3.9*	27.8 *1.8*	53.8 *2.4*	13.6 *0.6*	5.5 *0.1*	1.1 *0.1*	2.5 *0.3*	1.1 *0.11*
4	3, NPKS	99.9 *11.0*	161.2 *10.4*	31.3 *1.6*	45.0 *3.1*	13.0 *1.1*	5.1 *0.3*	0.8 *0.1*	1.4 *0.2*	1.0 *0.06*
5	1, none	204.6 *16.2*	219.8 *14.9*	65.1 *2.9*	62.8 *4.8*	15.6 *0.8*	40.7 *3.8*	1.5 *0.1*	5.6 *1.5*	1.3 *0.24*
5	2, N	92.1 *8.0*	112.6 *4.3*	37.5 *2.5*	38.2 *2.0*	31.4 *3.5*	28.7 *4.2*	0.8 *0.1*	5.0 *1.1*	1.0 *0.16*
5	3, NPK	113.8 *9.8*	147.9 *9.2*	30.3 *2.5*	54.4 *4.1*	16.9 *1.6*	6.1 *0.4*	1.2 *0.2*	2.2 *0.5*	1.1 *0.07*
5	3, NPKS	93.6 *9.4*	145.0 *6.6*	35.8 *3.0*	46.7 *3.8*	18.1 *2.3*	6.3 *0.7*	1.2 *0.2*	1.4 *0.2*	1.1 *0.16*
6	1, none	230.4 *48.4*	227.3 *64.6*	77.5 *11.2*	72.5 *16.7*	23.3 *5.2*	78.3 *43.4*	2.0 *0.4*	10.6 *2.8*	0.6 *0.09*
6	2, N	95.6 *8.3*	117.3 *5.3*	47.8 *3.3*	44.8 *1.3*	33.4 *1.5*	27.9 *1.4*	1.0 *0.1*	3.8 *0.8*	1.2 *0.11*
6	3, NPK	95.9 *n.d.*	110.1 *n.d.*	81.1 *n.d.*	37.2 *n.d.*	15.9 *n.d.*	0.4 *n.d.*	1.1 *n.d.*	1.2 *n.d.*	1.0 *0.00*
6	3, NPKS	164.4 *11.5*	158.9 *12.2*	66.4 *4.8*	53.1 *2.0*	22.1 *2.0*	1.1 *0.4*	1.0 *0.1*	1.6 *0.2*	1.4 *0.11*
7	1, none	211.6 *16.8*	193.7 *17.0*	69.0 *5.5*	61.0 *6.2*	24.3 *6.1*	60.2 *18.4*	1.4 *0.1*	4.5 *0.8*	0.5 *0.10*
7	2, N	116.1 *11.6*	142.0 *9.4*	52.9 *2.5*	34.0 *2.2*	16.4 *1.3*	1.8 *0.2*	0.8 *0.1*	1.1 *0.2*	0.9 *0.05*
7	3, NPK	175.2 *8.5*	169.8 *14.8*	39.2 *7.9*	53.2 *2.3*	18.0 *2.1*	1.1 *0.4*	0.9 *0.2*	2.8 *0.3*	1.3 *0.26*
7	3, NPKS	133.2 *13.2*	123.8 *7.0*	36.0 *5.9*	45.7 *4.1*	13.3 *1.4*	2.1 *0.7*	1.3 *0.4*	1.6 *0.4*	1.5 *0.38*
8	1, none	196.2 *15.4*	186.3 *12.6*	73.0 *3.6*	61.6 *4.1*	23.9 *4.0*	63.3 *12.6*	1.4 *0.1*	6.3 *1.2*	1.4 *0.26*
8	2, N	n.d.	n.d.	n.d.	n.d.	n.d.	n.d.	n.d.	n.d.	n.d.
8	3, NPK	147.7 *n.d.*	157.7 *n.d.*	64.4 *n.d.*	60.7 *n.d.*	15.6 *n.d.*	1.2 *n.d.*	1.1 *n.d.*	1.0 *n.d.*	1.3 *0.00*
8	3, NPKS	132.9 *10.0*	142.3 *6.7*	75.5 *4.0*	49.1 *3.5*	19.8 *2.7*	2.1 *0.7*	0.9 *0.1*	1.0 *0.2*	1.1 *0.06*
9	1, none	251.1 *76.9*	257.6 *75.8*	81.1 *24.2*	72.7 *21.2*	34.7 *11.0*	65.0 *20.5*	1.6 *0.4*	14.3 *6.7*	3.6 *1.20*
9	2, N	83.6 *5.5*	120.2 *3.9*	46.5 *2.9*	37.2 *1.5*	37.2 *1.8*	32.0 *1.5*	0.8 *0.1*	6.6 *1.5*	0.9 *0.08*
9	3, NPK	150.9 *n.d.*	149.9 *n.d.*	40.1 *n.d.*	48.0 *n.d.*	13.4 *n.d.*	1.3 *n.d.*	1.1 *n.d.*	2.8 *n.d.*	0.6 *0.00*
9	3, NPKS	107.8 *14.9*	116.2 *5.8*	33.0 *2.8*	39.4 *3.9*	19.9 *1.4*	1.9 *0.2*	0.7 *0.1*	0.5 *0.2*	0.6 *0.04*
Bethlehem	1, none	215.4 *16.4*	216.9 *14.4*	65.9 *5.4*	65.2 *4.3*	23.9 *1.0*	29.6 *3.3*	1.5 *0.1*	5.1 *0.7*	1.5 *0.16*
Bethlehem	2, N	108.9 *9.6*	187.0 *7.2*	32.5 *3.2*	46.5 *2.2*	18.7 *1.4*	6.3 *0.6*	1.1 *0.1*	3.9 *0.8*	1.1 *0.26*
Bethlehem	3, NPK	131.3 *7.1*	168.0 *7.4*	54.1 *4.6*	50.4 *2.9*	23.8 *1.3*	3.2 *0.1*	1.1 *0.1*	1.9 *0.2*	1.0 *0.06*
Bethlehem	3, NPKS	110.3 *7.2*	171.1 *7.3*	61.9 *2.6*	44.1 *1.6*	25.3 *3.5*	3.4 *0.3*	1.2 *0.1*	1.7 *0.2*	1.3 *0.10*
Dariel	1, none	220.4 *21.6*	248.9 *29.7*	69.1 *4.4*	69.6 *6.0*	19.7 *1.3*	42.2 *17.5*	2.1 *0.2*	7.6 *1.8*	2.0 *0.62*
Dariel	2, N	84.1 *8.4*	177.6 *8.8*	33.2 *3.0*	44.6 *2.3*	20.0 *1.3*	3.8 *0.2*	1.3 *0.1*	4.2 *1.0*	0.7 *0.07*
Dariel	3, NPK	169.5 *3.3*	169.6 *4.8*	41.5 *1.0*	69.0 *1.7*	26.2 *0.6*	3.7 *0.1*	1.8 *0.2*	3.1 *0.2*	1.4 *0.14*
Dariel	3, NPKS	125.7 *9.0*	184.5 *7.1*	50.9 *2.9*	53.6 *2.9*	26.1 *1.1*	3.5 *0.2*	1.5 *0.1*	1.4 *0.1*	1.0 *0.04*
Bareket	1, none	249.6 *12.3*	312.4 *11.7*	72.2 *3.5*	61.9 *3.2*	23.9 *1.4*	34.3 *3.6*	1.5 *0.1*	8.3 *2.2*	2.3 *0.70*
Bareket	2, N	107.7 *9.8*	222.6 *9.5*	46.6 *2.8*	44.3 *3.1*	16.6 *0.8*	10.1 *0.7*	0.9 *0.0*	3.6 *0.4*	0.8 *0.12*
Bareket	3, NPK	137.4 *10.5*	184.5 *7.8*	48.5 *2.3*	45.5 *5.5*	17.1 *1.2*	3.5 *0.7*	1.2 *0.2*	2.5 *0.6*	1.6 *0.29*
Bareket	3, NPKS	112.0 *11.2*	198.7 *12.4*	49.1 *2.7*	38.6 *3.0*	17.7 *2.2*	2.8 *0.2*	1.0 *0.2*	1.3 *0.1*	1.1 *0.12*

The ratio of a nutrient element to gram of N was either stable or variable (Table 2). Large differences existed in the amount of S in relation to a gram of N in the grains of different genotypes, and the ratio was affected by fertilization regime. During the 1st year, when no N was added (and was the yield limiting element) and S was not yet depleted, all genotypes attained a S/N ratio above 60 (except genotype 4 which, even under these growth conditions, produced grains with only 45 mg S/g N; Fig. 2). Nitrogen fertilization without addition of S usually reduced the S/N ratio to <50 mg/g (Fig. 2). The reduction of the S/N ratio was less after PK addition in genotypes 6, 8 and *T. aestivum* cv. Bethlehem (the 3rd year).

The Zn/N ratio can be stable, as shown for genotype 1 (Table 2), but it can also be variable (genotype 5 and *T. aestivum* cv. Bethlehem) or even unstable (genotypes 6 and 9). Common wheat grains contained more K per gram N than did wild emmer wheat grains, especially when nitrogen was applied. The highest K/N ratio that existed in genotype 1 was even lower than the minimum level of all common wheats. Under NPK fertilization, the Mg/N ratio in the grains of genotype 2 and *T. aestivum* cv. Dariel was as high as that without any fertilization (~65 mg/g). However, in genotype 3 and *T. aestivum* cv. Bethlehem, this ratio was the same as when N or NPKS was added (~50 mg/g). The Na/N ratio exhibited the greatest reduction (in overall minerals ratio) over the years in all genotypes. This was the result of Na depletion due to irrigation with deionized water. This shows that the available Na reserves in the soil were much lower in comparison with those of the other nutrients in the studied soils.

4. Discussion

Ionic content, as regulated by nutrient excesses or deficiencies, is one option for the development of tolerance to water shortage, salinity, and other stresses (Bohnert *et al.*, 1995). The ionic content in the grain might reflect the adaptation to the growth conditions of a specific habitat (Bonfil and Kafkafi, in preparation). In this greenhouse study, the growth conditions were free of stress ascribable to water or salinity. The mineral content in wheat grain is affected mainly by the minerals available in the soil and supplied by fertilizer. Depletion of S and all the micronutrients, and enrichment of N or NPKS after fertilization in the 2nd and 3rd years, respectively, led to the wide range in the mineral element-to-nitrogen ratio in the seeds (Table 2).

Different Triticeae species, as well as common and wild wheat, contain genes that control mineral uptake, as was shown for K/Na discrimination in the shoot (Gorham, 1993). This discrimination does not occur in *T. durum* or *T. dicoccoides*, since this trait gene is located on the long arm of chromosome 4D of the hexaploid wheat, on chromosome 4A of diploid wheat, and absent in both *T. durum* and *T. dicoccoides* (Gorham, 1993). However, Nevo *et al.* (1992) showed variation in Na uptake among *T. dicoccoides* genotypes, and Dvořák *et al.* (1994) reported that the K/Na discrimination trait (*Kna1* locus) can be transferred from *T. aestivum* to *T. durum*. The K/Na in grains is not the same as in leaves, as a result of different rates of K and Na translocation in the phloem (Marschner, 1986). Therefore, leaf and grain data cannot be compared directly. The present study showed that *T. aestivum, T. durum* and *T. dicoccoides* all reach similar K/Na ratio in their grains. However, this ratio can be changed. Our work demonstrates diversity in element content in the grains. Moreover, we show that nutrient content per unit of N in wheat grains varies among genotypes, and is not stable even within each genotype at different nutrient supply levels in the soil. Although genotype 1 exhibited the lowest K/N ratio in grains, K is the limiting factor for grain production of this genotype (Bonfil and Kafkafi, 1995, In preparation).

The high protein content and genetic diversity in wild wheat grain are attractive to wheat breeders (Ciaffi *et al.*, 1991, 1992; Feldman and Sears, 1981; Grama *et al.*, 1987). Table 2 and Figure 1 show diversity among wild wheat genotypes in their grain mineral content and, even more important, in the minerals' ratio to nitrogen in grains. Sulphur is the second most important element, after N, in its influence on seed storage protein content and composition (Bonfil *et al.*, 1997; Randall and Wrigley, 1986; Wrigley *et al.*, 1980). The lowest critical S/N ratio for good baking quality is approximately 60-65 mg S/ g N (Randall and Wrigley, 1986), and therefore it is necessary to prevent a reduction in the S/N ratio. Hence, the variability found in this work could be of help in finding a new genetic source from the wild type to insure a high S/N ratio in grain even under sulphur deficiency, as shown especially for genotypes 6 and 8 in Figure 2 and Table 2. Flour baking quality could be related to nutrient concentration in the grains (Bequette *et al.*, 1963; Douglas and Dyson, 1985) and to protein composition (Randall and Wrigley, 1986). The concentrations of macro- and micronutrients in wheat grains differ among cultivars (Boila *et al.*, 1993). In this study we showed that cultivars may also differ in the ratio of mineral content to nitrogen content in the grain within each genotype, in addition to changes in protein components composition (Bonfil *et al.*, 1997). It is concluded that there is genetic diversity in discrimination and mineral concentration in wheat seeds. This genomic diversity exists in both wild and common wheat. This diversity can be used in wheat quality breeding programs as well as for adapting new cultivars to specific nutrient deficient soils.

5. References

Bequette, R.K., Watson, C.A., Miller, B.S., Johnson, J.A., and Schrenk, W.G. (1963) Mineral composition of gluten, starch and water-soluble fractions of wheat flour and its relationship to flour quality, *Agron. J.* **55**, 537- 542.

Bohnert, H.A., Nelson, D.E., and Jensen, R.G. (1995) Adaptation to environmental stresses, *Plant Cell.* **7**, 1099-1111.

Boila, R.J., Campbell, L.D., Stothers, S.C., Crow, G.H., and Ibrahim, E.A. (1993) Variation in the mineral content of cereal grains grown at selected locations throughout Manitoba, *Can. J. Anim. Sci.* **73**, 421-429.

Bonfil, D.J., Czosnek, H., and Kafkafi, U. (1997) Wheat seed storage protein fingerprint changes due to soil types and their mineral content, *Euphytica* **95**, 209-219.

Bonfil, D.J. and Kafkafi, U. (1995) Plant genotype effects on efficient use of plant nutrients, *Proc. 24th Colloquium International Potash Institute, Chiang Mai, Thailand*, pp. 557-573.

Bonfil, D.J. and Kafkafi, U. Seeds mineral concentration as a measure for wild wheat adaptation in different soil ecosystems, *In preparation.*

Ciaffi, M., Benedettelli, S., Giorgi, B., Porceddu, E., and Lafiandra, D. (1991) Seed storage proteins of *Triticum turgidum* ssp. *dicoccoides* and their effect on the technological quality in *durum* wheat, *Plant Breeding* **107**, 309-319.

Ciaffi, M., Dominici, L., Lafiandra, D., and Porceddu, E. (1992) Seed storage proteins of wild wheat progenitors and their relationships with technological properties, *Hereditas* **116**, 315-322.

Dan, Y., Yaalon, D.H., Koyumdjisky, H., and Raz, Z. (1976) *The Soils of Israel (with map 1:500.000)*. Division of Scientific Publications, Pamphlet No. 159. Agric. Res. Organization (ARO), The Volcani Center, Bet Dagan, Israel.

Douglas, J.A. (1987) The effect of various fertilisers and the elemental components of wheat and flour on baking quality, *Proc. Agron. Soc. New Zealand* **17**, 115-120.

Douglas, J.A. and Dyson, C.B. (1985) Preliminary investigation of concentrations of minerals and nitrogen in wheat grain, and their relationship with baking quality and grain weight, *New Zealand J. Agric. Res.* **28**, 81-85.

Dvořák, J., Noaman, M.M., Goyal, S., and Gorham, J. (1994) Enhancement of the salt tolerance of *Triticum turgidum* by the *kna1* locus transferred from the *Triticum aestivum* chromosome 4D by homoeologous recombination, *Theor. Appl. Genet.* **87**, 872-877.

Feldman, M. and Sears, E.R. (1981) The wild gene resources of wheat, *Sci. Am.* **244**, 98-109.

Gorham, J. (1993) Genetics and physiology of enhanced K/Na discrimination, in P.J. Randall *et al.* (eds.), *Genetics Aspects of Plant Mineral Nutrition,* Kluwer Academic Publishers, Dordrecht, The Netherlands, pp. 151-158.

Grama, A., Porter, N.G., and Wright, D.S.C. (1987) Hexaploid wild emmer wheat derivatives grown under New Zealand conditions. 2. Effect of foliar urea sprays on plant and grain nitrogen and baking quality, *New Zealand J. Agric. Res.* **30**, 45-51.

Jones, J.B. Jr. and Case, V.W. (1990) Sampling, handling, and analyzing plant tissue samples, in R.L. Westerman (ed.), *Soil Testing and Plant Analysis. 3rd ed.* SSSA Book Series, no. 3. Madison, WI, USA, pp. 389-427.

Marschner, H. (1986) *Mineral Nutrition of Higher Plants*, Academic Press, New York, NY.

Moss, H.J., Wrigley, C.W., MacRitchie, F., and Randall, P.J. (1981) Sulfur and nitrogen fertilizer effects on wheat. II Influence on grain quality, *Aust. J. Agric. Res.* **32**, 213-226.

Nevo, E. (1986) Genetic resources of wild cereals and crop improvement: Israel, a natural laboratory, *Isr. J. Bot.* **35**, 255-278.

Nevo, E. and Beiles, A. (1989) Genetic diversity of wild emmer wheat in Israel and Turkey. Structure, evolution, and application in breeding, *Theor. Appl. Genet.* **77**, 421-455.

Nevo, E., Gorham, J., and Beiles, A. (1992) Variation for ^{22}Na uptake in wild emmer wheat, *Triticum dicoccoides* in Israel: Salt tolerance resources for wheat improvement, *J. Exp. Bot.* **43**, 511-518.

Rahman, S., Shewry, P.R., Forde, B.G., Kreis, M., and Miflin, B.J. (1983) Nutritional control of storage-protein synthesis in developing grain of barley (*Hordeum vulgare* L.), *Planta* **159**, 366-372.

Randall, P.J., Thomson, J.A., and Schroeder, H.E. (1979) Cotyledonary storage proteins in *Pisum sativum*. IV Effects of sulfur, phosphorus, potassium and magnesium deficiencies, *Aust. J. Plant Physiol.* **6**, 11-24.

Randall, P.J. and Wrigley, C.W. (1986) Effects of sulphur supply on the yield, composition, and quality of grain from cereals, oilseeds, and legumes, *Adv. Cereal Sci. Technol.* **8**, 171-206.

Shewry, P.R., Franklin, J., Parmar, S., Smith, S.J., and Miflin, B.J. (1983) The effects of sulphur starvation on the amino acid and protein compositions of barley grain, *J. Cereal Sci.* **1**, 21-31.

Shewry, P.R., Napiar, J.A., and Tatham, A.S. (1995) Seed storage proteins: structures and biosynthesis, *Plant Cell* **7**, 945-956.

Smith, V.R. (1980) A phenol-hypochlorite manual determination of ammonium-nitrogen in Kjeldahl digests of plant tissue, *Commun. Soil Sci. Plant Anal.* **11**, 709-722.

Wrigley, C.W., du Cros, D.L., Archer, M.J., Downie, P.G., and Roxburgh, C.M. (1980) The sulphur content of wheat endosperm proteins and its relevance to grain quality, *Aust. J. Plant Physiol.* **7**, 755-766.

UPTAKE OF NITRATE, AMMONIUM AND SULPHATE BY SEEDLINGS OF TWO CONIFER SPECIES: THE DECIDUOUS LARCH AND THE EVERGREEN SCOTS PINE

A. BOTTACIN, A.M. DAL CANAL, M. MALAGOLI,
S. QUAGGIOTTI and G. FERRARI
Department of Agricultural Biotechnology, University of Padua
Agripolis, Strada Romea 16, 35020 Legnaro, Italy.

Abstract

Larix decidua and *Pinus sylvestris* demonstrated a preferential uptake of NH_4^+ over NO_3^- as appeared by the 5-fold higher NH_4^+ uptake rate and the lower Cmin in comparison with NO_3^- induced or non-induced uptake and Cmin. The NH_4^+ uptake rate was unaffected by previous nitrogen supply, while it was reduced (-25%) by sulphur starvation. Vmax for SO_4^{2-} was 410 and 625 nmol g^{-1}fw h^{-1} and Km 68 and 182 μM, respectively in larch and pine. In the comparison between the two species, larch demonstrates a better capacity of adaptation to the paucity of nutrient availability as appeared by the lower Cmin for both NH_4^+ and NO_3^-, the higher NH_4^+ and NO_3^- uptake rates (2 folds) and finally the higher affinity of SO_4^{2-} transport system.

Introduction

European larch (*Larix decidua* Mill.) and Scots pine (*Pinus sylvestris* L.), two conifers naturally occurring and widespread in Alpine mountains, are both valuable for ecological and economic aspects. They grow at altitudes ranging from 700 m till the timberline, depending on climatic pattern, where the edaphic conditions of soils are usually very poor and highly heterogeneous both spatially and temporally due to the low mineralization rate, the rocky background and the water deficit caused by frost temperature of soil for several months of the year. Despite mineral nutrition has long been recognised to play an essential role in biomass production of plants, only recently nutrient uptake (mainly N) by conifers has been taken into account for the increasing exposure of forest plants to stress conditions. Among these, the heterogeneous mineral distribution in soil (George *et al.*, 1997), nitrogen and sulphur deposition (Flaig and Mohr, 1992; Bauer *et al.*, 1997) and the increase of NO_3^- to NH_4^+ ratio after forest disturbance (Kronzucker *et al.* 1997) are considered the major nutritional stresses. In this view, the knowledge of the physiological utilisation of ions by conifers appears

G. Gissel-Nielsen and A. Jensen (eds.), Plant Nutrition – Molecular Biology and Genetics, 35–38.

particularly interesting not only in forest-grown trees but also in young plants or seedlings cultured under defined laboratory conditions.
In the present investigation we evaluated the uptake rate of nitrate, ammonium and sulphate by nonmycorrhizal larch and pine seedlings grown in sterile sand, then transferred to aerated hydroponic solutions. The measures of NO_3^-, NH_4^+ and ^{35}[S]-SO_4^{2-} uptake were carried out in plants previously pre-treated with either complete nutrient solution or deprived of each ion (NO_3^-, NH_4^+, SO_4^{2-}). The effect of S-starvation on ammonium uptake was also determined.

Materials and methods

Seeds of *Larix decidua* and *Pinus sylvestris* were surface-sterilised (30% v/v H_2O_2 for 30 min) and germinated at 24°C and 90% RH in sterile moist sand. Seedlings were grown with a 14 h daylength, under 200 μmol m^{-2} s^{-1}, day/night temperature of 24/18 °C, 50/70% RH and watered every 3 d with 1:50 strength Arnon and Hoagland (1940) nutrient solution. After four weeks plants were transferred to an aerated nutrient solution containing 100 μM of either NO_3^-or NH_4^+or no nitrogen and pre-treated for 3 d in each solution. In N-deprived solutions $NO_3^-$and NH_4^+ were respectively replaced by Cl^- and K^+. Uptake rate of $NO_3^-$and NH_4^+ was calculated by depletion of N-ions from the experiment solution (100 μM NO_3^-or NH_4^+) and measured colorimetrically by the method of Cawse (1967) and of Weatherburn (1967), respectively. Sulphate uptake was determined in seedlings pre-treated either in condition of steady supply or after a 4 d period of SO_4^{2-}-deprivation. S-free solution was obtained replacing $MgSO_4$ with $MgCl_2$. ^{35}[S]-SO_4^{2-} uptake (100 μM SO_4^{2-}, 5 MBq $mmol^{-1}$ specific activity) was evaluated in the digested tissue (130% v/v H_2O_2 at 85 °C) of whole plant by liquid scintillation analyser.

Results and discussion

Plants previously starved of NO_3^-, NH_4^+ or SO_4^{2-} exhibited different behaviour in the uptake rate after the exposure to solutions containing each ion. The uptake of NO_3^- by non-induced and induced seedlings was delayed 6-9 h and 4-5 h, respectively, following the exposure to nitrate. On the other hand, the onset of NH_4^+ and SO_4^{2-} uptake appeared immediately after the exposure to solution containing ammonium and sulphate, respectively. In NO_3^- deficient plants the uptake rate of nitrate was lower than in NO_3^- sufficient ones, the decrease being only 5% in larch and 45% in pine. The induction of NO_3^- transport system by nitrate is a common feature in plants, including *P. sylvestris* but seems not involve *L. decidua*. Similar differences were reported for white spruce, characterised by NO_3^- inducible transport, and aspen requiring no previous NO_3^- treatment (Kronzucker *et al.*, 1997). NH_4^+ uptake rate was irrespective to N-pre-treatment (nitrate, ammonium or none), suggesting that NH_4^+ transport doesn't require induction (Kronzucker *et al.*, 1997), but it was reduced by 25% by

SO_4^{2-} deprivation in both species. A co-regulation of N-S nutrition on their reciprocal transport systems was reported by Clarkson *et al.* (1989) in barley where a depression of NH_4^+ intake by sulphur starvation was observed. The two conifers demonstrated a preferential uptake of NH_4^+ over NO_3^-, as appeared by the higher NH_4^+ uptake rate (1.06 larch and 0.55 pine μmol g^{-1}fw h^{-1}, induced; 0.92 larch and 0.54 pine μmol g^{-1}fw h^{-1}, non-induced) and the lower Cmin (0 larch and 7μM pine) in comparison with NO_3^- induced (0.20 and 0.18 μmol g^{-1}fw h^{-1}) or non-induced (0.19 and 0.10 μmol g^{-1}fw h^{-1}) transport system and Cmin for nitrate ranging from 21 to 43 μM (Tab. 1). The 5-fold higher uptake rate of ammonium and the lower Cmin with respect to nitrate observed in larch is of similar magnitude as in spruce (Marschner *et al.*, 1991; Kronzucker *et al.*, 1996; Kronzucker *et al.*, 1997). Sulphate uptake was derepressed by S-starvation by 74% and 55%, respectively in larch and pine, as in most other species (Clarkson and Saker, 1989; Malagoli *et al.*, 1994). The kinetic parameters of sulphate uptake in S-deprived plants were Vmax 410 and 625 nmol g^{-1}fw h^{-1} and Km 68 and 182 μM, respectively in larch and pine (Tab. 1). The same order of Km magnitude was observed in beech (Kreuzwieser *et al.*, 1996). Even if SO_4^{2-} uptake by larch was lower than pine, after S-starvation larch exposed to SO_4^{2-} showed a 3 times higher internal pool of sulphate with respect to pine (Tab. 1), demonstrating a higher ability of recovery from sulphur shortage. In the comparison between the two species, larch demonstrated a better capacity of adaptation to the paucity of nutrient availability and to the increase of NO_3^- input, as appeared by the lower Cmin for both NH_4^+ and NO_3^-, the higher NH_4^+ and NO_3^- uptake rates (2 folds), the higher affinity of SO_4^{2-} transport system and finally the better response to sulphur deprivation.

TABLE 1. Biochemical parameters of nitrate, ammonium and sulphate uptake by European larch and Scots pine seedlings. Uptake rate of NO_3^-, NH_4^+, SO_4^{2-} and Vmax are expressed in nmol g^{-1}fw h^{-1}; Cmin and Km are expressed in μM; SO_4^{2-} internal pool in nmol g^{-1}fw. Values are averages of 4-5 independent experiments ± standard error. Seedling pre-treatment is indicated in brackets.

Biochemical parameters	Pre-treatment	Larch	Pine
NO_3^- uptake	($+NO_3^-$)	200±18	180±14
NO_3^- uptake	($-NO_3^-$)	190±15	100±8
NH_4^+ uptake	($+NO_3^-$)	1060±42	550±24
NH_4^+ uptake	(-N)	925±33	548±35
NH_4^+ uptake	($+NH_4^+$)	1045±38	553±27
NH_4^+ uptake	($-SO_4^{2-}$)	812±22	424±18
Cmin NO_3^-	($+NO_3^-$)	21±1.0	35±2.2
Cmin NO_3^-	($-NO_3^-$)	29±1.5	43±3.1
Cmin NH_4^+	($+NH_4^+$)	00±00	7±0.5
SO_4^{2-} uptake	($-SO_4^{2-}$)	260±21	340±27
SO_4^{2-} uptake	($+SO_4^{2-}$)	150±13	220±19
SO_4^{2-} pool	($-SO_4^{2-}$)	350±25	120±12
Vmax SO_4^{2-}	($-SO_4^{2-}$)	410±15	625±33
Km SO_4^{2-}	($-SO_4^{2-}$)	68±9.0	182±15

Acknowledgement

We gratefully acknowledge M. Cagnin for his skillful technical and computing assistance and Dr. F. Gorian of the Ministry for Agriculture and Forestry Resources for supplying seeds. The research was financially supported by the National Research Council.

References

Arnon, D.J. and Hoagland, D.R. (1940) Crop production in artificial culture and in soil with special reference to factors influencing yields and absorption of inorganic nutrient, *Soil Sci.* **50**,463-483.

Bauer, G., Schulze, E.-D. and Mund, M. (1997) Nutrient contents and concentrations in relation to growth of *Picea abies* and *Fagus sylvatica* along European transect, *Tree Physiol.* **17**,777-786.

Cawse, P.A. (1967) The determination of nitrate in soil solution by ultraviolet spectrophotometry, *Analyst.* **92**,311-315.

Clarkson, D.T. and Saker, L.R. (1989) Sulphate influx in wheat and barley roots becomes ore sensitive to specific protein-binding reagents when plants are sulphate-deficient, *Planta.* **178**,249-257.

Clarkson, D.T., Saker, L.R. and Purves, J.V. (1989) Depression of nitrate and ammonium transport in barley plants with diminished sulphate status. Evidence of co-regulation of nitrogen and sulphate intake, *J. Exp. Bot.* **40**,953-963.

Flaig, H. and Mohr, H. (1992) Assimilation of nitrate and ammonium by Scots pine (*Pinus sylvestris*) seedling under conditions of high nitrogen supply, *Physiol. Plant.* **84**,568-576.

George, E., Seith, B., Schaeffer, C. and Marschner, H. (1997) Response of *Picea, Pinus* and *Pseudotsuga* roots to heterogeneous nutrient distribution in soil, *Tree Physiol.* **17**,39-45.

Kreuzwieser, J., Hersbach, C. and Rennenberg, H. (1996) Sulfate uptake and xylem loading of non- mycorrhizal excised roots of young *Fagus sylvatica* trees. *Plant Physiol.* **34**,409-416.

Kronzucker, H., Siddiqui, M.Y. and Glass, A.D.M. (1996) Kinetics of NH_4^+ influx in spruce, *Plant Physiol.* **110**,773-779.

Kronzucker, H., Siddiqui, M.Y. and Glass, A.D.M. (1997) Conifer root discrimination against soil nitrate and the ecology of forest succession, *Nature.* **385**,59-61.

Malagoli, M., Trettenero, A. and Ferrari, G. (1994) Response of root mRNA population to sulphate shortage in maize seedlings, *Plant Soil.* **162**,309-313.

Marschner, H., Häussling, M. and George, E. (1991) Ammonium and nitrate uptake rates and rhizosphere pH in non-mycorrhizal roots of norway spruce (*Picea abies* L. Karst), *Trees* **5**,14-21.

Weatherburn, M.W. (1967) Phenol-hypoclorite reaction for determination of ammonia, *Anal. Chem.* **39**,971-974.

CONTROL OF PROTEIN CONTENT IN THE RICE PHLOEM SAP

A. FUKUDA, Y. ISHIWATARI, K ABE, M CHINO, T. FUJIWARA and H. HAYASHI
Department of Applied Biological Chemistry, The University of Tokyo, Yayoi, Bunkyo-ku, Tokyo 113-8657, Japan

1. Abstract

We describe the alteration of protein contents in the phloem sap of rice plants by introduction of transgenes. The coding sequences of rice thioredoxin h (TRXh) and oryzacystatin-I (OC-I) were fused in sense and antisense orientations to the promoter region of the rice TRXh gene. These proteins are present in rice phloem sap. The contents of TRXh and OC-I were reduced in the phloem sap of the transgenic plants carrying the corresponding antisense cDNAs. The content of OC-I protein increased in the phloem sap of transgenic plants carrying the TRXh promoter sense OC-I cDNA. This is the first demonstration of the control of the phloem protein contents in rice plants by analysis of phloem sap of transgenic plants.

2. Introduction

In vascular plants, phloem is the route of long distant transport of photoassimilates and signals. Over a hundred proteins have been detected in wheat and rice phloem sap collected through cut ends of insect stylets (Fisher et al. 1992; Nakamura et al. 1993). Wheat phloem proteins have been shown to constantly turn over (Fisher et al. 1992). Sieve elements, the main component of sieve tubes, lose their nuclei and most ribosomes during the course of differentiation (Cronshaw 1981). Thus, the sieve elements are considered to be incapable of protein synthesis. Therefore proteins in the sieve elements are assumed to be transported from neighboring companion cells. Sieve

G. Gissel-Nielsen and A. Jensen (eds.), Plant Nutrition – Molecular Biology and Genetics, 39–45.

elements and companion cells are connected by specialized plasmodesmata (Lucas et al. 1993). Upon microinjection into *Cucurbita* and tobacco mesophyll cells, some phloem proteins have been shown to dilate and traffic through plasmodesmata (Balachandran et al. 1997; Ishiwatari et al. 1998). As the mRNA of some phloem proteins were mainly detected in companion cells (Dannenhoffer et al. 1997; Ishiwatari et al. 1998), phloem proteins are likely synthesized in companion cells and transported to sieve elements through the plasmodesmata.

To examine whether or not alteration of protein contents in companion cells leads to changes in protein contents in phloem sap in rice plants, we introduced rice thioredoxin h (TRXh) and oryzacystatin-I (OC-I) cDNA under the cauliflower mosaic virus 35S RNA promoter (p35S) and the rice TRXh promoter (pTRXh) into rice plants. Rice TRXh, a 13kDa protein, is one of the major proteins in rice phloem sap (Ishiwatari et al. 1995), and TRXh mRNA is mainly found in companion cells in the mature leaves of rice plants (Ishiwatari et al. 1998). OC-I, a 11kDa protein, is a cystein protease inhibitor present in rice seeds (Abe et al. 1987) and in phloem sap. We report the alteration of contents of TRXh and OC-I in the phloem sap of transgenic rice plants.

3. Experimental procedure

3.1. CONSTRUCTION OF PLASMIDS

3.1.1. *The constructs carrying p35S-TRXh cDNA (sense or antisense orientation)*

The cDNA fragment of TRXh (Ishiwatari et al., 1995), was digested with *Not*I and subcloned into the pBluescriptII SK(+) (Stratagene, CA, U.S.A). Two types of plasmids were obtained; pλ3121 which has the TRXh cDNA in a sense orientation to lacZ, and pλ3122 which has the insert in an antisense orientation. The DNA fragment containing p35S, β-glucuronidase gene (*uid*A) and nopaline synthase polyadenylation signal (tNOS) was excised from pBI221 (Clontech, CA, U.S.A) using *Eco*RI and *Hind*III, and inserted into the same sites of pBluescripII SK(+) of which the *Sac*I and *Xba*I fragment was eliminated by *Sac*I and *Xba*I digestion, blunting, and re-ligation prior to subcloning of the fragment. The *uid*A fragment of the resulting plasmid was replaced with the *Sac*I-*Xba*I fragment of TRXh cDNA excised from pλ3121 or pλ3122. The fragments containing p35S-TRXh cDNA (antisense or sense orientation)-tNOS were excised using *Spe*I and *Hind*III, and inserted into the corresponding sites of pTF339 vector, which contains p35S-hygromycin phosphotransferase gene (HPH)-tNOS fragment in place of the kanamycin selectable marker of pBIN19 (Bevan 1984, Fig1-A).

3.1.2 *The construct carrying pTRXh-TRXh cDNA (antisense orientation)*

An 1.0 kbp fragment containing 5' promoter region of the TRXh gene (pTRXh) (Genbank accession No.D26547, Ishiwatari et al., 1997) was excised using *Bam*HI and *Eae*I, and inserted into the *Bam*HI-*Not*I sites of pλ3121 to ligate the pTRXh and TRXh cDNA in antisense orientation. The pTRXh-TRXh cDNA (antisense orientation) fragment was excised using *Bam*HI and *Sac*I, and inserted into the same sites of pBI221. The *Bam*HI-*Eco*RI fragment containing pTRXh-TRXh cDNA (antisense orientation)-tNOS was then subcloned into the same sites of pBluescriptII SK(+). Then the pTRXh-TRXh cDNA (antisense orientation)-tNOS fragment was excised using *Xba*I and *Hin*dIII, and inserted into the same site of pTF338, which contains p35S-HPH-tNOS fragments in reverse orientation to that of pTF339 (Fig.1-B).

3.1.3. *The constructs carrying pTRXh-OC-I cDNA (sense or antisense orientation)*

The promoter region of TRXh from -1015 to -16 bp from the translation start site was amplified by a polymerase chain reaction using two primers:

5' CTAGTGGATCCATCTAAAATGGGAATAGAG 3';

3' CTAGAGCAGCGGCTCCTTAAGCCTAGGCTC 5'

The amplified fragment was digested with *Bam*HI and *Eco*RI, and subcloned into pBluescriptII SK(+). The resulting plasmid was digested with *Eco*RI and *Kpn*I, and ligated with the *Eco*RI-*Kpn*I fragment of OC-I cDNA (Abe et al. 1987). The pTRXh-OC-I cDNA fragment was excised using *Sac*I, and inserted into pUC19 which contains the tNOS fragments in the *Sac*I-*Bam*HI sites. The OC-I cDNA fragment was isolated by *Eco*RI and recloned into the same site in the antisense orientation. The pTRXh-OC-I cDNA (sense or antisense orientation)-tNOS fragments were excised using *Xba*I and inserted into the same site of pTF338 (Fig.1-C).

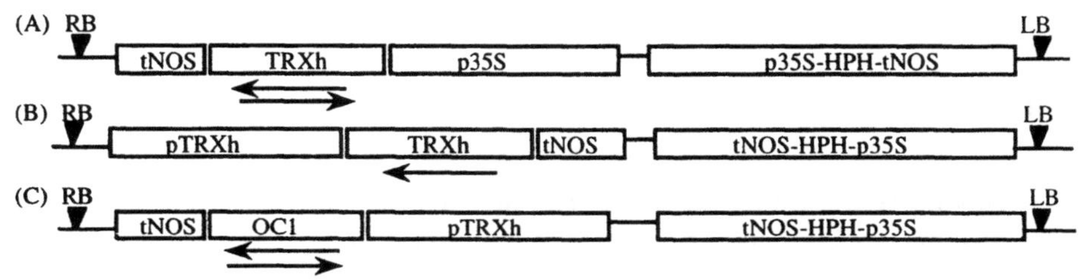

Fig.1. Structure of the chimeric genes used for rice transformation
(A): p35S-TRXh cDNA (sense or antisense orientation), (B): pTRXh-TRXh cDNA (antisense orientation), (C): pTRXh-OC-I cDNA (sense or antisense orientation). The arrows indicate sense direction of fragments. RB, T-DNA right border; LB, T-DNA left border; p35S, cauliflower mosaic virus 35S RNA promoter; tNOS, nopaline synthase terminator; HPH, hygromycin phosphotransferase; pTRXh, promoter region of thioredoxin h; TRXh, thioredoxin h cDNA; OC-I, oryzacystatin-I cDNA.

3.2. TRANSFORMATION OF RICE PLANTS

The plasmids were introduced into *Agrobacterium tumefaciens* strain C58C1Rif^R (pGV2260) (Deblaere et al. 1985) by triparental mating using *E.coli* strain DH1 (pRK2013) as a helper (Figurski et al. 1979). Rice (*Oryza sativa* L. cv. Tsukinohikari) transformation was according to Hiei et al (1994).

3.3. SOUTHERN ANALYSIS OF GENOMIC DNA

Total DNA was extracted from leaf tissue with cetyl trimethyl ammonium bromide (Murray and Thompson. 1980). Genomic DNA was digested with *Kpn*I and *Pst*I and then applied to a 0.6% agarose, 1xTris-acetate-EDTA gel. The separated DNA was transferred to Zeta probe blotting membranes (BIO RAD, CA, U.S.A). Hybridization was performed using Random Primer Fluorescein Labeling Kit (Dupont, MA, U.S.A.).

3.4. COLLECTION OF PHLOEM SAP AND WESTERN ANALYSIS

Rice phloem sap was collected from hydroponically cultured transgenic rice plants of 7th- or 8th-leaf stages by the insect laser technique (Kawabe et al. 1980). The phloem sap was separated by SDS-PAGE (Nakamura et al. 1993). Separated proteins were transferred to PVDF membranes. Immunoblotting was performed as described elsewhere (Ishiwatari et al. 1995) with antiserum against TRXh (Florencio et al. 1988) and OC-I (Kondo et al. 1989). The antiserum against TRXh were kindly provided by Dr. B. B. Buchanan (University of California, Berkeley, CA, U.S.A.).

4. Result

Transformants are shown in Table 1. Phloem saps of hygromycin resistant T2 transgenic plants were collected and subjected to western blot analysis. T2 plants used for the analysis were not necessarily homozygous. Contents of TRXh and OC-I were visually judged based on the intensities of the corresponding bands.

Table 1 contains the protein content results and Fig. 2 shows examples of such analysis. Decrease in TRXh protein was evident in two independent lines of transformants carrying p35S-antisenseTRXh and two independent lines of transformants carrying pTRXh-antisenseTRXh. The TRXh levels in phloem sap of three independent

Table1. The lines of transgenic plants and the results of western blot analysis

Introduced foreign gene	Number of independent T1 transformants	Names of transgenic lines used for experiments	TRXh and OC1 contents in phloem sap (a) (number of Hyg^R T2 plants)		
			TRXh contents		
			low	unchanged	high
vector control	2	VC1	0	1	0
		VC2	0	1	0
p35S -senseTRXh	3	p35S-senseTRX1	0	3	0
		p35S-senseTRX2	0	1	0
		p35S-senseTRX3	0	2	0
p35S -antisenseTRXh	3	p35S-antiTRX1	0	3	0
		p35S-antiTRX2	2	0	0
		p35S-antiTRX3	3	0	0
pTRXh -antisenseTRXh	2	pTRX-antiTRX1	2	0	0
		pTRX-antiTRX2	1	1	0
			OC1 contents		
pTRXh -senseOC-I	6	pTRX-senseOC1	0	0	1
		pTRX-senseOC2	0	0	2
		pTRX-senseOC3	0	0	2
		pTRX-senseOC4	0	0	1
		pTRX-senseOC5	0	0	2
		pTRX-senseOC6	0	0	1
pTRXh -antisenseOC-I	4	pTRX-antiOC1	2	0	0
		pTRX-antiOC2	0	1	0
		pTRX-antiOC3	0	1	0
		pTRX-antiOC4	0	1	0

(a) Phloem sap were collected from hygromycin resistant T2 plants of each line.

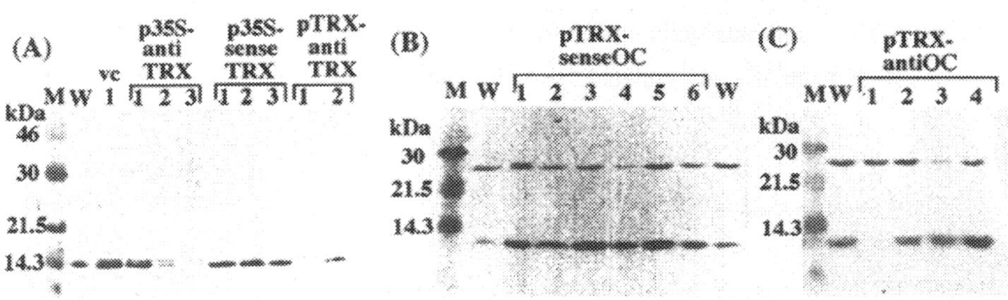

Fig. 2. Western blot analysis of TRXh and OC-I of phloem sap from transgenic plants

(A). Western blot analysis of TRXh. TRXh are the lower bands. The weaker band of a high molecular weight may be another thioredoxin h-like protein (Ishiwatari et al. 1995). The size-markers (M), one μl of the phloem sap from wild type (W), a vector control (VC1), p35S-antisenseTRXh lines (p35S-antiTRX), p35S-senseTRXh lines (p35S-senseTRX), pTRXh-antisenseTRXh lines (pTRX-antiTRX) were analyzed.

(B). Western blot analysis of OC-I. OC-I band is that with a molecular weight of 11kDa . The band with higher molecular weight (ca 25kDa) is not identified. Three μl of the phloem sap from wild type (W), pTRXh-senseOC-I lines (pTRX-senseOC) were analyzed.

(C). Western blot analysis of OC-I. Three μl of the phloem sap from wild type (W), pTRXh-antisenseOC-I lines (pTRX-antiOC) were analyzed.

lines of transgenic plants with p35S-sense TRXh and vector control, which has T-DNA region of pTF339, were similar to that of the wild type plant.

OC-I protein level was increased in all lines of sense transgenic plants. OC-I protein was not detected in one out of four lines of antisense transgenic plants.

Southern analyses were performed on the same plants used for the collection of phloem sap to examine the presence of the introduced trans genes (data not shown). pTRX-antiOC2 had a single copy, pTRX-antiOC1 had two or more copies of foreign OC-I genes. The antisense line in which OC-I contents in phloem sap were decreased had more copies of insertions compared to that of the plants with unchanged OC-I contents, suggesting that antisense inhibition is gene dosage dependent in our case.

5. Discussion

The results of present study showed that it was possible to control the protein contents in phloem sap of rice plants by transgene introduction. It has been shown that snowdrop lectin (GNA) was present in the phloem sap of transgenic tobacco which express GNA gene in phloem region driven by the rice sucrose synthase-1 promoter. The existence of GNA was shown by the analysis of honeydew of the aphid which feed on the transgenic tobacco plants (Shi et al. 1994).

Both TRXh promoter and CaMV 35S promoter reduced the TRXh contents in the phloem sap of some of the transgenic rice plants carrying antisense TRXh cDNA. TRXh promoter was active mainly in companion cells (Ishiwatari et al. 1997), and CaMV 35S promoter was active throughout the leaf tissue including companion cells. Gene expression in companion cells is likely to be required for control of the protein contents in phloem sap.

Although OC-I content in phloem sap was increased in all transgenic plants carrying sense OC-I cDNA, the TRXh content was not changed in the transgenic plants carrying sense TRXh cDNA. It is possible that the foreign TRXh gene was not active in the transgenic plants. Another explanation is that the sieve elements can accept only a limited amount of protein.

This work was supported by a Grant-in-Aid for Scientific Reseasch on Priority Areas (no.09274I01 and no.09274102) from the Ministry of Education, Science, Sports and Culture, Japan.

6. Reference

Abe, K., Emori, Y., Kondo, H., Suzuki, K., and Arai, S. (1987) Molecular cloning of a cysteine proteinase inhibitor of rice (oryzacystatin). *J. Biol. Chem.* **262**, 16793-16797.

Balachandran, S., Xiang, Y., Schobert, C., Thompson, G. A., and Lucas, W. J. (1997) Phloem sap protein from *Cucurbita maxima* and *Ricinus communis* have the capacity to traffic cell-to-cell through plasmodesmata. *Proc. Natl. Acad. Sci. USA* **94**, 14150-14155.

Bevan, M.W. (1984) Binary *Agrobacterium* vectors for plant transformation. *Nucleic Acids Res.* **12**, 8711-8721.

Cronshaw, J. (1981) Phloem structure and function. *Ann. Rev. Plant Physiol.* **32**, 465-484.

Dannenhoffer, J. M, Schulz, A., Skaggs, M. I., Bostwick, D. E., and Thompson, G. A. (1997) Expression of the phloem lectin is developmentally linked to vascular differentiation in cucurbits. *Planta* **201**, 405-414.

Deblaere, R., Bytebier, B., De Greve, H., Deboeck, F., Schell, J., Van Montagu, M., and Leemans, J. (1985) Efficient octopine Ti plasmid-derived vectors for *Agrobacterium*-mediated gene transfer to plants. *Nucleic Acids Res.* **13**, 4777-4788.

Figurski, D. H., and Helinski, D. R. (1979) Replication of an origin-containing derivative of plasmid RK2 dependent on a plasmid function provided in trans. *Proc. Natl. Acad. Sci. USA* **76**, 1648-1652.

Fisher, D. B., Wu, Y., and Ku, M. S. B. (1992) Turnover of soluble proteins in the wheat sieve tube. *Plant Physiol.* **100**, 1433-1441.

Florencio, F. J., Yee, B. C., Johnson, T.C., and Buchanan, B. B. (1988) An NADP/thioredoxin system in leaves: purification and characterization of NADP-thioredoxin reductase and thioredoxin h from spinach. *Arch. Biochem. Biophys.* **266**, 496-507.

Hiei, Y., Ohta, S., Komari, T., and Kumashiro, T. (1994) Efficient transformation of rice (*Oryza sativa* L.) mediated by *Agrobacterium* and sequence analysis of the boundaries of the T-DNA. *Plant J.* **6**, 271-282.

Ishiwatari, Y., Fujiwara, T., McFarland, K. C., Nemoto, K., Hayashi, H., Chino, M., and Lucas, W. J. (1998) Rice phloem thioredoxin h has the capacity to mediate its own cell-to-cell transport through lasmodesmata. *Planta* **205**, 12-22.

Ishiwatari Y (1997) Cell-to-cell movement and gene expression of rice phloem thioredoxin h. Ph. D. Thesis the Univ. Tokyo.

Ishiwatari, Y., Honda, C., Kawashima, I., Nakamura, S., Hirano, H., Mori, S., Fujiwara, T., Hayashi, H., and Chino, M. (1995) Thioredoxin h is one of the major proteins in rice phloem sap. *Planta* **195**, 456-463.

Kawabe, S., Fukumorita, T., and Chino, M. (1980) Collection of rice phloem sap from stylets of homopterous insects severed by YAG laser. *Plant Cell Physiol.* **21**, 1319-1327.

Kondo, H., Abe, K., and Arai, S. (1989) Immunoassay of oryzacystatin occurring in rice seeds during maturation and germination. *Agric. Biol. Chem.* **53**, 2949-2954.

Lucas, W. J., Ding, B., and van der Schoot, C. (1993) Plasmodesmata and the supercellular nature of plants. *New Phytol.* **125**, 435-476.

Murray, M. G., and Thompson, W. F. (1980) Rapid isolation of high molecular weight plant DNA. *Nucleic Acids Res.* **8**, 4321-4325.

Nakamura, S., Hayashi, H., Mori, S., and Chino, M. (1993). Protein phosphorylation in the sieve tubes of rice plants. *Plant Cell Physiol.* **34**, 927-933.

Shi, Y., Wang, M. B., Powell, K. S., Van-Damme, E., Hilder, V. A., Gatehouse, A. M. R., Boulter, D., and Gatehouse, J. A. (1994) Use of the rice sucrose synthase-1 promoter to direct phloem-specific expression of beta-glucuronidase and snowdrop lectin genes in transgenic tobacco plants. *J. Exp. Bot.* **45**, 623-631.

Effect of Different Irrigation Programs with Nitrogen Fertilizer Application on Nitrogen Use Efficiency and Fruit Quality in Tomato

M.Rüştü KARAMAN[1], Hikmet GÜLEÇ[2], Sabit ERŞAHİN[3]

[1] *Gaziosmanpaşa Univ., Agricultural Faculty, Department of Soil Sci., Tokat-Turkey.*
[2] *Research Institute of Rural Services, Tokat-Turkey.*
[3] *Depart. of Crops and Soils, 215 J. Hall Washington State Univ. Pullman W.A. 99164, USA.*

Abstract

A lysimeter study was conducted based on completely randomized block design at the Agricultural Research Institute of Tokat in 1996-98. The lysimeters were filled with soils taken from A_p, C_1 and C_2 horizons of the Kazova plain in Tokat-Turkey. Three seedlings of tomato variety H-2274 were transplanted to each lysimeter. Nitrogen fertilizer at the rates of 0 and 140 kg $N.ha^{-1}$ was side dressed in two split applications. The lysimeters were irrigated based on programs C 0.75, 1.00, 1.25 and 1.50, C referring to class A-Pan evaporation coefficients. The maximum yield was obtained with irrigation programs C 0.75 and C 1.00 together with 140 kg $N.ha^{-1}$ treatment. Significant negative correlations ($r = -0.66$ for 1996 and $r = -0.70$ for 1997) were found between Class A pan coefficients (C) and fertilizer-N-use. In general, N-uptake decreased with utilization of C 1.00, C 1.25 and C 1.50 programs possibly due to the maximum N-leaching losses occured with these practices. Considering fruit quality, the water soluble dry matter, protein and nitrate contents were decreased, however, total acidity and ascorbic acid were increased with increasing irrigation practices. The results further revealed that the concentrations of NO_3^- and NO_2^- in drainage water were closely related to leaching which were considerably higher with programs C 1.25 and C 1.50.

Introduction

Plant growth and yield are closely related to the internal water status of the plants (Begg and Turner,1976). Any drop in soil water potential below a critical level leads to the development of water deficit in leaves causing a decrease in reduced productivity (Meidner and Mansfield,1968). Different irrigation programs together with nitrogen fertilization had considerable effect on fertilizer-N-use efficiency. Availability and uptake of mineral nutrients are modified by soil water status (Begg and Turner,1976). In general, N-uptake and N-use efficiency decrease with high amount of irrigation possibly due to the maximum N-leaching losses occured with these practices. However, N-uptake and N-use efficiency decrease with reduced irrigation (Hegde and Srinivas, 1990). Irrigation and N-fertilization practices affect not only yield and N-use efficiency but also fruit quality. Quality properties of fruit may be affected with different irrigation programs together with N-fertilization. For example, WHO and FAO reports have recommended a 1000 mg NO_3 and 16 mg NO_2 for per 60 kg body weight of human (Corre and Breimer, 1979). Oxalic acid causes acute toxicity if taken more than 5

G. Gissel-Nielsen and A. Jensen (eds.), Plant Nutrition – Molecular Biology and Genetics, 47–51.

grams.day^{-1} by human. Below this dose it will react with Ca to make precipitation to form stones in Kidneys (Adrianse and Robbers,1970). Other quality properties may be varied with different irrigation practices and N-fertilization.

Materials and Methods

A lysimeter study was conducted based on completely randomized block design with three replications at the Agricultural Research Institute of Tokat in 1996-98. The lysimeters were 2 m high CTP cylinders having 0.21 m inner diameter. Each lysimeter was filled up to 0.5 m with coarse gravel and sand mixture, and the remaining 1.5 m was filled with Ap, C_1 and C_2 horizons of a fluvent from the Kazova Plain, Tokat-Turkey. The properties of experimental soil are given in Table 1. Three seedlings of tomato variety H-2274 were transplanted to each lysimeter. Nitrogen fertilizers, at the rate of 140 kg N ha^{-1} (N_1), were side dressed in two split applications, first half as ammonium sulfate and other half as ammonium nitrate on 4th and 21st days after transplanting, respectively. In addition, 45 kg P ha^{-1} as TSP was applied to each treatment. The lysimeters were irrigated on Class A-Pan evaporation coefficients (C) of 0.75 (I_1), 1.00 (I_2), 1.25 (I_3) and 1.50 (I_4), respectively. In control treatment (I_2N_0), irrigation programme I_2 was applied without nitrogen fertilizer. In this study, Class C_2S_1 (Ec=0.26 dS m^{-1}, SAR=0.5) water, containing 3.96-4.63 mg l^{-1} NO_3^-, were used for irrigation. Tomato yields were recorded, and leaf and fruit samples were taken for analysis. Drainage waters from the lysimeters were collected during the growth period. In 1996, experimental site had 12.3 °C av. temparature, and had 616 mm av. precipitation of which 5.7, 0.15, 2.06 and 7.40 % were in Jun., Jul., Aug. and Sept., respectively. In 1997, annual average temperature was 11.2 °C and an av. precipitation was 350 mm of which 3.7, 0.77, 7.71 and 1.80 % fell in Jun., Jul., Aug. and Sept., respectively. Total N (Chapman and Pratt, 1961), water soluble dry mat. (Anon., 1986), sugar (Anon.,1990), protein (Lowry et al., 1951), total acidity and ascorbic acid (Anon., 1978), oxalic acid (Adrianse and Robbers, 1970), nitrate (Anon., 1991), nitrite (Barnes and Folkard, 1951) were determined in fruit samples.

TABLE 1. Some characteristics of experimental soil

	Soil depth, cm				
Parameters	0-30	30-60	60-90	90-120	120-150
Horizon	Ap	C_1	C_2	C_2	C_2
Sand, %	30.5	18.2	22.0	22.1	18.0
Silt, %	28.5	34.5	39.6	35.0	35.0
Clay, %	41.0	47.3	38.4	42.9	47.0
Texture	C	C	CL	C	C
Field capacity, %	39.98	39.99	39.19	36.85	39.74
Wilting point, %	28.56	28.59	27.44	26.60	28.59
CEC, cmol.kg^{-1}	48.9	42.0	36.3	29.5	28.8
$CaCO_3$, %	9.3	9.6	12.6	14.8	15.9
EC, ds.m^{-1}	0.52	0.50	0.65	0.51	0.45
pH, 1:2 W/v	7.88	8.05	8.25	8.47	8.35
Avail. P, mg.kg^{-1}	32.56	21.50	11.49	11.49	11.00
Avail. K, mg.kg^{-1}	203.8	136.1	131.1	119.1	97.53
Total C, gr.kg^{-1}	16.8	15.7	13.3	11.0	3.50
Total N, gr.kg^{-1}	0.67	0.73	0.62	0.68	0.47

RESULTS and DISCUSSION

Yield and N-uptake

The highest fruit yield of 152 ton.ha^{-1} was obtained at I_1N_1 (irrrigation and fertilization) treatment in 1996, whereas the highest yield of 106 ton.ha^{-1} was obtained at I_2N_1 treatment in 1997 (Table 2). Although the fruit yields did not show a definite trend under various irrigation programs, the highest yield was obtained at the lowest level of irrigation during 1996. Fien and Kerchev (1976) also report similar findings. Hegde and Srinivas (1990) have found that the differences between -25, -45 and -65 kPa treatments were not significant. In our study, the Class A pan coefficients of 0.75-1.00 appears to be sufficient for tomato production under experimental conditions.

TABLE 2. Yield and N uptake of tomato depending on different irrigation programs with N-fertilization.

Irrig. and fertiliz.	*Fruit yield t.ha^{-1}*	*Fruit D.M.Y t.ha^{-1}*	*Fruit N-cont. %*	*N-uptake by fruit kg.ha^{-1}*	*Leaf+ Stem D.M.Y. t.ha^{-1}*	*L+S N-cont. %*	*N-uptake by L+S kg.ha^{-1}*	*Total N uptake kg.ha^{-1}*	*N-use effic. % #*
1996									
Control	128 b	6.40	1.49	95.4	2.26	1.53	34.6	130.0	-
I_1N_1	152 a	7.60	1.55	117.8	2.64	2.54	67.1	184.9	39.21
I_2N_1	119 b	5.95	1.58	94.0	2.41	2.63	63.4	157.4	19.57
I_3N_1	122 b	6.10	1.55	94.6	2.38	2.39	56.9	151.5	15.36
I_4N_1	139 ab	6.95	1.51	104.9	2.47	2.31	57.1	162.0	22.86
1997									
Control	84 bc	5.01	1.33	66.6	2.12	1.46	30.0	96.6	-
I_1N_1	93 b	5.58	1.48	82.6	2.38	2.11	50.2	132.8	25.86
I_2N_1	106 a	6.36	1.48	94.1	2.53	2.09	52.9	147.0	36.00
I_3N_1	96 b	5.76	1.45	83.5	2.42	2.00	48.4	131.9	25.21
I_4N_1	74 c	4.44	1.47	65.3	2.29	2.02	46.3	111.6	10.71

N-use efficiency.= Total N upt. (for 140 kg N.ha^{-1}) -Tot. N upt. (at 0 kg N.ha^{-1}) / Appl. N rate x 100

Total N uptake was the highest under I_1N_1 and I_2N_1 treatments, possibly due to the production of maximum amount of biomass under this program (Table 2). Fruit, leaf and stem-N contents tended to increase with N applications. However, the significant increase in yield was obtained at 140 kg N.ha^{-1} as compared to 0 kg N.ha^{-1} treatment with irrigation program I_1 and I_2. Total amounts of N taken up by plants also followed a similar pattern. These results are in agreement with the findings of Hegde and Srinivas (1990). Significant negative correlations (r = -0.66 for 1996 and r = -0.70 for 1997) were also found between Class A pan evaporation coefficients (C) and fertilizer-N-use efficiencies (Figure 1). Different irrigation programs had considerable effect on fertilizer-N-use efficiency. The highest N-use efficiency, based on dry weights, was obtained with irrigation programs I_1 and I_2 together with 140 kg N.ha^{-1} application in 1996 and 1997, respectively. These results are in agreement with the findings of Hills et al. (1983).

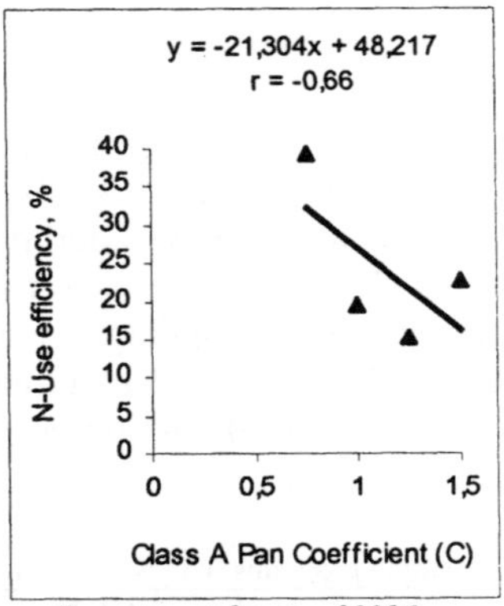

Experimental year of 1996

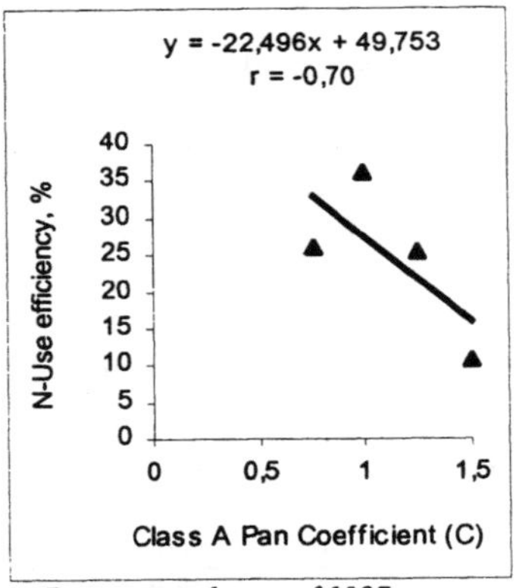

Experimental year of 1997

Figure 1. Relationship between Class A pan evaporation coefficients (C) and N-use efficiency.

Quality aspects

In general, water soluble dry matter, protein, nitrate and nitrite contents of fruit were decreased, whereas total acidity and ascorbic acid were increased with increasing irrigation practices (Table 3 and 4). Hegde and Srinivas (1990) have also found that fruit firmness tended to decline and acidity of the juice tended to increase with increasing irrigation frequency. The oxalic acid, protein, nitrate, and nitrite content of fruit increased,water soluble dry matter and ascorbic acid were decreased with N application.

TABLE 3. Some quality properties of tomato in 1996.

Properties	Control	$I_1 N_1$	$I_2 N_1$	$I_3 N_1$	$I_4 N_1$
Water soluble dry mat., %	5.00	5.03	5.11	4.94	4.87
Soluble sugar, g.100 ml^{-1}	4.15	4.11	4.12	4.34	4.23
Total acidity, g.100 ml^{-1}	0.51	0.51	0.48	0.54	0.62
Ascorbic acid, mg.100 g^{-1}	17.6	16.2	19.6	17.6	19.5
Oxalic acid, mg.kg^{-1} (fw)	29.5	48.1	39.4	51.2	46.5
Protein, %	9.31	9.69	9.88	9.69	9.44
Nitrate, mg.kg^{-1} (fw)	23.7	36.9	41.5	32.1	33.5
Nitrite, mg.kg^{-1} (fw)	0.28	0.58	0.55	0.56	0.54

TABLE 4. Some quality properties of tomato in 1997.

Properties	Control	$I_1 N_1$	$I_2 N_1$	$I_3 N_1$	$I_4 N_1$
Water soluble dry mat., %	5.03	4.90	4.77	4.80	4.70
Soluble sugar, g.100 ml^{-1}	4.01	4.31	4.59	4.55	4.26
Total acidity, g.100 ml^{-1}	0.61	0.57	0.66	0.66	0.68
Ascorbic acid, mg.100 g^{-1}	16.3	15.0	17.9	18.7	17.3
Oxalic acid, mg.kg^{-1} (fw)	50.4	48.3	61.1	54.6	55.1
Protein, %	8.88	15.7	15.3	13.7	13.3
Nitrate, mg kg^{-1} (fw)	29.9	33.8	34.0	32.5	29.1
Nitrite, mg kg^{-1} (fw)	0.67	0.67	0.58	0.70	0.59

The nitrate and nitrite contents were found less than critical levels reccommended for human health. The oxalic acid content have been found less than the critical levels of

5 grams.day^{-1} (Corre and Breimer, 1979). Similar results were obtained by Michalik and Szwonek (1987) and Karaman, (1996). Hegde and Srinivas, (1990) have also found that N-fertilization increased the acidity and fruit firmness.

Nitrogen leaching losses

There was a significant polynomial relationship between NO_3^--N loss and Class A pan evaporation coefficients (C) (r = 0.99, P < 0.01). Concentrations of both NO_3^- and NO_2^- in drainage water were above the critical levels (45 mg l^{-1} for NO_3^- and 0.20 mg l^{-1} for NO_2^- according to WHO and FAO) for drinking water (USEPA,1989). Leaching part of this paper was presented in detail in other symposium (Güleç et al.,1997).

As a result; Nitrogen uptake and crop quality was varied with varying irrigation programs together with N-fertilization. The uptake efficiency of fertilizer-N was found very sensitive to excess irrigation. The results showed that the optimal irrigation program with nitrogen fertilization, needed to provide critical N-uptake and the highest yield, will lead to the maximum N-use efficiency and minimum groundwater pollution.

References

Adrianse A. and Robbers I.E., (1970) Uber eÝne modifi-zierte gessamtoxalat. *Z.Lenesm-unters.* 141:158.

Anonymous, (1972) The determination of total acidity in fruit and vegetables. The Standart Institute of Turkey. *Republic Press,* No: 1125. Ankara.

Anonymous, (1986) The determination of water soluble dry matter in fruit and vegetables. The Standart Institute of Turkey, *Republic Press,* No:4890. Ankara.

Anonymous, (1990) Zeiss-sugar tester. *Carl zeiss Jena Operating Instructions.* Publication No: 32-G.

Anonymous, (1991) Measurement procedure by electrode.*Orion Application Information.* Nitrate in Plants, Waters and Fertilizers Method.

Bar-Yosef, B. and Sagiv, B., (1982) Response of tomatoes to N and water applied via a trickle irrigation system. II. Water. *Agronomy Journal,* 74: 637- 639.

Begg, J.E. and Turner, N.C., (1976) Crop water deficits. *Advances in Agronomy,* 28:161-170.

Barnes, H. and Folkard, A.R., (1951) The determination of nitrites. *Analyst,* 76:599-603, London.

Chapman,H.D. and Pratt,P.T.,(1961) *Methods of Anal. for plants.* Univ. of California Agr. Sci.U.S.A.

Corre,W.J. and Breimer,T., (1979) *Nitrate and nitrite in vegetables.* Centre for Agricultural Publishing and Documentation. 1-85 pp. Wageningen.

Fien, T. and Kerchev, V., (1976) The effect of different soil water potantiels on the growth and productivity of tomatoes. *Agrochemia,* 11:13-22.

Güleç, H., Saltalı,K., Karaman, M.R., Balçın, M. and Derici, M.R., (1997) A Lysimeter Study on the Effects of Different Irrigation Programs on Tomato Yield and Nitrogen Leaching, *Proceeding of 11th World Fert. Cong. of CIEC,* Fertilization for Sustainable Plant Product, 7-13 Sept., Gent-Belgium.

Hegde, D.M. and Srinivas, K., (1990). Effect of irrigation and nitrogen fertilization on yield, nutrient uptake and water use of tomato. *Gartenbauwissenschaft,* 55 (4). 173-177.

Hills, F.J., Broadbent, F.E. and Lorenz, O.A., (1983). Fertilizer nitrogen utilization by corn, tomato and sugarbeet. *Agronomy Journal,* 75:423-426.

Karaman, M.R., (1996). Influence of N-fertilizers on the quality of tomato in green-maturity, semi-red maturity and red-maturity stages. *Proceeding of ESNA XXVIth Meeting,* Sept. 12-16. Romania.

Lowry, O.H., Rosenbrough, N.S., Farr, A.L. and Randall, R.J., (1951) Protein measurement with folin-fenol reagent. *J. Biol. Chem,* 6:193-365.

Meidner,H. and Mansfield,T.A.,(1968) *Physiology of stomata.* Mc Graw Hill Publishing Comp.,.London.

Michalik, H. and Szwonek, E., (1987) Influence of nitrogen fertilization on nitrate content of vegetables. *Adv. in Agronomy.* 28:71-114.

U.S.E.P.A., (1989) *Federal register.* 54 FR 22062, USEPA, Washington, DC.

Assesment of Periodical Nitrogen Use of Tomato Using a Computer Program

M.Rüştü KARAMAN[1], Sabit ERŞAHİN[2], Hikmet GÜLEÇ[3] and M.Rıfat DERİCİ[4]

[1] *Gaziosmanpaşa Univ., Agricultural Faculty, Department of Soil Sci.,Tokat-Turkey*
[2] *Depart. of Crops and Soils, 215 J. Hall Washington State Univ. Pullman W.A. 99164,USA.*
[3] *Research Institute of Rural Services, Tokat-Turkey*
[4] *Çukurova Univ. Agricultural Faculty, Dept. of Soil Sci., Adana, Turkey*

Abstract

In this study, conducted with lysimeters, we wanted to test the performance of LEACHN computer model to simulate the nitrogen uptake of plants at experimental sites. Nitrogen fertilizer at the rates of 0, 80, 160 and 240 kg N.ha^{-1} was side dressed in two split applications, first half as ammonium sulphate and the other half as ammonium nitrate. The lysimeters were irrigated based on program C 1.00, C referring to class A-Pan evaporation coefficients. Yield and N-use efficiency increased with increasing nitrogen rates up to 160 kg N.ha^{-1}, however, they decreased drastically under 240 kg N.ha^{-1}. The results also releaved that some fruit qualities were significantly affected by N-rates. The ability of LEACHN model to simulate crop-N-use was tested using Model parameters for experimental site. Significant correlations (r = 0.95 for 1996 and r = 0.94 for 1997) were found between measured and simulated values for the plant N-uptake, suggesting that the LEACHN model can be confidently used to estimate the crop-N consumption.

Introduction

Fruit production requires that all the involved growing factors be optimized. Among these factors, nitrogen fertilization and irrigation plays an important role. Fertilizer and water management practices affect not only yield and quality but also N recovery and nitrogen leaching (Hegde and Srinivas, 1990). Excess nitrogen, especially in the form of nitrate, in the root zone must be avoided at times when the soil nitrogen is vulnerable to leaching by excess rainfall or irrigation. Ideally, all sources of plant-available N are taken into account and are just sufficient to maximize crop yield and quality. Computer simulation models are useful tools to predict the risk of agricultural chemicals to contaminate groundwater, and to fallow the translocation of chemicial materials in soil-water-plant systems (Wagenet and Hutson, 1989). The computer program LEACHN, a version of LEACHM, was developed for farmers, extension personnel and action agencies (Hutson and Wagenet, 1992). Indeed, it is a deterministic model for simulating the movement of chemical materials in soil-water-plant systems, and not intended to predict crop yields (Jabro et al. 1994). The fertilizer-N-use in crops in any agricultural system may be simulated using LEACHN computer model.

G. Gissel-Nielsen and A. Jensen (eds.), Plant Nutrition – Molecular Biology and Genetics, 53–57.

Materials and Methods

Lysimeter experiments were conducted at the Agricultural Research Institute of Tokat during May, 1996 through October, 1996 and May, 1997 through October, 1997. A completely randomised block design with three replications was adopted using lysimeters, 2 m high CTP cylinders with 0.21 m inner diameter. They were filled with 0.2 m coarse over 0.3 m coarse gravel and sand mixture at their bottoms and the remaining 1.5 m were filled with Ap, C_1 and C_2 horizons of a fluvent from the Kazova Plain in Tokat-Turkey. The properties of experimental soil are given in Table 1. Three seedlings of tomato variety H-2274 were transplanted to per lysimeter for both experiments. Nitrogen fertilizers at the rates of 0, 80, 160 and 240 kg $N.ha^{-1}$ were side dressed in two split applications, first half as ammonium sulphate and the socond half as ammonium nitrate. In addition, 45 kg $P.ha^{-1}$ as TSP were applied to each treatment. Lysimeters were irrigated on a program based on class A pan evaporation coefficient of 1.00 with C_2S_1 quality water. Yelds were recorded and fruit samples were taken 83th day and 93th day after transplanting in 1996 and 1997, respectively. In 1996, experimental site had an average annual temparature of 12.3 °C, and had an average precipitation of 616 mm of which 5.7, 0.15, 2.06 and 7.40 % were in Jun., Jul., Aug. and Sept., respectively. In 1997, annual average temperature was 11.2 °C and an average precipitation was 350 mm of which 3.7, 0.77, 7.71 and 1.80 % fell in Jun., Jul., Aug. and Sept., respectively. Total N (Chapman and Pratt, 1961), oxalic acid (Adrianse and Robbers,1970) and nitrate (Anonymous,1991) were determined in fruit samples. Other analyses were made using standard methods. LEACHN computer model requires a variety input data. The meteorogical data for each year were collected at a weather station near experimental site. Initial N-transformation parameters used to calibrate the model were presented in Table 2. Other input parameters required by the model were obtained from direct laboratory and field measurements or through the calibration process (Jabro et al., 1994; Jalal et al.,1995). Modified addiscott's capacity model was used to simulate water flow (Addiscott and Bailey, 1990). The saturated hydraulic conductivity of soil ranged from 0.2 to 12.0 $cm.d^{-1}$ at 0.30-0.60 m, 0.05 to 2.1 $cm.d^{-1}$ at 0.60-0.90 m, and 0.04 to 1.2 $cm.d^{-1}$ at 0.90-1.50 m depths.

TABLE 1. Some characteristics of experimental soil

	Soil depth, cm				
Parameters	0-30	30-60	60-90	90-120	120-150
Horizon	Ap	C_1	C_2	C_2	C_2
Soil water content, m^3 m^{-3}					
at -0.03 MPa	0.39	0.41	0.39	0.43	0.38
at -1.5 MPa	0.27	0.26	0.25	0.26	0.27
Sand, %	30.5	18.2	22.0	22.1	18.0
Silt, %	28.5	34.5	39.6	35.0	35.0
Clay, %	41.0	47.3	38.4	42.9	47.0
Texture	C	C	CL	C	C
Bulk dencity, $kg.dm^{-3}$	1.31	1.35	1.35	1.49	1.29
CEC, $cmol.kg^{-1}$	48.9	42.0	36.3	29.5	28.8
EC, $ds.m^{-1}$	0.52	0.50	0.65	0.51	0.45
pH, 1:2 W/v	7.88	8.05	8.25	8.47	8.35
Avail. P, $mg.kg^{-1}$	32.56	21.50	11.49	11.49	11.00
Avail. K, $mg.kg^{-1}$	203.8	136.1	131.1	119.1	97.53
Total C, $gr.kg^{-1}$	16.8	15.7	13.3	11.0	3.50
Total, N, $gr.kg^{-1}$	0.67	0.73	0.62	0.68	0.47

TABLE 2. Initial input parameter s used for calibtration of LEACHN model.

Parameter	Input value	Parameter	Input value
Partition coefficient, NH_4^+-N ($L.kg^{-1}$)	3.0	Partition coefficient, NO_3^--N ($L.kg^{-1}$)	0.0
Nitrification rate constant (day^{-1})	0.04-0.4	Denitrification rate constant (day^{-1})	0.02-0.1
Litter mineraliz. rate constant (day^{-1})	0.01	Manure mineraliz. rate constant (day^{-1})	0.02
Humus mineralization rate const. (day^{-1})	0.15	Ammonia volatilization rate const. (day^{-1})	0-0.4
C:N ratio for biomass and humus	10	Q10 factor	3.0

Results and Discussion

Yield and N-use efficiency

A significant increase in yield (157 $ton.ha^{-1}$ in 1996 and 113.8 $ton.ha^{-1}$ in 1997) was obtained at 160 kg $N.ha^{-1}$ both years. Total amounts of N taken up by plants followed a similar pattern (Table 3). These results are in agreement with the findings of Hegde and Srinivas (1990). N application rates had considerable effect on fertilizer use efficiency. The highest fertilizer use efficiencies of 36.38 % and 38.34 %, based on dry weights, were obtained with the 160 kg $N.ha^{-1}$ rate both years. This type of insufficient response to high N applications was expected (Hills et al., 1983).

TABLE 3. Yield and N-use efficiency of tomato depending on nitrogen rates.

N kg.ha⁻¹	*Fruit yield t.ha⁻¹*	*Fruit D.MY. t.ha⁻¹*	*Fruit N-cont. %*	*N-upt. by fruit kg.ha⁻¹*	*Leaf+ Stem D.M.Y. t.ha⁻¹*	*Leaf+ Stem N-cont. %*	*N-upt. by L+S kg.ha⁻¹*	*Total N-upt. kg.ha⁻¹*	*N-use effic. (%)* ♣
1996									
0	128 b	6.40	1.49	95.4	2.26	1.53	34.58	130.0	-
80	117 b	5.85	1.61	94.2	2.21	2.04	45.08	139.3	11.63
160	157 a	7.85	1.65	129.5	2.54	2.31	58.67	188.2	36.38
240	127 b	6.35	1.73	109.9	2.30	2.52	57.96	167.9	15.79
1997									
0	83.5 b	5.01	1.33	66.6	2.12	1.46	30.95	97.6	-
80	102.3 ab	6.14	1.36	83.5	2.15	1.61	34.62	118.1	25.68
160	113.8 a	6.83	1.58	107.9	2.34	2.18	51.01	158.9	38.34
240	79.6 b	4.78	1.68	80.3	2.28	2.57	58.60	138.9	17.22

♣ *N-use efficiency.=Total N upt. (for 80,160,240 kg $N.ha^{-1}$)-Total N upt. (at 0 kg $N.ha^{-1}$)/appl. N ratex100*

Quality aspects of nitrogen in tomato

In general, water soluble dry matter and ascorbic acid content of fruit were decreased, whereas, oxalic acid, nitrate and nitrite contents were increased. with increasing N-rates (Table 4-5). The results aggree to the findings of (Karaman, 1996). Hegde and Srinivas, (1990) have also found that N-fertilization increased the acidity and fruit firmness. The nitrate and nitrite contents of fruit have been found to be lower than critical levels. Oxalic acid contents of fruit have also been found to be lower than the critical levels of 5 $grams.day^{-1}$ (Adrianse and Robbers, 1970). Total acidity, fruit and leaf nitrate content were decreased, whereas water soluble dry matter, sugar, ascorbic acid and oxalic acid contents were increased by maturity of tomato. The highest water soluble dry mat., sugar, ascorbic acid and oxalic acid in tomato fruit were found at red maturity stage. Similar results have also been found by others (Kaynas and Sürmeli,1994; Karaman, 1996)

TABLE 4. Some quality properties of tomato (at green maturity stage).

Properties	Nitrogen, kg N.ha^{-1} (1996)				Nitrogen, kg N.ha^{-1} (1997)			
	0	80	160	240	0	80	160	240
Water soluble dry matter, %	4.21	4.35	4.46	4.42	4.37	4.20	4.35	4.28
Soluble sugar, g.100 ml^{-1}	3.82	3.65	3.94	3.83	3.75	3.86	3.91	3.82
Total acidity, g.100 ml^{-1}	0.59	0.61	0.62	0.58	0.58	0.65	0.61	0.52
Ascorbic acid, mg.100 g^{-1}	14.1	13.2	14.4	13.6	13.3	14.3	14.8	12.2
Oxalic acid, mg.kg^{-1} (fw)	28.1	27.5	30.7	30.1	30.1	36.2	35.0	41.5
Fruit nitrate cont.mg.kg^{-1}, fw.	42.6	47.3	49.4	57.0	50.3	49.1	56.3	57.7
Fruit nitrite cont. mg.kg^{-1}, fw.	0.52	0.71	0.64	0.67	0.63	0.65	0.62	0.76
Fruit N content, %	1.88	2.02	2.24	2.41	2.03	2.34	2.36	2.57
Leaf + Stem N content, %	2.51	2.61	2.86	3.01	2.87	3.03	3.01	3.24

TABLE 5. Some quality properties of tomato (at red maturity stage).

Properties	Nitrogen, kg N.ha^{-1} (1996)				Nitrogen, kg N.ha^{-1} (1997)			
	0	80	160	240	0	80	160	240
Water soluble dry matter, %	5.00	5.10	5.00	4.90	5.03	4.90	4.83	4.73
Soluble sugar, g.100 ml^{-1}	4.15	4.28	4.13	4.19	4.01	4.11	4.13	4.09
Total acidity, g.100 ml^{-1}	0.51	0.59	0.56	0.58	0.61	0.65	0.64	0.65
Ascorbic acid, mg.100 g^{-1}	17.6	19.5	16.3	15.0	16.3	18.8	17.6	18.0
Oxalic acid, mg.kg^{-1} (fw)	29.5	37.1	35.0	40.1	50.4	43.8	49.2	52.2
Fruit nitrate cont,mg.kg^{-1}, fw.	23.6	33.1	38.8	44.5	29.8	32.5	39.3	53.6
Fruit nitrite cont. mg.kg^{-1}, fw.	0.28	0.42	0.58	0.69	0.67	0.58	0.70	0.59
Fruit N content, %	1.49	1.61	1.65	1.73	1.42	1.44	1.44	1.55
Leaf + Stem N content, %	1.53	2.04	2.31	2.52	1.84	2.05	2.22	2.60

Application of LEACHN model to estimate nitrogen uptake

Plant N-uptake under under 0, 80, 160 and 240 kg N.ha^{-1} nitrogen application rates was predicted with the LEACHN model in 1996 and 1997. Simulated values for each nitrogen rate indicated that significant increase in nitrogen uptaken by crops was obtained up to 160 kg N.ha^{-1} and it decreased with 240 kg N.ha^{-1} application (Figure 1). Periodical distribution of simulated values during the growing season was showed in Figure 2. There was close association between simulated and measured values for plant N-uptake ($P < 0.04$ and $r^2 = 0.90$ for 1996, and $P < 0.05$ and $r^2 = 0.89$ for 1997),which suggest that the LEACHN model can be confidently used to estimate the crop-N consumption (Table 6).

TABLE 6. Statistical comparison of the cumulative measured and simulated values

Calibration year	d#, kg N ha^{-1} 0-N	80-N	160-N	240-N	Intercept	Slope	r^2
1996	-15.0	3.02	-14.2	-8.32	5.90	0.91	0.90**
1997	10.0	25.2	9.0	27.06	17.77	0.99	0.89**

 # d (difference) = simulated value - measured value at applied nitrogen rate

As a result; the LEACHN model adequately predicted crop N-uptake component. Adopting these models to estimate N-uptake will be beneficial to calculate N-balance, and to recommend rational fertilizer N-rate for large agricultural areas. Both simulated and measured values revaled that 160 kg N.ha^{-1} rate should not be exceeded for optimal tomato production, fruit quality, and N-use efficiency under the experimental conditions.

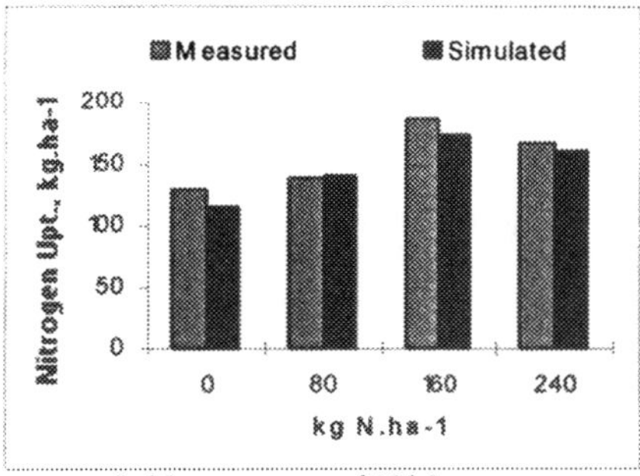

Calibration year of 1996

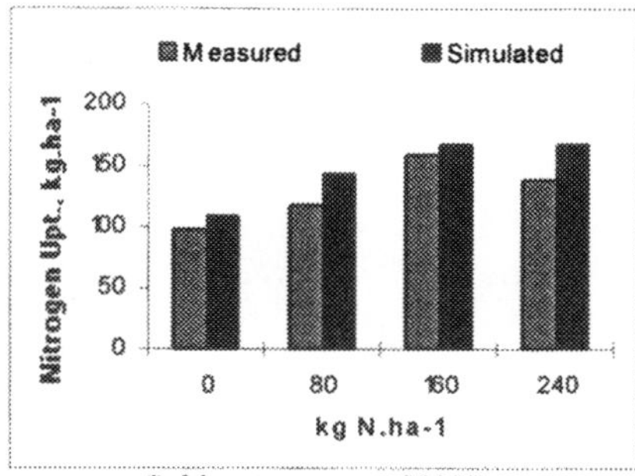

Calibration year of 1997

Figure 1. Measured and simulated values of the cumulative nitrogen uptake

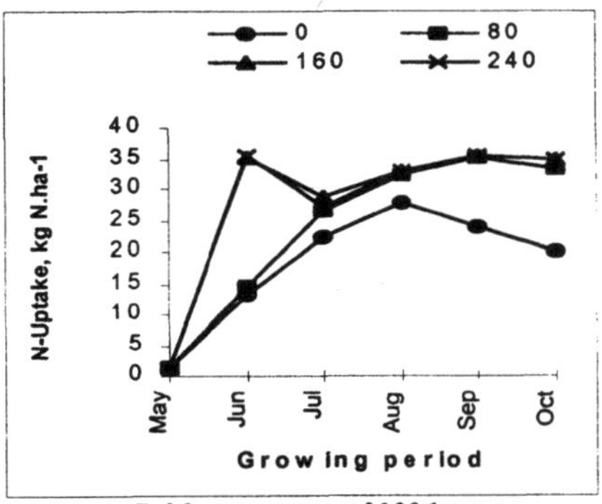

Calibration year of 1996

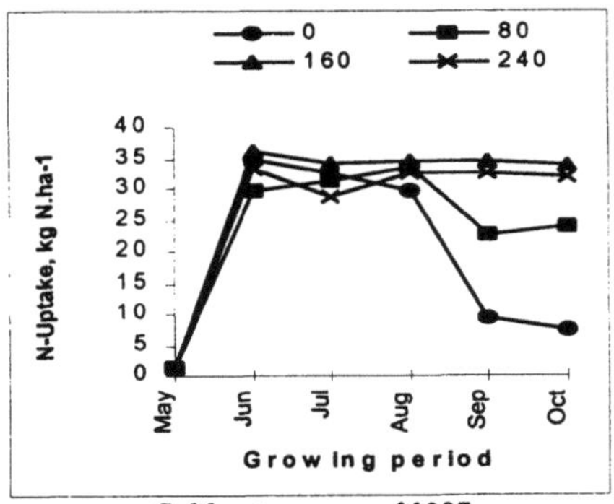

Calibration year of 1997

Figure 2. Periodical distribution of simulated values during the tomato growing season.

References

Addiscott,T.M., and Bailey,N.J.,(1990) Relating the parameters of a leaching model to the percent of clay and other soil components.. K. Roth (ed). *Monte Verita Birkhauser Verlag*, 209-221.

Adrianse, A. and Robbers, I.E. (1970) Uber eine modifi-zierte gessamtoxalat, *Z.Lenesm.* U.Fors. 141:158.

Chapman,H.D. and Pratt,F.P.(1961) *Methods of Anal. for plants.*Univ.of California. Agr. Sci.U.S.A.

Hegde, D. M. and Srinivas, K. (1990) Effect of irrigation and nitrogen fertilization on yield, nutrient uptake and water use of tomato. *Gartenbauwissenschaft.* 55 (4). 173-177.

Hills, F.J., Broadbent, F.E. and Lorenz, O.A. (1983) Fertilizer N utilization by tomato *Agron. J.* 75:423.

Hutson, J.L., and Wagenet, R.J. (1992) *LEACHM, Leaching estimation and chemistry model*, Version 3. Department of Agr., Ithaca, N.Y.

Jabro, J.D., Jemison, J.M., Fox, R.H., and Frittion, D.D. (1994) Predicting bromide leaching under field conditions using SLIM and MACRO. *J. Soil Sci.* 157:215-223.

Jalal,D.J., John, D.T., Zhengxia, D., Richard, H.F., and Daniel, D.F. (1995) Evaluation of nitrogen version of LEACHM for predicting N leaching. *J. Soil Sci.* Vol.160, 209-217 p., USA.

Karaman, M.R. (1996) Influence of nitrogenous fertilizers on the quality of tomato in green, semi-red and red-maturity stages. *The Proceedings of ESNA XXVIth Annual Meeting,*, Romania.

Kaynas, K. and Sürmeli, N. (1994) Some chemical properties of tomato at the different maturity stages. *Tr. J. of Agricultural and Forestry*, 18 (2), 71- 79.

Wagenet, R.J. and Hutson, J.L. (1989) LEACHM. Leaching estimation and chemistry model, *Continuum,* Vol.2. Water Res. Inst., Univ., Ithaca, NY.

GENETIC MANIPULATION OF THE LOCATION OF NITRATE REDUCTION AND ITS EFFECTS ON PLANT GROWTH.

M. LEXA
Masaryk University Brno, Faculty of Science
Laboratory of Plant Molecular Physiology
Kotlarska 2, 611 37 BRNO
Czech Republic

J. M. CHEESEMAN
University of Illinois at Champaign-Urbana
Department of Plant Biology
505 S. Goodwin Ave, URBANA, 618 01 IL
USA

1. Abstract

Nitrate reductase (NR) activity can be found in both roots and shoots of most plant species. However, the activity is very variable and the proportion of nitrate reduced in roots and shoots varies among species and even among individual plants. Using a mathematical model, we have shown that changes in the location of nitrate reduction might cause changes in biomass and nitrogen partitioning between shoots and roots. In a previous experiment we grafted wild-type and mutant pea plants to shift nitrate reduction from shoots to roots or vice versa. In this paper we lay out possible advantages or disadvantages of root and shoot reduction. We also report preliminary results of *Arabidopsis* transformation with tobacco *nia2* cDNA. The experiments are expected to alter the location of nitrate reduction in the transgenic plants.

2. Introduction

2.1. BACKGROUND

Nitrate reductase activity can be found in both roots and shoots of most plant species. However, the activity is very variable and the proportion of nitrate reduced in roots and shoots varies among species and even among individual plants (Table 1). The activity

G. Gissel-Nielsen and A. Jensen (eds.), Plant Nutrition – Molecular Biology and Genetics, 59–63.

is highly regulated at different levels (expression, modulation by phosphorylation) and depends on factors such as light, nitrate and water availability, developmental stage, salinity etc. For example, increasing nitrate availability to plants will cause a shift from root reduction to shoot reduction in many species. Often this is accompanied by an increase in shoot activity (because of substrate induction), but in theory an increase in shoot reduction could be observed even if shoot NR activity was already fully induced before the increase in nitrate availability.

Although many of the observed differences are based on genetic variation among species, very little is known about how the differences evolved and what competitive advantages or disadvantages they may have for the species involved. In our research we began to search for answers to these questions. We have set out to engineer plants with altered location of nitrate reduction and measure their growth and nitrogen assimilation characteristics. We have formulated hypotheses about the consequences of such alterations, some of which were based on a compartmental simulation model of nitrate uptake and assimilation (Figure 1). In our previous research we used grafting of wild-type and NR-deficient pea plants to acheive the desired effect (Lexa, Cheeseman, 1997). We found considerable plasticity in these plants and only very little change in overall growth.

Table 1 - Shoot nitrate assimilation in various species (adapted from Andrews et al., 1992).

SPECIES	% SHOOT ASSIMILATION	
	low N	high N
Pinus nigra Arnold	0	0
Pyrus malus L.	0	80
Lupinus albus L.	4-21	15-69
Pisum sativum L.	4-44	54-81
Vicia faba L.	7-49	28-83
Zea mays L.	38	58-91
Hordeum vulgare L.	42-79	62-98
Xanthium strumarium L.	59-89	80-85
Ricinus communis L.	76	64
Glycine max (L.) Merr	58-86	69-100
Phaseolus vulgaris L.	62-96	58-79
Solanum tuberosum L.	91	88

This poster presents some of the possible advantages and disadvantages of root or shoot reduction of nitrate. We hope we could provoke our readers to come up with their own hypotheses, perhaps related to their field of expertise. We also describe our first efforts in changing the location of nitrate reduction by genetic engineering, using *Arabidopsis thaliana*, *nia2* gene from Nicotiana tabacum and several presumably organ-specific promoters.

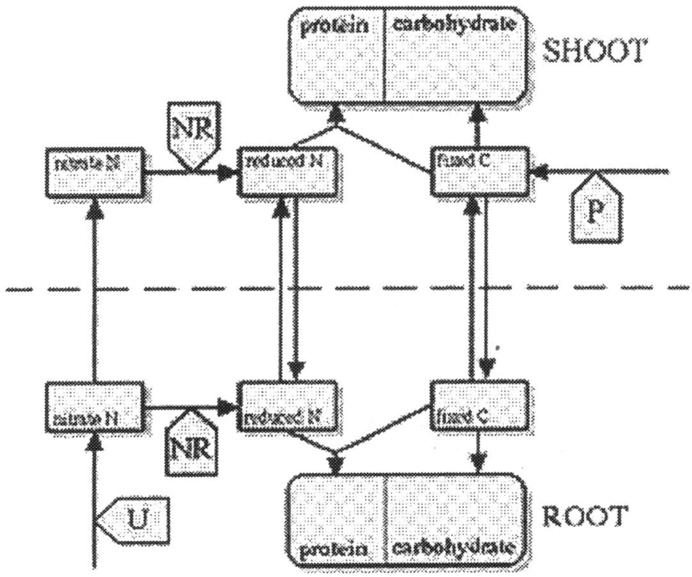

Figure 1 - Block sheme of the compartmental simulation model of nitrate assimilation and growth. U - nitrate uptake; P - photosynthetic carbon uptake; NR - nitrate reduction.

2.2. THE POSSIBLE SIGNIFICANCE OF THE LOCATION OF NITRATE REDUCTION

- Many plant shoots are a source of food for animals, including humans. A plant species subject to grazing or harvesting may have faster regrowth rates if nitrate reduction is done in the roots.

- Low activity of NR in the shoot may provide for higher nitrate concentrations in the leaves. Some animals are known to avoid species known to have high nitrate levels in their leaves.

- For plants with very low shoot NR activity, nitrate transported to the shoot is physiologically lost, because of the extremely low shoot to root transport of nitrate. NR activity in the shoot determines the rate at which it can be utilized.

- Plants using ammonium as their nitrogen source assimilate most of their nitrogen in roots. It is possible that many of the same enzymes can be used for nitrate assimilation if nitrate is reduced in the roots.

- Reduction in roots may be able to respond faster to rapid fluctuations in nitrate availability.

- Reduction in the shoot may be able to respond better to changes in light intensity because of the proximity of the photosynthetic apparatus. Nitrate reduction in shoot may be regulated by molecules that are not present in the root in sufficient or meaningful quantities.

- Shoot reducers may use extra reducing power from photosynthesis to reduce nitrate.

- Shoot reducers need special mechanisms to maintain balance beteween reduction and uptake of nitrate, such as malate signalling.

- Under certain conditions plants growth may be indifferent to the location of nitrate reduction. Reduced N incorporated into amino acids can be rapidly recycled between roots and shoots, regardless of its root or shoot origin.

3. Methods

3.1. PREPARATION OF CONSTRUCTS

To change the location of nitrate reduction in Arabidopsis we prepared DNA constructs of organ-specific promoters and a tobacco *nia2* cDNA clone. We used a full length CaMV35S promoter (for preparation of our construct pLX524), a rolD promoter from Agrobacterium rhizogenii (for pLX661) and its deletion that has been shown to drive root-specific expression of GUS (for pLX1163), and *cab* promoter from *Oryza sativa* which has been shown to drive leaf-specific expression of foreign genes in tobacco (for pLX1077). The constructs were made using available restriction sites and finaly cloned into pBIN19 and transformed into *Agrobacterium tumefaciens* GV3101 pMP90.

3.2. TRANSFORMATION OF *ARABIDOPSIS*

The constructs described above were introduced into wild-type (WT) and NR-deficient *Arabidopsis* (CS2356 *nia1*,*nia2*) plants by vacuum infiltration. Seeds were collected from the infiltrated plants at maturity and germinated on a kanamycin-cntaining MS agar medium. Seedlings forming true leaves were selected as resistant and grown in soil. T2 seeds were collected from these plants for further characterisation of the transformants.

4. Results and Discussion

We obtained about 100 transformants (Table 2) and are currently characterizing them. Activity measurements in the T1 generation of CS2356 transformants showed that none of the plants have higher activity of NR in the leaves. However, several transformed lines are showing increased growth compared to untransformed controls or controls transformed without a NR gene. Of these, some have significantly higher seed yields. One line is characterized by increased root growth when germinated on agar MS plates. Several lines of WT transformants have lower NR activity in their leaves and show signs of co-suppression, such as chlorosis. Other plants show signs of increased growth and loss of apical dominance. Our results are not complete enough yet to draw any final conclusions about the characteristics of these transformants. Selection of appropriate T3 seed populations is under way and data on their overall growth response to nitrate and NR activity in roots and leaves will be collected soon.

Table 2 - Putative transformants obtained after first screening on kanamycin

CONSTRUCT	HOST PLANT	KAN RESISTANT PLANTS
pMON530	CS2356	4
pLX524	CS2356	3
pLX524	WT	28
pLX661	CS2356	5
pLX661	WT	18
pLX1077	CS2356	0
pLX1077	WT	35
pLX1163	CS2356	27
pLX1163	WT	16

5. References

Andrews, M. et al.. (1986). The partitioning of nitrate assimilation between root and shoot of higher plants. *Plant, Cell and Environment* 9, 511-519.

Lexa, M. and J. M. Cheeseman (1997). Growth and nitrogen relations in reciprocal grafts of wild-type and nitrate reductase-deficient mutants of pea (Pisum sativum L. var. Juneau). *Journal of Experimental Botany* **48**, 1241-1250.

THE STRATEGY OF THE WHEAT PLANT IN THE PRODUCTION OF GRAINS AT REDUCED NITROGEN AVAILABILITY

PETTER OSCARSON
Department of Plant Breeding, SLU, S-268 31 Svalöv, Sweden

1. Introduction

The wheat plant somehow sense and adapt its production of grains to prevailing nutrient conditions. The mechanisms of this acclimation process to limiting nutrient availability is however poorly investigated. To be able to distinguish between broad environmentally imposed responses and fine-tuned genetically derived responses we need to culture plants under fully defined and reproducible conditions. Culturing in soil gives only marginal control of the flow of nutrients to the root surface and the amounts of nutrients available for the plant. Culturing in hydroponics with external concen- tration of nutrients as driving variable is as inadequate as soil culturing in this respect. Unless the external concentration of nutrients is kept constant, extremely low and under full control, there are only two options in concentration controlled cultures; optimal supply or acute deficiency. Neither conditions are relevant in mimicking field culturing of cereals where there is a more or less continuous but usually growth limiting flow of nutrients to the root surface. Growth and grain production is then acclimated to the changes in nutrient availability that occurs over the season.

In the experiment presented here the plants were grown to full maturity with relevant ontogeny under fully controlled conditions and under various levels of (N) nitrogen availabilty. So, not only relevant conditions are produced but also the possibilty of studying acclimation processes over a gradient of nutrient availability.

2. Materials and Methods

Two cultivars of spring wheat (*Triticum æstivum* L.) were grown to full maturity in hydroponic culture. The cultivar Drabant is relatively stable concerning yield and grain protein concentration whereas Sport gives a lower yield but higher grain protein concentration under favourable conditions, but sometimes, of not yet identified reasons, gives a lower baking quality compared to Drabant.

Culturing procedures has been presented elsewhere (Oscarson 1996). Nitrogen was added daily at growth limiting rates. The size of the daily doses of nitrogen was calculated as $N_t - N_o$ from the equation:

$$N_t = N_o * e^{RAt} \qquad (1)$$

where N_t and N_o denotes nitrogen content at day t and day 0 respectively, and RA represents the preset rate at which the N was added to the plants. In order to produce growth limiting N availability conditions similar to field conditions, the RA was

G. Gissel-Nielsen and A. Jensen (eds.), Plant Nutrition – Molecular Biology and Genetics, 65–68.

stepwise decreased as time progressed; day 14 - RA 0.12, day 24 - RA 0.11, day 34 - 0.10, day 44 - RA 0.05, day 54 - RA 0.02, day 64 - RA 0.01, day 74 - RA 0.005, day 84 - RA 0.0025, day 94 - RA 0.0013, day 104 - RA 0.0006, day 114 - RA 0.0003. The plants were growth limited by the availability of nitrogen at all times through development.

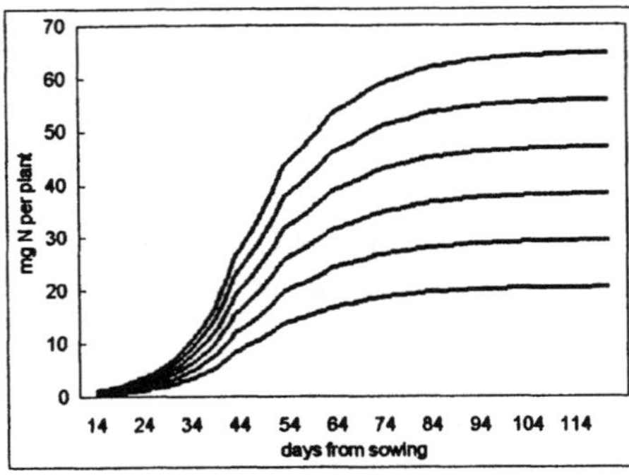

Figure 1. Cumulated N additions. Six different levels of total N availability were obtained by varying the N_0-value (0.35, 0.5, 0.65, 0.8, 0.95 and 1.1 mg/plant). All plants were thus exposed to the same relative rates of N increment throughout development but the final amounts of N added varied from 20 to 65 mg nitrogen per plant.

3. Results

The mathematic definition of the N accumulation curves was the same but the curves had different amplitudes (Fig. 1). All N added was taken up as there was full agreement between N added and N recovered in the plant (not shown).
Increased N availability resulted in increased weight of all vegetative parts (Fig. 2). Most of the increase, however, was found in the almost linear increase in tiller dry weight which represent 50% of vegetative dry matter at the highest N level. The contribution of the root to total vegetative dry matter was constant throughout, showing that the relative growth of the different organs were in phase.

The increased weight of tillers was mostly explained by an increased number of tillers per plant with increasing N levels (Fig. 3). The availability of N strongly affected the initiation of tiller growth, but more than half of the formed tillers never produced ears (Fig. 3). Tiller production was the most important single factor responsible for increased yield. At all N levels the plant produced more tillers than actually ended up setting grain. It is possible that this is one the strategies of the plant concerning acclimation to nutrient availability; to produce slightly more tillers than at the moment seem possible to support. If more nutrients appear to be available, these extra tillers can also be included in the overall support scheme, or if no extra nutrients are available, the support to these extra tillers is withdrawn and the nutrients re- mobilised for use in the rest of the plant.

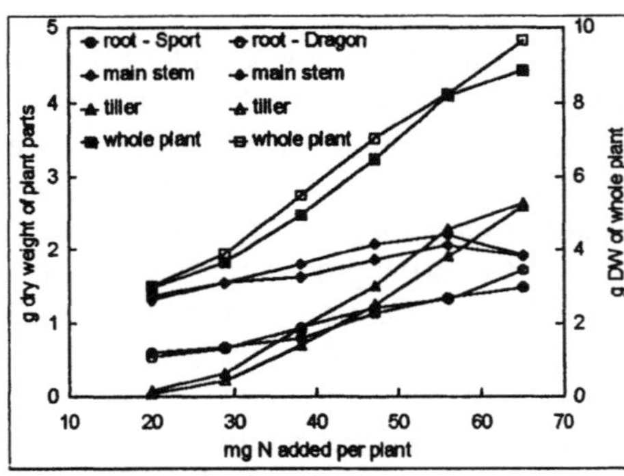

Figure 2. Dry weights of different plant parts at maturity. Linearisation of plots of DW vs. N content is often used as a measure of nitrogen productivity and this constant was 0.141 g DW/mg N in Sport and 0.152 g DW/mg N in Drabant.

The tiller spikes produced more grains per spikelet with increased N levels which was not the case in the main stem spike (Fig. 5). There seem to be a larger plasticity in tiller spike development compared to the main stem. The main stem spikes were formed earlier and at that time the nutritional differences may have been smaller compared to when tiller spike meristems were initiated. It seems likely, with respect to the overall results presented here that most of the production plasticity mechanism towards N availability lies in the the formation and reduction of tillers. As was observed with main stem development, the patterns of tiller response to increased N level was essentially similar in both genotypes.

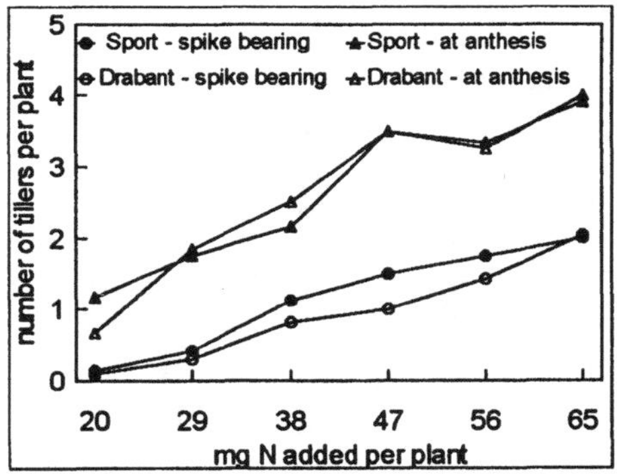

Figure 3. The number of tillers per plant at anthesis and final number of spike bearing tillers at maturity.

Increased grain yield per plant was the result of increased yields of main stem and tillers (Fig. 4). Grain weight was more or less constant over the whole range of N regimes (Fig. 4). The number of grains per spikelet was also relatively constant (Fig. 5), so the increased yield of the main stem spike could only be explained by an increased number of spikelets (Table 1), and thus grains, per spike.

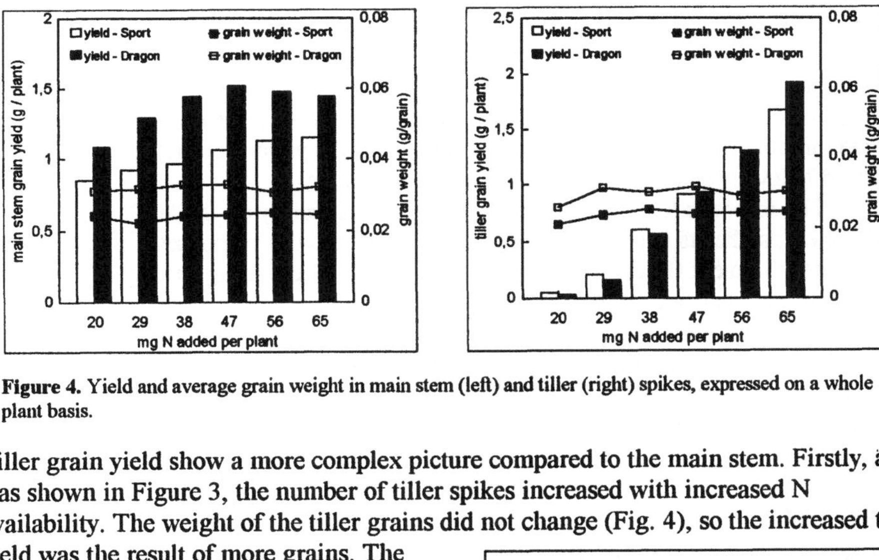

Figure 4. Yield and average grain weight in main stem (left) and tiller (right) spikes, expressed on a whole plant basis.

Tiller grain yield show a more complex picture compared to the main stem. Firstly, as was shown in Figure 3, the number of tiller spikes increased with increased N availability. The weight of the tiller grains did not change (Fig. 4), so the increased tiller yield was the result of more grains. The number of grains per spike increased with N level, and this was brought about by both increased number of spikelets (Table 1) and of number of grains per spikelet (Fig. 5). Tiller yield was thus the result of three varying components, number of tiller spikes, number of spikelets per spike and number of grains per spikelet.

Table 1. Number of spikelets in main stem and tiller spikes.

	Sport		Dragon	
N-level	main stem	tiller	main stem	tiller
20	16,5	14,7	14,6	15,0
29	16,5	16,6	14,5	15,6
38	18,2	17,9	16,5	16,3
47	20,0	19,1	17,3	17,9
56	20,8	20,7	18,6	19,3
65	19,1	19,5	17,0	18,0

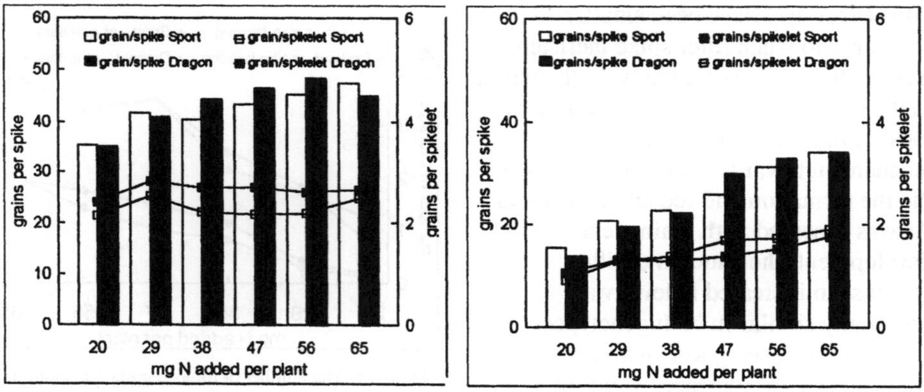

Figure 5. The number of grains per spike and grains per spikelet in main stem (left) and tiller (right) spikes.

4. Summary and Conclusion

Increased availability of N resulted in increased vegetative and generative production of the plant. No single trait was responsible for the full response but instead several processes were integrated. The more N, the longer the main stem with a spike with slightly more spikelets and more grains but with constant grain weight. The number of tillers increased with increased N availability. The number of spikelets per tiller spike also increased as did the number of grains per spikelet. Most measured parameters differed quantitatively between the two genotypes but the pattern of the response to increased N level was essentially similar in both genotypes.

The response to changes in N availability was mainly on the level of meristem development; spikelets per spike and tillers per plant. Similar responses has been seen in e.g. root development (Drew 1975, Agrell et al. 1994). The sensing mechanism is as yet unclear but there has been reports on N effects on the control of the cell cycle (Gould et al. 1981) and this could be the onset or offset of meristem activity. Either nitrate or another N containing compound, e.g. an amino acid, could act as a regulator of cell proliferation and differentiation, acting at the gene level as explained by Chasan (1995). Deficiency or low levels of this compound would inevitably lead to cessation of meristem growth.

5. References

Agrell, D, Oscarson, P, Larsson, C-M. 1994. Translocation of N to and from barley roots: its dependence on local nitrate supply in split-root culture. Physiol Plant 90:467-474.

Chasan, R. 1995. STARTing the cell cycle. The Plant Cell 7:1-3

Drew, MC. 1975. Comparison of the effects of a localised supply of phosphate, nitrate, ammonium and potassium on the growth of the seminal root system, and the shoot, in barley. New Phytol 75:479-490.

Gould, AR, Everett, NP, Wang, TL & Street, HE. 1981. Studies on the control of the cell cycle in cultured plant cells I. Effects of nutrient limitation and nutrient starvation. Protoplasma 106:1-13.

Oscarson, P. 1996. Transport of recently assimilated ^{15}N nitrogen to individual spikelets in spring wheat grown in culture solution. Ann Bot 78:479-488.

REGULATION OF AMMONIUM DISTRIBUTION IN PLANTS

JAN K. SCHJOERRING, JØRGEN FINNEMANN, SØREN HUSTED, MARIE MATTSSON, KENT H. NIELSEN & J. NICHOLAS PEARSON

Plant Nutrition Laboratory, Department of Agricultural Sciences, The Royal Veterinary and Agricultural University, Thorvaldsensvej 40, DK-1871 Frederiksberg C, Copenhagen, Denmark

Abstract

Ammonium is constantly generated from a variety of processes in plant nitrogen metabolism. Re-fixation of the liberated ammonium is catalysed by the enzyme glutamine synthetase (GS). Cytosolic GS isoforms dominate in the roots, while a chloroplastic GS isoform is the quantitatively most important in green leaves. Despite the central role of NH_4^+ as an intermediate in plant nitrogen metabolism very little is known about the processes regulation NH_4^+ distribution between cell organelles and plant organs.

The extent of NH_4^+ translocation from roots to shoots in oil-seed rape plants (*Brassica napus* L.) depended on the external N supply and the activity of GS isoforms in the root. Addition of either NO_3^- or NH_4^+ to N-starved induced both cytosolic GS isogene expression and GS activity in the roots. In N-replete plants, root GS isogene expression and activity were repressed causing enhanced NH_4^+ translocation to the shoots. Xylem NH_4^+ concentrations increased linearly both with time of exposure to NH_4^+ and with increasing external NH_4^+ concentration. The maximum xylem NH_4^+ concentration in NH_4^+-fed plants was 8 mM, corresponding to 11% of the nitrogen

G. Gissel-Nielsen and A. Jensen (eds.), Plant Nutrition – Molecular Biology and Genetics, 69–82.

translocated in the xylem, while in NO_3^--fed plants the xylem NH_4^+ concentration was around 0.6 mM, constituting 2% of the nitrogen translocated in the xylem.

Regulation of apoplastic NH_4^+ concentration in leaves of oil-seed rape was studied using a vacuum infiltration technique allowing controlled manipulations of the apoplastic solution. The apoplastic solution constituted a highly dynamic NH_4^+ pool. Ammonium was constantly added to this pool *via* NH_3 efflux from the mesophyll cells. The NH_4^+ was retrieved by a system involving a transporter with channel-like properties which was able to respond very rapidly to perturbations in apoplastic NH_4^+ concentration, thereby maintaining apoplastic NH_4^+ homeostasis. More detailed investigations into $^{15}NH_4^+$ uptake into *B. napus* leaf protoplasts revealed the presence of an additional high-affinity NH_4^+ uptake system, similar in kinetics to the *AMT1* transporter observed in *Arabidopsis*, tomato and rice. Primers based on conserved regions of the *AMT1* transporter were used in RT-PCR reactions to identify and subsequently clone a homologue in *B. napus (BnAMT1)*. The *Bnamt1* gene was found to be strongly expressed under all N conditions but regulated by N supply.

Introduction

Ammonium (NH_4^+) is constantly generated from a variety of anabolic and catabolic processes in plant leaves such as photorespiration, nitrate reduction, protein turnover and lignin biosynthesis (Table 1; Joy, 1988). Substantial amounts of NH_3/NH_4^+ generated from photorespiration are released by the mitochondria, representing up to 10 times that involved in primary N assimilation in the plant (Leegood et al., 1995). Lignin biosynthesis also contributes significantly to NH_4^+ release and occurs directly in the leaf apoplast (Nakashima et al., 1997). Refixation of NH_4^+ mainly takes place in the chloroplasts and is catalysed by the chloroplastic isoform of glutamine synthetase (GS_2; Leegood et al., 1995). Very few studies have so far been undertaken to improve the understanding of the regulation of NH_4^+ transport between organelles in plant cells. Indeed little has been reported on mechanism controlling whole-plant distribution of NH_4^+, in particular the factors regulating root to shoot translocation of NH_4^+. Furthermore, a greater understanding of mechanism controlling NH_4^+ homeostasis between the apoplast and symplast of the leaf cell is required. Current research in our laboratory is trying to reveal the dynamics and physiological implications of NH_4^+ transport and turnover in plants. Some of the recent advances in this research are presented below.

Table 1. Processes generating and assimilating NH_4^+ in plants

NH_4^+ generating processes	NH_4^+ assimilating processes
NH_4^+ uptake	Glutamine synthesis
	GS/GOGAT pathway
NO_3^- reduction	NADH-GDH (?)
Photorespiration	
Phenylpropanoid pathway	Carbamoyl phosphate synthesis
Senescence induced protein degradation	
Utilization of transport nitrogen compounds in sink organs	

NH_4^+ translocation from roots to shoots

It is often assumed that NH_4^+, whether absorbed directly from the external solution or generated by NO_3^- reduction, is assimilated in the roots and not translocated to the shoots (Sechley et al., 1992; Oaks, 1994; Glass and Siddiqi, 1995; Marschner, 1995; Lam et al., 1996). However, the generality and validity of this dogma is questioned by the fact that widely different techniques have shown substantial differences in NH_4^+ concentrations in the cytoplasm (3-8 mM in maize root cells, Lee and Ratcliffe, 1991; up to 40 mM in rice and spruce root cells, Wang et al., 1993; Kronzucker et al., 1995; and up to 2 mM in xylem sap of maize and barley plants, Cramer and Lewis, 1993; Mattsson and Schjoerring, 1996).

The extent of NH_4^+ appearance in the xylem sap is dependent upon the synchronisation of N uptake, NO_3^- reduction and NH_4^+ assimilation. Ammonium is assimilated in an amidation reaction condensing NH_4^+ and glutamate into glutamine with the consumption of ATP (Lea et al., 1992). In higher plants several isoenzymes of glutamine synthetase (GS) catalyse the reaction. The cytoplasmic form of GS (GS_1) predominates in the roots (Mäck, 1995; Lam et al., 1996). In roots of oilseed rape, two isoforms of cytosolic GS have been identified (Ochs et al., 1995, EMBL database accession number X76736 and X82997). The two isoforms may be localised in different cell types as shown by *in situ* hybridisation of two GS_1 isoforms in root cells of *Nicotiana tabacum* L. (Dubois et al., 1996). Assimilation of NH_4^+ in the roots requires co-ordination between GS_1 activity and that of glutamate synthase (GOGAT), which catalyses the formation of glutamate, the substrate for GS_1, by transamination of 2-oxoglutarate with the amino group from glutamine. This imposes a requirement for a steady supply of photosynthates from the shoots in order to generate sufficient amounts of 2-oxoglutarate for subsequent NH_4^+ assimilation. Differences in carbon skeleton availability (Raab and Terry, 1995), and the C/N ratio of the cell (Hoelzle et al., 1992) may be the reason for conflicting reports on effects of nitrogen supply on GS activity. NH_4^+ has been found to be both inhibitory (Ratajczak et al., 1981; Schmidt et al., 1989) and stimulatory (Vollbrecht et al., 1989; Mäck and Tischner, 1990) to GS activity.

Assimilation of NH_4^+ translocated to the shoots must occur in the chloroplasts as the dominating GS isoform (GS_2) is located there (Kamachi et al., 1992). The main role of GS_2 is the assimilation of NH_4^+ originating from photorespiration and NO_3^- reduction in leaves (Wallsgrove et al., 1987). In the shoot, the cytosolic isoform GS_1 seems only to be located in phloem companion cells and is mainly involved in the synthesis of transport amides during senescence (Kamachi et al., 1992; Bauer et al., 1997; Buchanan-Wollaston and Ainsworth, 1997).

We have studied the translocation of NH_4^+ in relation to the expression of three glutamine synthetase (GS) isogenes and total GS activity in roots and leaves of oilseed rape (*Brassica napus*). The concentration of NH_4^+ in the stem xylem sap of NO_3^- fed plants were 0.65 mM, which was 60% higher than that in plants deprived of external N for 2 days. In NH_4^+ fed plants, xylem NH_4^+ concentrations increased linearly both with time of exposure to NH_4^+ and with increasing external NH_4^+ concentration. The maximum xylem NH_4^+ concentration was 5 mM, corresponding to 11% of the nitrogen and 20% of positive charge equivalents translocated in the xylem. Addition of either NO_3^- or NH_4^+ to N-starved plants induced both root cytosolic *gs* isogene expression

(Fig. 1) and GS activity (not shown). In N-replete plants, *gs* isogene expression was repressed, probably due to increased levels of glutamine (data not shown), thereby protecting the roots against excessive drain of photosynthates. Repressed *gs* isogene expression and GS activity under N-replete conditions caused an enhanced NH_4^+ translocation to the shoots.

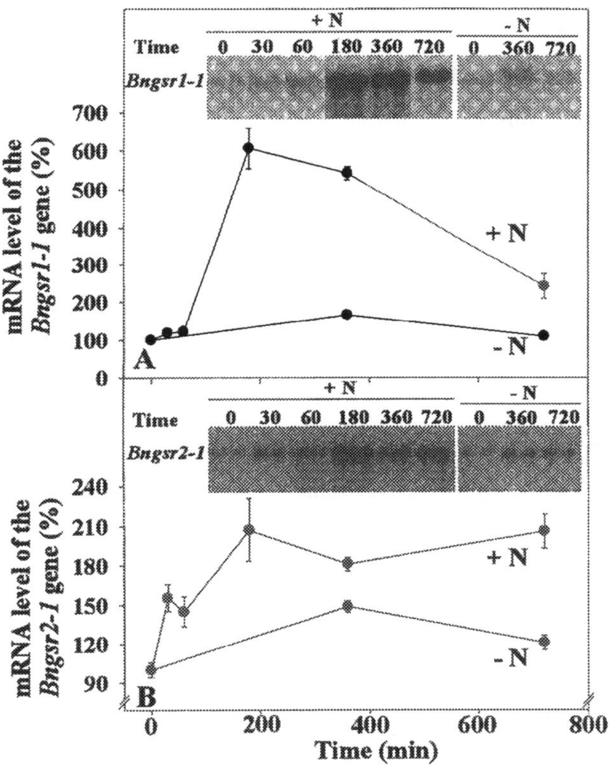

Figure 1. Northern blot analysis of the time-dependent expression of isogenes encoding cytosolic GS in roots of 5-week-old *B. napus* plants. The plants had been growing at 3 mM NO_3^- for 32 d, thereafter without external N for 3 d. At time zero of the experiment, 10 mM NH_4^+ was added to the nutrient solution. (A) Top: Northern blot of the root specific *Bngsr1-1* gene. Bottom: Graph illustrating the quantitated relative transcription levels of the root specific *Bngsr1-1* gene. (B) Top: Northern blot of the root specific *Bngsr2-1* gene. Bottom: Graph illustrating the quantitated relative transcription levels of the *Bngsr2-1* (n = 4, mean ± SE). Transcription level of 100% refers to the 3 mM NO_3^- treatment.

The leaf apoplastic solution constitutes a highly dynamic NH_4^+ pool

The rapid turnover of NH_4^+ in plant leaves leads to the establishment of a finite NH_4^+ concentration in the leaf apoplastic solution (Husted and Schjoerring, 1995). The concentration of NH_4^+ and H^+ determines the size of the NH_3 compensation point or the NH_3 mole fraction in the air within the substomatal cavities (Farquhar et al., 1980; Husted and Schjoerring, 1996). The NH_3 compensation point ranges between 0.1 and 20 nmol mol^{-1} air and is thus of the same order of magnitude as the naturally occurring atmospheric NH_3 concentrations (Sutton et al., 1994). At an NH_3 compensation point of, e.g., 5 nmol mol^{-1} this would under conditions of equilibrium correspond to an apoplastic NH_4^+ concentration of 2 mM at 20 °C and pH 5.8 (Husted and Schjoerring, 1996). The existence of an NH_3 compensation point infers that vegetation has a major influence on the transport and budgets of atmospheric ammonia, a pollutant with damaging environmental impacts (Langford and Fehsenfeld, 1992; Dentener and Crutzen, 1994; Sutton et al., 1995).

The concentration of NH_4^+ in leaf apoplastic solution is very sensitive to leaf nitrogen status and external N supply. Thus, the apoplastic NH_4^+ concentration may be about ten times higher in high-N leaves than in leaves of low-N treated oil seed rape plants (Husted and Schjoerring, 1996). Barley plants having access to NH_4^+ in the root medium have higher apoplastic NH_4^+ concentrations than plants supplied with NO_3^-. Leaf apoplastic NH_4^+ concentrations increase directly with the NH_4^+ concentration in the root medium (Mattsson and Schjoerring, 1996). Inhibition of GS leads to a rapid and very substantial increase in apoplastic NH_4^+ (Husted and Schjoerring, 1995) and barley mutants with reduced GS activity have increased apoplastic NH_4^+ relative to wild type plants (Mattsson et al., 1997).

The apoplastic solution in leaves infiltrated with an NH_4^+-free solution re-equilibrated and was maintained at 0.8 mM NH_4^+ in less than 1.5 min (Fig. 2). No changes in apoplastic NH_4^+ concentrations were observed upon infiltration with a solution containing 0.8 mM NH_4^+ (Fig. 2).

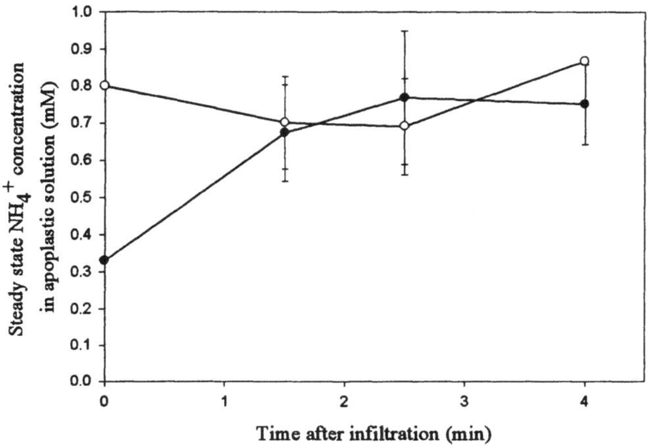

Figure 2. Time course of apoplastic NH_4^+ concentrations in leaf disks infiltrated with either 0 mM (●) or 0.8 mM (○) NH_4 Cl in buffered solutions. Values at time 0 corrected for NH_4^+ already present in the apoplast. Values are means ± SE (n = 4).

The stable isotope ^{15}N was used to assess the contribution of bi-directional NH_3/NH_4^+ transport to the maintenance of apoplastic NH_4^+ homeostasis. Infiltration with 1.0 mM 98% ^{15}N-enriched NH_4^+ resulted in an initial ^{15}N excess of 72 atom% in the apoplastic NH_4^+ pool. Five min later the ^{15}N excess was diluted to 20.0 atom% and after 25 min to 8.4 atom% without any changes in apoplastic NH_4^+ concentration (0.8 mM) (Fig. 3) or total apoplastic N (7.1 mM - data not shown). The relative dilution of ^{15}N with ^{14}N during the 0-25 min experimental period was 75% at the start to 12% after 25 min. The calculated exchange rate of ^{15}N and ^{14}N over the plasma membrane was 29.4 μmol g^{-1} FW h^{-1} between 0 and 5 min and 3.9 ± 0.7 μmol g^{-1} FW h^{-1} between 15 and 25 min after infiltration (Table I).

These results clearly demonstrate the dynamic nature of the apoplastic solution in *B. napus*. Ammonia constantly effluxes from the cell into the apoplast solution with highly efficient transporters operating to maintain NH_4^+ homeostasis.

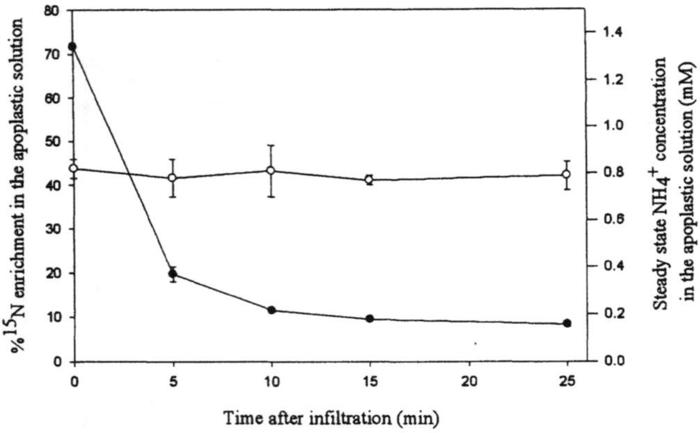

Figure 3. Time course of ^{15}N excess (●) and NH_4^+ concentration (O) in the apoplast solution following infiltration with 1.0 mM $^{15}NH_4Cl$. Values are means ± SE (n = 6).

Table 2. Exchange rate of $^{15}N/^{14}N$ over the plasma membrane determined by measuring the dilution of ^{15}N with ^{14}N in the apoplast. Calculated from Fig. 3.

Interval	Exchange rate of $^{15}N/^{14}N$
min	µmol g^{-1} fresh wt h^{-1}
0-5	29.4 ± 4.0
5-10	19.9 ± 4.4
10-15	6.2 ± 1.8
15-20	3.9 ± 0.7

Transport mechanisms for NH_4^+ entry into leaf cells

Despite the importance of control of leaf apoplastic NH_4^+ concentration for NH_4^+ recovery and plant-atmosphere NH_3 exchange very little information is available concerning the transport of NH_4^+ across the leaf cell plasma membrane. In leaf discs of *Phaseolus vulgaris,* Raven and Farquhar (1981) observed that uptake of methylammonium (an NH_4^+ analogue) could not be accounted for by passive diffusion, but seemed to be mediated by some kind of energy requiring transport system. In roots of various plant species, as well as in *Chara,* biphasic uptake systems have been reported. Ammonium was found to enter root cells via high-affinity transport systems displaying Michaelis-Menten kinetics with a K_m around 15-40 μM, and via low-affinity transport systems displaying linear dependence to external NH_4^+ supply (Ritchie, 1987; Glass et al., 1997). Ninnemann et al. (1994) isolated and characterised a gene for a high affinity NH_4^+ transporter from Arabidopsis that was highly expressed in both roots and leaves and has now been found in rice and tomato (Lauter et al., 1996). The low affinity transport system has been proposed to be a uniport with fluxes driven by the electrochemical gradient across the plasma membrane (Wang et al., 1994). This uniport may be a specific NH_4^+ channel, a K^+ channel or a shared cation channel, for example a K^+/NH_4^+ channel as indicated by Avery et al. (1992) and Schachtman et al. (1992). Considering the relatively high concentrations of NH_4^+ (0.5-1.5 mM) frequently encountered in the leaf apoplastic solution of oilseed rape plants (ref), the low affinity system appears to be central in NH_4^+ transport.

A further complicating factor of NH_4^+ transport in leaves relative to that in roots is the possible existence of a large efflux component due to diffusion of dissolved NH_3. This efflux is due to the fact that even under conditions where the intracellular NH_4^+ concentration is 10-100 times lower than the extracellular, a high pH in the cytoplasm (pH 7.0-7.5; Martin et al., 1982) and in the chloroplasts (pH around 8 in light) relatively to that in the apoplastic solution (pH around 6.0) may maintain a gradient of dissolved NH_3 directed towards the apoplast.

Experimental work in our laboratory has identified some of the transport mechanism for NH_4^+ entry into leaf cells of *B. napus*. Using an apoplast infiltration technique, the net uptake of NH_4^+ over a 4 min period responded linearly to increasing NH_4^+ concentrations up to 10 mM (Fig. 4). At higher concentrations the net NH_4^+ uptake began to saturate, with saturation at concentrations above 40 mM NH_4^+ (Fig. 4). $^{15}NH_4^+$ uptake into isolated protoplasts from *B. napus* leaves closely paralleled these results and further demonstrated a pH optimum of 5.5 (J.N. Pearson, J. Finnemann and J.K. Schjoerring, unpublished results). Current investigations will reveal whether low-affinity uptake of NH_4^+ occurs via a K^+ channel or via a novel, specific NH_4^+ channel.

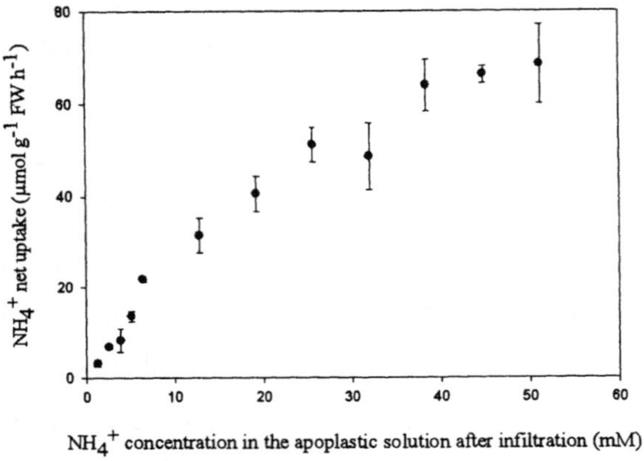

Figure 4. Net uptake of NH_4^+ from the apoplast solution in leaf disks infiltrated for 4 min with a range of concentrations of NH_4^+. Values are means ± SE (n = 4).

The ATPase inhibitor DES and the protonophore CCCP had little effect ($P>0.05$) on the net NH_4^+ uptake (Fig. 5a). Conversely, the unspecific channel blocker La^{3+} and the specific K^+-channel-blocker TEA-Cl reduced ($P<0.05$) the NH_4^+ net uptake (Figs 4c,d). The reduction caused by La^{3+} amounted to 30%, while that of TEA-Cl was 47%. Both increasing concentrations of K^+ and Na^+ resulted in a decrease in NH_4^+ net uptake (Fig. 5b). The inhibition caused by K^+ was 50% at 100 mM KCl in the infiltration solution. A similar concentration of NaCl only resulted in a 20% decline of NH_4^+ net uptake.

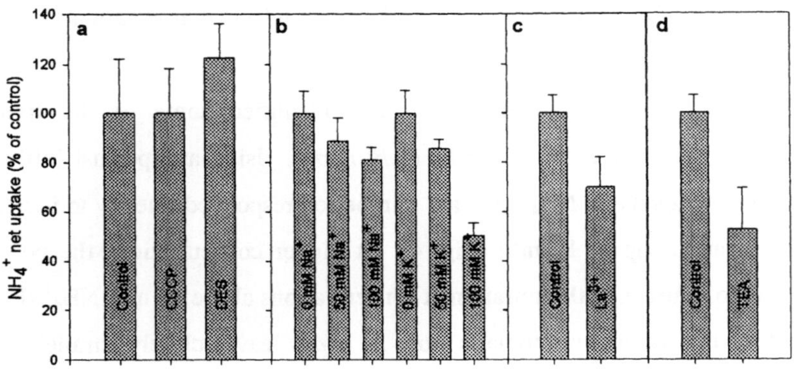

Figure 5. Influence of various inhibitors and competitors on net uptake of NH_4^+ from the apoplast. Leaves were infiltrated with 10 mM NH_4Cl for 4 min. Values are means ± SE (n = 4).

More detailed investigations into $^{15}NH_4^+$ uptake into *B. napus* leaf protoplasts revealed the presence of a high-affinity NH_4^+ uptake system, similar in kinetics to the *AMT1* transporter observed in *Arabidopsis*, tomato and rice. Primers based on conserved regions of the *AMT1* transporter were used in RT-PCR reactions to identify and subsequently clone a homologue in *B. napus (BnAMT1)*. The *Bnamt1* gene was found to be strongly expressed under all N conditions but regulated by N supply (J.N. Pearson, J. Finnemann and J.K. Schjoerring, unpublished results). Long-term exposure of plants to high NH_4^+ resulted in reduced levels of *BnAMT1* mRNA. On the other hand, short-term exposure of leaf cells to NH_4^+ concentrations above 0.2 mM stimulated *Bnamt1* expression in the leaves. *BnAMT1* appears to serve a dual function in the recovery of photorespiratory NH_4^+ from the apoplast and maintenance of symplastic NH_4^+ concentrations.

Both the high and low-affinity transport systems observed in the leaves of *B. napus* are involved in the maintenance of apoplastic NH_4^+ concentrations. The low-affinity transporter may serve an important function in apoplast homeostasis, in particular with regard to the rapid re-equilibration of apoplastic NH_4^+ levels. Variations in apoplastic NH_4^+ levels greatly influence the NH_3 compensation point of the apoplast solution and thus NH_3 emissions from the plant.

Future perspectives

Previous work in our laboratory has demonstrated the dynamic nature of the apoplastic NH_4^+ pool and the existence of specific transporters to recover NH_4^+ from the apoplast solution. However much remains unknown. In particular, the mechanism for the uptake of NH_4^+ into the chloroplast during primary N assimilation and re-assimilation of photorespiratory NH_4^+ is not known. The pH gradient across the chloroplast envelope precludes the passive influx of NH_3 as the main mechanism. Similarly, the mechanism for the release of NH_3/NH_4^+ from the mitochondria during photorespiration is not known. Much more however is known of the low-affinity NH_4^+ uptake system on the plasma membrane. Current work in our laboratory is attempting to isolate the transport protein responsible for low-affinity NH_4^+ uptake in *Arabidopsis*. It remains to be seen as to whether the protein is a novel transporter or whether low-affinity uptake occurs via an already identified transporter, such as a K^+ channel.

References

Avery, S.V., Codd G.A. and Gadd G.M. (1992) Caesium transport in the cyanobacterium *Anabaena variabilis*: kinetics and evidence for uptake via ammonium transport systems. *FEMS Microbiol. Lett.* **95**, 253-258.

Bauer, D., Biehler, K., Fock, H., Hirel, B., Migge, A. and Becker, T.W. (1997) A role for glutamine synthetase in the remobilization of leaf nitrogen during water stress in tomato. *Physiol. Plant.* **99**, 241-248.

Buchanan-Wollaston, V. and Ainsworth, C. (1997) Leaf senescence in Brassica napus: cloning of senescence related genes by subtractive hybridisation. *Plant Mol. Biol.* **33**, 821-834.

Cramer, M.D. and Lewis, O.A.M. (1993) The influence of nitrate and ammonium nutrition on growth of wheat (*Triticum aestivum* L.) and maize (*Zea mays* L.) plants. *Ann. Bot.* **72**, 359-365.

Dentener, F.J. and Crutzen, P.J. (1994) A three dimensional model of the global ammonia cycle. *J. Atmos. Chem.* **19**, 331-369.

Dubois, F., Brugière, N., Sangwan, R.S. and Hirel, B. (1996) Localization of tobacco cytosolic glutamine synthetase enzymes and the corresponding transcripts shows organ- and cell-specific patterns of protein synthesis and gene expression. *Plant Mol. Biol.* **31**, 803-817.

Farquhar, G.D., Firth, P.M., Wetselaar, R. and Weir, B, (1980) On the gaseous exchange of ammonia between leaves and the environment: Determination of the ammonia compensation point. *Plant Physiol.* **66**, 710-714.

Glass, A.D.M. and Siddiqi, M.Y. (1995) Nitrogen absorption by plant roots, in H.S. Srivastava and R.P. Singh (eds.), *Nitrogen nutrition in higher plants*, Associated Publishing Co., New Delhi, pp. 21-56.

Glass, A.D.M., Erner, Y., Kronzucker, H., Schjoerring, J.K., Siddiqi, M.Y. and Wang, M.-Y. (1997) Ammonium fluxes into plant roots: energetics, kinetics and regulation. *J. Plant Nutr. Soil Sci.* **160**, 261-268.

Hoelzle, I., Finer, J.J., McMullen, M.D., Streeter, J.G. (1992) Induction of glutamine synthetase activity in nonnodulated roots of *Glycine max*, *Phaseolus vulgaris* and *Pisum sativum*. *Plant Physiol.* **100**, 525-528.

Husted, S. and Schjoerring, J.K. (1995) Apoplastic pH and ammonium concentration in leaves of *Brassica napus* L. *Plant Physiol.* **109**, 1453-1460.

Husted, S. and Schjoerring, J.K. (1996) Ammonia flux between oilseed rape plants and the atmosphere in response to changes in leaf temperature, light intensity and relative air humidity. Interactions with stomatal conductance and apoplastic NH_4^+ and H^+ concentrations. *Plant Physiol.* **112**, 67-74.

Joy, K.W. (1988) Ammonia, glutamine, and asparagine: A carbon-nitrogen interface. *Can. J. Bot.* **66**, 2103-2109.

Kamachi, K., Yamaya, T., Hayakawa, T., Mae, T. and Ojima, K. (1992) Vascular bundle-specific localization of glutamine synthetase in rice leaves. *Plant Physiol.* **99**, 1481-1486.

Kronzucker, H.J., Siddiqi, M.Y. and Glass, A.D.M. (1995) Compartmentation and flux characteristics of ammonium in spruce. *Planta* **196**, 691-698.

Lam, H.-M., Coschigano, K.T., Oliveira, I.C., Melo-Oliveira, R. and Coruzzi, G.M. (1996) The molecular-genetics of nitrogen assimilation into amino acids in higher plants. *Annu. Rev. Plant Physiol. Plant Mol. Biol.* **47**, 569-593.

Langford, A.O., Fehsenfeld, F.C. (1992) Natural vegetation as a source or sink for atmospheric ammonia: A case study. *Science* **255**, 581-583.

Lauter, F.-R., Ninnemann, O., Bucher, M., Riesmeier, J.W. and Frommer, W.B. (1996) Preferential expression of an ammonium transporter and two putative nitrate transporters in root hairs of tomato. *Proc. Natl. Acad. Sci. USA* **93**, 8139-8144.

Lea, P.J., Blackwell, R.D. and Joy, K.W. (1992) Ammonia assimilation in higher plants, in K. Mengel and D.J. Pilbeam (eds.), *Nitrogen metabolism of plants*, Oxford Science Publications, Oxford, pp. 153-186.

Lee, R.B., Ratcliffe, R.G. (1991) Observations on the subcellular distribution of the ammonium ion in maize root tissue using *in-vivo* ^{14}N-nuclear magnetic resonance spectroscopy. *Planta* **183**, 359-367.

Leegood, R.C., Lea, P.J., Adcock, M.D. and Häusler, R.E. (1995) The regulation and control of photorespiration. *J. Exp. Bot.* **46**, 1397-1414.

Marschner, H. (1995) *Mineral nutrient of higher plants*. Ed. 2, Academic Press, London.

Martin, J.-B., Bligny, R., Rebeille, F., Douce, R., Leguay, J.J., Mathiey, Y. and Guern, J. (1982) A ^{31}P nuclear magnetic resonance study of intracellular pH of plant cells cultivated in liquid medium. *Plant Physiol.* **70**, 1156-1161.

Mattsson, M. and Schjoerring, J.K. (1996) Ammonia emission from young barley plants: influence of N source, light/dark cycles, and inhibition of glutamine synthetase. *J. Exp. Bot.* **47**, 477-484.

Mattsson, M., Häusler, R.E., Leegood, R.C., Lea, P. and Schjoerring, J.K. (1997) Leaf-atmosphere ammonia exchange in barley mutants with reduced activities of glutamine synthetase. *Plant Physiol.* **114**, 1307-1312.

Mäck, G. (1995) Organ-specific changes in the activity and subunit composition of glutamine-synthetase isoforms of barley (*Hordeum vulgare* L.) after growth on different levels of NH_4^+. *Planta* **196**, 231-238.

Mäck, G. and Tischner, R. (1990) Glutamine synthetase oligomers and isoforms in sugarbeet (*Beta vulgaris* L.). *Planta* **181**, 10-17.

Nakashima, J., Awano, T., Takabe, K., Fujita, M. and Saiki, H. (1997) Immunocytochemical localization of phenylalanine ammonia-lyase and cinnamyl alcohol dehydrogenase in differentiating tracheary elements derived from *Zinnia* mesophyll cells. *Plant Cell Physiol.* **38**, 113-123.

Ninnemann, O., Jauniaux, J.-C. and Frommer, W.B. (1994) Identification of a high affinity NH_4^+ transporter from plants. *EMBO J.* **103**, 3464-3471.

Oaks, A, (1994) Primary nitrogen assimilation in higher plants and its regulation. *Can. J. Bot.* **72**, 739-750.

Ochs, G., Schock, G. and Wild, A. (1995) Purification and characterization of glutamine synthetase isoenzymes from leaves and roots of *Brassica napus* (L.). *J. Plant. Physiol.* **147**, 1-8.

Raab, T.K. and Terry, N. (1995) Carbon, nitrogen and nutrient interactions in *Beta vulgaris* L. as influenced by nitrogen source NO_3^- versus NH_4^+. *Plant Physiol.* **107**, 575-584.

Ratajczak, L., Ratajczak, W. and Mazurowa, H. (1981) The effect of different carbon and nitrogen sources on the activity of glutamine synthetase and glutamate dehydrogenase in lupin embryonic axes. *Physiol. Plant* **51**, 277-280.

Raven, J.A. and Farquhar, G.D. (1981) Methylammonium transport in *Phaseolus vulgaris* leaf slices. *Plant Physiol.* **67**, 859-863.

Ritchie, R.J. (1987) The permeability of ammonia, methylamine and ethylamine in the charophyte *Chara corallina* (*C. australis*). *J. Exp. Bot.* **38**, 67-76.

Schatchtman, D.P., Schroeder, J.I., Lucas, W.J., Anderson, J.A. and Gaber, R.F. (1992) Expression of an inwardly-rectifying potassium channel by the *Arabidopsis* KAT1 cDNA. *Science* **258**, 1654-1658.

Schmidt, S. and Mohr, H. (1989) Regulation of the appearance of glutamine synthetase in mustard (*Sinapis alba* L.) cotyledons by light nitrate and ammonium. *Planta* **177**, 526-534.

Sechley, K.A., Yamaya, T. and Oaks, A. (1992) Compartmentation of nitrogen assimilation in higher plants. *Int. Rev. Cyt.* **134**, 85-163.

Sutton, M., Asman, W.A.H. and Schjoerring, J.K. (1994) Dry deposition of reduced nitrogen. *Tellus* **46**, 255-273.

Sutton, M., Schjoerring, J.K. and Wyers, P. (1995) Plant-atmosphere exchange of ammonia. *Phil. Trans. R. Soc. London* **351**, 261-278.

Vollbrecht, P., Klein, E. and Kasemir, H. (1989) Different effects of supplied ammonium on glutamine synthetase activity in mustard (*Sinapis alba*) and pine (*Pinus sylvestris*) seedlings. *Physiol. Plant* **77**, 129-135.

Wallsgrove, R.M., Turner, J.C., Hall, N.P., Kendall, A.C. and Bright, S.W.J. (1987) Barley mutants lacking chloroplast glutamine synthetase - biochemical and genetic analysis. *Plant Physiol.* **83**, 155-158.

Wang, M.-Y., Siddiqi, M.Y., Ruth, T.J. and Glass, A.D.M. (1993) Ammonium uptake by rice roots. I. Fluxes and subcellular distribution of $^{13}NH_4^+$. *Plant Physiol.* **103**, 1249-1258.

Wang, M.-Y., Glass, A.D.M., Shaff, J.E. and Kochian, L.V. (1994) Ammonium uptake by rice roots. III. Electrophysiology. *Plant Physiol.* **104**, 899-906.

REGULATION OF THE EXPRESSION OF TWO NEW MEMBERS OF THE *BCH* FAMILY OF GENES, *BCH3* AND *BCH4*, WHICH ENCODE INDUCIBLE HIGH AFFINITY NITRATE TRANSPORTERS IN *HORDEUM VULGARE*, WITH RELATION TO NITRATE INFLUX.

J. JOHN VIDMAR[1-3], DEGAN ZHUO[1], M. YAESSH SIDDIQI[1], JAN K.SCHJOERRING[2], BRUNO TOURAINE[3] and ANTHONY D. M. GLASS[1]

[1]Department of Botany, University of British Columbia, Vancouver, British Columbia, V6T 1Z4, Canada; [2]Plant Nutrition Laboratory, Department of Agricultural Sciences, Royal Veterinary and Agricultural University, Thorvaldsensvej 40, DK-1871 Frederiksberg C, Copenhagen, Denmark; [3]Biochimie et Physiologie Végétales, Ecole Nationale Supérieure d'Agronomque de Monpellier/Institut National de la Recherche Agronomique/ Centre National de la Recherche Scientifique Unité de Recherche Associée 2133, 34060 Montpellier Cedex 1, France

Abstract

We isolated two new cDNAs, *BCH3* and *BCH4* (barley *CRNA* homologues), which are members of the *BCH* multigene family from barley. *BCH* genes are thought to encode putative inducible high affinity nitrate transporters. *BCH3* was isolated using RACE-PCR with specific primers to the BCRNA fragment (Trueman *et al.*, 1996), while *BCH4* was isolated using RT-PCR. *BCH3* and *BCH4* are 1822 and 1705 bp, encode putative polypeptides of 507 amino acids, with a predicted m.w. of 54.6 kDa. Predicted BCH3 and BCH4 proteins are members of a nitrate/nitrite subfamily of major facilitator superfamily. These cDNAs are related to *Aspergillus nidulans CRNA*,

G. Gissel-Nielsen and A. Jensen (eds.), Plant Nutrition – Molecular Biology and Genetics, 83–94.

Chlamydomonas rienhardtii NAR3 and *NAR4*, *Hordeum vulgare BCH1* and *BCH2*, and *Nicotinia plumbaginifolia NRT2Np*. In barley, southern analysis indicates 7-10 copies of *BCH* genes (Trueman *et al.*, 1996). *BCH1, BCH2, BCH3* and *BCH4* are highly homologous (greater than 87% identity). In Northern blot analysis, we found that NO_3^- supply to N-deprived plants increased both the abundance of *BCH* transcripts and $^{13}NO_3^-$ influx. We analyzed the expression pattern of the four *BCH* genes using oligonucleotides direct to the 3'UTR, and found that all four genes are co-ordinately up-regulated in response to NO_3^- treatment. We assayed the effect of various external concentrations of NO_3^- and found that 50 μM NO_3^- grown plants showed the highest *BCH* transcript abundance and $^{13}NO_3^-$ influx. These results provide indirect evidence that the *BCH* genes encode proteins which are involved in inducible high-affinity transport in plants. We analyzed the effect of amino acids feeding on *BCH* transcript levels, when co-supplied with nitrate, and found that asparagine, aspartate, glutamate and glutamine dramatically decrease transcript levels (>60%). Concurrently, we measured influx using $^{13}NO_3^-$ and found a different pattern for the amino acid treatments. Glutamine and asparagine decreased influx by 40 and 20% respectively, while aspartate and glutamate decreased influx by 78 and 65%, respectively. The effect of inhibitors of N assimilation were assayed. Tungstate (an inhibitor of nitrate reductase) increased the level of expression of *BCH* transcripts, whereas there was a slight decrease for methionine sulfoximine (an inhibitor of glutamine synthetase), compared to control (NO_3^- treatment for 6h), but NO_3^- influx decreased by 50%. Azaserine (an inhibitor of glutamate synthase) decreased the levels of *BCH* transcript by 95% of the control, while NO_3^- influx decreased by >95 %. Amino-oxyacetate (an aspartate amino-transferase and glutamate decarboxylase inhibitor) decreased both the *BCH* transcripts abundance and NO_3^- influx by 65%.

Introduction

The uptake of nitrate in terrestrial plants is mediated by specific transporters. It is thought to be a thermodynamically active process, since transport occurs against the electrochemical potential gradient of the cell, and transport is sensitive to metabolic inhibitors (Clarkson, 1986). At a physiological level, nitrate transport is mediated by at least three systems namely CHATS, IHATS and LATS (see review by Glass and Siddiqi, 1995).

In plants, genes encoding putative inducible high-affinity transporters (IHATs) have been cloned from barley (Trueman *et al.*, 1996), *Arabidopsis thaliana* (Z97058), *Nicotinia plumbiginifolia* (Quesada *et al.*, 1997), and soybean (Amarasinge *et al.*, 1998). In barley, a putative IHATs is encoded by a multigene family of 7-10 members (Trueman *et al.*, 1996). To date, two members of the *BCH* family have been isolated (Trueman *et al.*, 1996). *BCH* cDNAs encode polypeptides of 507-509 amino acids, with a m.w. of 54-55 kDal, with 12 hydrophobic domains, and belong to the Major Facilitator Superfamily (MFS). Their amino acid sequence contains the conserved motif AG(W/L)GNMG, which is specific to the nitrate/nitrite transporters subgroub of MFS. It has been shown that the mRNA levels of these IHAT genes increase due to the provision of nitrate, a process referred to as «induction» (Trueman *et al.*, 1996; Quesada *et al.*, 1997; Amarasinge *et al.*, 1998). Nitrate influx has been demonstrated at a physiological level to be induced by the provision of NO_3^- to plant that were N-Starved (Hole *et al.*, 1990; and Siddiqi *et al.*, 1989).

In *N. plumbaginifolia,* Krapp *et al.* (1998) demonstrated that genes involved in NO_3^- assimilation, *NpNRT2* (a nitrate transporter), a *NIA* (nitrate reductase) and *NII* (nitrite reductase) were co-ordinately expressed, also that a deficiency in nitrate reductase lead to increase abundance of *NpNRT2* mRNA.

The down-regulation of IHATS, is thought to be mediated by NO_3^- or product(s) of N-assimilation. At a physiological level down-regulation of NO_3^- influx has been attributed to NO_3^- itself (King *et al.*, 1993; Doddema *et al.*, 1978), NH_4^+ (Aslam *et al.*, 1996) and amino acids (Doddema and Otten 1979; Muller and Touraine, 1992).

In this communication, we report the isolation of *BCH*3 and *BCH4* cDNAs, which encode putative inducible high-affinity nitrate transporters. We characterize the regulation of *BCH* transcript abundance and $^{13}NO_3^-$ influx in barley seedlings in response to nitrate, amino acids (glutamine (Gln), glutamate (Glu), aspargine (Asn), and aspartate (Asp)), and N-assimilation inhibitors (tungstate, methionine sulfoximine (MSO), azaserine (AZA), and amino-oxyacetate (AOA)).

Materials and methods

PLANT MATERIAL

Seven day old seedlings of *Hordeum vulgare* cv. Klondike were used in all experiments. Plants were grown as reported by Siddiqi *et al.* (1989), on N-free 1/10 modified Johnson's solution (Siddiqi *et al.*, 1989). N was supplied in the form of NO_3^- at 1 or 10 mM depending on experimental design; amino acids were supplied at a 1 mM concentration. Inhibitors tungstate, MSO, AZA, and AOA were supplied at the concentrations 0.5, 1.0, 0.25 and 0.5, respectively. Potassium concentration was monitored daily and other nutrient including potassium was resupplied accordingly. The pH of the solution was maintained at 6.2 ± 0.3 by the addition of $CaCO_3$ powder. The cultures were carried out in controlled environmental chamber with a 16 h/8 h light dark cycle at 20 ± 2°C and 70 % relative humidity. Light was provided by fluorescent tubes with a spectral composition similar to sunlight. The photon flux density at plant level was 300 μmol m^{-2} s^{-1}.

RNA ISOLATION AND cDNA SYNTHESIS

Total RNA was isolated using Trizol Reagent (Life Technologies), with two modifications. Firstly, after the tissue was ground in a mortar and Trizol reagent was added at a ratio of 0.2 g tissue/ 1 ml Trizol. The homogenate was centrifuged at 8000 *g* for 30 min to remove cellular debris. Secondly, after the total RNA was isolated, it was again extracted with phenol: chloroform: iso-amyl alcohol (25:24:1), and precipitated with 0.3 M sodium acetate (final concentration), and two volumes ethanol. Fastrack mRNA isolation Kit (Invitrogen, Carlsbad, Ca) was used in the isolation of mRNA, as per manufacturers instructions. Messenger RNA isolated from roots of 7 day old barley seedlings treated for 2 and 6 h with 10 mM KNO_3 were used as template for cDNA was synthesis. Marathon cDNA synthesis kit (Clontech, Palo Alto, Ca) was used for the construction of a cDNA library.

NORTHERN BLOT ANALYSIS

Total RNA was separated on a denaturing 1.2% agarose gel as described in Sambrook *et al.* (1989). RNA was transferred and fixed on N+ nylon membrane (Amersham) according to the manufacturer's instructions. Radioactive probes were made with Prime-A-Gene kit (Promega, Madison, WI) using internal fragment from the *BCH3* gene on plasmid pBCH3 by digestion with *Eco*RV and *Afl*III. For control levels of total RNA, blots were probed with a fragment of the 25S gene, on plasmid pV25S, by digestion with *XhoI*. Membranes were washed according to the manufacturer's instructions.

NITRATE INFLUX

Nitrate influx experiments were carried out essentially as described by Siddiqi *et al.* (1989), with the following modifications, influx measurements were carried out in 50 $\mu M\ NO_3^-$, for a period of 5 min. Production of $^{13}NO_3^-$ was as described by Kronzucker *et al.* (1995).

Results and Discussion

ISOLATION AND CHARACTERIZATION OF *BCH3* AND *BCH4* cDNAS

Using RT-PCR and RACE-PCR we isolated two new full length cDNAs, *BCH3* and *BCH4,* which are members of the *BCH* gene family in barley. They are 1822 and 1705 bp and encode proteins of 507 amino acids, with a m.w of 53.4 kDa. They contain 12 membrane spanning domains, and the concensus amino acid sequence hallmark of the MFS and the consensus sequence for the nitrate/nitrate subgroup of this family (Trueman *et al.,* 1996). The two protiens are highly homologous with BCH1 and BCH2 proteins (greater than 89% identity between the four polypeptides). The encoded polypeptides are 32.5-33.6% identical to CRNA protein from *A.nidulans,* 48.3-49% to NAR3 protein from *Chlamydomonas reinhardtii*, 70.2-71.4% to NRT2Np protein from *N. plumbaginifolia*, and 72.4-72.7% to NRT2.1At protein from *A.thaliana.*

TIME PROFILE OF NO_3^- INDUCTION ON THE ABUNDANCE OF *BCH* TRANSCRIPT AND $^{13}NO_3^-$ INFLUX

The dramatic increase of NO_3^- uptake due to NO_3^- exposure has been referred to as nitrate induction (Jackson *et al.*, 1973; Goyal and Huffaker, 1986). This exposure can increase net uptake rates by 5 to 10 fold (Warren and Huffaker, 1989) and $^{13}NO_3^-$ influx by 30 fold (Siddiqi *et al.*, 1989). In our experiments using 1 mM NO_3^- to induce NO_3^- influx, there was an increase in both influx rates and *BCH* transcript levels which peaked at 3 h and 6 h, respectively. The two parameters followed the same pattern, displaying a sharp increased at the onset of treatment, followed by steady decrease (figure 1). All four mRNA of *BCH* cDNAs increased due to the provision of NO_3^- (data not shown). The *BCH* transcripts were not detected in shoot tissues. The earliest appearance of transcript was observed within 30 min of the onset of NO_3^- treatment (data not shown). The decrease of *BCH* transcript accumulation in long term experiments (greater than 12 h) has also been shown for other cDNAs encoding putative a IHATs (*NRT2Np* (Quesada *et al.*, 1997)). This response is thought to be a function of N-metabolite repression. It is noteworthy that by 24 and 48 h of exposure to 1 mM NO_3^- *BCH* transcripts abundance was reduced to a level of the uninduced plants. By contrast, $^{13}NO_3^-$ influx at these times was still substantially higher than that of uninduced plants. This may point to the contribution of other nitrate transporters in the observed fluxes or that these transporters have a long half-life.

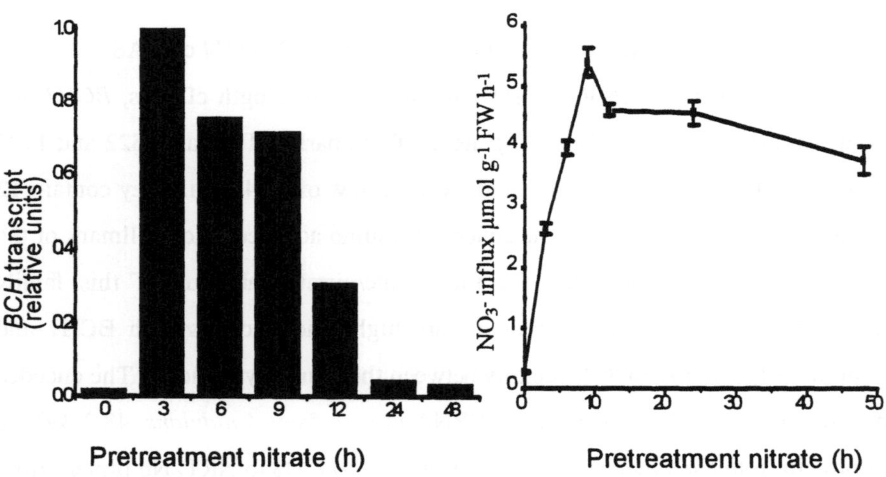

Figure 1. Time course of NO_3^- treatment on BCH mRNA accumulation and NO_3^- influx. 7 day old seedlings of barley grown in N-free Johnson's solution then supplied with 1 mM NO_3^- a) Northern blot analysis of 20 µg of total RNA quantified by phosphoimager (standardized by 25S transcript) and b) $^{13}NO_3^-$ influx.

EFFECT OF EXTERNAL NO_3^- ON THE ABUNDANCE OF *BCH* TRANSCRIPT AND $^{13}NO_3^-$ INFLUX

We investigated the effect of constant external NO_3^- concentration of 0, 10, 50, 100, or 500 µM, on $^{13}NO_3^-$ influx and *BCH* transcript levels, in order to determine if parameters were correlated (figure 2a and figure 2b). Both abundance of *BCH* transcript and $^{13}NO_3^-$ influx were highest at the 50 µM external NO_3^- treatment. The reported K_m values for IHATS vary between 10 to 100 µM, and these variations can be caused by intra and inter species differences. In *Hordeum vulgare* the K_m and V_{max} of IHATS has been shown to vary as a function of nitrate pretreatment (Siddiqi *et al.*, 1990).

The correlation between *BCH* transcript levels with influx rates in the time profile and NO_3^- concentration experiments, provide indirect evidence that *BCH* genes encode for a component of the inducible high-affinity transport systems.

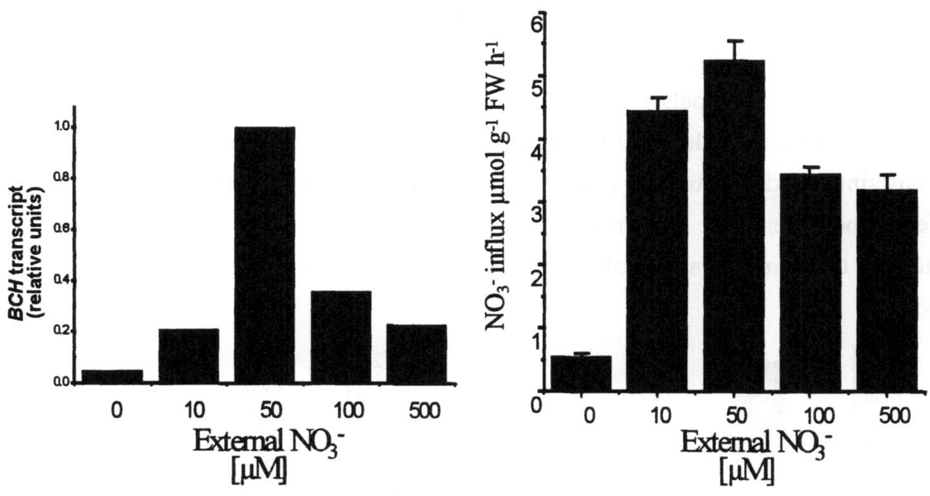

Figure 2. Effect of External NO_3^- concentration on *BCH* mRNA accumulation and NO_3^- influx. 7 day old seedlings of barley grown for 4 d in N-free Johnson's solution containing 0, 10, 50, 100, or 500 µM $KJNO_3^-$. a) Northern blot analysis of 20 µg of total RNA quantified by phosphoimager (standardized by 25S transcript) and b) $^{13}NO_3^-$ influx.

EFFECT OF AMINO ACID PRETREATMENT ON THE ABUNDANCE OF *BCH* TRANSCRIPT AND $^{13}NO_3^-$ INFLUX

Pretreatment of intact roots with glutamine (Gln), glutamate (Glu), aspartate (Asp), and asparagine (Asn) showed that the four amino acids tested decreased *BCH* transcript levels (figure 3a) and in roots NO_3^- influx (figure 3b) to varying degrees. The amino acid effect shows the following responses on abundance of *BCH* transcript Glu=Asp>Asn>Gln (but the decrease was greater than 50 % in all cases), while $^{13}NO_3^-$ influx followed a different pattern Glu>Asp>Asn>Gln. When feeding plants with amino acids, one must take into account different rates entry into the root and different rates of metabolism. Feeding plants with amino acids may mimic the entry in the root of shoot-originated signals that control uptake (Imsande and Touraine, 1994). It has been demonstrated that external amino acids can decrease NO_3^- uptake, in *Arabidopsis thaliana* (Doddema and Otten, 1979), in *Phaseolus vulgare* (Breteler and Arnozis, 1985) and in wheat (Rodgers and Barneix, 1993). But the negative feedback effect of amino acid treatment was preceded by a lag period greater than 3 h (in the above cases). This indicates that the amino acid effect on NO_3^- uptake is not direct or allosteric (Breteler and Arnozis, 1985). Since amino acid can be metabolized upon entry in the cell, the pools of amino acids may vary drastically depending on treatment. Our results indicate that there is a down-regulatory effect by amino acids on *BCH* transcript abundance, but the specific N-pools responsible for these effects could not be determined. Hence we designed a new set of experiments using inhibitors of specific steps of the N-assimilation pathway, which would result in altered of the amino acid pools.

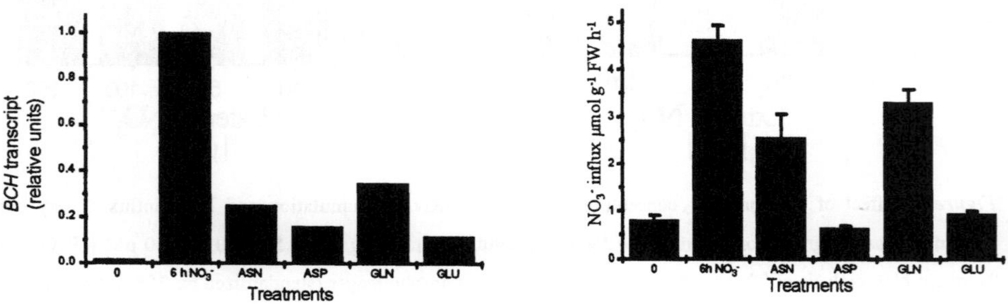

Figure 3. Effect of amino acid pretreatment on *BCH* mRNA accumulation and NO_3^- influx. 7 day old seedlings of barley were grown in N-free media then supplied for 6 h with 10 mM NO_3^- alone or together with each of the following amino acids (1 mM concentration): Asn, Asp, Gln, and Glu. a) Northern blot analysis of 20 µg of total RNA quantified by phosphoimager (standardized by 25S transcript) and b) $^{13}NO_3^-$ influx.

THE EFFECT OF N-ASSIMILATION INHIBITORS ON NITRATE INFLUX AND *BCH* TRANSCRIPT ACCUMULATION

The inhibitor treatments consisted in the addition of 0.5 mM tungstate, 1 mM methionine sulfoximine (MSO), 0.25mM azaserine (AZA) or 0.5 mM amino oxyacetate (AOA), all co-supplied with 10 mM NO_3^-. In the roots of tungstate treated plants, *BCH* transcript levels increased in comparison to control plants (approximately 20 %) (figure 4a). Conversely, we measured $^{13}NO_3^-$ influx uptake after 6 h of NO_3^- and tungstate co-supply and found a 50% decrease in influx (figure 4b). With regards to the effect of NO_3^- in down regulation of influx, the work by Ingermarsson *et al.* (1987) in *Lemna* showed that tungstate increased NO_3^- concentration in tissue, which resulted in decreased NO_3^-uptake. These workers concluded that NO_3^- accumulation was responsible for down-regulating $^{13}NO_3^-$ influx. Also using NR double mutants of barley King *et al.* (1993) concluded that NO_3^- itself may regulate influx. The experimental evidence may thus indicate that NO_3^- itself acts as a regulator of influx, but at a post-transcriptional level.

The effect of MSO co-supplied with NO_3^- on *BCH* transcript levels showed a slight decrease in comparison to control plants (approximately 5 %) (figure 4a). However, $^{13}NO_3^-$ influx decreased 50% after 6 h of co-supply. (figure 4b). NH_4^+ has been shown to have a multitude of effects on NO_3^- uptake. It is thought to affect influx (King *et al.*, 1993) and/or efflux (Aslam *et al.*, 1996). At the molecular level, Quesada *et al.* (1997) demonstrated that NH_4^+ application can down-regulate the abundance of the *NpNRT2* (IHAT gene), but it was unclear if the effect was from NH_4^+ *per se* or due to an assimilate of NH_4^+ (amino acid). Feeding plants with NO_3^- and MSO (an inhibitor of glutamine synthetase) should increase cytoplasmic NH_4^+ concentrations (Lee *et al.*, 1992). Since MSO blocks the conversion NH_4^+ to glutamine (thus further assimilation to other amino acids) there should be a lose of negative feedback by amino acids. The MSO treatment resulted in the same pattern of transcription, as that of NO_3^- treated control plants, thus indicating that NH_4^+ may act as a signal for negative feeback. In the corresponding $^{13}NO_3^-$ influx experiment we found a 50% decrease in influx, and considered that this effect may be due to down regulation of nitrate uptake at a post-transcriptional level by increased [NH_4^+].

AZA is an inhibitor of glutamate synthase (GOGAT) and asparagine synthetase (Sechley *et al.,* 1992). In our treatments, it affected both NO_3^- influx and *BCH* transcript accumulation. Nitrate influx decreased dramatically to levels lower that N-starved (uninduced) plants (figure 4a), while transcript abundance of *BCH* transcript decreased >95% in comparison to the control plants treated only with NO_3^- for a 6 h of period(figure 4b). As expected from the inhibition of GOGAT, AZA has been shown to elevate Gln levels and decrease Asn and Glu levels in plant tissue. These results indicate that Gln is a major regulator of BCH transcription.

AOA is an inhibitor of aspartate amino transferase and glutamate decarboxylase (Sechley *et al.,*1992). This inhibition should result in increased Glu levels in the tissue. We found that this treatment inhibited both *BCH* transcript accumulation and $^{13}NO_3^-$ influx by 65%.

The results of amino acid feeding and inhibitor treatments (AZA and AOA) indicate that (1) down-regulation of nitrate influx at the high affinity range and *BCH* transcript levels are correlated; and (2) that amino acids are possible signals for this regulation. Analyzing specifically the effect of Gln and Glu, we report that varying their level within the root tissue has pronounced effects on nitrate influx and *BCH* transcript levels.

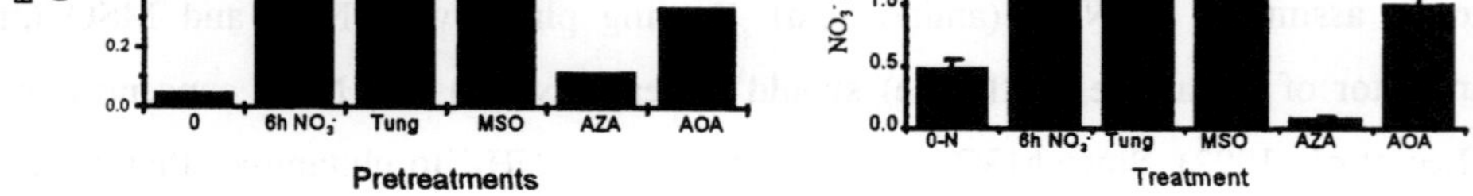

Figure 3. Effect of N-assimilation inhibitors on *BCH* mRNA accumulation and NO_3^- influx. 7 day old seedlings of barley grown in N-free media then supplied with 10 mM NO_3^- with Tungstate (0.5 mM), MSO (1.0 mM), AZA (0.25 mM) or AOA (0.5 mM). a) Northern blot analysis of 20 µg of total RNA quantified by phosphoimager (standardized by 25S transcript) and b) $^{13}NO_3^-$ influx.

SUMMARY

From our study, we draw the following conclusions

1) *BCH* genes products are involved IHATS in barley.
2) NO_3^- is signal that induces *BCH* accumulation and NO_3^- influx.
3) High levels of tissue nitrate may down-regulate high affinity transport of NO_3^- at post-transcriptional level.
4) Ammonium plays a role in transcriptional regulation of *BCH* transcript levels
5) Feeding Asn, Asp, Gln, or Glu exogenously decreased both influx and *BCH* mRNA accumulation, but due to possible differential rates of uptake and/or metabolism of these amino acids the specific effects can not be distinguished.
6) Gln or Glu, seem to play an important role in down regulation of transcript abundance of *BCH* and NO_3^- influx as inferred from comparing the response to amino acids, and to AZA and AOA treatments. However, this does not preclude the involvement of other amino acids in the regulation of inducible high affinity NO_3^- transporters.

References

Amarasinghe, B.H.R.R., DeBruxelles, G.L., Braddon, M., Onyeocha, I., Forde, B.G. and Udvardi, M.K (1998). Regulation of GmNRT2 expression and nitrate transport in roots of soybean (Glycine max) Planta In press

Aslam, M., Travis, R.L., Rains, D.W. and Huffaker, R.C. (1996) Effect of ammonium on the regulation of nitrate and nitrite transport systems in roots of intact barley (*Hordeum vulgare* L.) seedlings. *Planta* **200**, 58-63.

Breteler, H., and Arnozis, P.A. (1985) Effect of amino compounds on nitrate utilization by roots of dwarf bean. *Phytochemistry*. **24**, 653-657.

Breteler, H. and Siegerist, M. (1984) Effect of ammonium on nitrate utilization by roots of dwarf bean. *Plant Physiol.* **75**, 1099-1103.

Clarkson, D.T. (1988) Movement of ions across roots. In Baker DA, Hall JL eds. *Solute Transport in Plant Cells and Tissues*. Harlow: Longman Scientific and Technical, 251-304.

Doddema, H., and Otten, H. (1979) Uptake of nitrate by mutants of Arabidopsis thaliana, disturbed in uptake or reduction of nitrate. III. Regulation *Physiol. Plant.* **45**, 339-346.

Doddema, H., Hofstra, J.J., and Feenstra, W.J. (1978), Uptake of nitrate by mutants of Arabidopsis thaliana, disturbed in uptake or reduction of nitrate. I. Effect of nitrogen source during growth on uptake of nitrate and chlorate. *Physiol. Plant.* **43**, 343-350.

Glass, A.D.M., Siddiqi, M.Y., Ruth, T.J., and Rufty, T.W. (1990) Studies of the uptake of nitrate in barley. II. Energetics. *Plant Physiol* **93**, 1585-1589.

Glass, A.D.M., and Siddiqi, M.Y. (1995) Nitrogen absorption by plant roots. In Srivastava HS and Singh RP, eds, *Nitrogen Nutrition in Higher Plants,* (Associated Publishing Co., New Delhi, India), pp. 21-56.

Goyal, S.S., and Huffaker, R.C. (1986) The uptake of NO3-, NO2-, and NH4+ by intact wheat (Triticum aestivum) seedlings. *Plant. Physiol.* **82,** 1051-1056.

Hole, D.J., Emran,A., Farres, Y., and Drew, M.C., (1990) Induction of nitrate transport in maize roots and kinetics of influx measured with nitrogen-13. *Plant Physiol.* **93,** 642-647.

Imsande, J., and Touraine, B. (1994) N Demand and the regulation of nitrate uptake. *Plant Physiol.* **105,** 3-7.

Ingemarsson, B., Oscarson, P., Ugglas, M., and Larsson, C-M. (1987) Nitrogen utilization in Lemna II. Studies of nitrate uptake using $^{13}NO_3$. *Plant physiol.* **85,** 860-864.

Jackson, W.A., Flesher, D., and Hageman, R.H. (1973) Nitrate uptake by dark-grown corn seedlings. Some characteristics of apparent induction. *Plant Physiol* **51:** 120-127.

King, B.J., Siddiqi, M.Y., Ruth, T.J., Warner, R.L. and Glass, A.D.M. (1993). Feedback regulation of nitrate influx in barley roots by nitrate, nitrite and ammonium. *Plant Physiol.* **102,** 1279-1286.

Krapp, A., Fraisier, V., Scheible, W.-R., Quesada, A., Stitt, M., Gojon, A., Caboche, M., and Daniel-Vedel, F. (1998) Expression studies of *Nrt2:1Np,* a putative high affinity transporter: evidence for its role in nitrate uptake. *Plant Journal* in press

Kronzucker, H.J., Siddiqi, M.Y., and Glass, A.D.M. (1995). Compartmentation and flux characteristics of nitrate in spruce. Planta. **196,** 674-682.

Lee, R.B., and Drew, M.C. (1989) Rapid reversible inhibition of nitrate influx in barley by ammonium. *J. Exp. Bot.* **40,** 741-752.

Lee, R.B., Purves, J.V., Ratcliffe, R.G., and Saker, L.R. (1992) Nitrogen assimilation and the control of ammonium and nitrate absortion by maize roots. *J. Exp. Bot.* **43,** 1385-1396.

Muller, B., and Touraine, B. (1992) Inhibition of NO_3^- uptake by various phloem-translocated amino acids in soybean seedlings. *J.Exp. Bot.* **43,** 617-623.

Quesada, A., Galvan, A., and Fernandez, E. (1994) Identification of nitrate transporter genes in *Chlamydomonas reinhardttii. Plant J.* **5,** 407-419.

Quesada, A., Krapp, A., Trueman, L.J., Daniel-Vedele, F., Fernandez, E., Forde, B.G., Caboche, M. (1997) PCR-identification of a *Nicotiana plumbaginifolia* cDNA homologous to the high-affinity nitrate transporters of the crnA family. *Plant Mol. Biol.* **34,** 265-274.

Rodgers, C.O., and Barneix, A.J. (1993). The effect of amino acids and amides on the regulation of nitrate uptake by wheat seedlings. *Journal of Plant Nutrition* **16,** 337-348.

Sambrook, J., Fritsch, E.F. and Maniatis T.(1989) Molecular Cloning: A Laboratory Manual. (Cold Spring Harbor, NY: Cold Spring Harbor Laboratory Press).

Sechley, K.A., Yamaya, T., and Oaks, A. (1992) Compartmentation of nitrogen assimilation in higher plants. *Int. Rev. Cytol.* **134,** 85-163.

Siddiqi, M.Y., Glass A.D.M., Ruth, T.J., and Fernando, M. (1989) Studies of the regulation of nitrate influx by barley seedlings using $^{13}NO_3$ *Plant Physiol* **90,** 806-813

Siddiqi, M.Y., Glass A.D.M., Ruth, T.J., and Rufty, T. (1990) Studies of the uptake of nitrate in barley. I. Kinetics of $^{13}NO_3$ influx. *Plant Physiol* **93:** 1426-1432

Siddiqi, M.Y., Glass A.D.M., Ruth, T.J. (1991) Studies of the uptake of nitrate in barley. III. Compartmentation of NO_3. *J Exp Bot* **42:** 1455-1463.

Trueman, L.J, Richardson A., and Forde, B.G. (1996) Molecular cloning of higher plant homologues of the high-affinity nitrate transporters of *Chlamydomonas reinhardtii* and *Aspergillus nidulans. Gene* **175,** 223-231.

Warner , R.L., and Huffaker R.C. (1989) Nitrate transport is independent of NADH and NAD(P)H nitrate reductases in barley seedlings. *Plant Physiol* **91,** 947-953.

MOLECULAR REGULATION OF PHOSPHATE ACQUISITION IN PLANTS

K.G. RAGHOTHAMA
Department of Horticulture, Purdue University, West Lafayette, IN. 47907.

Abstract

Phosphorus availability is considered as one of the major growth limiting factors for plants in many natural ecosystems. In response, plants have developed several physiological, biochemical and molecular adaptations to acquire phosphate (Pi). Enhanced ability to acquire Pi and altered gene expression are the hallmarks of plant adaptation to Pi deficiency. There is an increasing number of genes known to be activated under Pi starvation. High-affinity Pi transporter genes have been cloned and characterized from plants. They code for integral membrane proteins consisting of twelve membrane spanning domains separated into two groups of six by a large charged hydrophilic region.

It is presumed that regulated expression of plant Pi transporters is an adaptive response to varying levels of Pi in the rhizosphere. The observed increase in the rate of Pi uptake with little or no change in the apparent Km upon starvation has been proposed to be due to increased synthesis of high-affinity carrier molecules. Studies indicate that an increase in the transcripts under Pi starvation correlates directly with a concurrent increase in the transport protein, suggesting a transcriptional regulation for Pi acquisition. This molecular response is very specific to Pi deficiency and reversible upon resupply of the nutrient. The Pi transporters are enriched in the plasma membrane of Pi starved roots. The localization of the transporters to the plasma membranes, and their expression in the root epidermal cells and root hairs, provide a strong evidence for their proposed role in Pi acquisition. The intricate mechanisms involved in maintaining Pi homeostasis is a reflection of the complexity of Pi acquisition and translocation in plants. It is becoming clear that internal phosphorus levels play a role in the regulation of Pi transporters expression. It is likely that internal signals arising in response to Pi deficiency would lead to a Pi-starvation induced response including activation of Pi transporters.

G. Gissel-Nielsen and A. Jensen (eds.), Plant Nutrition – Molecular Biology and Genetics, 95–103.

Phosphorus a limiting nutrient!

Phosphorus plays an important role in photosynthesis, respiration, regulation of enzymes, biosynthesis of nucleic acids and membranes (39). Phosphate (Pi) deficiency is a major constraint for crop production in many parts of the world. Pi deficiency is widespread in the nature and the available Pi seldom exceeds 10 μM even in fertile soils (2). Inorganic phosphate is attached to surface active sesquioxides and oxyhydrates of clay minerals or precipitated as calcium salts (39). Due to these interactions up to 80% of applied Pi may be fixed in the soil, forcing farmers to use up to four times the fertilizer necessary for crop production (20). At the current usage rate of P fertilizer, readily available sources of phosphate rocks will be exhausted over the next 60 to 90 years (51). The Pi reserves (capital) in tropical soils are depleting at a faster rate due to lack of fertilizer or organic matter applications. In contrast, parts of temperate regions are faced with excessive soluble P in the soil that is deleterious to the environment.

RESPONSE OF PLANTS TO Pi STARVATION.

A steep gradient in available phosphorus between the soil solution (μM) and cytoplasm (mM) of plant cells has essentially led to numerous morphological, physiological, biochemical and molecular adaptations by plants to survive in the nature. Increased root:shoot ratio is a typical response of plants to Pi deficiency. This enhances the total root surface area available for soil exploration and Pi acquisition (36). In general Pi efficiency can be attributed to either high root-shoot ratios (e.g. rye and wheat) or high influx rates (e.g. rape and spinach) (16). Furthermore, both the production and elongation of root hairs increase under Pi deficiency. The significance of root hairs in Pi acquisition is well documented (17). Plants also exhibit remarkable phenotypic plasticity to maximize Pi uptake from the Pi rich regions of soil. An 80% increase in uptake by roots growing in enriched soil patches is a clear sign of physiological plasticity exhibited by plants (24). This flexibility is crucial in compensating for a lack of uptake by major portions of root system under Pi deficiency. Phosphorus efficient bean genotypes have highly branched, actively growing root system as compared to the root system of P inefficient genotypes suggesting that root architectural traits strongly influence P acquisition (37).

Phosphorus deficiency not only enhances root proliferation, root hair production and elongation, but also modify the biochemical processes of the roots. Plants such as white lupin adapt to Pi deficiency by formation of proteoid roots that are highly efficient in synthesis and secretion of organic acids to the rhizosphere (10). These acids help in the release of Pi from Ca, Fe, and Al phosphates. Proteoid roots have also been shown to secrete more phosphatases (18), and absorb Pi at a faster rate than non-proteoid roots. The oil seed crop rape is very efficient in solubilizing rock phosphate by secreting organic acids into the rhizosphere (23). Interestingly, secretion of organic acids is targeted to the region of the root that is directly in contact with rock phosphate. Increased secretion of organic acids is presumed to be due to activation or synthesis of enzymes involved in organic acid production, and anion transporters or channels to enhance secretory processes.

Production of phenolic compounds such as pisidic acid (p-hydroxybenzyl tartaric acid) by pigeon pea allows them to extract Pi associated with iron (1).

Detailed analysis of altered biochemical processes during Pi starved plants have been discussed in a recent review (47). The induction of acid phosphatases is a distinctive response of higher plants to Pi starvation (13). In view of considerable amount (20 -80%) of P in soil is found in the organic form (25) the role of phosphatases in P nutrition cannot be ignored. Production of both extracellular and intracellular phosphatases increases during Pi starvation (13). The extracellular phosphatases are likely to be involved in recapturing P from organic phosphorus compounds in the extracellular matrix. The extracellular phosphatases produced in shoots and leaves may also be involved in scavenging P from xylem-derived phosphocholine (PC), which is a major phosphorylated compound in xylem sap (47). Changes in specific isoforms of phosphatases under Pi starvation are commonly observed (57). A concurrent increase in phosphatases, phytase and RNases was observed in the absence of exogenous supply of Pi (4). It is becoming apparent that production of acid phosphatases is a part of an adaptive mechanism by which plants overcome the deficiency of Pi in soils. In addition to increased synthesis of phosphatases and RNases, the levels of nucleotides are reduced during Pi starvation (47). The enzymes of "bypass reactions" that circumvent Pi and adenylate requiring steps in glycolysis are activated in response to Pi starvation (12). Phosphate starvation often results in activation of alternate respiratory pathway, reduction in photosynthesis and stomatal conductance (6,50,52).

Another major biological adaptation to Pi deficiency is formation of mycorrhizal associations to enhance Pi acquisition. The recent advances in molecular biology of mycorrhizal fungi leading to cloning a Pi transporter gene and other symbiosis specific genes (5,22) have allowed the researchers not only to evaluate the contribution of mycorrhizal fungi to Pi acquisition, but also the interaction between plant and fungal transporters. Northern analyses of transcripts show that colonization of roots by mycorrhizal fungi and expression of the fungal transporters leads to repression of the plant Pi transporters expression (34). These data indicate a coordinated interaction between plant roots and mycorrhizal fungi in obtaining and transferring Pi to plants in the natural ecosystem.

Gene expression is altered during Pi staration.

The number of genes known to be expressed under Pi starvation is increasing rapidly (49). The list includes genes with known function such as high affinity Pi transporters (43), RNases (21), Ca2+ATPase (42), purple acid phosphatase (46), PEPcase, vegetative storage protein (53), enolase and pyruvate formate-lyase (14), b-glucosidase (38) and many novel genes such as TPSI1 (32,33) and Mt4 (5). The TPSI1 gene is induced rapidly in response to Pi starvation and temporally and specifically regulated by changes in Pi concentrations. Expression of TPSI1 gene increases within hours of Pi starvation in cultured tomato cells. Furthermore TPSI1 promoter sequences contain conserved cis elements found in many PHO-regulon genes of yeast. The complex pattern of induction of several genes during Pi starvation suggests there is a highly regulated molecular response mechanism in plants to alleviate Pi deficiency. The evidence for an intricate gene regulation system in plants similar to the yeast PHO regulon is increasing, in addition the complexity of plant morphology and biochemistry point to the existence of other regulatory mechanisms that respond to changes in cellular Pi levels.

Multiple transporters are involved in phosphorus homeostasis

Plants tend to maintain constant cytoplasmic concentrations of nutrients such as N, P, K, and Cl in spite of large fluctuations in the external concentrations of those ions (19). The vacuoles serve as the storage organs for Pi in plants. A relatively constant level of cytoplasmic Pi is maintained at the expense of stored Pi in the vacuoles. The role of the vacuole in Pi homeostasis is extensively studied by NMR techniques (29). 31P NMR studies showed that cytoplasmic P levels remained relatively constant and vacuolar Pi levels vary significantly in response to Pi supply. The transport of Pi across the tonoplast is energized by ATP and requires cytoplasmic alkalization (54). At present the exact nature of the Pi transport across the tonoplast is unclear, however the possibility of a Pi specific channel or a symporter operating at the tonoplast is likely. A complex pattern of Pi movement is also observed at the whole plant level. Under Pi limiting conditions, transfer of Pi from roots to shoots is reduced, Pi in older leaves is remobilized to younger growing leaves and other active sinks (3,40). Pi efflux is another mechanism by which plants regulate Pi concentrations in the cytoplasm (3,7). This may be a significant factor in maintaining Pi homeostasis particularly when plants are exposed to unlimited supply of Pi. The proportion of Pi excreted increased significantly with increased concentration of Pi in the media, from 28% of influx at 50 μM to 90% at 5 mM Pi (7). In a recent study using double isotopes (33P and 32P) the rate of efflux was determined to be ten-fold higher for the +P treatment (15). This study also suggested the possible involvement of anion channels in Pi efflux.

High affinity Pi transporter and plant nutrition

The concentration of Pi in cytosol is generally in the millimolar range and therefore, roots have to acquire Pi against a steep concentration gradient (1000 fold or higher) across the plasmamembrane. A number of studies have shown dual uptake patterns consisting of a high affinity transporter operating at low (μM) concentration and a low affinity transporter functioning at high concentration (mM) of Pi in plants (2,58). However, in the context of μM concentration of available Pi in soil solution the high affinity Pi transport should be considered as the primary mechanism of uptake for plants under natural conditions. An energy mediated co-transport process, driven by protons generated by a plasma membrane H+ATPase, has been proposed for Pi uptake in plants (54,58). This process may require 2 to 4 H+ for every phosphate molecule transported.

High affinity Pi transporters are membrane associated proteins translocating Pi from an external media containing μM concentrations to the cytoplasm. Recently, several high affinity Pi transporter genes have been cloned from plants

(8,26,30,31,34,35,43,56). Phosphate transporters are encoded by a small family of genes in the genome (43). Some of the genes have been used to complement yeast mutants deficient in high affinity Pi uptake (8,26,30,35,43), and enhance the uptake of Pi by tobacco cells (41). Plant Pi transporters exhibit a pH optimum of 4.5 to 5.0 in the yeast complementation studies, and are inhibited by chemicals which dissipate pH-gradients across membranes (8,30).

The deduced amino acid sequences of plant Pi transporters share a significant similarity with those of yeast (Saccharomyces cerevisiae), Neurospora, and the mycorrhizal fungi (Glomus versiforme) (43). The Pi transporters are integral membrane proteins consisting of 12 membrane spanning regions a common feature shared by members the Major Facilitator Super family (45). The members of this family include facilitated carriers capable of transporting small solutes such as sugars, ions, antibiotics and amino acids utilizing the chemiosmotic ion gradients. The intricate movement of Pi within the cell, and between different organs point to the existence of additional Pi transporters in plants. The thoroughly characterized Arabidopsis mutants (pho1 and pho2) with impaired loading of Pi in to xylem (48) and defective regulation of Pi loading into shoots (9) provide genetic evidence for the existence of other transporters involved in Pi distribution in plants. Some of these transporters are likely to be regulated in a temporal manner and exhibit spatial specificity. The plant growth and development will also affect the function of transporters involved in retranslocation of Pi from older, senescing parts of the plant to younger parts.

Phosphorus deficiency regulates the expression of Pi transporters.

Phosphate starvation leads to preferential expression of Pi transporters in the roots of Pi starved plants (8,31,43). All the Pi transporters examined in plants are induced in a temporal and concentration- mediated manner in roots and cell cultures (8,28,30,34,43). An increase in Pi transport activity observed in many of the earlier works is likely due to increased Pi transporters. In tomato plants, induction of transcripts and accumulation of Pi transporter proteins were observed within 24 of imposition of Pi deficiency (31,44). This may be the primary reason for observed increases in Vmax without any apparent changes in Km of Pi uptake in many plant species (11,55). It is becoming clear that the capacity for Pi uptake is regulated by increasing the total number of transporter molecules (49). Induction of Pi transporter gene expression is not only a rapid response to Pi starvation but also reversible upon replenishment of P (31). An earlier study showed that transferring tobacco cells to a Pi deficient media led to a gradual increase in the Vmax reaching the maximum at 20 hours. When the starved cells were replenished with P, the uptake rate decreased gradually to reach the normal uptake rates after 5 hours (55). Similarly, enhanced Pi uptake rates in P-starved tomato plants returned to normal 30 hours after Pi replenishment (27). Expression of Pi transporters correlate with the observed increases in Pi uptake.

Most of cloned Pi transporter genes are preferentially expressed in roots under Pi starvation (30,31,43). The preferential expression of these genes in roots is a strong indication of their role in acquiring Pi from soil. Tomato Pi transporter genes (LePT1 and LePT2) are expressed predominantly in the root epidermis and root hairs of Pi starved plants (8,31). Epidermal expression of high affinity Pi transporters is indicative of specific targeting of the transporters to epidermis, which are generally exposed to relatively higher concentrations of Pi. A similar argument could also be made for higher expression of Pi transporters in root hairs of Pi deficient plants. Since the contribution of root hairs to Pi acquisition is significant, particularly in Pi deficient soils, expression of Pi transporters in root hairs should result in better exploration of the rhizosphere and rapid uptake of Pi. In addition the transcripts for tomato (LePT1) and potato (StPT1) Pi transporters were also observed in leaves, stem, tubers and flowers (30,31). These transporters may also be involved in the intercellular movement of Pi or reloading Pi leaked into the apoplast. Detailed in situ hybridization and immunolocalization studies of different members of the Pi transporter family should help in understanding the complexity of P nutrition in higher plants.

Increased Pi uptake following Pi starvation has been shown to require de novo protein synthesis (55). The enhanced uptake could be due to increased synthesis of a carrier system under Pi deprivation (11). Immunological studies with antibodies specific to tomato Pi transporter (LePT1) revealed the accumulation of Pi transporters in roots of Pi starved plants (44). These studies indicated that Pi starvation results in concurrent accumulation of the transcripts and Pi transporters in the plasma membranes, suggesting the Pi uptake is likely to be regulated at the level of transcription. There is a continuing debate on the regulation of Pi uptake by plants. Divided root system plant experiments have revealed that shoot Pi levels play a role in the regulation of its uptake by roots (11). Analysis of transcripts isolated from tomato plants with divided roots exposed to P+ and P- conditions indicated that Pi transporter mRNA levels remained comparable to that of Pi sufficient plants (31). It appears that once the Pi requirements of plants are satisfied, the level of Pi transporter gene expression is down regulated in the entire root system. These observations support the notion that internal Pi status of a plant can regulate the rates of Pi uptake by roots. This type of coordinated regulation of gene expression in different parts of the roots by internal Pi levels will allow plants to obtain the required amount of the nutrient.

Given that Pi deficiency is the norm under natural conditions, the level of Pi transporter expression and protein may always be high, and repression of gene expression is likely to be the regulating factor of Pi acquisition. An examination of gene expression in roots of plants grown in soils with differing Pi fixation/buffering capacity is required to confirm this hypothesis. Furthermore, under natural conditions plant growth rate is normally adjusted to the availability of the nutrient. Reduced rate of growth leads to redistribution of Pi within the plant, thus allowing plants to overcome brief periods of Pi starvation without a major effect on growth.

Concluding remarks

Our understanding of the molecular regulation of processes associated with Pi acquisition and Pi starvation responses is increasing. However, the task of evaluating how plants adapt to Pi stress is still a challenging and open area of research. The 1990s, have been an exciting time for plant nutrition because of isolation and characterization of genes for many nutrient transporters. The recently emerging research area of functional and structural genomics will allow us to decipher global changes in gene expression during Pi deficiency. The potential benefits of this technology will be to identify molecular determinants of Pi acquisition and adaptation to Pi stress. This will allow us to generate plants that are efficient in utilizing phosphorus and perform better under nutrient limiting conditions.

Acknowledgments

Research in my laboratory is supported by a grant from United States Department of Agriculture.

References

1. Ae, N., Arihara ,J., Okada,, K, Yoshihara T, Johansen, C. (1990) Phosphorus uptake by pigeon pea and its role in cropping systems of Indian subcontinent. *Science* **248**, 477-80.
2. Bieleski, R.L. (1973) Phosphate pools, phosphate transport, and phosphate availability. *Ann. Rev. Plant Physiol.* **24**, 225-52.
3. Bieleski ,R.L., Ferguson, I.B. (1983) Physiology and metabolism of phosphate and its compound. In A Lauchi, RL Bieleski (eds.), *Encyclopedia of Plant Physiology; Inorganic Plant Nutrient*, Springer-Verlag, Berlin, pp. 422-29.
4. Bosse ,D., Kock ,M. (1998) Influence of phosphate starvation on phosphohydrolases during development of tomato seedlings. *Plant Cell Environ.* **21**, 325-32.
5. Burleigh, S.H, Harrison, M.J. (1997) A novel gene whose expression in Medicago truncatula roots is suppressed in response to colonization by vesicular-arbuscular mycorrhizal (VAM) fungi and to phosphate nutrition. *Plant Mol. Biol.* **34**, 199-208.
6. Clarkson, D.D., Scattergood, C.B. (1982) Growth and phosphate transport in barley and tomato plants during development of, and recovery from, phosphate stress. *J. Expt. Bot.* **33**, 865-75.
7. Cogliatti, D.H., Santa Maria, G.E. (1990) Influx and efflux of phosphorus in roots of wheat plants in non-growth-limiting concentrations of phosphorus. *J. Exp. Bot.* **41**, 601-07.
8. Daram, P., Brunner ,S., Amrhein, N., Bucher, M. (1998) Functional analysis and cell-specific expression of a phosphate transporter from tomato. *Planta* **206**, 225-33.
9. Delhaize, E., Randall, P.J. (1995) Characterization of a phosphate-accumulator mutant of Arabidopsis thaliana. *Plant Physiol* **107**, 207-13.
10. Dinkelaker, B., Hengeler, C., Marschner, H. (1995) Distribution and function of proteoid roots and other root clusters. *Bot. Acta.* **108**, 183-200.
11. Drew, M.C., Saker, L.R. (1984) Uptake and long-distance transport of phosphate, potassium and chloride in relation to internal ion concentrations in barley: evidence of non-allosteric regulation. *Planta* **160**, 500-07.
12. Duff, S.M.G., Moorhead, G.B.G., Lefebvre,. DD., Plaxton, W.C. (1989) Phosphate starvation inducible "bypasses" of adenylate and phosphate dependent glycolytic enzymes in Brassica nigra suspension cells. *Plant Physiol.* **90**, 1275-78.
13. Duff ,S.M.G., Sarath, G., Plaxton, W.C. (1994) The role of acid phosphatase in plant phosphorus metabolism. *Physiol. Planta.* **90**, 791-800.
14. Dumont, F., Joris, B., Gumusboga, A., Bruyninx, M., Loppes, R. (1993) Isolation and characterization of cDNA sequences controlled by inorganic phosphate in Chlamydomonas reinhardtii. *Plant Sci.* **89**, 55-67.
15. Dunlop, J., Phung, T. (1998) Efflux and influx as factors determining the relative abilities of ryegrass and white clover to compete for phosphate. Presented at Sixth International Symposium on Genetics and molecular biology of plant nutrition. Elsinore, Denmark. Abstract
16. Fohse, D., Claassen, N., Jungk, A. (1988) Phosphorus efficiency of plants I. External and internal P requirement and P uptake efficiency of different plant species. *Plant Soil* **110**, 101-09.

17. Gahoonia, T.S., Nielsen, N.E. (1998) Direct evidence on participation of root hairs in phosphorus (32P) uptake from soil. *Plant Soil* **198**, 147-52.
18. Gilbert, G.A., Allan, D.L., Vance, C.P. (1997) Phosphorus deficiency in white lupin alters root development and metabolism. In H.E. Flores, J.P. Lynch, D. Eissenstat (eds.) *Radical Biology: Advances and Perspectives on the Function of Plant Roots,* Amer. Soc. Plant Physiol, pp. 92-103.
19. Glass, A.D.M., Siddiqi, M.Y. (1984) The control of nutrient uptake rates in relation to the inorganic composition of plants. *Adv. Plant Nutr.* **1**, 103-47.
20. Goldstein, A.H. (1992) Phosphate starvation inducible enzymes and proteins in higher plants. In: JL Wray (ed.), : Inducible Plant Proteins, Society for Experi-mental Biology Seminar Series 49, Cambridge University Press, pp. 25-44.
21. Green, P.J. (1994) The ribonucleases of higher plants. Annu. Rev. Plant Physiol. *Plant Mol. Biol.* **45**, 421-45.
22. Harrison, M.J., van Buuren, M.L. (1995) A phosphate transporter from the mycorrhizal fungus Glomus versiforme. *Nature* 378, 626-29.
23. Hoffland, E., Findenegg, G.R., Nelemans, J.A. (1989) Solubilization of rock phosphate by rape. *Plant Soil* **113**, 155-60.
24. Jackson, R.B., Manwaring, J.H., Caldwell, M.M. (1990) Rapid physiological adjustment of roots to localized soil enrichment. *Nature* **344**, 58-60.
25. Jungk, A., Seeling, B., Gerke, J. (1993) Mobilization of different phosphate fractions in the rhizosphere. *Plant Soil* **155/156**, 91-94.
26. Kai ,M., Masuda, Y., Kikuchi, Y., Osaki ,M., Tadano, T. (1997) Isolation and characterization of a cDNA from Catharanthus roseus which is highly homologous with phosphate transporter. *Soil Sci Plant Nutr* **43**, 227-35.
27. Katz, D.B., Gerloff, G.C., Gabelman, W.H. (1986) Effects of phosphate stress on the rate of phosphate uptake during resupply to deficient tomato plants. *Physiol. Plant.* **67**, 23-28.
28. Kim, D.H., Muchhal, U., Raghothama, K.G. (1998) Tomato phosphate transporters respond to altered phosphorus levels in cell cultures. *Plant Physiol.* pp. 136. (abstract)
29. Lee, R.B., Ratcliffe, R.G. (1993) Nuclear magnetic resonance studies of the location and function of plant nutrients in vivo. *Plant Soil* **155/156**, 45-55.
30. Leggewie, G., Willmitzer, L., Riesmeier, J.W. (1997) Two cDNAs from potato are able to complement a phosphate uptake-deficient yeast mutant: Identification of phosphate transporters from higher plants. *Plant Cell* **9**, 381-92.
31. Liu, C., Muchhal, U.S., Mukatira, U., Kononowicz, A.K., Raghothama, K.G. (1998) Tomato phosphate transporter genes are differentially regulated in plant tissues by phosphorus. *Plant Physiol.* **116**, 91-99.
32. Liu ,C., Muchhal, U.S., Raghothama, K.G. (1997) Differential expression of TPSI1, a phosphate starvation-induced gene in tomato. *Plant Mol. Biol.* **33**, 867-74
33. Liu, C., Raghothama, K.G. (1995) A practical method for cloning cDNAs generated in a mRNA differential display. *Biotechniques* **20**, 576-79.
34. Liu, H., Trieu, A.T., Blaylock, L.A., Harrison, M.J. (1998) Cloning and characterization of two phosphate transporters from Medicago truncatula roots . Regulation in response to phosphate and to colonization by arbuscular mycorrhizal (AM) fungi. *Mol. Plant Microbe. Inter.* **11**, 14-22.
35. Lu, Y.P., Zhen ,R.G., Rea, P.A. (1997) AtPT4: A forth member of the Arabidopsis phosphate transporter gene family (accession no U97546). *Plant Physiol.* **114**, 747.
36. Lynch, J.P. (1995) Root architecture and plant productivity. *Plant Physiol.* **109**, 7-13.
37. Lynch ,J.P., Beebe, S.E. (1995) Adaptation of beans (Phaseolus vulgaris L.) to low phosphorus availability. *HortSci.* **30**, 1165-1171.
38. Malboobi, M.A., Lefebvre, D.D. (1997) A phosphate-starvation inducible b-glucosidase gene (psr 3.2) isolated from Arabidopsis thaliana is a member of a distinct subfamily of the BGA family. *Plant Mol. Biol.* **34**, 57-68.
39. Marschner, H. (1995) Mineral nutrition in plants. 2nd ed. Academic Press, San Diego, CA
40. Mimura ,T. (1995) Homeostasis and transport of inorganic phosphate in plants. *Plant Cell Physiol.* **36**, 1-7.
41. Mitsukawa, N., Okumura, S., Shirano, Y., Sato, S., Kato, T., Harashima, S., Shibata ,D. 1(997) Overexpression of an Arabidopsis thaliana high-affinity phosphate transporter gene in tobacco cultured cells enhances cell growth under phosphate-limited conditions. *Proc. Natl. Acad. Sci.* **94**, 7098-102.
42. Muchhal, U.S., Liu, C., Raghothama, K.G. (1997) Calcium-ATPase is differentially expressed in phosphate starved roots of tomato. *Physiol. Planta* **101**, 540-44.
43. Muchhal, U.S., Pardo, J.M., Raghothama, K.G. (1996) Phosphate transporters from the higher plant Arabidopsis thaliana. *Proc. Natl. Acad. Sci. USA* 93, 10519-23.

44. Muchhal, U.S., Varadarajan, D., Damsz, B., Raghothama, K.G. (1998) Tomato phosphate transporter (LePT1) is localized in plasmamembranes. In Lynch, J.P., Deikman, J. (eds.) *Phosphorus in Plant Biology, Regulatory roles in molecular, cellular, organismic, and ecological process*, ed. 1998 Penn State Summer Symposium in Plant Physiology,. (in press)
45. Pao, S.S., Paulsen, I.T., Saier, M.H. (1998) Major facilitator superfamily. Microbiol. *Mol. Biol Rev.* **62**, 1-34.
46. Patel, K., Lockless, S., Thomas, B., McKnight, T.D. (1995) A secreted purple acid phosphatase from Arabidopsis. *Plant Physiol (suppl)* **111**, 270. (abstract)
47. Plaxton, W.C., Carswell, M.C. (1998) Metabolic aspects of the phosphate starvation response in plants. In Lerner, H.R (ed.) *Plant Responses to Environmental Stresses: From Phytohormones to Genome Reorganization,*. Dekker .(in press)
48. Poirier, Y., Thoma, S., Somerville, C., Schiefelbein, J. (1991) A mutant of Arabidopsis deficient in xylem loading of phosphate. *Plant Physiol.* **97**, 1087-93.
49. Raghothama, K.G., Muchhal, U.S., Kim, D.H., Bucher, M. (1998) Molecular regulation of plant phosphate transporters. In Lynch, J.P., Deikman, J.(eds.) *Phosphorus in Plant Biology, Regulatory roles in molecular, cellular, organismic, and ecological process*, 1998 Penn State Summer Symposium in Plant Physiol. (in press)
50. Rao, M., Terry, N. (1995) Leaf phosphate status, photosynthesis, and carbon partitioning in sugar beet. IV. Changes with time following increased supply of phosphate to low-phosphate plants. *Plant Physiol.* **107**, 1313-21.
51. Runge-Metzger, A. (995) Closing the cycle: Obstacles to efficient P management for improved global food security. In H Tiessen (ed.) *Phosphorus in the Global Environment: Tansfers, cycles and Management*, John Wiley and Sons, NY, pp. 27-42.
52. Rychter, A.M, Misulska, M. (1990) The relationship between phosphate status and cyanide-resistant respiration in bean roots. *Physiol.Plant.* **79**, 663-67.
53. Sadka, A., DeWald, W.B., May, G.D., Park, W.D., Mullet, J.E. (1994) Phosphate modulates transcription of soybean VspB and other sugar-inducible genes. *Plant Cell* **6**, 737-49.
54. Sakano, K. (1990) Proton/phosphate stoichiometry in uptake of inorganic phosphate by cultured cells of Catharanthus roseus (L.) G. *Don. Plant Physiol.* **93**, 479-83.
55. Shimogawara, K., Usuda, H. (1995) Uptake of inorganic phosphate by suspension-cultured tobacco cells: Kinetics and regulation by Pi starvation. *Plant Cell Physiol.* **36**, 341-51.
56. Smith, F.W., Ealing, P.M., Dong ,B., and Delhaize, E. (1997) The cloning of two Arabidopsis genes belonging to a phosphate transporter family. *Plant J.* **11**, 83-92.
57. Trull, M.C., Guiltinan, M.J., Lynch, J.P., Deikman, J. (1997) The response of wild-type and ABA mutant Arabidopsis thaliana plants to phosphorus starvation. *Plant Cell Environ.* **20**, 85-92.
58. Ullrich-Eberius, C.I., Novacky, A., van Bel, A.J.E. (1984) Phosphate uptake in Lemna gibba G1: energetics and kinetics. *Planta* **161**, 46-52.

EFFLUX AND INFLUX AS FACTORS IN THE RELATIVE ABILITIES OF RYEGRASS AND WHITE CLOVER TO COMPETE FOR PHOSPHATE

JAMES DUNLOP AND THAI PHUNG
AgResearch Grasslands, Private Bag 11008, Palmerston North, New Zealand.

Abstract

Influx and efflux of phosphate has been measured for roots of white clover and perennial ryegrass. Efflux was consistently and substantially lower in ryegrass. For plants deprived of phosphate for 5 days, the initial rates of influx were significantly higher for ryegrass. It is concluded that higher rates of efflux may be a significant factor in the poorer ability of white clover to compete for phosphate.

Introduction

Pastures containing mixtures of perennial ryegrass and white clover can produce high quality forage without nitrogen fertiliser but require generous inputs of phosphate (PO_4) because of the poor ability of white clover to compete for PO_4 (Dunlop & Hart, 1987). The difference in competitiveness between the two species may arise from differences in the acquisition of PO_4 from the soil solution. Mouat (1983) demonstrated that, in this respect, clover was inferior to the low fertility grass *Agrostis tenuis* Sibth. but did not confirm that this also applied to ryegrass. This paper reports aspects of the acquisition of PO_4 by ryegrass and white clover from nutrient solutions.

The kinetics of nutrient ion uptake has been frequently studied using plants grown with minimal nutrients and radioactive tracers which minimise the effects of efflux. However net flux rather than the influx determines the amount of nutrient obtained by the plant and consequently efflux can be significant (Bieleski, 1973). In ryegrass Macklon and Sim (1992) found net absorption of PO_4 was only 40% of unidirectional influx. In this study we use $^{32}PO_4$ and $^{33}PO_4$ to simultaneously determine influx and efflux.

Materials and Methods

Seeds of white clover (*Lolium perenne* L.) and perennial ryegrass (*Trifolium repens* L.) were sterilised in 2% chlorine bleach and grown in sterile conditions on filter paper

G. Gissel-Nielsen and A. Jensen (eds.), Plant Nutrition – Molecular Biology and Genetics, 105–110.

supports in glass boiling tubes containing 50 ml of full strength Hoaglands solution, one seed per tube. After 14 days growth for ryegrass and 21 days for white clover, solutions were changed to carrier free ^{32}P labelled Hoaglands solutions modified to have either nil or normal PO_4 (-P or +P), at either pH 5.5 or 7.0 and the plants grown for a further 5 days.

Roots of the plants were rinsed in PO_4-free Hoagland's for 30 min, and then suspended in a 30 ml tube with the roots bathed in 10 ml of ^{33}P-labelled 5μM KH_2PO_4 and shaken on an orbital shaker. Samples of 20 μl were taken periodically and the contents of ^{32}P and ^{33}P were determined by scintillation counting. The initial rates of influx are the slopes of straight lines fitted to the data for the first 4 h. The kinetic constants KM and V_{max} were derived from functions fitted to the depletion curves (Shimogawara and Usuda, 1995).

Plants were dissected into root and shoot and the ^{32}P content determined by counting Cerenkov radiation of the fresh material. Roots and shoots were dried, weighed, wet ashed and total P content measured by spectrophotometry (Reay and Waugh, 1981). To calculate PO_4 efflux, the specific activity of ^{32}P was determined from the activity of the root samples and their total P content.

Results

INFLUX OF ^{33}P LABELLED PO_4

Typical examples of the depletion curves are given in Fig. 1. The -P treatments removed $^{33}PO_4$ more rapidly than the corresponding +P treatments. A stable minimum concentration of $^{33}PO_4$ was reached only by the -P treatment of ryegrass. The initial rates of ^{33}P influx derived from the depletion data are shown in Fig. 2. Influx for all -P treatments were higher ($p < 0.001$) than the corresponding +P treatments. Influx was more rapid ($p < .01$) for ryegrass than for white clover for the -P treatments (132% and 199% greater for pH 5.5 and 7.0 respectively). However there were no corresponding

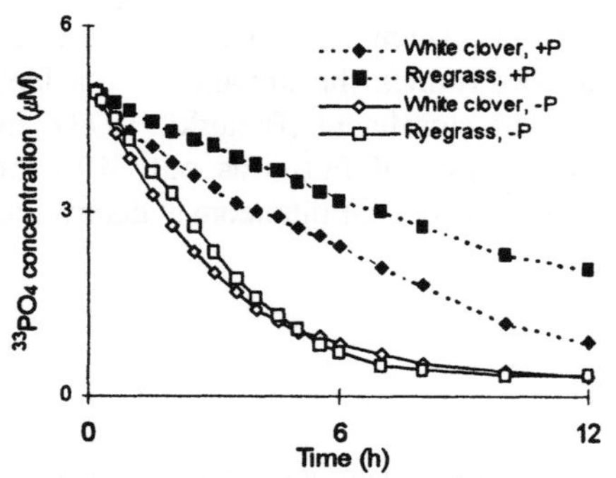

Fig. 1. Depletion curves for absorption of $^{33}PO_4$ by roots of intact white clover and ryegrass plants at pH 5.5.

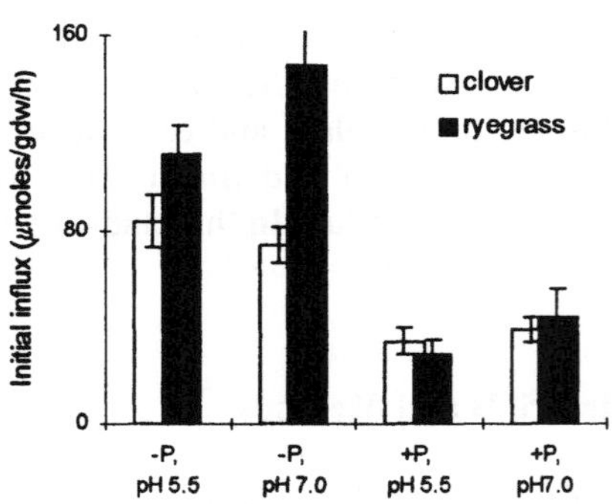

Fig. 2. Initial rates of $^{33}PO_4$ influx obtained from depletion curves. Data are means of 4 reps ± s.d.

differences between species for the +P treatments. If the influx data is expressed per plant instead of dry weight, there was no difference between species. Differences in pH resulted in differences in influx only for the -P treatment of ryegrass ($p < .001$).

Values for the kinetic constant K_M, derived from curves fitted to the depletion data, are given in Fig. 3 The only significant difference in K_M between species, was for the $-PO_4$, pH 7.0 treatment with clover having twice the value for ryegrass. Values for K_M for the other -P treatments were in the range 2 - 6 μM. For +P treatments the values were 16 to 23 μM. The values for pH 7.0 tended to be higher than the values for pH 5.5 but this was only significant for $-PO_4$ treatments of clover.

Similarly, V_{max}, Fig. 4, expressed on the basis of root dry weight, did not vary significantly between species. In contrast to K_M, differences in V_{max} relating to P status were not significant. Values for V_{max} tended to be higher at pH 7.0 than at pH 5.5.

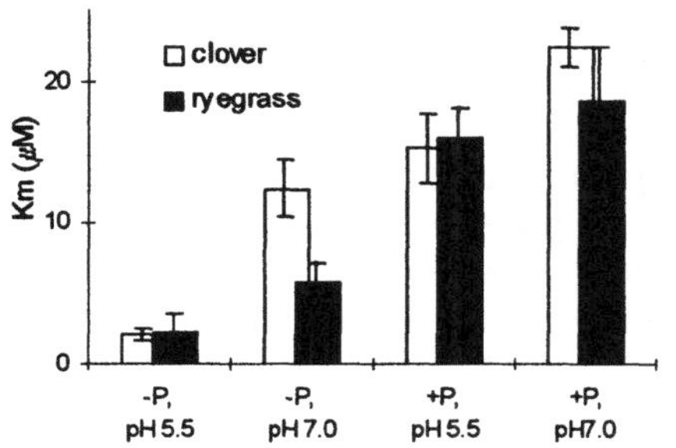

Fig. 3. Values for Km derived from depletion curves. Data are means for 4 reps ± s.d.

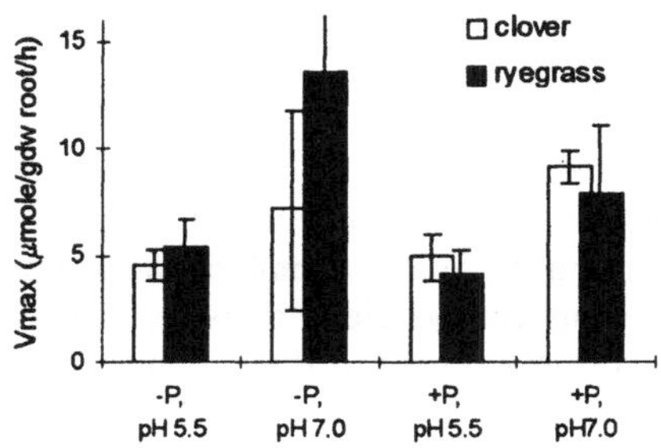

Fig. 4. Values for Vmax derived from depletion curves. Data are means for 4 reps ± s.d.

EFFLUX OF PREVIOUSLY ABSORBED $^{32}PO_4$

The efflux of ^{32}P from the root is illustrated in Fig. 5. Over the first 3 to 6 h efflux proceeded at an initially rapid but declining rate. The accumulation of ^{32}P in the solution was fitted to an exponential function of the form –

$$a + b\mathrm{e}^{-kt} \qquad (1)$$

where e is the exponent; t = time; *a, b,* and *k* are constants. The fitted values for the *a*, which quantify maximum efflux of ^{32}P, are given in Fig. 6. Log-transformed values for *a* were subjected to an analysis of variance. Values for +P treatments were significantly ($p<0.001$) greater than for the -P treatments by factors ranging from 5 to 10. The values for clover were significantly greater ($p<0.02$) than values for ryegrass by a factor of about three times except for the +P, pH 7.0 treatment where ryegrass exceeded clover. The values for *k*, the rate constant, did not vary significantly. Values for *b* were not significantly different from the corresponding values for *-a*, indicating that the intercepts at zero time were not significant.

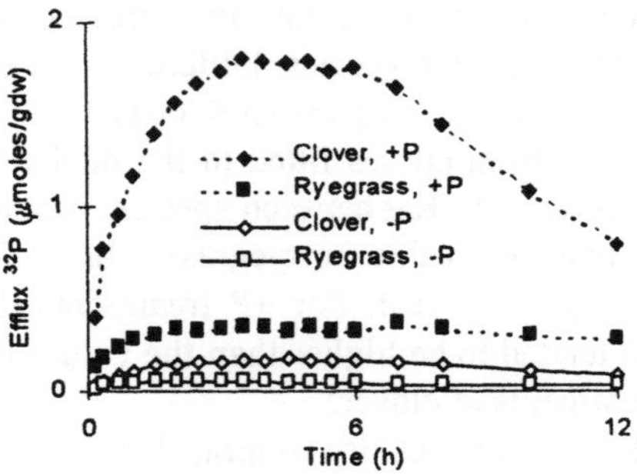

Fig. 5. Time course of $^{32}PO_4$ efflux from intact roots of white clover and ryegrass at pH 5.5.

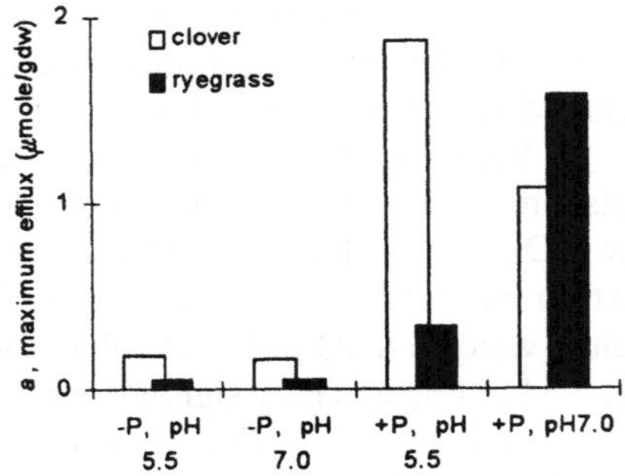

Fig. 6. Values for maximum efflux, *a*, from the equation $a + b e^{-kt}$. Data are means for 4 reps.

Discussion

The solution depletion method for determining KM and V_{max} was used in this study because the required numbers of samples of intact plants are not impracticably large. Solution depletion resembles the agricultural situation where, rather than solutions of constant composition, roots experience declining concentrations as they deplete the soluble phosphate in their vicinity. The values obtained for KM for PO_4 influx by -P plants, 2-13 μM, are consistent with values for other species (Bieleski, 1973; Mouat, 1983). The higher values for pH 7.0 compared with pH 5.5 may arise from the lower relative concentration of monovalent orthophosphate ($H_2PO_4^-$) at the higher pH. Higher values for KM resulting from a high P status, as found here, are not often reported but have been found previously (Cogliatti and Clarkson 1983; Dunlop et al., 1997).

The higher rates of influx for ryegrass are associated with lower root dry weights. For white clover the mean root dry weight did not differ between the +P and -P treatments whereas for ryegrass the -P treatment was only 0.73 the value of the +P treatment. This indicates that growth of the ryegrass roots but not the white clover roots was significantly reduced by the absence of phosphate during the 5 day pretreatment. This characteristic would offset the higher specific rate of PO_4 influx for ryegrass.

The progress of efflux found here is similar to a number of other studies (see Bieleski and Lauchli, 1992). The reabsorption of $^{32}PO_4$ after 3-9 h, evident in some treatments, results from the experiments being designed to measure influx and efflux simultaneously and to avoid exposing roots to transfer shock arising from frequent solution changes (Mouat 1983). To reduce the influence of $^{32}PO_4$ reabsorption on our results we used only the first 3 or 4 h of data to fit exponentials for efflux. The validity of this was tested by comparing the fits obtained for data for 1, 3, and 5 h and finding only minor differences. The significantly lower efflux for -P plants found in this study has also been reported for other species (Bieleski and Lauchli, 1992).

This study was undertaken to investigate the relative influences of influx and efflux in determining the differences in P acquisition between white clover and ryegrass. The largest and most consistent differences between the two species were the extent of efflux with ryegrass having lower values than white clover. Therefore efflux may be significant in determining the relative PO_4 acquisition abilities of the two species. Mouat (1983) found that C_{min}, the concentration of PO_4 at which efflux equals influx, was substantially lower for ryegrass than white clover which would impose a significant disadvantage on white clover when competing with ryegrass. This study extends Mouat's findings by identifying efflux as important in determining the relative values of C_{min}. Contrary to expectations, there were few differences in K_M and V_{max} between the two species. Notwithstanding this, the initial rates of influx for the -P treatments were faster for ryegrass.

In applying these findings to the P nutrition of pastures, it should be borne in mind that these experiments were conducted using seedlings with seminal roots. In pasture, roots systems of ryegrass and white clover are principally nodal. Furthermore the hydroponic system used provides an environment that is different from roots growing in soil. In particular, if sites of rapid influx in roots are remote from the sites of efflux, the potential for losing previously absorbed PO_4 could be amplified. We are now endeavouring to map the distribution of PO_4 influx and efflux along roots. Other factors likely to be important in the competition between clover and ryegrass include the relative morphology and growth rates of roots (Silberbush and Barber, 1983) which are more favourable for ryegrass (Dunlop and Hart, 1987). However in established ryegrass-white clover swards, the two species tend to segregate spatially into separate patches, a phenomenon possibly arising from the complexity of nitrogen nutrition (Schwinning and Parsons, 1996). This patchiness may reduce the importance of root growth and morphology in the competition between the two species. Clearly the competition involves many factors with the relative significance of each being as yet undetermined

References

Bieleski, R.L. (1973) Phosphate pools, phosphate transport, and phosphate availability. Annual Reviews of Plant Physiology, **24**, 225-252.

Bieleski, R.L. and Lauchli, A. (1992) Phosphate uptake, efflux and deficiency in the water fern *Azolla*. Plant, Cell, and Environment, **15**, 665-673.

Cogliatti, D.H. and Clarkson, D.T. (1983) Physiological changes and phosphate uptake by potato plants during development of, and recovery from phosphate deficiency. Physiologia plantarum, **58**, 287-294.

Dunlop, J. and Hart, A.L. (1987) Mineral nutrition. in M.J. Baker and W.M. Williams (eds.), *White Clover*, C.A.B. International, Wallingford, pp153-183.

Dunlop, J., Phung, H.T., Meeking, R., and White, D.W.R. (1997) The kinetics associated with phosphate absorption by *Arabidopsis* and its regulation by phosphorus status. Australian Journal of Plant Physiology, **24**, 623-629.

Macklon, A.E.S. and Sim, A. (1992) Modifying effects of a non-toxic level of aluminium on phosphate fluxes and compartmentation in root cortex cells of intact ryegrass seedlings. Journal of Experimental Botany, **43**, 1483-1490.

Mouat, M.C.H. (1983) Phosphate uptake from extended soil solutions by pasture plants. New Zealand Journal of Agricultural Research, **26**, 483-487.

Reay, P.F. and Waugh, C. (1981) Mineral-element composition of *Lupinus albus* and *Lupinus angustifolius* in relation to manganese accumulation. Plant and Soil, **60**, 435-444.
Schwinning, S. and Parsons, A.J. (1996) A spatially explicit population model of stoloniferous N-fixing legumes in grazing systems. Journal of Ecology, **84**, 799-813.
Shimogawara, K. and Usuda, H. (1995) Uptake of inorganic phosphate by suspension-cultured tobacco cells: kinetics and regulation by Pi starvation. Plant Cell Physiology, **36**, 341-351.
Silberbush, M. And Barber, S.A. (1983) Sensitivity of simulated phosphorus uptake to parameters used by a mechanistic-mathematical model. Plant and Soil, **74**, 93-100.

GENETIC VARIATION IN PHOSPHORUS UPTAKE OF CEREAL CULTIVARS

T. S. GAHOONIA and N. E. NIELSEN

The Royal Veterinary and Agriculture University, Department of Agriculture Sciences, Plant nutrition and Soil fertility laboratory, Thorvaldsensvej 40, DK-Frederiksberg C, Denmark

Abstract

Low phosphorus (P) availability in soils and exhausting P reserves emphasize the need to create plants that are more efficient P users. Studies of the mechanisms involved in P acquisition efficiency of the existing cereal varieties may serve as basis for improving soil P use by genetic means. We studied P depletion in thin slices (0.2 mm) of rhizosphere soil of cereal cultivars (for example, wheat: Kosack and Kraka; spring barley: Canut, Alexis). They differed considerably in P depletion of rhizosphere soil under laboratory conditions. The differences in P uptake were confirmed under field conditions at three P levels (no-P, 10 and 20 kg P ha^{-1}). Both in laboratory and field conditions, the variation between the cultivars was greatest in soil without P fertilizers (no-P) for about 30 years. The variation in P uptake was markedly reduced when 10 kg P ha^{-1} was applied.

Root length did not differ between the barley cultivars (Canut = 119±10; Alexis 112±15 m g^{-1} dry root), but there was wide, consistent variation in their root hairs, regardless whether they were grown in nutrient solution or soil. The cultivars with longer root hairs depleted more P from the rhizosphere and absorbed more P in the field, especially the barley cultivars in low P soil. To find out whether root hairs really participated in the P uptake, we measured the uptake of P via root hairs separately from that of remainder of the root surface. Only root hairs were made to grow into labelled soil so that they were the only pathways of ^{32}P uptake from soil. The presence of substantial amounts of ^{32}P in the plant shoot provided direct evidence on the role of root hairs in P uptake.

G. Gissel-Nielsen and A. Jensen (eds.), Plant Nutrition – Molecular Biology and Genetics, 111–119.

The wide variation in root hairs of the cereal cultivars together with the direct evidence of their role in P uptake suggest that selection for root hairs may be one of the criteria for improving soil P use. This way we may find P efficient genotypes, able to grow better both in widespread low-P soils (developing countries) and desired lower-P soils (developed countries) for environmental friendly agriculture.

Introduction

Cereals are among the important crops, requiring large phosphorus (P) applications. A crop usually utilises only 10 to 20 % of the fertiliser P in the year of application. The rest remains in soils mainly as non-soluble and also as soluble P, which may create environmental problems, depending upon P fixation capacity of soils. Limited P resources and many economical and environmental issues (leading to tighter legislation of fertiliser applications), emphasises the need to produce more with less P supply. Although better P cycling in agro-ecosystems may partly reduce exhaustion of P reserves, it is not known whether modern high yield varieties will sustain yield and product quality when grown with reduced P inputs. However, there are indications that some cereal varieties absorb more P than others (Nielsen and Schjørring, 1983; Gahoonia and Nielsen, 1996). This suggests that among the existing cereal varieties, there are varieties, which are superior to others in P uptake (unconscious selection of P efficiency). Identification of P efficient varieties can provide valuable genetic resource for maintaining yields with reduced P supply. The knowledge on the extent of genetic variation in P uptake and the mechanisms causing the variation are the primary steps for selecting and breeding more P efficient varieties.

Because of low mobility, major part of plant P is absorbed from soil close to roots (rhizosphere soil). From this widely accepted view, we may say that a cultivar is more P-efficient than others if it absorbs more P by exploiting more soil volume and/or by making P available by increasing solubility of P locally in the rhizosphere soil (P-acquisition efficiency). The extent of variation in P uptake between cultivars can be assessed from the P depletion profiles in rhizosphere soil. They can also provide information on the mechanisms causing the variation in P uptake. Such P depletion profiles can be obtained by analysing the thin slices of rhizosphere soil (Gahoonia and Nielsen, 1991; Gahoonia and Nielsen, 1996).

We investigated the ability of selected cereal cultivars to absorb P from rhizosphere soil in laboratory conditions and also measured their P uptake in field. The observed differences in P uptake is reported in this paper.

Materials and Methods

The following methods and procedures were applied to obtain results reported in this paper.

a) For rhizosphere studies, the method of Gahoonia and Nielsen (1991) was applied. In this method, rhizosphere soil is frozen in liquid nitrogen and then sliced with microtome into thin layers (0.2 mm) to obtain soil samples of known distances from the roots (a root mat of surface area 24.6 cm^2). The soil samples can be analysed for various fractions of phosphorus. The method has the advantage that confounding variation due to many environmental and nutritional factors is minimised. This makes it easier to assess genetic variation.

b) Root and root hair measurements using solution, soil columns and in the field (Gahoonia and Nielsen, 1997).

c) Direct evidence of the role of root hairs in phosphorus uptake from soil was obtained as described by Gahoonia and Nielsen (1998). Here only root hairs were allowed to grow in ^{32}P labelled soil, so that they were the only pathways of P uptake from the soil.

d) Field experiments were conducted using complete randomised design with 2 replicates. The plot size was 1.5 x 10 m^2. There were three phosphorus treatments (OP, 10P and 20P). The OP plots received no P fertiliser since 1966. Nitrogen (N), phosphorus (P) and Potassium (K) fertilizers applications (kg ha^{-1}) to plots were as follow:

OP plot: *60N, OP, 60K*; **10P plot**: *60N, 10P, 60K*; **20P plot**: *120N, 20P, 120K.*

Statistical analyses were performed using Statistical Analysis System (SAS Institute, 1989) and Microsoft Excel as found appropriate.

Results and Discussion

The P depletion profiles in the rhizosphere of barley cultivars varied significantly (p>0.05). For example, spring barley cultivar Canut depleted 64 % more P than Alexis in soil without P for about 30 years (Figure. 1). The variation was persistence at three P levels but it became gradually less with increase in P levels. At the highest P level, Canut depleted 7 % more P than Alexis from rhizosphere soil (Gahoonia *et. al.,* 1997). Also in the field, Canut and Alexis differed to produce shoot dry matter (DM) and in P uptake. At 0P levels, Canut was superior to Alexis in P uptake (Figure. 2). With 10P there was no difference.

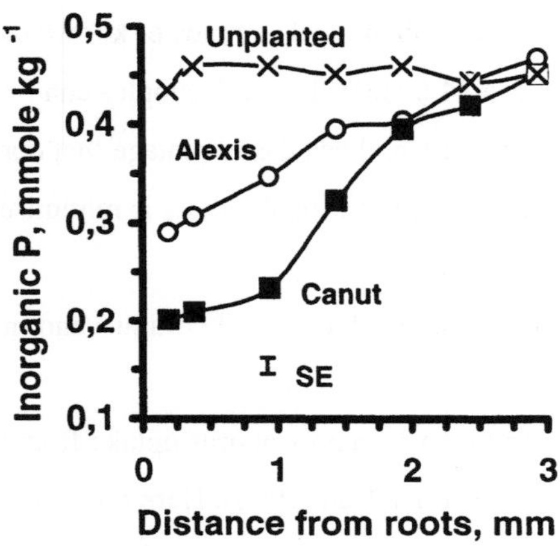

Figure 1. Phosphorus (extracted with 0.5 *M* $NaHCO_3$) depletion in the rhizosphere of two spring barley cultivars in 14 days. No P was applied to this soil for about 30 years.

At higher input levels (120N, 20 P, 120K), close to normal agriculture practices, Alexis produced more DM than Canut in the field. This is in agreement with the performance of Alexis in many agriculture soils, when high fertiliser inputs are applied. These results show that cultivars performing better under high- input conditions may not perform similar in low-input conditions. Therefore, selection for low-P-input has to be done under low P conditions, as also suggested by Ceccarelli and Grando (1991).

Application of 10 kg P ha^{-1} increased shoot biomass (DM) of the cultivars. However, this increase was nearly equal to the difference between the cultivars in 0P plots. The amount of DM produced by Alexis in 10P plot was nearly equal to that of Canut in 0P plot. Thus, in low P soil, the choice of cultivar Canut could help to produce same amount of DM with 10 kg ha^{-1} less P fertiliser application than to Alexis. These results suggest the added advantage of selection and breeding for P efficiency over amendment of soil by addition of high amounts of P fertilizers.

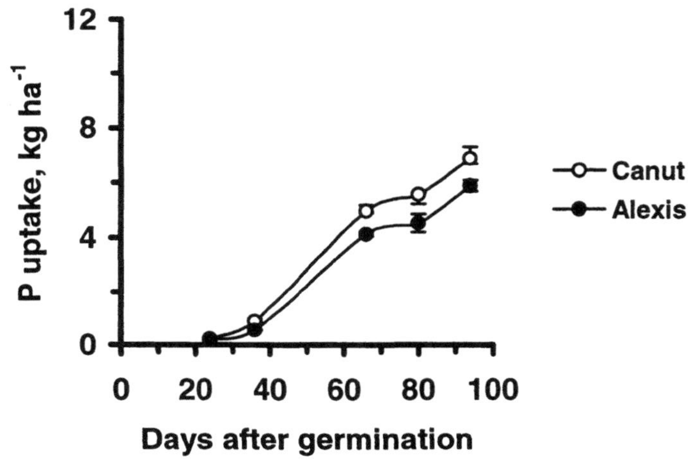

Figure 2. Phosphorus uptake of two spring barley cultivars in the field. Fertilizers : No P for about 30 years, 60 kg N and 60 kg K ha^{-1}.

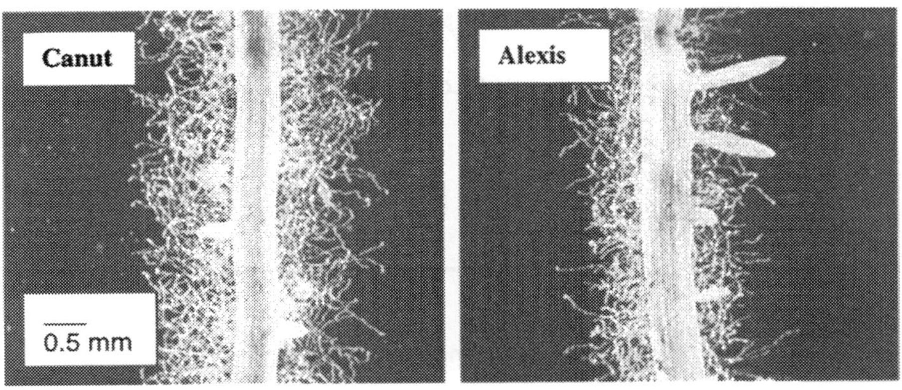

Figure 3. Root hairs of spring barley cultivars (Canut and Alexis). Root hairs were extracted from soil using ultrasound treatment (120W, 47 K Hz).

Wheat cultivars also differ to uptake P from rhizosphere soil in low P soil (Figure. 4). For example, the concentration of P near the root mat of Kosack decreased up to 0.28 mmoles kg^{-1} and that of Kraka up to 0.31 mmoles kg^{-1}. At one mm from the root mat P concentration in case of Kosack was 0.38 mmoles kg^{-1} whereas with Kraka it remained nearly unchanged (Figure. 4). Thus, Kosack was able to reduce soil P concentration near the roots to lower level, whereas Kraka extended the effective volume of soil close to the roots. This variation in P depletion profiles suggests that the two cultivars possess different strategies for acquiring P from rhizosphere soil. The root hairs of Kraka (1.26±0.17 mm) were longer than Kosack (0.59±0.16 mm). This explains why the P depletion profiles of Kraka extended further than Kosack. Kraka depleted nearly 20 % more P than Kosack from the rhizosphere soil. Under field conditions, Kosack was, however, slightly superior to Kraka, despite longer and denser root hairs of Kraka.

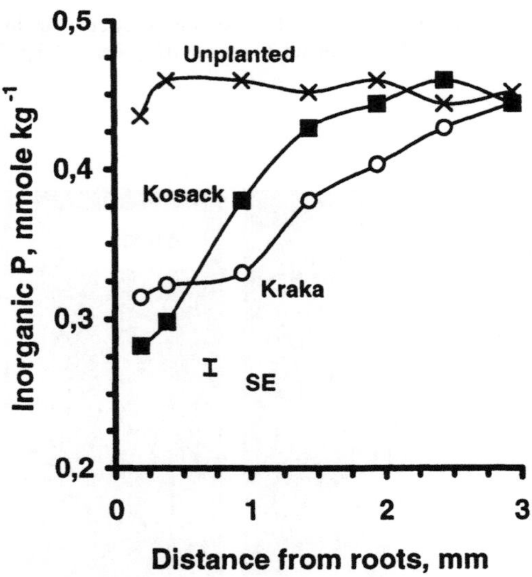

Figure 4. Phosphorus (extracted with 0.5 *M* $NaHCO_3$) depletion in the rhizosphere of two winter wheat cultivars in 14 days. No P was applied to this soil for about 30 years.

Soil P concentration in the immediate vicinity of roots of Kosack was lower (0.28 mmoles kg^{-1}) than Kraka (0.31 mmoles kg^{-1}) as can be seen in Figure 4. This suggests that Kosack might have released more P from soil under field conditions in the immediate vicinity of roots either due to higher root induced dissolution of soil P or it has lower C_{min} (minimum effective concentration) than Kraka as observed with barley cultivars in previous study (Nielsen and Schjørring, 1983).

The results showed that there is wide variation in root hairs of cereal cultivars, for example Canut and Alexis (Figure 3). Especially barley cultivars with longer and denser root hairs absorbed more P from rhizosphere soil and also in the field.

In previous studies (Gahoonia and Nielsen, 1997), barley cultivars (Salka and Zita) differed in net uptake of P from a soil-based system, e.g. the rhizosphere soil and in the field. They did not differ when grown in nutrient solution culture, despite similar variation in their root hairs in both cases. In solution culture, P diffuses freely to the root surface and the added advantage of having longer root hairs is lost (Clarkson, 1996). In soil, diffusion of P to the root surface is rate limiting and therefore, zone of soil close to the root is depleted uniformly, as observed with Canut in Figure. 1 and with Kraka in Figure. 4.

To find out whether root hairs really participated in the P uptake, we measured the uptake of P via root hairs separately from that of remainder of the root surface. Only root hairs were allowed to grow into soil labelled with radioisotope phosphorus (P) tracer ^{32}P (185 kBq g^{-1} dry soil) so that they were the only pathways of ^{32}P uptake from soil. The results (Gahoonia and Nielsen, 1998) showed that when 70 percent of the root hairs grew into the labelled soil, they contributed to 63 percent of the total P uptake. With decreasing number of root hairs growing into the ^{32}P labelled soil, the quantity of ^{32}P in plant shoot decreased. Therefore, this study provides strong direct evidence on the substantial participation of root hairs in uptake of phosphorus from soil.

The results show that there is wide genetic variation in phosphorus uptake of cereals, especially in low P soil. The supply of P to plant roots is diffusion limited (Lewis and Quirk, 1967). Due to the desired reduction in P fertiliser inputs, diffusion may become even more limiting for P uptake in the future. Under such conditions root hairs can help, because they become longer and denser with P limitation (Bates and Lynch, 1996). Under low P conditions, plants invest more carbon in root development (Lynch and Beebe, 1995). Although the metabolic costs (increased root respiration) associated with longer and denser root hairs are not known, the root dry weight in root hairs fraction is about 2 % (Röhm and Werner, 1987). With this little carbon investment, they can increase the effective root surface area three times.

The knowledge of the genetics of root hair formation (Schiefelbein and Somerville; 1990; Wada *et al.*, 1997) and P transporter genes (Muchhal *et al.*, 1996) of higher plants has started to emerge. There are indications that the P transporter genes are expressed mainly in roots and root hairs (Bucher *et al.*, 1997). All these studies support that it might be worthwhile and feasible to enhance the phosphorus efficiency of cereals by selection and breeding. Because of wide variation in root hairs of cereal cultivars and their substantial role in P uptake, root hairs can be one of the selection criteria.

Acknowledgement

The financial support of the Danish Ministry of food, Agriculture and Fisheries, under the Research Programme: *Development of Future Crop Plants* is thankfully acknowledged.

References

Bates, T. and Lynch, J. (1996) Stimulation of root hair elongation in *Arabidopsis thaliana* by low phosphorus availability. *Plant Cell and Environment* **19**, 529-538.

Bucher, M., Brunner, S., Daram, P. and Amrhein, N. (1997) Phosphorus acquisition in tomato Roots. In: H.E. Flores, J.P. Lynch, and Eissenstat, D. (eds.), *Radical Biology: Advances and Perspectives on functions of Plant roots.* American Society of Plant Physiologist.

Ceccareli, S. and Grando, S. (1991) Environment of selection and type of germplasm in barley breeding for low input conditions. *Euphytica* **57**, 207-219.

Clarkson, D.T. (1996) Root structure and site of ion uptake. In: Y. Waisel, A. Eshel and U. Kafkafi (eds.), *Plant Roots : The Hidden Half.* Marcel Dekker, Inc. New York, Basel, Hong Kong. pp 483-510.

Gahoonia, T.S. and Nielsen, N.E. (1991) A method to study rhizosphere processes in thin soil layers of different proximity to roots. *Plant and Soil* **135**, 143-146.

Gahoonia, T.S. and Nielsen, N.E. (1996) Variation in acquisition of soil phosphorus among wheat and barley genotypes. *Plant and Soil* **178**, 223-230.

Gahoonia, T.S. and Nielsen, N.E. (1997) Variation in root hairs of barley cultivars doubled phosphorus uptake from soil. *Euphytica* **98**, 177-182.

Gahoonia, T.S., Care, D. and Nielsen, N.E. (1997) Root hairs and acquisition of phosphorus by wheat and barley cultivars. *Plant and Soil* **191**, 181-188.

Gahoonia, T. S. and Nielsen, N. E. (1998) Direct evidence of participation of root hairs in phosphorus (^{32}P) uptake from soil. *Plant and Soil* **198**, 147-152.

Lewis, D.G. and Quirk, J.P. (1967) Phosphate diffusion in soil and uptake by plants. III. ^{31}P movement and uptake by plants as indicated by ^{32}P auto-radiography. *Plant and Soil* **27**, 445-453.

Lynch, J.P. and Beebe, S.E. (1995) Adaptation of beans to low soil phosphorus availability. *Hort. Sci.* **30**, 1165-1171.

Muchhal, U., Padro, J. and Raghothama, K. (1996) Phosphoate transporters from the higher plants *Arabidopsis thaliana. Proc. Natl.Acad. Sci.* USA **93**: 10519-10523.

Nielsen, N.E. and Schjørring, J.K. (1983) Efficiency and kinetic of phosphorus uptake from

soil by barley genotypes. *Plant and Soil* **72,** 225-230.
Röhm, M. and Werner, D. (1987) Isolation of root hairs from seedlings of *Pisum sativum.* Identification of root hair specific proteins by *in situ* labelling. *Physiol. Plant.* **69,** 129-136.
SAS Institute Inc., (1989) SAS/STAT. *User Guide.* Version 5. SAS Institute Inc., Cary, NC, USA.
Schiefelbein, J.W. and Somerville, C. (1990) Genetic control of root hair development in *Arabidopsis thaliana. Plant Cell* **2,** 235-243.
Wada, T., Tachibana, T., Simura, Y. and Okada, K. (1997) Epidermal cell differentiation in *Arabidopsis* determined by a *Myb* Homolog, CPC. *Science* **277,** 1113-1115.

LOW PHYTIC ACID MUTANTS AND HIGH PHYTASE CROPS: TWO STRATEGIES TO IMPROVE THE AVAILABILITY OF PHOSPHATE

F. HATZACK, K.S. JOHANSEN and S.K. RASMUSSEN
Plant Biology and Biogeochemistry Department
Risø National Laboratory, PO box 49, 4000 Roskilde, Denmark

1. Introduction

The nutritional availability of phosphate (P) from cereal grains is restricted because a high percentage (60-80%) of the total P content is stored as phytic acid (*myo*-inostol hexakisphosphate), which can not be efficiently digested by monogastric animals (Raboy, 1990). As a consequence, additional phosphate and/or phytases have to be supplied to the fodder. Furthermore, environmental problems arise when undigested phytate is released with animal manure. Microbial degradation of excreeted phytate then increases phosphate levels in the environment, which leads to eutrophication of freshwater reservoirs.

Other negative effects of the high phytate content in grain based fodder arise from the chemical nature of phytic acid itself. The high density of negatively charged phosphate groups makes phytic acid a strong chelator for nutritionally important mineral ions like $Fe^{2+/3+}$ and Zn^{2+}, which are consequently depleted from the fodder. Thus, lowering phytate levels and increasing the content of nutritional available P have become important goals in the development of novel crops with an improved nutritional value. To meet these goals we are pursuing two strategies: (1.) Identification and characterization of low phytate mutants and (2.) purification and cloning of wheat bran phytase with the prospect of overexpressing the recombinant enzyme in cereal grains.

(1.) Recently, chemically induced low phytate mutants were identified in the barley near-isogenic line Pallas-P01. Using thin layer chromatography for initial analysis, two mutant phenotypes could be distinguished, which differed in the amounts of grain localized inorganic phosphate and phytate (Rasmussen and Hatzack, 1998). Grains from A-type plants appeared to contain large amounts of inorganic phosphate and very little phytate, whereas B-type grains seemed to contain slightly increased levels of inorganic phosphate and somewhat less phytate than the wildtype. Here we present first quantitative data on the grain phosphate composition in low phytic acid mutants in barley, which in regard of the limited number of analysed plants have to be considered as a pilotstudy.

(2.) Transgenic expression of phytases represents an alternative approach to improve the nutritional availability of phosphate from cereal grains. Towards this

G. Gissel-Nielsen and A. Jensen (eds.), Plant Nutrition – Molecular Biology and Genetics, 121–124.

objective wheat bran phytase was selected because of its high specific activity and relative heatstability (Lim and Tate, 1973). Once purification of the enzyme and cloning of the respective gene are accomplished, suitable tranformation vectors will be constructed to overexpress this phytase in cereals. The optimal result of this strategy would be an engineered crop whose grain localized phytase activity is released during digestion in the animal intestine. In this context heatstability of the recombinant enzyme would be appreciable since commercial feedpellets have to be heatprocessed to avoid contamination from bacteria like *Salmonella*.

2. Results and Discussion

2.1. PHOSPHATE COMPOSITION IN GRAINS FROM LOW PHYTIC ACID MUTANTS

To obtain data on the grain phosphate composition in the low phytic acid mutants PLP1A-1 and PLP1B-2, we measured phytic acid P, inorganic P (Pi) and total P. Twentyfive grains per plant were ground and extracted with TCA as described in Rasmussen and Hatzack (1998). Inorganic P and phytic acid P was measured photometrically after TCA-extraction (Engelen *et al.* ,1994) and phytic acid P was determined by MDD-HPLC analysis (Mayr, 1988). To quantify total P, flour was subjected to wet oxidation in concentrated HNO_3:$HClO_4$ (4:1, v/v, Murphy and Riley, 1962).

The P-profiles in Fig. 1 show that phytic acid P was reduced in both mutants. In PLP1A-1 phytic acid P a reduction of ca. 86% in comparison to the wildtype controls was observed, whereas a more moderate reduction of ca. 25% was measured in PLP1B-2. In both mutants the decrease in phytic acid P was accompagnied by a proportional increase of inorganic P, showing that an impaired capacity to synthesize phytic acid resulted into accumulation of the phosphate substrate . Moreover, a strongly increased level of cellular P in PLP1A-1 indicated that phosphate can be stored in other phosphorylated compounds than phytic acid when phytic acid synthesis is partially blocked. Chromatograms obtained by MDD-HPLC showed that PLP1A-1 contained increased amounts of lower inositol phosphates, which could represent such alternative P-storage compounds (data not shown). A detailled characterization of these phosphoinositols has been initiated.

Taken together, our present data show that grain phosphorus composition in A- and B-type low phytate barley mutants can differ considerably from the wild type and that the percentage of those P-compounds, which are potentially available for animal nutrition, can be increased.

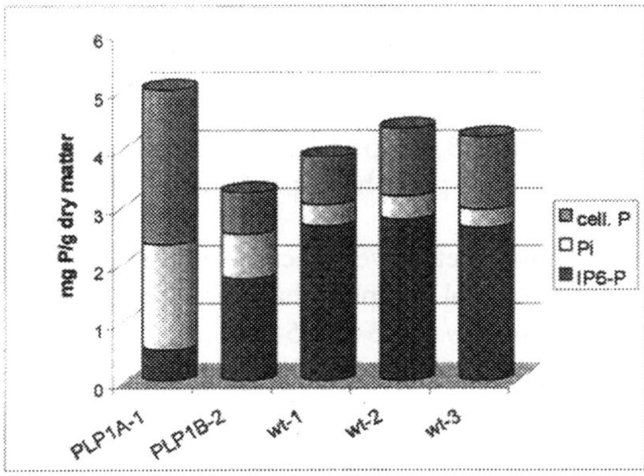

Figure 1. Phosphorus composition in grain material from low phytic acid mutants (PLP1A-1 and PLP1B-2) in comparison with wildtype controls (wt1, wt2 and wt3). Cellular P (cell. P) was calculated by substracting inorganic P (Pi) and phytic acid P (IP6-P) from the total P content.

2.2. PURIFICATION OF WHEAT BRAN PHYTASE

Phytase activity from wheat bran was enriched by a sequence of chromatographic steps, including ionic- and hydrophic interaction chromatography. Two proteins with apparent molecular weights of 31.000 and 56.000 copurified with phytase activity and were therefore assumed to represent wheat bran phytase(s). This assumption was further supported by western blot analysis, showing the two proteins crossreacting with both maize and mung bean phytase antibodies (Fig. 2). Amino-terminal sequences were obtained from both proteins, allowing the design of oligonucleotides, so that cloning of wheat bran phytase gene(s) could be initiated.

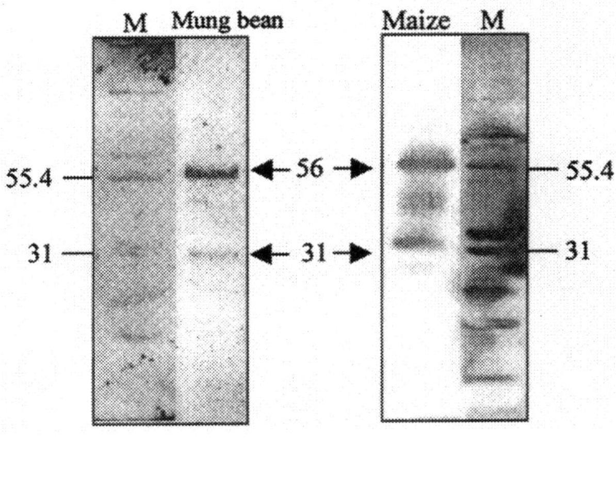

Figure 2. Western blot analysis of wheat bran phytase. Samples with peak phytase activity were resolved by SDS-PAGE, blotted and probed with either mung bean or maize phytase antibodies. Proteins with apparent molecular masses of 31 kd and 56 kd were detected in both experiments.

3. References

Engelen, A.J., Van der Heeft, F.C., Randsdorp, P.H.G., and Smit, E.L.C. (1994) Simple and rapid determination of phytase activity. *J. AOAC International* **77**,760-764.

Lim, P.E. and Tate, M.E. (1973) Properties of phytase fractions F1 and F2 from wheat bran and the myo-inositol phosphates produced by fraction F2. *Biochim. Biophys. Acta*, **302**, 316-328.

Mayr, G.W. (1988) A novel metal-dye detection system permits picomolar-range h.p.l.c. analysis of inositol polyphosphates from non-radioactively labelled cell or tissue specimens. *Biochem. J.* **254**, 585-591.

Murphy, J. and Riley, J.P. (1962) A modified single solution method for the determination of phosphate in natural waters. *Anal. Chim. Acta* **27**, 31-36.

Raboy, V. (1990) Biochemistry and genetics of phytic acid synthesis, in D.J. Morre, W.F. Boss and F.A. Loewus (eds.), *Inositol metabolism in plants*, John Wiley-Liss & Sons, Inc., New York. pp. 55-76.

Rasmussen, S.K. and Hatzack, F. (1998) Identification of two low-phytate barley (*Hordeum vulgare* L.) grain mutants by TLC and genetic analysis. *Hereditas* **129**, vol. 2, 107-112.

IDENTIFICATION AND CHARACTERISATION OF TWO CDNAS ENCODING PHOSPHATE TRANSPORTERS FROM *SOLANUM TUBEROSUM*

GEORG LEGGEWIE[1], LOTHAR WILLMITZER[2] and JÖRG W. RIESMEIER[3]

[1]*IACR Rothamsted, Harpenden, Hertfordshire AL52JQ, United Kingdom, Email: georg.leggewie@bbsrc.ac.uk*

[2]*Max Planck Institut für molekulare Pflanzenphysiologie, Karl-Liebknecht-Strasse 25, 14476 Golm, Germany*

[3]*PlantTec Biotechnologie GmbH, Hermannswerder 14, 14473 Potsdam, Germany*

Phosphate is an important macronutrient of higher plants, the molecular transport mechanisms of which have not been investigated until recently. Genetical optimizing of arable crops in their phosphate acquisition efficiency is highly desirable, in order to take advantage of their full genetic yield potential. Phosphate greatly influences the yield of crop plants because its lack has detrimental effects on growth, flowering, fruit setting and ripening. Genetic modification however, requires knowledge of the molecules involved in this process. To this end we isolated three cDNAs from potato that showed homology to a high affinity phosphate transporter from *Saccharomyces cerevisiae*. This paper summarizes the characterisation of two of these clones so far, using yeast as an heterologous expression system and analysis of the genes at the transcriptional level. The results will be discussed in respect to more recent publications in this field. This summary is part of an oral presentation given at the "Sixth International Symposium on Genetics and Molecular Biology of Plant Nutrition" in Elsinore, Denmark.

Introduction

Phosphate transport in higher plants is mainly driven by a proton phosphate symport mechanism utilising the proton motive force at the plasma membrane. Kinetic studies suggested an inducible high affinity and a constitutive low affinity phosphate uptake system (reviewed in Schachtman et al., 1998). Induction occurs during a period of Pi deprivation. This makes the inducible component of the high affinity phosphate transport system similar to the sulphate uptake system in higher plants (Smith et al., 1997) and different to the inducible high affinity transport system of nitrate (iHATS), which requires replenishment with the nitrate ion in order to be fully induced (Forde and Clarkson, in press). A sodium coupled symport system, as found in animals, has not yet been demonstrated in higher plants even though plant derived cDNAs with similarity to such transporters are known (Poirier and Theunissen, 1995). A similar situation is found in yeast, where two uptake systems have been described, one operating at a Km of 8 μM and the other at a Km of 770 μM. It seems that *PHO84*

G. Gissel-Nielsen and A. Jensen (eds.), Plant Nutrition – Molecular Biology and Genetics, 125–135.

(Bun-ya et al., 1991) encodes a component of the high affinity system. However, the recent discovery of several other genes, some of which encode membrane proteins, indicates that a more complicated mechanism may be responsible for high affinity uptake in yeast. Pho86p, Pho87p (Bun-ya et al., 1996), Pho88p (Yompakdee et al., 1996) and a GTP-binding protein, Gtr1p (Bun-ya et al., 1992), have been shown to be transcriptionally regulated by the PHO regulatory system and they are assumed to play a role in phosphate transport. Very recently, a gene encoding a sodium phosphate symporter has also been described in baker's yeast (*PHO89*, Martinez and Persson, 1998). The mode of interaction of these proteins is unknown. Although such interacting proteins have not yet been identified in higher plants, recent results indicate that in plants there are also multiple transporters responsible for phosphate uptake. So far in *Arabidopsis thaliana*, four genes (AtPT1 and AtPT2, Muchal et al., 1996; APT1 and APT2, Smith et al., 1996, with the coding region of APT2 being identical to AtPT1, and AtPT4, Lu et al., 1997; the cDNA PHT1 reported by Mitsukawa et al., 1997, is identical to AtPT1 and the coding region of APT2), in tomato two (LePT1, Daram et al., 1998, Liu and Muchal et al., 1998 and LePT2, Liu and Muchal et al., 1998), in potato three (StPT1, StPT2 and StPT3, Leggewie et al., 1997 and this report), in *Medicago truncatula* two (MtPT1 and MtPT2, Liu et al., 1998) and in *Catharanthus roseus* one gene (PIT1, Kai et al., 1997) have been published. As reported at this symposium, more genes can be expected to be cloned from both, Arabidopsis (up to 7) and barley (three so far). In our group one more cDNA, StPT3, has been isolated recently. The astonishingly high number of genes involved in phosphate transport opens up questions for their spatial and developmental regulation as well as their kinetic features.

Results and Discussion

Cloning of StPT1, StPT2 and StPT3

Using the *PHO84* sequence from yeast as a probe to screen the translated expressed sequencing tag database, an EST-clone (134M11T7) was identified which had 35% identity at the amino acid level to the fungal transporter. Subsequently, the cDNA of this EST-clone (1 Kb in length) was used to screen a root specific potato cDNA library. The roots were harvested from plants that were grown on tap water for three days to induce the Pi-uptake capacity. So far we have isolated 3 different clones, two of which (StPT1 and StPT2) have been characterized at the transcriptional level and in the yeast expression system. StPT1 and StPT2 are predicted to contain an ORF of 540 and 527 amino acids, respectively. This matches the ORFs observed from all other cDNAs isolated so far which range from 524 (AtPT1) to 542 (PIT1). The hydropathy plots of the deduced amino acid sequence fit well with that published for Pho84p, having 12 membrane spanning domaines separated by a hydrophilic loop after the sixth hydrophobic domain. All of the above mentioned cDNAs seem to be members of a well conserved family of transporters, because they share position and spacing of the membrane spanning domains as well as putative regulatory sites. These are a protein kinase C site, exposed to the cytosol due to its location at the beginning of the large hydrophilic central loop, and a casein kinase II site immediately after the last membrane spanning domain at the C-terminus, also facing the cytosol. Interestingly, these sites are shared with fungal Pi-transporters such as Pho84p (Bunya et al., 1991) and GvPT (Harrison and van Buuren, 1995). An N-glycosylation site is found after the tenth transmembrane domain. Table 1 summarizes the homology of StPT1 and StPT2 to all the cDNAs encoding phosphate transporter proteins in higher plants and fungi published so far.

Table 1. Percentage of amino acid identity of the potato and other phosphate transporters from plants and fungi, calculated by using the BESTFIT algorithm software of the Genetics Computer Group (GCG, Madison, WI).

	StPT1	StPT2	AtPT1	AtPT2	APT1	AtPT4	LePT1	LePT2	MtPT1
StPT1	100	78.5	80.8	83.7	80.5	81.4	95.5	80.8	78.5
StPT2	78.5	100	75.8	75.5	75.5	76.6	78.0	95.0	79.3

	MtPT2	PIT1	Pho84p	GvPT
StPT1	78.5	86.7	34.9	41.7
StPT2	79.8	76.3	36.0	41.8

Expression analysis of StPT1 and StPT2 in yeast

The *Saccharomyces cerevisiae* strain MB192 is defective in the high affinity uptake system due to an insertion into the *PHO84* gene. This results in retarded growth and a lower uptake rate for Pi. Additionally, the mutant is unable to incorporate Pi in sufficient quantities to supress the activity of an acid phosphatase, even when grown on Pi rich medium. This enzyme is secreted by the wildtype in case of Pi deprivation and its production is suppressed when sufficient Pi is in the medium. The activity of the acid phosphatase was used to monitor the Pi status of the mutant when complemented with StPT1 or StPT2. A staining assay allows the detection of the acid phosphatase activity, by producing a red dye on the yeast surface when the enzyme is secreted. This was the case with the non-complemented mutant, whereas expression of StPT1 or StPT2 did not result in staining. The non-staining indicated the suppression of the acid phosphatase formation which is due to a higher Pi incorporation after expression of StPT1 and StPT2. Expression of a PCR-product encoding the wildtype *PHO84* gene served as a positive control and gave a similar result (Figure 1). The same assay was sucessfully used to demonstrate complementation in case of AtPT1 and AtPT2 (Muchal et al., 1996) and the fungal phosphate transporter GvPT (Harrison and van Buuren, 1995).

A

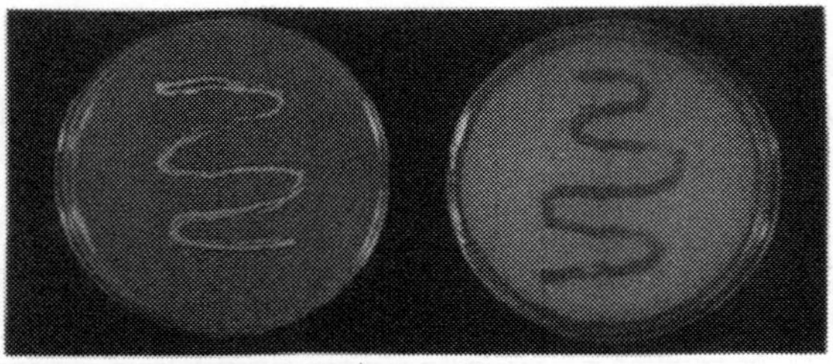

B

Figure 1 Staining test for acid phosphatase activity

(A) An acid phosphatase activity test was conducted with yeast mutant strain MB192 (middle) and transformants YStPT1 (left) and YStPT2 (right). The latter contain StPT1 and StPT2 cDNAs subcloned in the yeast expression vector p112A1NE. Phosphatase activity was detected by staining according to Bun-ya et al. (1991). Whereas mutant colonies stain red, the transformants remain pale.
(B) As a control, the *PHO84* polymerase chain reaction product in p112A1NE was transformed in MB192, yielding transformed strain YPHO84. YPHO84 (left) reveals the same staining phenotype as YStPT1 and YStPT2 and differs from the phenotype of MB192 (right).

To corroborate this finding we measured growth and uptake of Pi in the complemented mutants. In both cases however, complementation was observed which did not fully re-establish the wildtype characteristics as represented by complementation with the endogenous *PHO84* gene (Figure 2). The uptake of Pi was pH dependent, with a maximum of uptake at pH 4.5. The use of uncouplers such as carbonylcyanide *m*-chloro-phenylhydrazone (CCCP) and 2,4-dinitrophenol (DNP) confirmed the pH dependance of the Pi-uptake: Both substances destroy the proton gradient across the membrane in a dosage dependent way and hence their application is supposed to influence the proton driven flux of Pi across the yeast plasmamembrane. Indeed, 10 μM CCCP reduced the uptake to 75% (StPT1) and 55% (StPT2) and 100 μM DNP reduced the Pi-uptake to 58% (StPT1) and 55% (StPT2). This strongly indicates that StPT1 and StPT2 are Pi/proton symporters. An even stronger reduction by using the same inhibitors at comparable concentrations was recently reported by Daram et al. (1998) in the case of expression of the tomato LePT1 transporter in yeast. The yeast strain MB192 was then used for determination of the Km for the phosphate incorporation. The uptake followed Michaelis-Menten kinetics and had an apparent Km of 280 and

130 μM for StPT1 and StPT2, respectivly. A recent report on expression of the *Medicago truncatula* phosphate transporter MtPT1 in the yeast *PHO84* mutant strain NS219 gave a similar Km (192 μM) for the phosphate uptake (Liu. et al, 1998). These Km values differ very much from Km measurements that were obtained when fungal transporters were expressed in the *PHO84* mutant (GvPT, Harrison and van Buuren, 1995; *PHO84*, Bunya et al., 1991). In fact, the Pi concentration in soil solution is about 2 μM, hence a high affinity phosphate uptake system would be expected to operate in this range. This was indeed measured *in vivo* in several cases (Cogliatti and Clarkson, 1983).

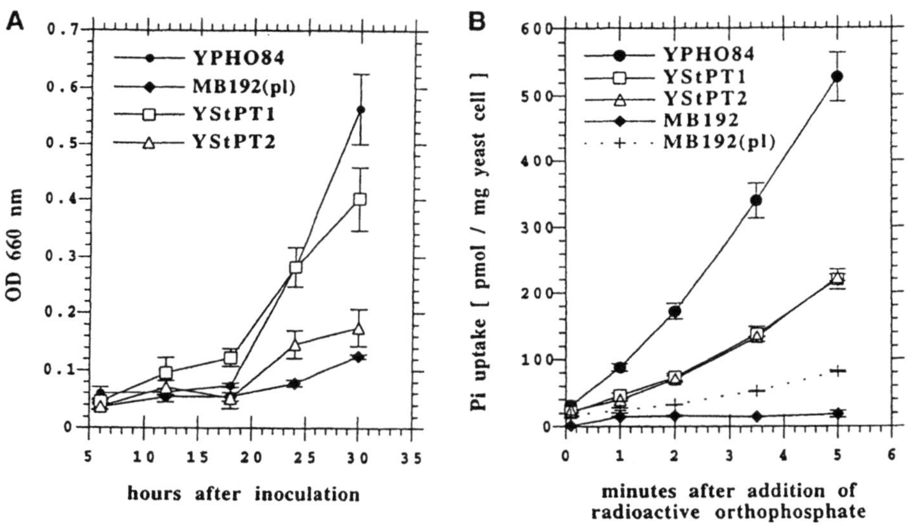

Figure 2 Phenotype of YStPT1 and YStPT2
(A) Growth of YStPT1 and YStPT2 compared with mutant strain MB192 (pl) carrying the yeast expression vector p112A1NE without an insert and YPHO84 carrying the *PHO84*-encoding PCR product.
(B) Uptake of radioactive orthophosphate by YStPT1, YStPT2, YPHO84, MB192(pl) and untransformed MB192.

This large discrepancy may be due to the expression in a heterologous expression system. Daram et al. (1998) reported an expression study of the tomato Pi transporter LePT1 in a yeast double mutant, where the *PHO84* gene and the *PHO89* gene were destroyed by insertion. They measured a Km for Pi-uptake into this particular yeast strain of 31 μM. Hence, the determination of this kinetic parameter may be very much dependent on the strain used for the measurements. However, even a Km of 31 μM is still 10-fold higher than the expected values. The use of other heterologous expression systems such as *Xenopus laevis* oocytes or the baculovirus-insect cell expression system did not give any functional expression at all when we used them for expression of either transporter (data not shown). In fact, the only approach that reproducibly allowed the measurement of Km values in the expected range of around 2-3 μM was a transgenic expression experiment in plants: Mitsukawa et al. (1997) reported that expression of PHT1 (identical with AtPT1 and the coding region of APT2) in tobacco using the 35S-promotor increased the uptake rate with an apparent Km of 3.1 μM, when uptake was measured in suspension cultured cells. Hence, it seems that only a homologous expression system is able to reveal the transporter as a component of a high affinity uptake system. It may be speculated that heterologous sytems are not able to provide auxillary factors that might be necessary for proper functioning of the transporters. The nature of these auxillary factors remains to be elucidated, but probably the genetic evidence from baker's yeast suggests that the additional presence of multiple subunits in higher plants is required for proper high affinity phosphate transport as well.

Expression analysis of StPT1 and StPT2 at the transcriptional level

Although StPT1 and StPT2 are almost 80% identical in their amino acid composition, their expression pattern at the transcriptional level is diverse. Expression of StPT1 in soil grown plants could be observed in roots, tubers, source leaves and floral organs. No expression was detected under these conditions in sink leaf and stem (Figure 3A). However, in these parts expression could be also detected when quartz sand cultures and 0.5x Hoagland solution were used to grow the plants. Hence, StPT1 seems to be constitutively expressed in all tissues even though an enhanced expression was detected under nutrient deprivation. Expression of StPT2 could not be detected at all in soil grown plants. Only when the plant was grown on tap water for a couple of days was a strong expression monitored in the roots (Figure 3B). This induction was not seen when plants were deprived of nitrogen or potassium, but was strong under Pi deprivation and detectable under sulphate deprivation (Figure 4). Therefore, StPT2 seems to be root specific and its expression strongly responds to the Pi-status of the plant. The inducibility of StPT2 upon S-deprivation is peculiar: Schröppel-Meier and Kaiser (1988) showed that sulfate deprivation does not have an influence on the Pi concentration in leaves from spinach and Alarcón et al. (1997) showed, in a nutrition study of tomato grown on rockwool, that sulphate uptake from the medium used is not related to phosphate uptake and vice versa.

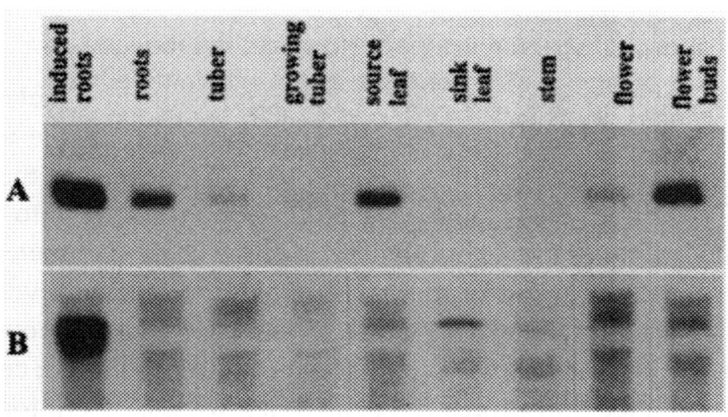

Figure 3

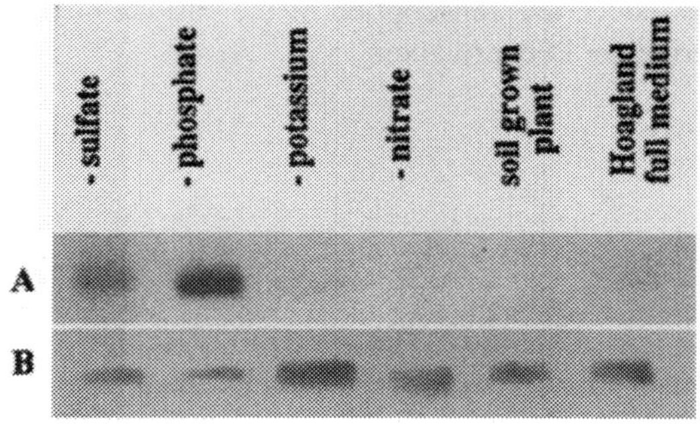

Figure 4

Figure 3 Gel blot analysis of potato RNA derived from different tissues
(A) Thirty micrograms of total RNA in each lane was hybridized with StPT1
(B) The same amount of total RNA as above was hybridized to StPT2
Hybridization shows the occurrence of a single RNA transcript at a length of 1.8 kb. Roots derived from nutrient-starved plants are termed induced roots.

Figure 4 RNA gel blot analysis of root-specific RNA derived from plants starved for phosphorus, sulfur, nitrogen, and potassium as well as plants grown under optimal nutrient conditions.
(A) Hybridization of StPT2 with 30 µg of total RNA shows the presence of an StPT2-specific mRNA with RNA from Pi-deprived plants. The presence of StPT2-specific mRNA was also detected to a lesser extent in sulfur-starved plants.
(B) To indicate the total amount of RNA present, hybridization was performed with 25S rRNA as a control.

The difference between phosphate and sulfate transport systems has been well documented at the physiological level (Clarkson et al., 1992) and the molecular level (Smith et al., 1995, 1996). However, a similar finding was reported by Daram et al. (1998) with LePT1 and to this symposium by F.W. Smith (CSIRO, Brisbane, Australia) with a Pi-transporter cDNA from barley (HvPT3). It may be that some high affinity phosphate transporters have a transport activity for sulphate as well. However, Daram et al. (1998) reported that LePT1 did not complement a sulphate uptake deficient yeast mutant and that sulphate does not compete with Pi in uptake studies using the phosphate uptake deficient yeast that expressed LePT1. Another possibility is the lack of a suppressor, the synthesis of which is hampered due to S-starvation and that would otherwise inhibit the transcription of the Pi-transporter encoding gene. In contrast, the effect of nitrogen on phosphate transport in roots is well documented: Smith and Jackson (1987 a,b) showed an enhanced uptake and translocation of Pi in maize seedlings in the presence of nitrate and ammonium. It will be interesting to see whether such treatments can induce the transcription of the various Pi transporters in a similar way to that observed under Pi deprivation.

Outlook

The identification of several phosphate transporters with different expression patterns suggests that the plant is able to cope with diverse nutritional conditions by using differential transcription as a tool. The availability of the cDNAs encoding these transporters allows us to answer various question that stem from this finding:

1) Under which condition is each transporter expressed and what are the locations of the different transporters in the plant tissue? This question may be answered by using promotor-GUS fusions, in situ hybridisations and immunogold labelling experiments. Initial experiments show mRNA of LePT1 in peripheral cell layers of roots such as rhizodermal and root cap cells but also in the central cylinder under Pi-deprivation. (Daram et al., 1998, Liu et al., 1998). Under these conditions Liu et al. (1998) could observe expression of LePT1 in palisade parenchyma and phloem cells of the leaves as well.

2) What proportion of the overall Pi uptake into the plant are these transporters responsible for ? To study this question, knock out mutants or antisense plants would be helpful. We have initiated the work by creating antisense plants in potato: Four lines have been isolated so far for both, the StPT1 and StPT2 genes, respectively. These show a 40-60% reduction of the amount of RNA hybridizing with each cDNA. The lines show a loss of apical dominance, displaying bushy growth and start to tuberize prematurely. This phenotype is enhanced under Pi-deprivation, but is also visible at sufficient Pi-supply though to a lesser extent. Premature tuberisation is often seen in wildtype potato as a response to environmental stress. Future analysis will show whether the root to shoot ratio and the Pi contents of these antisense plants are altered.

3) Is there allosteric control of the transporters, as suggested by several authors (e.g. Clarkson and Cogliatti, 1983)? If so, what is the nature of this control ? The data from the heterologous expression systems suggest an interaction with other proteins. Post-transcriptional modification may also change the kinetics of the transporter, because of the presence of conserved phosphorylation sites. These questions can be addressed by using phosphorylation assays and the yeast two hybrid system. Likewise a coexpression of the transporter cDNA and a plant cDNA-library in a phosphate uptake mutant is feasible. A similar strategy has been presented to the "11th International Workshop on Plant Membrane Biology" in Cambridge as a successful approach to identifying a kinase interacting with AKT1, the inward rectifying voltage gated potassium channel from Arabidopsis (Desbrosses-Fonrouge et al., 1998).

4) What are the phosphate starvation signals and how are they perceived and transduced downstream towards the final onset of the transcription of the phosphate transporters? Are there different phosphate receptors for the inner and the outer cell phosphate

concentration ? Liu and Muchal et al. (1998) demonstrate that in divided-root experiments, where the root system has been exposed to different Pi concentrations (0 and 250 µM Pi), expression of the Pi transporter follows the internal Pi concentration which was high in both halves of the root system. This corroborates early data from Cogliatti and Clarkson (1983) who showed in split root experiments that the activity of the Pi transport system is governed by the overall demand for phosphate. If the leaf phosphorus status regulates the uptake of Pi in the roots, as suggested by Marschner (1995), then an interesting signal transduction pathway has to be unravelled connecting the shoot and root parts of the plant.

5) Finally, how is the expression of these transporters influenced when roots are colonized by mycorrhizal fungi ? This fungal symbiosis occurs in most of the arable crop plants. The network of hyphae that extend out of the roots into the soil contribute significantly to the Pi absorption of the plant. Pioneering work in this field has been done by the group of Maria Harrison: Liu et al. (1998) showed the downregulation of the expression of phosphate transporters *from Medicago truncatula* upon infection with mycorrhiza which are otherwise highly expressed in roots under low Pi conditions. However, Sally E. Smith (University of Adelaide, Australia) reported to the symposium that LePT1 in tomato is not downregulated upon mycorrhizal infection. LePT1 is also expressed in organs of the plant other than the root, therefore again different phosphate transporters may undergo different regulation when the plant is colonized by mycorrhiza. Split root experiments with infected and non infected roots may demonstrate whether a downregulation is systemic or restricted to the infected part of the root system. It will also be interesting to test the expression of the plant transporters under different Pi conditions in the soil when the plant is infected with the fungus.

In conclusion, it seems that transcriptional regulation of the Pi-transporter genes is an important tool for the higher plant to cope with various Pi nutrition conditions. To understand this regulation is a crucial step in the attempt to use transgenic approaches in order to modify phosphate uptake efficiency in higher plants.

Acknowledgement

I like to thank the organizing commitee of the 6th International Symposium of Genetics and Molecular Biology of Plant Nutrition to invite me for giving a presentation of our recent work. It is also a pleasure to thank Risø National Laboratory for organising this most interesting and highly enjoyable conference in Elsinore. I thank the Biochemistry and Physiology Department of IACR Rothamsted for the financial support to attend the conference and my colleagues Marcel Bucher, Dave Carden and Steve Chatfield for carefully reading the manuscript and helpful comments.

Literature cited

Alarcón, A.L., Madrid, R, and Egea, C. (1997) Hydric and nutrient element nutrition of a tomato crop on rockwool: Ionic interrelationships, *J. Plant Nutr.* **20**, 1811-1828
Bun-ya, M., Nishimura, M., Harashima, S., and Oshima, Y. (1991) The *PHO84* gene of *Saccharomyces cerevisiae* encodes an inorganic phosphate transporter, *Mol. Cell. Biol.* **11**, 3229-3238.
Bun-ya, M., Harashima, S., and Oshima, Y. (1992) Putative GTP-binding protein, Gtr1, associated with the function of the PHO84 inorganic phosphate transporter in *Saccharomyces cerevisiae. Mol. Cell. Biol.* **12**, 2958-2966.
Bun-ya, M., Shikata, K., Nakade, S., Yompakdee, C., Harashima, S., and Oshima, Y. (1996) Two new genes, *PHO86* and *PHO87*, involved in inorganic phosphate uptake in *Saccharomyces cerevisiae. Curr.Genet.* **29**, 344-351.
Clarkson, D.T., Hawkesford, M.J., Davidian, J.-C., and Grignon, C. (1992) Contrasting responses of sulphate and phosphate transport in barley (*Hordeum vulgare* L.) roots to protein-modifying reagents and inhibition of protein synthesis, *Planta* **187**, 306-314.
Cogliatti, D.H., and Clarkson, D.T. (1983) Physiological changes in, and phosphate uptake by potato plants during development of, and recovery from phosphate deficiency, *Physiol. Plant.* **58**, 287-294.
Daram, P., Brunner, S., Persson, B.L., Amrhein, N., and Bucher, M. (1998) Functional analysis and cell-specific expression of a phosphate transporter from tomato, *Planta* **206**, 225-233.
Desbrosses-Fonrouges, A.-G., Chérel, I., Grignon, C., and Sentenac, H. (1998) Control of K^+ channel activity: utilization of yeast to identify proteins interacting with AKT1, in: 11th International Workshop on Plant Membrane Biology, Cambridge 9-14th August 1998, Abstractbook, p131.
Forde, B.G., and Clarkson, D.T. Nitrate and ammonium nutrition of plants: Physiological and molecular perspectives, *Adv. in Bot. Res.*, in press.
Harrison, M. and van Buuren, M.L. (1995) A phosphate transporter from the mycorrhizal fungus *Glomus versiforme*, *Nature* **378**, 626-629.
Kai, M., Masuda, Y., Kikuchi, Y., Osaki, M.,, and Tadano, T. (1997) Isolation and characterisation of a cDNA from *Catharanthus roseus* which is highly homologous with phosphate transporter, *Soil Sci. Nutr.* **43**, 227-235.
Leggewie, G., Willmitzer, L, and Riesmeier, J.W. (1997) Two cDNAs from potato are able to complement a phosphate uptake-deficient yeast mutant: Identification of phosphate transporters from higher plants, *Plant Cell* **9**, 381-392.
Liu, C., Muchal, U.S., Uthappa, M., Kononowicz, A.K., and Raghothama, K.G. (1998) Tomato phosphate transporter genes are differentially regulated in plant tissues by phosphorus, *Plant Physiol.* **116**, 91-99.
Liu, H., Trieu, A.T., Blaylock, L.A., and Harrison, M.J. (1998) Cloning and characterisation of two phosphate transporters from *Medicago truncatula* roots: Regulation in response to phosphate and to colonization by arbuscular mycorrhizal (AM) fungi, *MPMI* **11**, 14-22.
Lu, Y.P., Zhen, R.G., and Rea, P.A. (1997) AtPT4: A fourth member of the *Arabidopsis* phosphate transporter gene family (Accession No. U97546) (PGR 97-082) *Plant Physiol.* **114**, 747.
Martinez, P., and Persson, B.L. (1998) Identification, cloning and characterisation of a derepressible Na^+-coupled phosphate transporter in *Saccharomyces cerevisiae, Mol. Gen. Genet.* **258**, 628-638.
Marschner, H. (1995) Mineral nutrition in plants, Ed2, Academic Press, San Diego, CA.
Mitsukawa, N., Okumura, S., Shirano, Y., Sato, S., Kato, L., Harashima, S., and Shibata, D. (1997) Overexpression of an *Arabidopsis thaliana* high-affinity phosphate transporter gene in tobacco cultured cells enhances cell growth under phosphate-limited conditions, *Proc. Nat. Acad. Sci. USA* **94**, 7098-7102.
Muchal, U.S., Pardo, J.M., and Raghothama, K.G. (1996) Phosphate transporters from the higher plant *Arabidopsis thaliana, Proc. Natl. Acad. Sci. USA* **93**, 10519-10523.
Poirier, Y., and Theunissen, H. (1995) Analysis of two proteins in *Arabidopsis thaliana* showing homology to animal Na^+/Pi cotransporters, in: Program and abstracts, 10th International Workshop on Plant Membrane Biology, Regensburg, August 6-11, 1995, poster R34.
Schachtman, D.P., Reid, R.J., and Ayling S.M. (1998) Phosphorus uptake by plants: From soil to cell, *Plant Physiol.* **116**, 447-453.
Schröppel-Meier, G., and Kaiser, W.M. (1988) Ion homeostasis in chloroplasts under salinity and mineral deficiency, *Plant Physiol.* 87, 828-832.
Smith, F.W., and Jackson, W.A. (1987a) Nitrogen enhancement of phosphate transport in roots of *Zea mays* L. I. Effects of ammonium and nitrate pretreatment, *Plant Physiol.* **84**, 1314-1318.

Smith. F.W. and Jackson, W.A. (1987b) Nitrogen enhancement of phosphate transport in roots of *Zea mays* L. II. Kinetic and inhibitor studies, *Plant Physiol.* **84**, 1319-1324.
Smith, F.W., Ealing, P.M., Hawkesford, M.J. and Clarkson, D.T. (1995) Plant members of a family of sulphate transporters reveal functional subtypes, *Proc.Natl. Acad. Sci. USA* **92**, 9373-9377
Smith, F.W., Ealing, P.M., Dong, B., and Delhaize, E. (1997) The cloning of two *Arabidopsis* genes belonging to a phosphate transporter family, *Plant J.* **11**, 83-92.
Yompakdee, C., Ogawa, N., Harashima, S., and Oshima, Y. (1996) A putative membrane protein, Pho88p, involved in inorganic phosphate transport in Saccharomyces cerevisiae, *Mol. Gen. Genet.* **251**, 580-590

THE HAK1 AND HAK2 GENES OF BARLEY ENCODEHIGH-AFFINITY K^+ TRANSPORTERS THAT ARE ALSO PRESENTIN ARABIDOPSIS

G. SANTA-MARêA[a],F. RUBIO[a], J. DUBCOVSKY[b] AND A.RODRêGUEZ-NAVARRO[a]

[a]Departamento de Biotecnolog'a, E.T.S.I.Agr—nomos, 28040 Madrid, Spain

[b]Department of Agronomy and Range Science, Universityof California Davis, California, USA

Introduction

Potassium is an essential macronutrientrequired for plant growth and development (Flowers and LŠuchli, 1983).The K^+ concentration in the soil may be very low, at least in certain conditions,and plants accumulate the cation via high-affinity K^+transporters located in the plasma membrane of root cells. In barley,K^+ absorption exhibits a biphasic response to the increase of K^+ concentration in the external medium (Epstein *etal.*, 1963) showing a first saturable component in the micromolar range and asecond saturable component in the millimolar range.

Although the existence of high- andlow-affinity K^+ uptake systems in plants is an accepted paradigm, the number andcharacteristics of the transporters have yet to be determined. Many groupshave devoted their efforts to the isolation of plant genes

G. Gissel-Nielsen and A. Jensen (eds.), Plant Nutrition – Molecular Biology and Genetics, 137–143.

encodingK$^+$ transporters. As a result, two root-expressed cDNAs, *AKT1*from *Arabidopsis thaliana* (Lagarde *etal*., 1996; Sentenac *et al*., 1992) and *HKT1* from *Triticum aestivum* (Schachtman and Schroeder, 1994), have beenisolated by complementing yeast mutants defective in K$^+$uptake. AKT1 is a K$^+$ channel involved in high-affinityK$^+$ uptake (Hirsch *et al*., 1998) and HKT1 a Na$^+$-coupled K$^+$ transporter (Rubio *etal*., 1995) whose contribution to high-affinity K$^+$ uptake in plants remains unclear (Rubio *et al*., 1996;Walker *et al*., 1996). As an alternative to the procedureof complementing yeast mutants defective in K$^+$ uptake, we tried to identify plant transporters homologous to fungalK$^+$ transporters, because fungal K$^+$transport exhibits kinetic (Ramos *et al*., 1985) andmechanistic (Maathuis and Sanders, 1994; Rodriguez-Navarro *et al*.,1986) characteristics similar to those found for K$^+$ transport in plant roots. Using a method based on the polymerase chainreaction coupled with reverse transcription (RT-PCR), we identified inbarley, Arabidopsis and rice, genes that encode proteins homologous to the*Schwanniomyces occidentalis* HAK1 (Ba–uelos *et al*., 1995) and the*Escherichia coli* Kup (Schleyer and Bakker, 1993). A cDNAfrom barley, *HvHAK1* (Santa-Mar'a *etal*., 1997), and a cDNA from Arabidopsis, AtHAK1, conferred high-affinityK$^+$ uptake to a K$^+$-uptake deficientyeast mutant. HvHAK1, exhibited the hallmark characteristics of thehigh-affinity K$^+$ uptake described in barley roots.

Results

By performing RT-PCR with RNA obtained fromroots of barley seedlings grown in the absence of K$^+$,we isolated a 2.7-kb cDNA, *HvHAK1*,whose translated sequence showed high degree of homology to the *S. occidentalis* HAK1 (Ba–uelos *et al*., 1995)and the *E. coli* Kup (Schleyer and Bakker, 1993). The*HvHAK1* cDNA was expressed in a yeast mutant defective in K$^+$ uptake. Growth experiments revealed that HvHAK1 conferred on the yeastmutant the capacity to grow as fast as the wild-type strain in mineralmedium containing micromolar concentrations of K^+.Rb^+ uptake experiments carried out with the yeast strain expressing HvHAK1showed that the barley transporter mediated high-affinity Rb^+ uptake, and that it was competitively inhibited by micromolarconcentrations of K^+ and millimolar concentrations of Na^+ (Table 1).

TABLE 1. K_m value forHvHAK1-mediated Rb^+ uptake and $K_i$values for inhibition of Rb^+ uptake by K^+ and Na^+

Rb^+ K_m (μM)	K^+ K_i (μM)	$Na^+$$K_i$ (mM)
18	27	15

To define the conditions and organs inwhich *HvHAK1* is expressed in barley plants, we performedgel blot hybridizations with RNA from shoots and roots of plants grown inthe presence or absence of K^+. *HvHAK1* transcript was detectedexclusively in roots, and that its expression was enhanced at leastfivefold in K^+-starved plants relative to plants grownin medium with a 1 mM K^+ concentration. Gel blot hybridization of genomic barley DNA revealedthat several restriction fragments hybridized with *HvHAK1*, indicating that *HvHAK1* belongs to a large genefamily.

In addition to the *HvHAK1* cDNA described above, a second 2.8-kb cDNA, *HvHAK2*, and a cDNA fragment, *HvHAK3*, wereisolated from root RNA, whose translated sequence showed homology to the HAK1-Kup type oftransporters. Unlike HvHAK1, HvHAK2 did not complement the yeast mutantdefective in K^+ uptake although preliminary results suggest that HvHAK2 could mediatelow-afinity K^+ uptake. *HvHAK2* transcript could not be detected in total RNA isolated from barley plants.DNA gel blot hybridization showed that only one restriction fragmenthybridizes with *HvHAK2* in barley DNA, indicating that this gene is present in one single copy inthe barley genome. For HvHAK3, a full-length cDNA could not be obtained.

Following a similar approach on RNA from K^+-starved Arabidopsis roots, we isolated several cDNA fragments whosetranslated sequences showed homology to the HAK-Kup type of K^+ transporters. Sequence analyses demonstrated that we had isolated cDNAsfragments that corresponded to the previously described AtKT1/AtKup1 andAtKT2 (Fu and Luan, 1998; Kim *et al.*, 1998; Quintero and Blatt, 1997) and to two other Arabidopsis genes thathad not been described yet. The translated sequences of these Arabidopsis cDNA fragments showedhigh similarity to the barley HvHAK2. We also isolated two other cDNAfragments whose translated sequence showed high similarity to HvHAK1 andHvHAK3.

Full-length cDNAs were obtained for the Arabidopsis gene, AtHAK1, thatshowed the highest similarity to HvHAK1, and for AtKT2. The AtHAK1transporter expressed in yeast did not confer growth capacity to the yeastmutant, although Rb^+ uptake experiments performed on this strain showed that the Arabidopsistransporter promoted high-affinity Rb^+ uptake (Fig. 1). However, the low rates of transport mediated by AtHAK1preclude the growth of the yeast mutant at micromolar concentrations ofK^+. AtKT2 expressed in yeast did not confer growthcapacity nor high-affinity Rb^+ uptake to the yeast mutant.

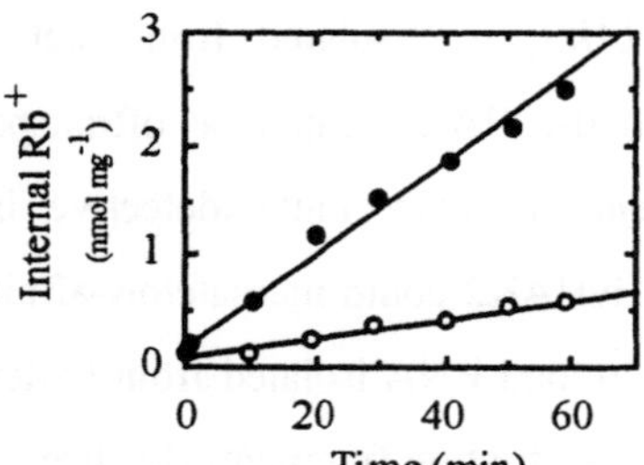

Figure 1. Rb^+accumulation in the yeast mutant defective in K^+ uptake(open symbols) and the same strain expressing AtHAK1 (filled symbols) froma 20 μM Rb^+ solution.

Inaddition to the barley and Arabidopsis genes described above, several ricecDNAs have been identified in the gene bank whose translated sequences showsimilarity to the HAK1-Kup transporters. We have expressed one of thesecDNAs, *OsHAK2*, in the yeast mutant defective in K^+ uptake, obtainingneither yeast growth at low K^+ concentrations norRb^+ uptake.

Discussion

We have identified several genes in barley,Arabidopsis and rice that belong to the HAK1-Kup type of K^+ transporters. HvHAK1 expressed in yeast shows high-affinityK^+ uptake with high rates of transport, allowing growth of the yeast mutantdefective in K^+ uptake in media with low K^+ concentrations. The kinetic characteristics and pattern of expression ofHvHAK1 suggest that this transporter could be a major contributor to thehigh-affinity K^+ uptake described in barley roots(Santa-Mar'a *et al.*, 1997). HvHAK2 is not expressed in yeast and its function could notbe studied.

In Arabidopsis, we have carried out anextensive search for HAK1-Kup type of transporters following an approachbased on RT-PCR performed on root RNA. As a result, we have identifiednovel genes that encode putative K^+ transporters homologous to the HAK1-Kup type ofK^+ transporters and to genes that either had beendescribed previously or corresponded to data bases entries. The ArabidopsisAtHAK1 mediates high-affinity K^+ uptake in yeast, although at very low rates oftransport. *AtHAK1* is probably exclusively expressed inArabidopsis roots, since we were unable to amplify by RT-PCR *AtHAK1* cDNA fragments from Arabidopsis shoot RNA. This results suggest thatAtHAK1 is the ortologue of HvHAK1. Recently, the Arabidopsis K^+-channel AKT1 has been demonstrated to play an important role inK^+ absorption from micromolar solutions (Hirsch *et al.*,1998). Therefore, the role of AtHAK1 and other K^+transporters belonging to the HAK1-Kup type of transporters in Arabidopsisremains to be determined.

The rice cDNA *OsHAK2*could not be expressed in yeast thus its function could not be studied.

The results described abovedemonstrate that the HAK1-Kup type of transporters are widely distributedamong plants. The number of putative HAK1- Kup type of K^+ transporters identified in plants is increasing continuously, suggestingthat this type of transporter fulfils important functions related toK^+ absorption and K^+ movements within the plant. The molecular characterization of this type oftransporters and the elucidation of their contribution to K^+movements in the plant will clarify our understanding ofK^+ nutrition in plants.

References

Ba–uelos, M. A., Klein, R. D., Alexander, S. J. and Rodriguez-Navarro, A.(1995) A potassium transporter of the yeast Schwanniomyces occidentalishomologous to the Kup system of Escherichia coli has a high concentrativecapacity. *EMBO J,* 14, 3021-3027.

Epstein, E., Rains, D. W. andElzam, O. E. (1963) Resolution of dual mechanisms of potassium absorptionby barley roots. *Proc. Natl. Acad. Sci. USA,* 49,684-692.

Fu, H.-H. and Luan, S. (1998) AtKUP1: A dual-affinity K^+ transporter from Arabidopsis. *Plant Cell,* 10,63-73.

Hirsch, R. E., Lewis, B. D., Spalding, E. P. and Sussman, M. R. (1998) Arole for the AKT1 potassium channel in plant nutrition. *Science,*280, 918-921.

Kim, E. J., Kwak, J. M., Uozumi, N. and Schroeder, J. I. (1998) *AtKUP1*: An Arabidopsis gene encoding high-affinity potassiumtransport activity. *Plan Cell,* 10, 51-62.

Lagarde, D., Basset, M.,Lepetit, M., Conejero, G., Gaymard, F., Astruc, S. and Grignon, C. (1996)Tissue-specific expression of *Arabidopsis AKT1* gene is consistent with a role in K^+ nutrition.*Plan J,* 9, 195-203.

Maathuis, F. J. M. and Sanders,D. (1994) Mechanism of high-affinity potassium uptake in roots of

Arabidopsis thaliana. Proc. Natl. Acad. Sci. USA, 91, 9272-9276.

Quintero, F. J. and Blatt, M. R. (1997) A new family of K transporters fromArabidopsis that are conserved across phyla. *FEBS Lett,* 415, 206-211.

Ramos, J., Contreras, P. and Rodriguez-Navarro, A. (1985) A potassiumtransport mutant of *Saccharomyces cerevisiae. Arch. Microbiol.,* 143, 88-93.

Rodriguez-Navarro, A., Blatt, M. R. and Slayman, C. L. (1986) Apotassium-proton symport in *Neurospora crassa. J. Gen. Physiol.,* 87, 649-674.

Rubio, F., Gassmann, W. and Schroeder, J. I. (1996) High-affinity potassiumuptake in plants. *Science,* 273, 977-979.

Rubio, F., Gassmann, W. and Schroeder, J. I. (1995) Na^+-driven K^+ uptake by the Plant K^+Transporter HKT1 and Mutations Conferring Salt Tolerance. *Science,* 270, 1660-1663.

Santa-Mar'a, G., Rubio, F.,Dubcovsky, J. and Rodriguez-Navarro, A. (1997) The Hak1 Gene of Barley is aMember of a Large Gene Family and Encodes a High-Affinity PotassiumTransporter. *The Plant Cell,* 9, 2281-2289.

Schachtman, D. P. and Schroeder,J. I. (1994) Structure and transport mechanism of a high-affinity potassiumuptake transporter from higher plants. *Nature,* 370,655-658.

Schleyer, M. and Bakker, E. P.(1993) Nucleotide sequence and 3«-end deletion studies indicate that theK^+ uptake protein Kup from *Escherichiacoli* is composed of a hydrophobic core linked to a large and partiallyessential hydrophilic C-terminus. *J. Bacteriol,* 175,6925-6931.

Sentenac, H., Bonneaud, N.,Minet, M., Lacroute, F., Salmon, J.-M., Gaymard, F. and Grignon, C. (1992)Cloning and expression in yeast of a plant potassium ion transport system.*Science,* 256, 663-665.

Walker, N. A., Sanders, D. andMaathuis, F. J. M. (1996) High-affinity potassium uptake in plants.*Science,* 273, 977-978.

REGULATION OF EXPRESSION OF GENES ENCODING PHOSPHATE TRANSPORTERS IN BARLEY ROOTS

SMITH, FRANK W., CYBINSKI, DAISY H., AND RAE, ANNE L.
CSIRO Tropical Agriculture,
Cunningham Laboratory,
306 Carmody Road, St Lucia Qld 4067. Australia

1. Abstract

A family of genes that encode phosphate transporters has been isolated from the barley genome. Three of these genes, *HVPT1*, *HVPT2* and *HVPT3* are expressed in the roots. The polypeptides encoded by *HVPT1*, *HVPT2* and *HVPT3* are similar in structure and show significant homology to phosphate transporters that have been isolated from dicotyledonous species. There is a high level of homology between *HVPT1* and *HVPT2* and these two genes are closely linked, lying on the same 20kB fragment of the genome. Although homologous in conserved regions and having a similar topology to HVPT1 and HVPT2, the polypeptide encoded by *HVPT3* differs significantly from HVPT1 and HVPT2.

The expression of these root phosphate transporter genes is regulated by the phosphate status of the plant. Using probes that are specific for *HVPT1*, *HVPT2* and *HVPT3* the effects of phosphorus deprivation and resupply on regulation of the expression of these genes in barley root and shoot tissue has been studied. These data, together with the effects of some other nutrient deficiencies on expression of these genes will be presented.

2. Introduction

The movement of phosphate ions across plant membranes is mediated by specific transporter proteins. Phosphate taken up from the soil solution and inorganic phosphate redistributed between various plant tissues passes through these transporter proteins embedded in the membranes. They are therefore key molecules, the number and activity of which controls the rates of movement of inorganic phosphate into plants and within plants. A number of the genes that encode these phosphate transporter proteins have recently been isolated from plants (Muchhal, Pardo and Ragothama, 1996; Leggewie, Willmitzer and Reismeier, 1997; Smith *et al.*, 1997b; Mitsukawa, Oumara and Shibata, 1997; Liu *et al.*,1998a, 1998b). This has permitted the amino acid sequence of the

G. Gissel-Nielsen and A. Jensen (eds.), Plant Nutrition – Molecular Biology and Genetics, 145–150.

transporters to be derived thus providing physiologists with new insights into the molecular mechanisms and regulation of phosphate transport in plants.

Following the isolation of genes encoding root phosphate transporters from the model plant *Arabidopsis* (Smith *et al.*, 1997b), our laboratory has turned its attention to the major cereal crops. A number of genes that encode phosphate transporters have been isolated from barley. Three of these, *HVPT1*, *HVPT2* and *HVPT3*, are expressed in roots. This paper reports on the effects of various nutrient stresses on the regulation of these genes and raises some issues regarding the function of transporter proteins involved in high-affinity phosphate transport in plant roots.

3. Materials and Methods

3.1 GROWTH OF PLANTS

Plants of *Hordeum vulgare* cv. Corvette were grown in 5 L containers in a nutrient solution containing 500 μM $Ca(NO_3)_2$, 250 μM $CaCl_2$, 250 μM KNO_3, 250 μM $MgSO_4$, 20 μM KH_2PO_4, 50 μM NaCl, 9 μM Na-Fe-EDTA, 45 μM H_3BO_3, 4.5 μM $MnCl_2$, 315 nM $CuCl_2$, 750 nM $ZnCl_2$ and 15 nM $(NH_4)_6Mo_6O_{24}$. Plants were grown at 28°C with a 16 h day length and harvested when 21 days old. Nutrient limitations were induced by transferring plants to modified solutions for 72 hours prior to harvest. For phosphate deprivation, KH_2PO_4 was omitted. For sulfate deprivation, $MgSO_4$ was replaced by $MgCl_2$. For nitrate deprivation, $Ca(NO_3)_2$ was replaced by $CaCl_2$ and KNO_3 by KCl. For potassium deprivation, KNO_3 was replaced by $NaNO_3$ and KH_2PO_4 by NaH_2PO_4.

3.2 EXTRACTION OF TOTAL RNA

Barley roots and shoots were ground in liquid nitrogen and RNA was extracted with guanidinium isothiocyanate and purified by ultracentrifugation through a CsCl cushion (Chirgwin et al., 1979).

3.4 NORTHERN HYBRIDISATIONS

Approximately 10 μg of total RNA was denatured and separated by the method of Smith *et al.* (1997a). RNA was transferred to Hybond-N$^+$ membrane (Amersham) by capillary blotting with 10x SSC. Membranes were hybridised overnight at 65°C with ^{32}P labelled probes prepared from 3' untranslated regions of *HVPT1*, *HVPT2* and *HVPT3*. Membranes were washed at 65°C with 2x SSC containing 0.5% SDS, 1x SSC containing 0.1% SDS and 0.1x SSC containing 0.1% SDS. As a control for equal loading, membranes were subsequently hybridised to a probe prepared from plasmid pTA71 which encodes wheat ribosomal RNA (Gerlach and Bedbrook, 1979). Membranes were exposed to phosphor screens for 1 to 7 days and detected with a Cyclone Phosphor-Imager (Packard).

4. Results and Discussion

The *HVPT*1, *HVPT*2 and *HVPT*3 genes were isolated from a genomic library of barley (Cybinski, Rae and Smith, personal communication). Corresponding cDNAs have been isolated from a cDNA library prepared from RNA extracted from barley roots that had been deprived of phosphate for 72 hours. *HVPT*1 and *HVPT*2 are clustered on the same 20kb genomic fragment and encode polypeptides that are identical except for the five carboxy-terminal amino acids. They therefore bear a striking similarity to the *APT*1 and *APT*2 genes isolated from *Arabidopsis* (Smith *et al.* 1997b). *HVPT*3 encodes a polypeptide that is 84% similar and 71% identical to HVPT1.

The HVPT1, HVPT2 and HVPT3 proteins have significant homology with the other plant phosphate transporters that have been reported. These key proteins have been highly conserved during evolution. They are comprised of 520 to 525 amino acids and contain twelve membrane spanning domains arranged in a 6+6 configuration with a long central cytoplasmic loop (Figure 1). This topology is characteristic of the major facilitator superfamily of proteins (Marger and Saier, 1993) that includes many solute transporters. In common with a number of the phosphate transporters that have been described HVPT1, HVPT2 and HVPT3 contain potential sites for phosphorylation and N-glycosylation (Figure 1). By analogy with other sequences, it is likely that they function as $H^+/H_2PO_4^-$ cotransporters.

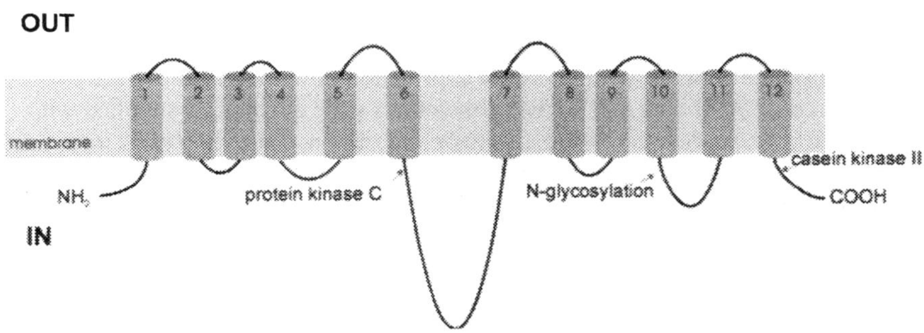

Figure 1. Topology of barley root phosphate transporters HVPT1, HVPT2 and HVPT3.

Labeled probes, specific to the 3' ends of each of the *HVPT*1, *HVPT*2 and *HVPT*3 transcripts, were used to differentiate between the three gene transcripts in northern blot studies. Within the detection limit of northern blot analyses, none of these genes appeared to have significant levels of expression in vegetative barley shoots (Figure 2). The steady-state levels of expression of *HVPT*3 in roots appeared to be less than those of either *HVPT*1 or *HVPT*2. The level of expression of all three genes was enhanced by deprivation of an external phosphate supply for 72 hours. This has also been observed with a number of the other genes encoding root phosphate transporters that have been isolated (Leggewie, Willmitzer and Reismeier, 1997; Smith *et al.*, 1997b; Liu *et al.*,1998a, 1998b). It is consistent with physiological measurements indicating that the capacity for phosphate uptake is considerably enhanced by phosphate starvation (Clarkson and Scattergood, 1982; Cogliatti and Clarkson, 1983).

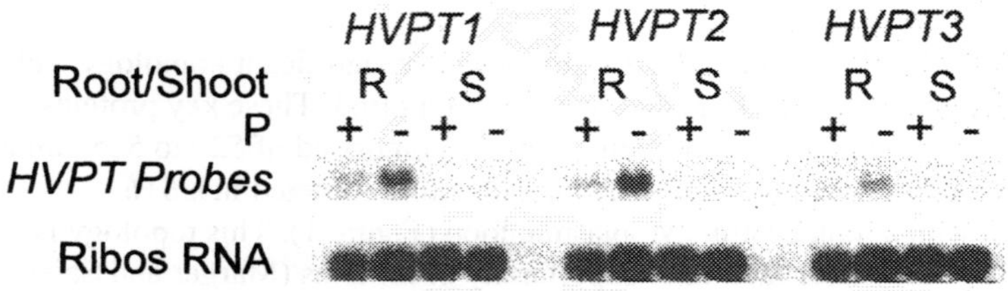

Figure 2. The effects of depriving barley plants of an external supply of phosphate for 72 hours on the steady state levels of transcripts corresponding to HVPT1, HVPT2 and HVPT3 in the root and shoot tissues.

Regulation of the expression of *HVPT*1 and *HVPT*2 seemed to respond in a similar manner to S, N, and K deprivation (Figure 3). Both sulfate and nitrogen deprivation for 72 hours restricted the enhanced expression of *HVPT*1 and *HVPT*2 due to phosphate deprivation. Potassium deprivation however enhanced the expression of *HVPT*1 and *HVPT*2 when the plants were adequately supplied with phosphate, but not when they were also deprived of phosphate. If *HVPT*1 and *HVPT*2 encode high affinity phosphate transporters, the observed effects of dual S/P and N/P deprivation on steady state levels of these transcripts may represent a mechanism which enables plants to maintain chemostasis during multiple deficiencies. Down-regulating these transporters would prevent accumulation of high levels of phosphate when nitrogen or sulfate was also limiting. Such an explanation cannot account for the observed effects of potassium deprivation however, where some more specific influence of potassium on phosphate transport appears likely. Interestingly, expression of *HVPT*3 responds similarly to potassium deprivation.

The enhancement of steady-state levels of *HVPT*3 due to phosphate deprivation is not reduced by sulfate or nitrogen deprivation – in fact it appears to be enhanced by dual S/P deprivation (Figure 3). Such differences from *HVPT*1 and *HVPT*2 in response to nutrient stresses suggests that *HVPT*3 has a different function in roots than that of *HVPT*1 and *HVPT*2.

Uptake of phosphate by roots from the low phosphate concentrations that commonly occur in soil solutions require high affinity phosphate transporters. It is of interest therefore to determine the kinetics of phosphate transport by HVPT1, HVPT2 and HVPT3. Complementation of a yeast mutant provided a valuable means of doing this for plant sulfate transporters (Smith *et al.* 1995, 1997a). However high affinity kinetics are not displayed when genes encoding root phosphate transporters are heterologously expressed in other organisms. Although it has been demonstrated that cDNAs derived from plants can complement the *pho84* mutant of yeast, which lacks a functional yeast high affinity phosphate transporter, kinetics derived from this system yield K_m values far higher than expected of a high affinity transporter (Leggewie, Willmitzer and Reismeier, 1997; Liu *et al.*,1998a). A similar gene expressed in cultured tobacco cells displayed a K_m of 3.1 μM (Mitsukawa *et al.*, 1997).

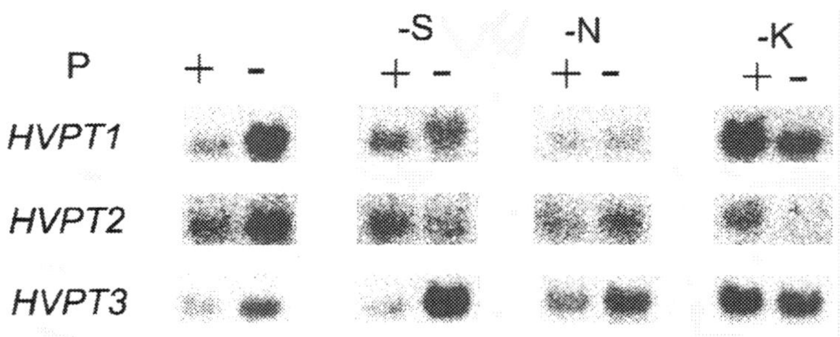

Figure 3. The effects of 72 hours deprivation of S, N, or K on the steady state levels of transcripts corresponding to HVPT1, HVPT2 and HVPT3 in barley roots. Barley plants were either maintained on an external P supply (+P) or deprived af an external P supply for 72 hours (-P). Ribosomal RNA controls (not shown) indicated that all lanes were loaded with a similar amount of barley root RNA.

The reasons for poor expression of root phosphate transporters in heterologous expression systems such as yeast and bacullo viruses need to be explored. A number of genes have been identified that encode products that interact with the PHO84 yeast high affinity phosphate transporter (Bun-ya, Harashima and Oshima, 1992; Bun-ya *et al.*, 1996). This raises the important question of whether there are similar genes in plants encoding products that interact with root high affinity phosphate transporters. There are also potential phosphorylation and glycosylation sites that are conserved on many phosphate transporter sequences (Figure 1) which suggest there may be regulation of the transporter by activation. Study of these points may identify additional molecular components that are required for a fully functional high affinity root phosphate transporter.

5. Acknowledgement

We are grateful to Mr Peter Vanden Berg for the technical assistance he provided to the research reported in this paper.

6. References

Bun-ya, M., Harashima, S. and Oshima, Y. (1992) Putative GTP-binding protein, Grtl, associated with the function of the Pho84 inorganic phosphate transporter in *Saccharomyces cerevisiae. Mol. Cell. Biol.*. **12**, 2958-2966.

Bun-ya, M., Shikata, K., Nakade, S., Yompakdee, C., Harashima, S. and Oshima, Y. (1996) Two new genes, PHO86 and PHO87, involved in inorganic phosphate uptake in *Saccharomyces cerevisiae. Curr. Genet.* **29**, 344-351.

Chirgwin, J.M., Pryzybyla, A.E., MacDonald, R.J. and Rutter, W.J. (1979) Isolation of biologically active ribonucleic acid from sources enriched in ribonuclease. *Biochemistry 18,* 5294-5299.

Clarkson, D.T, and Scattergood, C.B. (1982) Growth and phosphate transport in barley and tomato plants during development of and recovery from phosphate stress. *J. Exp. Bot.* **33**, 865-875.

Cogliatti, D.H. and Clarkson, D.T (1983) Physiological changes in phosphate uptake by potato plants during development of and recovery from phosphate deficiency. *Physiol. Plant.* **58**, 287-294.

Garlach, W.L. and Bedbrook, J.R. (1979) Cloning and characterisation of ribosomal RNA genes from wheat and barley. *Nuc. Acid Research* **7**, 1869-1885.

Leggewie, G., Willmitzer, L. and Reismeier, J.W. (1997) Two cDNAs from potato are able to complement a phosphate uptake-deficient yeast mutant: Identification of phosphate transporters from higher plants. *Plant Cell* **9**, 381-392.

Liu, C., Muchhal, U.S., Uthappa, M., Kononowicz, A.K. and Ragothama, K.G. (1998a) Tomato phosphate transporter genes are differentially regulated in plant tissues by phosphorus. *Plant Physiol.* **116**, 91-99.

Liu, H., Trieu, A.T., Blaycock, L.A., and Harrison, M.J. (1998b) Cloning and characterisation of two phosphate transporters from *Medicago truncatula* roots: Regulation in response to phosphate and to colonization by arbuscular mycorrhizal (AM) fungi. *Molecular Plant-Microbe Interactions* **11**, 14-22.

Marger, N.D. and Saier, M.H. (1993) A major superfamily of transmembrane facilitators that catalyse uniport, symport and antiport. *Trends Biochem. Sci.* **18,** 13-20.

Mitsukawa, N., Okumura, S. and Shibata, D (1997) High-affinity phosphate transporter genes of *Arabidopsis thaliana.* In T Ando *et al* (eds.) *Plant nutrition – for sustainable food production and environment,*. Kluwer Academic Publishers, pp. 187-190.

Mitsukawa, N., Okumura, S., Shirano, Y. Sato, S., Kato, T., Harashima, S., and Shibata, D (1997). Overexpression of an *Arabidopsis thaliana* high-affinity phosphate transporter gene in tobacco cultured cells enhanced cell growth under phosphate-limited conditions. *Proc. Natl. Acad. Sci. USA* **94**, 7098-7102.

Muchhal, U.S., Pardo, J.M., and Ragothama, K.G. (1996) Phosphate transporters from the higher plant *Arabidopsis thaliana. Proc. Natl. Acad. Sci. USA* **93**, 10519-10523.

Smith F.W., Ealing, P.M., Hawkesford, M.J., and Clarkson, D.T. (1995) Plant members of a family of sulfate transporters reveal functional subtypes. *Proc. Natl. Acad. Sci. USA* **92**, 9373-9377.

Smith F.W., Hawkesford, M.J., Ealing, P.M., Clarkson, D.T., Vanden Berg, P.J., Belcher, A.R. and Warrilow, A.G.S. (1997a) Regulation of expression of a cDNA from barley roots encoding a high affinity sulphate transporter. *Plant J.* **12** ,83-92.

Smith F.W., Ealing, P.M., Dong, B., and Delhaize, E. (1997b) The cloning of two *Arabidopsis* genes belonging to a phosphate transporter family. *Plant J.* **11,** 83-92.

PHYSIOLOGICAL AND GENETIC ASPECTS OF MICRONUTRIENT UPTAKE BY HIGHER PLANTS

XIAOE YANG[1] AND VOLKER RÖMHELD [2]
[1] Institute of Plant Nutrition and Fertilizers, Zhejiang University, Hangzhou 310029, China
[2]Institute of Plant Nutrition, Hohenheim University, 70593 Stutgart , Germany

Abstract

Currently, eight trace elements have been considered to be essential for higher plants, Fe, Zn, Mn, Cu, Ni, B, Mo, and Cl, which are termed micronutrients. Genotypes of plants vary widely in their adaptation to micronutrient deficiency stresses in soils. The adaptation of plants to micronutrient deficiency stresses appears to be associated, in most cases, with more efficiency in microniutrient acquisition from soils. The efficiency in acquisition of micronutrients from soils by plants is genetically controlled. This review briefly summarizes the current knowledge of physiological mechanisms of major micronutrient uptake by higher plants, genotypic differences in micronutrient acquisition from soils, and what is known of the genetic mechanisms of micronutrient efficiency in plants.

Key Words: Micronutrient efficiency, genotypic difference, genetic background, physiological bases

1. Introduction

There are 17 elements in the Periodic Table known to be required by all higher plants (Asher, 1991; Marschner 1995), of these essential elements, nine are macronutrients (C, H, O, N, K, Ca, Mg, P, and S) that normally have concentrations in plant tissues higher than 0.1%(dry weight). Eight micronutrients defined as essential elements that are required in plant tissues at concentrations of less than 100 ugg^{-1} dry weight) are now recognized as essential for all higher plants, B, Cl, Cu, Fe, Mn, Mo, Ni, Zn.

Interest in plant micronutrients has increased within the last decade because micronutrient deficiency stresses in soils are widespread and become one of the major limiting factors for crop production. The recent research demonstrates that micronutrients also play important roles in plant disease resistance and in root-stress resistance and moreover, plant foods are significant sources of these essential elements for animals and humans (Graham and Webb, 1991; Miller et al., 1991; Nielsen, 1992; van Campen, 1991). Micronutrient uptake process is a key factor controlling plant adaptation to both micronutrient deficiency and toxicity stresses. A great progress has been made in our understanding of Fe uptake mechanisms by

G. Gissel-Nielsen and A. Jensen (eds.), Plant Nutrition – Molecular Biology and Genetics, 151–186.

plants. However, considerably fewer investigations have been conducted into the uptake mechanisms of other micronutrients, with Mo in particular.

This review paper focuses on advancements in our understanding of physiological and genetical aspects of micronutrient uptake by higher plants. For the genetic aspects of micronutrient uptake, this review emphases micronutrient efficiency, i.e. the strategies for plant adaptation to micronutrient deficiency in soils. For basic mechanisms of micronutrient uptake, the readers can refer to reports elsewhere (Kochian, 1991; Mortvedt et al., 1991; Welch, 1995). Other reviews and books are available for those who are interested in more complete discussion of micronutrient nutrition in plants (Asher, 1991; Hewitt, 1983; Kabata-Pendias and Pendias, 1992; Manthey et al., 1994; Marschner, 1995; Mortvedt et al., 1991; Robb and Pierpoint, 1983; Shkolnik, 1984)

2. Cation Micronutrient Uptake

The differences between cation macronutrients and micronutrients must be considered before developing any hypothetical molecular models to study micronutrient uptake by higher plants. The major chemical characteristics cation micronutrients (i.e. Fe^{2+}, Mn^{2+}, Cu^{2+}, Zn^{2+}, and Ni^{2+}) compared to cation macronutrients (such as K^+) include: 1) more tendency to form complexes with organic ligands ; 2) more sensitivity to environmental redox change; 3) more sensitivity to pH change; 4) smaller chemical activities in soil solution. The negatively charged ion exchange sites and reactive ligand groups (e.g. carbonyl, hydroxyl, thiol, carboxylic, amide-N, amino-N, and phenolic) in cell wall constituents of root-cell apoplasmic spaces bind some micronutrient cations (e.g. Cu^{2+}) much more tightly than macronutrient cations such as K^+. If the apoplasmic pools of micronutrient ions from roots are not removed during kinetic studies of micronutrient uptake, values for influx rates of micronutrient cations by roots can be greatly exaggerated (Kochian, 1991).

In the last two decades, great progress has been made in understanding of Fe uptake by plants, and much more is known about microorganism Fe transport mechanisms than Fe absorption by higher plants. If aerobic Fe transport processes in microorganisms were conserved during the evolution of higher plants we could apply these processes to higher plant systems. The progress made in understanding of other cation micronutrient uptake by plants mainly based on our knowledge on Fe uptake. The mechanisms of Fe uptake have been used as models for other cation micronutrients.

2.1. IRON

Iron deficiency is a widespread problem in the world because of low Fe availability in aerobic environment and at biological pH, especially in calcareous soils which cover approximately one-third of the earth surface. To overcome iron deficiency, plants have evolved various adaptive mechanisms to acquire iron. As it is known that cereals developed a different approach to adapt Fe deficiency from other plant species, and several research groups have developed various Fe uptake models for dicot and nongraminaceous monocots and for graminaceous monocots (Bienfait, 1988, 1989; Bienfait and Luttge, 1988; Kochian. 1991; Römheld, 1987; 1991; R ömheld and Marschner,1990). Marschner and Römheld have termed these two Fe uptake models Strategy I (for dicots and nongrass monocots) and Strategy II (for grass). Bienfait (1989) has added the third model, termed strategy III, for uptake of microbial siderophores by plant roots.

2.1.1. Iron Uptake by Strategy I Plant Species

For Strategy I species, iron efficiency is a function of a number of induced responses in the rhizosphere by plant roots (Chaney et al., 1992; Römheld, 1987; Welkie and Miller, 1993; Korcak, 1987). These physiological responses primarily include:

1) Increased rates of electron transfer reactions at root surfaces, capable of driving Fe^{3+} reduction reactions;
2) Increased rates of rhizosphere acidification;
3) Increased release of phenolic compounds, i.e. caffeic and chlorogenic acid;
4) The accumulation of citric acid in plant roots;

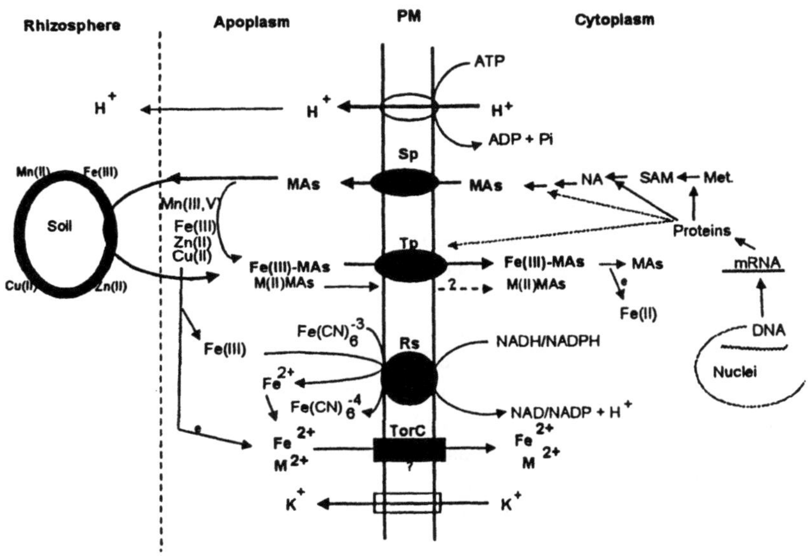

Figure 1. Iron uptake model for dicotyledonous and non-grass monocotyledonous plants (Strategy I)- from genetic to physiology. Symbols refer to: Ri-inducible reductase; Rc-constitutive reductase; Rs – standard reductase; TorC- transporter or channel.

Iron uptake model for dicots and nongraminaceous monocots is shown in Figure 1. For the Strategy I plant species, Fe^{2+} is the only soluble ionic Fe species in nutrient medium significantly absorbed by roots (Bienfait, 1989; Chaney, et al., 1972; Marschner et al., 1986). The reductants from roots account for little of the total Fe(III) reducing capacity of Fe deficiency-stressed roots unless excessive rhizosphere acidification occurs, causing root cell injury and increasing efflux of reductants (Bienfait, 1988b; Römheld and Marschner, 1986). Consequently, the focus for the understanding of Fe uptake by Strategy I species is the types, activities, and regulatory mechanisms controlling the membrane bound Fe(III) reductase that operates at the exterior surface of root-cell plasma membrane (Bienfait and Luttge 1988b; Crane and Barr, 1989; Holden et al., 1994 , Kochian and Lucas, 1991).

Three types of membrane bound Fe(III) reductase existing in root cell of Strategy I plants have been suggested (Fig. 1): a standard reductase (Rs), a constitutive Fe(III)-chelate reductase(Rc), and an inducible Fe(III)-chelate reductase (Ri) (Bienfait, 1988, 1989; Marchner et al., 1986; Römheld et al., 1984). The standard reductase occurs in plasma membrane of all higher plant species, and is generally not considered to be capable of reducing Fe(III) in synthetic Fe(III)-chelates in situ, but can reduce Fe(III) in other electron acceptor molecules such as ferri cyanide; while the constitutive Fe(III)-chelate reductase and the inducible Fe(III)-chelate reductase systems can reduce Fe(III) in various Fe(III)-chelates from different origins (Bienfait, 1988, 1989). Bruggeman et al. (1990) showed that the inducible reductase was the results of increased activity of the constitutive reductase under Fe-deficiency stress conditions. For the ferric reductase in tomato root plasma membranes, Holden et al., (1991, 1992) identified more than one electron transport systems in the root plasma membrane, one of which uses ferric chelates and ferri cyanide (constutive or inducible reductase?) as electron acceptors, while the other reduces only ferricyanide (standard reductase?). The electron donor for reduction of Fe(III)-chelates by the constitutive reductase system appears to be NAD^+/NADH couple rather than $NADP^+$/NADPH couple (Bruggeman et al., 1990; Buckhout et al., 1989; Holden et al., 1991; Schmidt et al., 1990;). The ability of the constitutive Fe (III)-chelate reductase to reduce Fe (III)-chelates is associated with redox stability of Fe(III) in Fe(III)-chelates and with redox potential in the reductase system. The roles of the reductases have been discussed in other review paper (Welch, 1995). After reduced from Fe (III), Fe^{2+} is mainly absorbed through ion channels in the membrane (Fig. 1). The existence of these channels has not yet been proven. More likely, the Fe^{2+} may be absorbed through the nonspecific divalent cation channel which is similar to those of Ca^{2+} channels identified in plant membranes of higher plants (Spalding et al., 1992; Tester, 1990). As Ca^{2+} channels can also transport other ions apart from Ca^{2+}, such as Mg^{2+}, Ba^{2+}, Rb^+, K^+, Na^+ (Tester, 1990), it is reasonable to suggest that Ca^{2+} channels may be membrane passages for Fe^{2+}, as well as for other divalent micronutrient cations (Welch, 1995)

2.1.2. Genetic Backgrounds for Iron Uptake by Strategy I Plant Species

Genetic study on Fe efficiency in dicots was dated as early as in 1940's. Table 1 lists some genetic backgrounds (or genes) related to Fe(III) reduction or ferric reductase of dicots plant species. Weiss (1943) found that Fe efficiency in soybean seemed to be controlled by a single gene. However, subsequent study on soybean showed that although one primary gene confers susceptibility to Fe deficiency, there are other genes which contribute to the trait (Fehr, 1984). There are two single-gene mutants related to Fe uptake in strategy I plant species, called the *chloronerva* mutant of tomato (Stephen and Grun, 1989) and the *brz* mutant of pea (Grusak et al., 1990a; Kneen et al., 1990; Welch and LaRue, 1990;). These mutants accumulate more iron than their wild-type parents. The *chloronerva* mutant is a nicotianamine auxotroph, which exhibits typical Fe-deficiency responses even with addition of Fe. As a result, it takes up more iron than the wild-type. Application of nicotianamine to roots or leaves of the *chloronerva* mutant led to recovery of the chlorosis (Scholz et al., 1985), probably due to the activation of an iron sensor or gene regulator by the "active" Fe form of Fe(II)-nicotianamine in this mutant (Pich et al., 1991). The *brz* mutant of pea can accumulate 50-fold more Fe in leaves than the parental plants (Kneen et al., 1990), which resulted from higher rates of ferric reduction, and from the release of more protons than the wild type. Another interesting mutant is the *fer* mutant of tomato, which is unable to respond to Fe deficiency (Brown et al., 1971). The wild type gene, FER, encodes the proteins of ferric reductases, and recently, two membrane proteins which are produced under Fe-deficiency conditions in the wild type have been identified (Bienfait, 1988). The mutants which are defective in ferric reductases signated as ys1 and ys3, have been identified in maize and used for iron uptake mechanisms (Brown and Jolley, 1989, Basso et al., 1994)

TABEL 1. Identified single-gene mutants, genes or clones in relation to ferric reduction in some Strategy I plant species

Species	Genes, clones, Single-gene mutants	Characteristics	Reference
Tomato	-	Ferric reductase, 35kDa; neither heam-containing, Nor glycosylated proteins	Holder et al., 1994
Tomato	Chloronerva	Nicotiamine auxotroph; express Fe deficiency regardless of Fe status in plant	Stephen and Grun, 1989; Pich et al., 1991
Tomato	Fer	Unable to response to Fe deficiency, lack of ferric reductase protein	Brown et al., 1971; Bienfait , 1988
Maize	Ys1, ys3	Defective in ferric reductase	Brown and Jolley, 1989; Basso et al., 1994
Pea	Brz	High rate of ferric reduction, excess Fe accumulator	Welch and LaRue, 1990; Crusak et al., 1990
Arabidopsis	•YY1, •YY2, •YY3	Totally or partially homologous to Yeast ferric reductase gene, FRE1	Yi et al., 1994
Arabidopsis	Froh C	Gene encodes deduced protein, falls into the same class as FRE1, FRP1 from yeast	Robinson et al., 1997

Some advances have been made in the understanding of ferric reductases at molecular level. The characterization of the plasma membrane enzymes responsible for the Fe(III) reduction catalysis in higher plants are reported (Holden et al., 1994, Yi et al., 1994). Holden et al

(1991, 1992) demonstrated that ferricyanide reductase activity in tomato has two different isoelectric points, only one of which was contiguous with the Fe reductase isoforms by means of preparative isoelectric focusing. Partial purification of Fe-chelate reductase and identification of a relevant polypeptide in tomato root plasma membrane indicated that the purified reductase has a 35-kDa polypeptide as a putative component of the reductase, and has neither heam-containing nor glycosylated components (Holden et al., 1994). However, the molecular nature of the electron transport function of the reductase is yet unknown. The tomato Fe-chelate reductase has an optimum pH of 6.0 and a low substrate specificity, but prefers organic acid-chelated Fe, such as citrate, to synthetics like EDTA (Holden et al., 1991). The partially purified yeast plasma membrane reductase utilized Fe-EDTA and ferrioxamine B equally well but did not use Fe-citrate (Lesuisse et al., 1990). A gene for yeast Fe-chelate reductase has been cloned (Dancis et al., 1990), which encodes for a 78-kDa polypeptide with some sequence similarity to cytochrome b 558, the component of a pagocyte membrane oxidoreductase (Dancis et al. 1992). Efforts have been made to screen for ferric reductase mutants in *Arabidopsis* and to determine the mode of inheritance as well as the dominant relationship of the mutant alleles (Yi et al., 1994). In yeast, it is clear that the yeast mutant genes, FRE-1 and FRP1 regulate both Fe(III) reduction and transport (Dancis et al., 1992, Roman et al., 1993). Yi et al., (1994) cloned *Arabidopsis* homologs using the yeast as a probe, and identified three different clones showing homology to FRE-1, termed λYY1, λYY2, and λYY3. The nucleotide sequence of the regions in λYY3 was found to be homologous to FRE-1, while for λYY1 and λYY2, partial sequence information is homologous to FRE-1. λYY3 carries an open reading frame, which encodes 286 amino acids. This induced amino acids sequence shares 17.5% identity and 49% similarity with the deduced FRE protein. The inserts carried by the three clones show approximately 40% identity to each other (Yi et al., 1994). Recently, Robinson et al. (1997) isolated a family of genes from *Arabidopsis thaliana*, designated *froh*, which encodes deduced proteins falling into the same class as FRE1, FRP1 from yeast, and gp91phox from human. The *froh C* transcripts accumulated in response to low Fe in both roots and leaves while the action transcripts (control) remained constant, and with other evidence showing that *Froh* C is more likely to be involved in iron-reduction both in roots and in leaves (Robinson et al., 1997). Zaharieva et al.(1997) studied the characterization of ferric chelate reductase in plasma membrane of cucumber roots. The Fe(III)-chelate reductase activity sharply decreased with time when pre-incubated with 0.03% Triton X-100 in the absence of substrates. Whereas, the inhibition of the activity was completely alleviated by adding NADH, FAD or both during pre-incubation with Triton X-100, only partly alleviated by Fe citrate, and not alleviated by FMN. The stimulation of Fe(III)-chelate reductase activity by FAD was concentration dependent with a maximum effect at 100 nM. These results indicated that a FAD containing protein may be important for the integrity of transplasma membrane reductase involved in Fe(III) reduction.

2.1.3. Iron uptake by Strategy II plant species

The Startegy II plants (graminaceous monocots) respond to iron deficiency by the production of phytosiderophores, instead of showing the responses what Stategy I plants have (Crowley et al., 1991, Marschner et al., 1989; Römheld, 1991; Römheld and Marschner, 1986). However, Strategy II plants possess plasma membrane-bound standard reductase that are capable of reducing electron donor molecules such as ferricyanide

(Bienfait and Luttge, 1988). The iron uptake model by Strategy II plants was shown in Figure 2. There are several processes involved in the regulation of Fe uptake by Strategy II plant species, including the biosynthesis of phytosiderophores in roots, secretion of phytosiderophores from root to rhizosphere and absorption of Fe(III)-phytosiderophores. Figure 2 depicted the relation of these three processes, too.

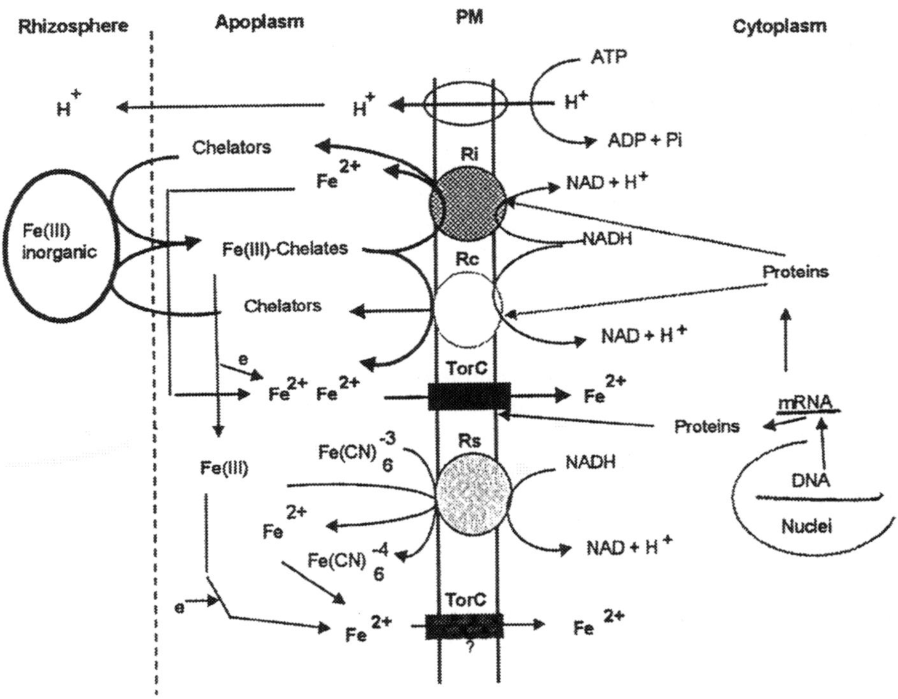

Figure 2. Iron uptake model for grasses (Strategy II) – from genetic to physiology. Symbols refer to: Tp-transporter protein; Rs – standard reductase; Sp –secretion passage or protein for MAs; TorC- transporter or channel; met – methionine; SAM- S-adenosyl-L-methionine; NA – nicotinamine; MAs – mugenic acids.

Great progress has been made in understanding the biosynthesis of phytosiderophores in recent years. Phytosiderophores named a series of mugeneic acids, including mugeneic acid (MA), avenic acid (AVA), 3-hydroxymugineic acid (HMA), 3-epihydroxymugineic acid (epiHMA); 2-deoxymugineic acid (DMA), and distichonic acid, and different grass species synthese and release different kinds of mugeneic acids (Nomoto et al., 1981, 1987). The methionine is shown to be the main precursor of mugeneic acid and the related phytosiderphores (Kawai et al., 1988; Mori and Nishizawa, 1987). The structure and biosynthesis pathway of MAs are understood (Kawait et al., 1988; Mori, 1994, 1997;

Shojima et al., 1989, 1990).

There are different regulation mechanisms of biosynthesis of MAS from those of the secretion of MAs. The secretion of MAS from cereals roots into rhizosphere is temperature and ATP dependent (Takagi, 1991), was suggested to be cotransported with K^+ through the operation of H^+, K^+-ATPase (Takagi, 1991). The secretion of MAs in barley under Fe deficiency conditions peaked after initiation of light for 1-2 hours (Takagi, 1991).

Central to the absorption of Fe by Strategy II plants is the operation of specific transport proteins in plasma membranes of roots for binding Fe(III)-phytosiderophores and transporting them across the plasmalemma into the cytosol. Marschner et al. (1989) reported that Fe(III)-3-HMA was transported mainly as dissociated Fe(III)-chelate across the root plasma membrane because the ^{14}C-HMA/^{59}Fe(III)-ratio in the root tissues after the absorption period remained near unity. When comparing the uptake of ^{59}Fe labelled D-enantiometric forms with that of the natural- occurring ^{59}Fe-labeled L-enantiomeric forms into the leaves of rice plants, Oida et al. (1989) found that L- enantiomeric forms significantly stimulated ^{59}Fe uptake into the leaves, whereas the D forms did not stimulate uptake obviously. There are other convincing evidences reported by Römheld and Marschner (1986), Mihashi and Mori (1989), Mahashi et al. (1991), that Fe(III)MAs was absorbed by plant roots through a highly specific-transporter-located in root cell membrane. Thus, all these evidences indicated that there exists specific Fe(III)-MAs binding sites in root cell plasma membranes. Further molecular research is needed to identify and characterize the Fe(III)-MAs transporter proteins in Startegy II plant roots.

The absorption of Fe(III)MAs by rice roots were found to be energy dependent (Takagi (1984). The activity of the Fe(III)MAs transporter can be induced during Fe deficiency, inhibited by an endogenous ATPase inhibitor DCCD (dicyclohexylcardodiimide), by a membrane proton conductor CCCP (carbonylcyanide m-chlorophenylhydrazone), and also by addition of the H^+ and K^+ ionophore (nigericin) (Mihashi and Mori, 1989). While addition of 10 mM K^+ enhanced Fe(III)MAs uptake, but this enhanced activity was completely canceled by addition of 5 microM valinomycin. These results implied that there existed close relationship between Fe(III)MAs uptake and the root cell plasma membrane proton motive force, which is operated by ATPase. The Fe(III)MAs uptake dependent on H^+ and K^+ gradients implied that K^+ might be co-transported with MA . Treatment of barley roots with 1 mM NEM (N-ethlmaleimide), a sulfhydryl binding reagent, almost completely inhibited the transporter of Fe(III)-MAs, implying an involvement of cysteine residue (i.e. sulfhydryl groups) at the active side of the Fe(III)-MAs transporter protein in plasma membrane.

The role of root apoplast in Fe (Zn) uptake has recently emphasized according to increasing evidence that apoplast Fe is important for Fe nutrition for barley and sorghum (V. Zhang et al., 1991; Wiren et al 1995). ^{59}Fe uptake rate from ^{59}Fe-rhizoferrin under -Fe preculture was 4-fold higher than that under + Fe preculture, implying that the loaded appoplast Fe enhanced exchange chelation (Yehuda et al., 1996). The concentration of apoplast Fe in roots of graminaceous plants is affected by the culture system. The root apoplast Fe grown in nutrient solution with 100 microM FeEDTA was 1300 mg Fe kg^{-1} D.W (v. Wiren et al., 1995), while the root apoplast Fe grown in soil (direct contact to soil) was as high as 11000 mg Fe/kg D.W (Mengel, 1994).

2.1.4 . Genetic Backgrounds of Iron Uptake for Strategy II Plant Species

In recent years, genetic and molecular studies to understand the regulation of Fe efficiency in grasses have been directed to MAs biosynthesis, and the transporter of Fe(III)-MAs. Table 2 lists the reported results of genetic backgrounds involved in Fe uptake by Strategy II plant species. The studies on chromosomal location of the genes which code the enzymes responsible for the pathway of MAs biosynthesis showed that the gene for the hydroxylase responsible for the conversion of DMA to MA is located on chromosome 4 of Betzes barley

TABLE 2. Identified genes, chromosomes and single-gene mutants in relation to biosynthesis of mugeneic acids or Fe(III)-MAs transporters in some Strategy II plant species.

Species	Genes, chromosomes, or single -gene mutants	Characteristics	Reference
Barley	Chromosome Add.4	Encoder of genes of MA Synthesis	Mori and Nishiza 1989
Wheat	R-5- L2.3	Fe efficiency ,Cu efficiency	Schlegel and Cakmak, 1997
Rye	R-5	A carrier of the genes for MA Synthetase and HMA Synthetase	Mori et al., 1990
Barley	SFD1	Fe deficiency specific clone, Encodes a novel protein likely Involved in uptake of Fe(III) or copper	Yamaguchi et al., 1997
Barley	Ids1	Fe deficiency specific clone, a plant MT gene, also responsive to excess Cu.	Okumura et al., 1992, Mori, 1997
Barley	Ids2	The genes expressed only in the -Fe roots, also under Mn and Zn, deficiency and high NaCl,	Okumura et al., 1994
Barley	Ids3	A clone specifically expressed in -Fe roots, a possible candidate for a MA synthetase gene.	Nakanishi et al., 1993; Mori., 1997
Barley	naat	Genes encode of nicotianamine aminotransferase, expressed under Fe deficiency.	Mori., 1997; Takahashi et al., 1997

(Mori and Nishizawa, 1989). Whereas, for Imperial rye, both genes for the hydroxylases responsible for the conversion from DMA to MA, and from MA to HMA are located on chromosome 5 (Mori et al., 1990). More recently, Schlegel and Cakmak (1997) demonstrated that the genes from the rye chromosome region 5RL2.3 might contribute to a higher Fe efficiency in wheat. There were other examples showing the genes located on nonhomologous chromosomes in barley and rye (Mori, 1997). With regards to the study of the genes that code for MAs biosynthesis, the enzymes of nicotianamine synthetase (NAS) and nicotianamine aminotransferase (NAAT) are already known to be induced under Fe deficiency, but SAM (S-Adenosyl-L-methionine) synthetase activity was not enhanced by Fe-deficiency (Mori, 1997). Nicotianamine aminotransferase and DMA-synthetase (3-keto acid reductase) are the most important enzymes and the gene code for the nicotianamine synthetase is known to exist both in dicots (Shojima et al., 1989) and in monocots (Nishizawa et al., 1989). The NAS has been purified, which has an optimum pH of 9.0 and molecular mass about 45 kD(Higuchi et al., 1994). Two isoforms of NAAT were found (Kanazawa et al., 1995), and the NAAT-I (molecular mass: 80 kD, Km: 4.0) was stronly induced under -Fe while the NAAT-II (molecular mass: 90 kD , Km: 0.9 mM) was detected both in the control and the Fe-deficient roots. The genes encode NAAT have been cloned (Mori, 1997). In addition, The genes encode alcoholdehydrogenase has also been cloned, which are considered to be related to MAs synthesis (Mori, 1997).

Mihashi and Mori (1989) purified the plasma membrane from both Fe-deficient and Fe-sufficient barley roots, and analysed the differences in the protein composition, consequently, they isolated three polypeptides (40, 28, and 14 kDa) induced under Fe deficiency conditions. Okumura et al., (1992) selected seven cDNA clones specific to Fe deficient barley roots by different hybridization. The longest subcloned mRNA, about 500 bp, was assumed to be of full length by northern hybridization and designated ids1 (iron deficiency-specific clone 1). It consisted of 503 nucleotides and contained a putative open reading frame encoding a protein of 74 residues (7500 Da) having two cysteine rich domains like animal (Okumura et al., 1992). These results implied that the transporter proteins might have the similar constraint of metal-binding characteristics of MTs. More recently, another two clones from the seven clones have been identified and characterized (Okumura et al., 1994, Nakanishi et al., 1993). One clone has the putative full-length mRNA of dioxygenase as judged by northern hybridization and named Ids2 (iron deficiency-specific clone 2). The predicted amino acid sequence of the Ids2 resembled that of 2-oxoglutarated-dependent dioxygenase. The Ids2 is expressed in the Fe-deficient barley roots but is not in the leaves, and also strongly expressed under Mn deficiency and weakly under Zn deficiency or excess NaCl (0.5%). The expression is repressed by the availability of Fe. The upstream 5'-flanking region of Ids2 has a root-specific cis element of the CaMV 35S promoter and a nodule-specific element of leghemoglobin (a metal regulatory element (MRE) and several Cu regulatory elements (UAS) of yeast metallothionein (CUP1)) (Okumura et al., 1994). Another clone, designated Ids3, was found to be specifically expressed in the roots of iron-deficient barley (Nakanishi et al., 1993). It consists of 1685 nucleotides encoded a putative protein of 169 amino acids and a M(g) of 18704. A search for homologies in a protein database (NBRF) revealed that the predicted protein product has a functional peptide domain that resembles that of 2-oxoglutarate-dependent dioxygenases. Northern hybridization and Western analysis with anti-Ids3-antibody was carried out and both expression and protein appearance were observed in the roots of Betze-barley and in the Add-4 as well, suggesting that ids3 is a possible candidate for a MA synthetase gene (Mori, 1997).

2.2. ZINC

Zinc is required at low levels in the plants, the critical deficiency concentrations in the shoots ranging from 10 to 15 mg kg^{-1} in most graminaceous species, and 20 to 30 mg kg^{-1} in most dicotyledenous species (Marschner, 1995). Zinc enters the plant primarily via root absorption of Zn^{2+} from the soil solution. The adequate concentrations of Zn in hydroponic solutions for plant growth are in the range of 0.05 to 0.25 microM, with Zn toxicities observed above concentrations of 3-6 microM. Because of the low concentration of Zn in soil solution, usually in the range of 0.03 to 0.5 microM, Zn supply by mass flow is limited and diffusion is the major process for Zn to reach the roots. Therefore, the characteristics of root morphology and vitality are crucial for the plant to efficiently explore Zn in soils.

2.2.1. Mechanisms of Zn absorption

Relatively much less work has been done on understanding the mechanisms of Zn uptake by higher plants compared to Fe uptake. Zinc uptake by higher plant appears to be mostly controlled by the transport of Zn across the plasma membrane, which is largely metabolism-dependent and genetically controlled (Loneragan, 1972; Mullins and Sommers, 1986; Santa Maria and Cogliatti, 1988;). The speculated mechanisms of Zn uptake in the plant include: 1) thermodynamic transport of Zn driven by electrochemical potential gradient across the membrane; 2) transported through an ion pump operated by H^+-ATPase; 3) the involvement of Zn-chelate transport system; and 4) the involvement of ion channels.

Thermodynamic Transport. The electrochemical potential of an ion thermodynamically indicates the free energy for that ion available for performing work (Nobel, 1983). For an ion species, j, the chemical potential is given as:

$$Fj = Fj^* + RT \ln aj + zjFE$$

Where, Fj^* is the chemical potential of ion j in its standard state, R is the gas constant, T is the temperature in degrees Kelvin, aj is the chemical activity of j, zj is the valence, F is the Faraday constant, and E is the electrical potential. When an ion is in passive equilibrium across a membrane, the electrical potential difference with respect to the ion j between inside and outside the membrane can be described by a Nernst equation:

$$E^i_j - E^o_j = E^N_j = RT/zjF \ln a^o_j/a^i_j$$

Where E^i_j and E^o_j are the electrical potential of inside and outside the membrane, respectively, corresponding to the activity of j ion inside and outside membrane at a^i_j and a^o_j, respectively. The above equation can be simplified as follows:

$$E^N_j = 59.2/zj \log a^o_j/a^i_j \quad \text{mV at } 25\,^0\text{C}$$

Root cells often maintain membrane potentials across the plasma membrane in the range of -120 to -180 mV, with each -29 mV balancing 10-fold accumulation gradient for divalent cation. Therefore, a membrane potential of -120 mV could balance an accumulation ratio of 10000:1 for Zn in the form of free ions. This mechanism may play an important role in Zn uptake at moderate to high concentrations in soil solution.

Active Transport Systems. Active transport systems such as H^+-ATPase, Zn-chelate, and ion channels have been reported to involve in Zn uptake (Kastrup et al., 1996; Kochian et al., 1991; Welch et al., 1993; Zhang et al., 1989, 1991). Kastrup et al. (1996) demonstrated that Zn as $ZnATPase^{2-}$ serves as a substrate of the plasma membrane H^+-ATPase. The native H^+-ATPase was activated by divalent metal ions in the order of Mg>Mn>Zn at pH 6.5 and in the order of Zn>Mn >Mg at pH 5.5 for total metal ion concentrations not higher than 3.0 mM. A Zn: H^+-ATP ratio of 0.6 at pH 5.5 gave more than 100% of the Mg-dependent ATPase activity (Mg: ATP ratio of 1, pH 5.5). By contrast, Free Zn^{2+} is an effective inhibitor of ATP hydrolysis and H^+ transport with a strong dependence upon the pH of medium. Both activation by $ZnATPase^{2-}$ and inhibition by free Zn^{2+} of the native plasma membrane H^+-ATPase and a partially purified enzyme were strongly affected by the pH of the medium (Kastrup et al., 1996). It appears that Zn, in combination with the pH of the cytoplasm plays a role in regulating the plasm membrane H^+-ATPase and thus in nutrient uptake of plant systems. In grasses, Fe or Zn deficiency could enhance release of phytochelate or phytometallophore from plant roots to the rihzosphere, which can effectively chelate Zn^{2+} and increase solubility of Zn from the soil, thus leading to an enhanced uptake of Zn (Zhang et al., 1991; v. Wiren et al., 1996). The mechanism by which these metal-chelate complexes are transported into root cell is poorly understood. It is assumed that both Zn-chelate and Fe-chelate are transported into the root cell via the same transport protein, followed by binding of the complex to certain recognition site (v. Wiren et al., 1996). The transport system would facilitate the transport of either complex into the cell, as has been shown by Crowley et al (1991) that both the Zn-mugeneic acid and Fe-mugeneic acid complexes were absorbed at approximately the same rate. Elaborate study is needed to elucidate the mechanism of Zn-chelate transport into root cell.

The mechanism of Zn influx was investigated using giant algal cells, which is similar to plant root cells (Reid et al., 1996). This study showed that most of the cell-associated Zn was in the cell walls. The cytoplasmic and vacuolar Zn concentration was 56 and 32 mM, respectively for cells grown in a Zn concentration at 0.1 mM. Influx of Zn was linear over several hours, with rapid transfer to the vacuole, but only slow efflux. Influx occurred in a biphasic manner, which might be attributed to the operation of two separate systems, a high-affinity system, which is saturated at 0.1 mM, and a low-affinity system, which showed a linear dependence on concentration up to at least 50 mM. Influx of Zn through the low-affinity system depended on external pH with an optimum around 7 and stimulated by cytoplasmic acidification (Reid et al., 1996). It was sensitive to metabolic inhibition, but not affected by either blocking of Ca^{2+} and K^+ channels or plasma membrane potential. These results suggest that active transport processes are dominant in the uptake of Zn even at relatively high concentration.

Via Ion Channels . Recent studies indicate that cation channels might be involved in the absorption of a number of micronutrient cations, including Zn^{2+}, Fe^{2+}, Cu^{2+}, and Mn^{2+} (Kochian et al., 1991, Welch et al., 1993). Fe deficiency was observed to remarkably enhance the absorption of Zn^{2+}, Cu^{2+}, and Mn^{2+} in addition to Fe^{2+} by peas. The plasma membrane reductase has been speculated to gate ion channels that mediate the uptake of certain divalent cations, including Zn^{2+} (Kochian et al., 1991). Some evidence showed that the ferric reductase is not regulated only by plant Fe status, but also induced by Zn^{2+} deficiency (Jolly and Brown, 1991, Kochian, 1993). How the gating of cation channel is related to the plasma membrane reductase is unclear. When the reductase is induced by Fe or Zn deficiency, critical sulfhydryl groups in the cation channel are reduced, thus opening the channel and allowing Zn^{2+} to enter the cell (Kochian, 1993).

2.2.2. Physiological Bases for Zn Efficiency

Zinc deficiency is one of the worldwide nutritional problems in soils affecting crop production. It has been estimated that Zn deficiency is the most widespread micronutrient deficiency decreasing crop production and quality in cereals, such as wheat, rice and other crop species. Genotypes of plants vary widely in their tolerance to Zn-deficient soils. Tolerance to Zn-deficient soils, as a genetic trait, is usually called Zn efficiency and defined as the ability of a cultivar (species *etc.*) to grow and yield well in soils too deficient in Zn for a standard cultivar (Graham, 1984). Growing Zn-efficiency plants on Zn-deficient soils represents the strategy of "tailoring the plants to fit the soil", which could be a cost-effective approach in improving crop yield in Zn-deficient soils widespread in the world.

Genotypic differences in Zn efficiency have been identified for a number of crop species. The crop species include: spinach (*Spinacea oleracea*), potato (*Solanum tuberosum*), bean (*Phaseolus vulgaris*) tomato (*Lycopersicon esculentum*), pearl millet (*Pennisetum americanum*), sorghum (*Sorghum valgare*), maize (*Zea mays*), oats (*Avena sativa*), wheat (*Triticum aestivum*), rice (*Oryzal satival*), and others (Ambler and Brown, 1969; Brown and McDaniel, 1978; Parker et al, 1992; Sharma and Grewal, 1990; Shukla and Raj, 1974; Shukla et al., 1973; Takkar et al., 1988; Yang et al., 1993, 1994a). A number of possible mechanisms may operate in different species and different genotypes of a species. These mechanisms may be operational at several levels of plant organization (molecular, physiological, structural, or developmental) and/or soil environments (Graham, 1993). Different mechanisms may be responsible for Zn-efficiency for different species or genotypes of the species, and more than one mechanism may be applied to one species. Briefly, the following physiological mechanisms of Zn efficiency have been suggested:

Root Exudation of Organic Acids and Chelators. Under Fe, P or Zn deficiency, exudation of organic acids, like citric acid or H^+, by roots of some plant genotypes can be greatly enhanced, which results in an increased Zn uptake through rihzosphere acidification and /or Zn complexation reactions (Chairidchai and Ritchie, 1993; RÖmheld, 1987). Graminaceous species respond to Fe or Zn deficiency by the release of non-proteinogenic amino acids, called phytosiderophores or phytometallophores (Kochian, 1991; RÖmheld, 1987; Zhang, et al., 1989). These compounds form stable chelates with Zn and are as effective as DTPA in mobilizing Zn from calcareous soils. Under cacareous soil conditions, Fe- deficient barley roots mobilized more Zn from Zn(II)-MA (699 nmoles Zn g^{-1}soil) than that from ZnDTPA (656 nmoles g^{-1}) (Treeby et al., 1989). In most cases, efficient species and genotypes are characterized by a greater acquisition from soils. Differences in Zn efficiency between plant species are probably related to inherent difference in rihzosphere pH and exudation (Marschner, 1993). For instance, the Zn-inefficient Durati grows poorly on soils low in Zn, in contrast to the Zn efficient Aroona (Graham et al., 1992). In agreement with this, under Zn deficiency, release of phytosiderophores in much higher in Aroona than in Durati (Cakmak et al., 1993).

Inherent Tolerance to Inhibitory Effect of Bicarbonate. Bicarbonate has inhibitory effect on plant growth in calcareous soil. Rice cultivars differ much in their sensitivity to Zn deficiency, especially when growing in calcareous soils, and bicarbonate has been regarded as one of the major factors in inducing Zn deficiency for the susceptible rice cultivars (Forno et al., 1975; Yang et al., 1993). In low Zn soil, bicarbonate inhibited remarkably the uptake of not only Zn, but also Fe and Mn for the Zn-inefficient rice cultivars, whereas no obvious effects were noted for the Zn-efficient cultivars, implying that bicarbonate has no specific inhibition on Zn uptake (Yang et al., 1993). Zinc efficiency in lowland rice is causally related to high tolerance of the plant to elevated bicarbonate concentrations. Even concentration of 5-10 mM bicarbonate inhibits root growth of the Zn inefficient cultivar, but slightly stimulate root growth of the efficient cultivars (Yang et al., 1994b). Recently, we have demonstrated that the root growth of a wide range of Zn efficient rice cultivars was stimulated by bicarbonate, whereas the root growth of Zn efficient wheat cultivars was severely inhibited by bicarbonate. However, there is no such close relationship between the inhibition of root growth by bicarbonate and Zn efficiency in wheat (Yang et al. unpubl. data). Thus, the impairment of root growth and activity of the Zn-inefficient rice cultivars by bicarbonate appears to be the early effect of bicarbonate induced Zn deficiency in rice. Root organic acids concentrations, malate and citrate in particular, considerably increased with increasing of bicarbonate levels for both Zn-efficient and Zn-inefficient rice cultivars, but to a greater extent for the Zn-efficient cultivars (Yang et al., 1994b). The results from the compartmentation of organic acids in roots and the distribution of $H^{14}CO_3^-$-assimilates showed that Zn efficiency in rice is closed associated with the efficient compartmentation of organic acids. These are synthesized through dark fixation with the operation of PEP-case, in root cells and fast translocation of $H^{14}CO_3^-$-assimilates to the shoots (Yang et al., unpubl. data). It could be concluded that, for lowland rice, the Zn efficiency is highly related to the inherent tolerance to bicarbonate inhibition on root growth.

Infection of Vesicular-arbuscular (VA) Mycorrhizae. Infection of VA mycorrhizae can enhance uptake of Zn by the host plants through the increased root surface from the external hyphae. The effect are usually more distinct in soils low in extractable Zn or low Zn mobility and in plants with coarse root systems, such as fruit trees including peaches, citrus and apple, and tree legumes (George et al. 1994; Gilmore, 1971; Marschner, 1995; Runjin, 1989). In graminaceous species such as maize and wheat gown in soils with low Zn contents, VA mycorrhizae increase Zn uptake and shoot contents in the dry matter, despite an increase in shoot biomass (Faber et al., 1990; Sharma et al., 1994; Swamvinathan and Verma, 1979;). The Zn uptake rates of mycorrhizal corn roots were around 10-fold greater than nonmycorrhizal roots at very low zinc concentrations (below l mmol m-3) (Sharma et al., 1994). Obviously, mycorrhizal plant species are expected to be more Zn-efficient than the non-mycorrhizal species. More work is needed to address the relationship between Zn-efficiency and mycorrhizal infection.

Root Geometry Differences. Plant species or cultivars that produce more smaller roots with diameter (<3.0 mm) can explore a larger volume of soil and hence more efficiently scavenge of the small amounts of immobile Zn ion, rendering it Zn-efficient. For example, Excalibur, the Zn-efficient wheat genotype, develop smaller roots than the cv. Gatcher, the Zn-inefficient wheat genotype (Graham, 1993). In addition, plant species with longer root system is expected to be higher Zn-efficient, as it has deeper rooting zone, by which to explore Zn more efficiently in subsoil (Grewal et al., 1997).

Absorption and Root to Shoot (or Seed) Transport. Zn-efficient genotype appears to have a better absorption and root to shoot transport, probably due to a more effective transport system such as ion channel or ion pump than the Zn-inefficient genotypes (Grewal et al., 1997; Khan et al., 1998). The efficient chickpea genotypes (CTS-60543, CTS-11308 and T-1587) have been reported to a higher Zn accumulation per plant and higher Zn uptake per g of root dry weight than the inefficient genotypes (Tyson and Dooen). The CTS-11308 and T-1587 transported more than 70% of the total absorbed Zn to the shoot compared with the Tyson and Dooen (Khan et al., 1998). For oilseed rapes, Zn-efficient genotype (CSIRO-1) has a higher Zn concentration and content in seed than the inefficient genotype (Zhongyou 821), suggesting that the efficient genotype has a superior Zn transport mechanism from source (roots) to sink (seed) (Grewal et al., 1997).

TABLE 3. Some identified chromosomes and/or genes in relation to the efficiency of Zn, Mn, or Cu in plants

Element	Plant species	Chromosome/genes	Characteristics	Reference
Zn	*Agropyron*	Chromo. L1	Zn efficiency	Schlegal et., 1997; Cakmak et al., 1997
	Haynaldia	Chromo. V2, V7	Related to Zn efficiency	Schlegal et al., 1997; Cakmak et al., 1997
	Rye	Chromo. 7R, 1Rs	Related to Zn efficiency	Schlegal and Cakmak, 1997
	Rice	Multiple genes	Control Zn efficiency	Majunder et .al., 1990
	Arabidopsis	ZIP1, ZIP2, ZIP3, ZIP4	Genes expressed in response to Zn deficiency	Crotz et al., 1998
Mn	Rye	Chromo. 2R	Related to Mn efficiency	Graham, 1988
	Wheat	A single dominant Gene	Related to Mn efficiency	Lonnecker et al., 1990
	Barley	A single dominant Gene	Related to Mn efficiency	Rengel et al., 1994
Cu	Rye	Chromo. 5RL2.3	Encode genes for Cu efficiency	Schlegal and Cakmak, 1997

2.2.3. Genetics of Zn Efficiency

Although some physiological mechanisms involved in Zn efficiency have been documented, a little information is available on the genetic control of these mechanisms and identification of molecular backgrounds or genes responsible for Zn efficiency. Table 3 listed reported genetic backgrounds with relation to Zn, Mn and Cu efficiency in higher plants. Studies of addition lines have shown that Cu, Zn and Mn efficiency in rye were independent traits and carried on different chromosomes (Graham, 1984). Recently, Schlegel et al., (1997) found that rye chromosome 1R and 7R seem to carry genes affecting Zn efficiency. By using the disomic wheat-rye addition lines to study the roles of rye chrosomes on Zn efficiency, Cakmak et al. (1997) confirmed that rye is inherent for high Zn efficiency, and its chrosome 1R and 7R carry the genes controlling Zn efficiency. It seemed that the Zn-efficient genes are transferable into wheat. F*or Agropyron*, chromosome L1seemed to be responsible for Zn efficiency, while for *Haynaldia*, chromosome V2 and V7 carry the genes controlling Zn efficiency (Schlegel et al., 1997). Similarly, a few genes may be involved in Zn efficiency in rice. The genetic effects responsible for Zn efficiency in rice were suggested to be mostly additive, and to a lesser extent dominant (Majumder et al., 1990). Different responses of soybean genotypes to Zn fertilizer appeared to result from differential efficiency of Zn absorption, the distribution of F_3 lines from the cross between Zn-efficient and Zn-inefficient genotypes suggested that only a few genes control the Zn efficiency trait in soybean (Hartwig et al., 1991)

As mentioned above, zinc might be absorbed by plant roots via transporting proteins located on the plasma membrane. Crotz et al. (1998) reported the first zinc transporter genes cloned from *Arabidopsis* thaliana. They found that the expression of three genes, designated ZIP1, ZIP2 and ZIP3, in yeast conferred zinc uptake activities. In the plant, ZIP1 and ZIP3 are expressed in roots in response to zinc deficiency, suggesting that they encode the proteins for zinc transport across root membranes. Although expression of ZIP2 has not been detected, a fourth related *Arabidopsis* gene identified by genome sequencing, ZIP4, is induced in both shoots and roots of zinc-deficiency plants. In *Escherichia* coli, the genes responsible for the high-affinity uptake for Zn have been cloned, and the complementing gene, zur, showed 27% sequence identity with the iron regulator Fur (Patzer and Hantke, 1998). In giant algal cells, Zn uptake is considered to be controlled by two separate transport systems, a high-affinity system and a low affinity system (Reid et al., 1996). Future research should direct on identifying those uptake systems in higher plants. Future study is needed to clarify the molecular bases of Zn efficiency in higher plants.

2.3. UPTAKE OF OTHER CATION MICRONUTRIENTS

The Cu^{2+} has been considered to be major Cu form for plant absorption. The Cu^{2+} ion has a high affinity for peptide N and S, and binds quite strongly to the proteins, especially proteins high in cysteine residues. Also there is a strong and specific adsorption of Cu^{2+} to cell walls that is not easily desorbed (Harrison et al., 1978), due to the high affinity of Cu^{2+} for carboxyl, sulfhydryl, and phenolic groups in cell walls. As Cu(II) can be easily reduced to Cu(I) in the range of physiological redox potentials, the significance of the absorption of Cu(I) by roots should not be ignored (Graham, 1981).

Manganese is absorbed into the plant mainly as the free Mn^{2+} ion, but it exist in soils as different reduced forms including Mn(III) and Mn(IV) oxides. The stimulated acidification and reductase due to Fe deficiency could increase reduction of Mn(III,IV) to Mn(II) (Marschner, 1988). However, unlike Fe, physiological and genetic mechanisms of Mn uptake by plants are not fully understood. Future researches should direct more on those mechanisms.

Nickel has been shown to be essential for higher plants (Brown et al., 1987, Eskew et al., 1984). The Ni^{2+} is the major form absorbed into the plants. Nickel levels in Ni-adequate plant tissues are extremely low, ranging between 0.2 and 5 microM, an adequate growth for most plants can be maintained in solutions containing Ni at concentrations below 1 microM (Brown et al., 1987; Welch, 1981). Most of the Ni uptake studies have come from concerns about heavy-metal pollutants, and have been conducted with relation to Ni toxicity.

2.3.1. Speculative Absorption Mechanisms

Little work has been done on the mechanisms of Cu, Mn, and Ni absorption by plants in recent decades. Many of the speculative uptake mechanisms are from Fe uptake models, including thermodynamic transport driven by electrochemical potential gradient; through ion channels; and the involvement of active transporter systems. As the principle of thermodynamic transport, which has been discussed for Zn uptake, is similar for all divalent micronutrient cations, only the later two speculative mechanisms are discussed here for the uptake of Cu, Mn and Ni.

Via Cation Channels. As has been described previously for other divalent micronutrient cations, if free Cu^{2+} ion is the dominant species absorbed by the root, there exist a large electrochemical potential gradient for passive Cu^{2+} uptake into plant cells, possibly via ion channel (Graham, 1981). There are considerable evidences that free Cu^{2+} is the absorbed species (Kochian, 1991). More recently, Welch (1995) proposed that Cu absorption by higher plants may have similar strategies as Fe uptake. For dicots and nongramineous monocots, Cu^{2+} might be absorbed through the cation ion channel in coupled with membrane-bound reductases. Recent evidence has shown that Cu deficiency also induces plasma membrane Fe reduction (Cohen, et al., 1997). Of the nutrient deficiencies (K, Mg, Ca, Mn, Zn, Fe, and Cu), only Cu and Fe deficiencies induced ferric reductase activity in roots of peas seedlings. Cu deficiency induced the reductase activity and rhizosphere acidification rates were similarly intermediate between those of control and Fe-deficient plants. These results support the hypotheses that the stimulation of the reductases in the membranes results in the reduction of sulfhydryl groups involved in gating a divalent cation channel that mediate the influx of Cu^{2+}, Mn^{2+}, Ni^{2+} and Zn^{2+}. The estimated redox potential for the standard reductase in plant root membrane is about +0.15 V, if so, then the standard reductase system might be involved in destabilizing Cu and Mn metal complexes at the plasma membrane surface thus releasing Cu(II) and Mn(III, IV) from their complexes as free Cu^{+}, and Mn^{2+} ions, and play a role in the subsequent transport of these micronutrient into root cells (Welch, 1995).

Kinetic analysis by Cataldo et al., (1978) indicated that both Cu^{2+} and Zn^{2+} competitively inhibited Ni^{2+} uptake, which suggested, quite speculatively, that all three ions could be transported by the same (or similar) transport system. The results from the studies of Ni uptake by different crop species at toxic levels by Yang et al., (1996 ab) showed that uptake of Zn, Cu and other cation ions were competitively inhibited by high Ni levels in the dicots plant species used, while such inhibition was not observed in the monocots plants species used. These results suggested that Ni absorption in dicots might be through a similar cation channel as that for other divalent cation ions (e.g. Ca^{2+}, Zn^{2+}, Mn^{2+}, Mg^{2+}) (Yang et al., 1996b). The expression of the ion channels or transporters seemed to be related to Fe-deficiency induced response (Cohen et al., 1998). Expression studies with the Fe^{2+} transporter cloned from *Arabidopsis*, IRT1, indicated that Fe deficiency induced the expression of this transporter, which might facilitate the transport of divalent cations such as Zn^{2+}, Cd^{2+}, in addition to Fe^{2+} (Cohen et al., 1998). As mentioned above, although Ca^{2+} channels can transport Ca^{2+} ions across the plasma membrane, they can also pass other ions such as Mg^{2+}, Ba^{+}, Rb^{+}, K^{+}, and Na^{+}. So it could be suggested that micronutrient cation channels in the plasma membrane might be similar or the same as Ca^{2+} channels. Future research is needed to address this possibility.

Through M(nI)-chelate Transporter. It has been generally considered that membrane transport of divalent cations can be mediated by proteinaceous ion channels. Based on the thermodynamic considerations. However, because of the tendency for Cu^{2+} to bind rather strong to proteins, Graham (1981) has suggested that Cu^{2+} uptake could be facilitated by a carrier-type protein (transporter), which across the plasma membrane. Welch (1995) speculated that the uptake of other micronutrient cations (e.g. Cu, Mn, Zn and Ni) into the root cells of grasses might be in a similar way as Fe (III)MAs through metal-phytometallophores transporters across the plasma membrane. The term of phytometallophores is based on broad meaning "plant metal bearer" or "plant claw", including phytosiderophores, phytochelatins, and low molecular metal chelators. Both copper and manganese can form complexes in the rhizophere with organic legends of plant and microbial origin, which may increase the mobility of Mn and Cu in the rhizosphere. (Facilitating diffusion up to the root-cell plasma membrane). It has been reported that Mn and Cu absorption can be enhanced by phytosiderophores produced in response to Fe deficiency (Marschner, 1988, Marschner et al., 1989). Zhang et al., (1989, 1991) demonstrated that phytosiderophores not only mobilize Fe but also mobilize Zn from both the rhizosphere and root-cell apoplasmic spaces. However, Mn deficiency did not induce phytosiderophore release in roots of Mn-efficient barley genotypes (Webb, 1994), more reductants of microorganism source were found to exist in the rhizosphere of Mn-efficient wheat genotype (Rengel, 1997). Recently, from Ni-tolerant ryegrass mutant, we found that Ni uptake by the mutant plant was closely related to Fe-deficiency induced phytosiderophore release (Yang and Römheld, 1997). More evidences are needed to be provided that these metal ions are absorbed as metal-phytometallophore complex. It could be speculated that there might exist transporters for cation micronutrient ions complexed with phytometallophores in the plasma membrane, similar to the Fe(III)-MAs transporter. Cation micronutrient uptake by grass might follow similar model as that for Fe uptake (Stategy II) by grass. Whereas, till now, there is no direct evidence showing M(nI)-chelate complexes transporter proteins or genes except for some evidence showing the presence of Fe(III)-MAs transporters in root plasma membranes. Further researches are needed to address those possibilities.

2.3.2. Genetic Aspects of Mn and Cu Uptake

Physiological and Genetic Bases for Mn Efficiency. Manganese deficiency in crops occurred in sand soils and calcareous soils, calcareous soil conditions favor chemical and microbial oxidation and immobilization of soluble Mn, resulted in low availability of Mn although they contain generally high total Mn. On calcareous soils, application of Mn fertilizers has less responsive which make it unsuitable for correction of Mn deficiency. An alternative strategy is to breed crop cultivars of high efficiency for Mn grown in these soils. A Mn-efficient genotype, in an agronomic sense, is one that is able to grow and yield well without added Mn fertilizer in a soil, which has insufficient available Mn for another. Great genotypic differences in Mn efficiency in soils have been widely recognized since Mn deficiency was first identified in the 1920s, and has been reported in wheat, durum wheat (*Triticum turgidium* L. Var. *Durum*), oat and barley (*Hordeum vulgare* L.) (Bansal et al., 1991; Graham, 1988; Huang and Graham, 1997; Kaur et al., 1989; Marcar and Graham, 1987; Nyborg, 1970; Saberi et al., 1997). As Mn deficiency is the most intractable trace element problem in crops grown in most South Australian soils, genetic improvement of Mn efficiency in cereals has been attracted much interested by the Australian scientists. A wide range of germplasm with wide differences in Mn efficiency is already available which could provide great possibility to improve genetically modern cultivars in Mn efficiency. Further improvement would be possible through transfer of Mn efficiency from rye (Rengel et al., 1994).

The physiological mechanisms of Mn efficiency in plants have not yet been fully understood. Mn-efficient genotypes absorb more Mn from soils and thus can survive and yield better in a soil with low Mn availability than do Mn-inefficient genotypes (Bansal et al., 1991; Graham, 1988; Huang and Graham, 1997; Marcar et al., 1989). Mn-efficient barley genotype took up more Mn independent of Mn supply levels both in solution (Huang et al., 1994) and in the soil (Huang and Graham, 1997), suggesting that efficient Mn uptake in barley is a constitutive system.

Using chelate-buffered system, it was showed that the Mn absorption rate is linear in the range from deficiency to adequacy (Webb et al., 1993), suggesting that the capacity for Mn absorption is not a limiting factor. Clarkson (1988) has commented that plants have a high capacity for Mn absorption; a capacity which is much greater than their needed. However, Mn absorption by young roots is relatively slow and increased rapidly with age before reaching an apparent plateau at about 20-day old (Webb, 1994). Huang et al (1993) used a chelate buffering technique to maintain a constant activity of Mn over a range of pH levels, and showed that increasing pH (without decreasing Mn activity) increased the absorption rate of Mn by barley plants. This further supports the suggestion that H^+ in solution inhibits Mn absorption. Generally high pH makes Mn less available in soils but low pH inhibits Mn absorption from solution because of H^+ competition (Marschner, 1988).

Manganese can form complexes in the rhizophere with organic legends of plant and microbial origin, which may increase the mobility of Mn in the rhizosphere (facilitating diffusion up to the root-cell plasma membrane). It has been reported that Mn absorption can be enhanced by phytosiderophores produced in response to Fe deficiency (Marschner, 1988; Marschner et al., 1989). Whereas, The results from the mixture culture experiments with Mn-efficient and Mn-inefficient barley genotypes showed that Mn acquisition by barley is not enhanced by the release of some Mn mobilizing compound as has been shown for Fe acquisition in wheat (Zhang et al., 1989). Rengel (1997) found that rhizosphere of wheat genotypes contained an increased proportion of Mn reducers under Mn deficiency compared to Mn-sufficiency conditions. When grown on a soil with low Mn availability, some wheat Mn-efficient genotypes (like cv. Aroona) had a greater ratio of Mn-reducers to Mn-oxidisers in the rhizosphere compared to the Mn-inefficient genotypes (Rengel, 1997).

However, until recently, a little was known about the genetics and molecular backgrounds for Mn efficiency in higher plants (Table 3). As it is well known that rye is outstandingly Mn-efficient in comparison with wheat. The study with rye addition lines, showed that efficiency is carried on the 2R chromosome (Graham, 1988). Recent experiments with barley suggest the involvement of simple mechanisms of Mn efficiency controlled by single, major, dominant gene (Longnecker et al., 1990; McCarthy et al., 1988). In the study of 72 barley genotypes from a world collection, the pedigree relationships within the most efficient group and within the inefficient group were consistent with single, major-gene inheritance. In the study of the progeny from cross between Mn-efficient and Mn-inefficient barley genotypes, the distribution of F_2 individual followed a 3:1 ratio characteristic of a single dominant gene for Mn efficiency in barley, with a narrow-sense heritability of 71%(Rengel et al., 1994).

Genetics of Cu efficiency. Copper deficiency often occurs in soils either inherently low in total copper (e.g. ferallitic and ferruginous coarse textured soils, or calcareous soils derived from chalk) or on soils high in organic matter where copper is complexed with organic substance (Alloway and Tills, 1984). Plant species differ considerably in sensitivity to copper deficiency. Some plant species like wheat (*Triticum aestivum*), oats (*Avena sativa*), and spinach (*Spinacea oleracea* L.) are more sensitive to Cu deficiency than the other species like pea (*Pisum sativum* L.), rye (*Avena sativa*), and rape (*Brassica napus* L.)(Alloway and Tills, 1984). However, very little work has been on the physiological and genetic mechanisms of Cu efficiency in higher plants (Table 3), because, like Ni, most of research on Cu uptake are related to mechanisms of plant tolerance to Cu toxicity, which is beyond the scope of this review. Copper efficiency in rye appears to be controlled by a single major gene (Graham, 1984), and is linked to the chromosome 5R. Recently, Schelegel et al., (1993) revealed that the gene for Cu efficiency is linked to a dominant hairy neck character from rye and to two rye-specific leaf esterase loci, all of which are assumed to map to the distal part of 5RL. By utilization of amphiliploids, wheat-alien addition, substitution and translocation lines it was demonstrated that genes from rye chromosome region of 5RL2.3 contributed to a high Cu efficiency (Schlegel and Cakmak, 1997). Great efforts have been made to transfer Cu-efficient genes from rye to wheat and oats. Future research should direct more on physiological and molecular bases of Cu efficiency in higher plants.

3. Anion Micronutrient Uptake

3.1. BORON

3.1.1. Mechanisms of B Absorption

Boric acid is a very weak acid in aqueous solution, so B occurs mainly as boric acid,

$B(OH)_3$ in soil or nutrient solution at physiological pH range (Raven, 1980). The estimated permeability coefficient of plant cell membrane to boric acid is around 10^{-6} cm s^{-1} using the linear relationship between permeability coefficient and the ether-water partition coefficient (Raven, 1980). From the theoretical calculation, Raven (1980) suggested that B absorption would be a passive process.

Whether B absorption is passive or active, there are many conflicting results in the literature. Boron uptake, as suggested by most investigators, is not necessarily related to metabolic process, based on plant responses to changes in B concentration or temperature as well as respiratory inhibitors (Brown and Hu; 1994; Martini and Thellier, 1993; Thellier et al., 1979). Brown and Hu (1994) used suitable B isotopes to study B absorption by tobacco cells, sunflower roots and squash roots, respectively. They found that B absorption rates were linear over a range of B concentrations (0-0.2 mM or 0.2 – 10 mM) and there was no indication of saturation kinetics, no inhibition by DNP and KCN either. By using isotopes of ^{10}B and ^{11}B to study $B(OH)_3$ flux in duckweed (*Lema minor*), Thellier et al (1979) suggested that there existed four compartments within cell, i.e. free space, cytoplasm, vacuole and the cell wall. Each of these compartments contained higher B than the external solution (0.16 mM), but they explained that this internal higher B concentration than external was due to the formation of borate mono- or di-ester within the various compartments and subsequent exchange of B isotopes. As most of the researchers used relative high B concentration for uptake study, which could not represent the real conditions in soils, especially in B deficient soils. More recently, by changing the B concentration (1 or 100 microM) of preculture and treatment for short term B uptake, Pfeffer et al (1997) found that the B concentrations in root cell sap and xylem exudate varied remarkably with external B concentrations used for preculture and/or treatment (Fig. 3). In addition, the B uptake into the root cell sap was significantly inhibited by low temperature (7°C) or the treatment with metabolic inhibitor (DNP) only at low external B supply during preculture (Fig.3). Different changes of B in xylem exudates vs. external pH at low (1 μM) and high (100μM) B were studied for sunflower (Dannel et al., 1997). At low external B (1 μM), about 25-fold higher B in xylem sap than that in external solution was found, and the B concentrations in xylem sap changed with increasing of pH in a similar way to the pH-dependent concentration of undissociated boric acids. All these evidences implied that at high B supply, B uptake seems mainly through a passive process, whereas at low supply, there is a mechanism to concentrate B against external level and a short term active energy dependent uptake. Research with more elaborate techniques is needed to further understand B uptake by plants.

Another important property of boric acid should be kept in mind when developing any model to study B uptake by higher plants. That is its ability to form cis-diol complexes with many organic compounds, such as mannitol in celery (Hu et al., 1997), sorbitol, glycerol, ribose, apiose, nicotiamid-adenine-dinucleotide (NAD) and fructose (Loomis and Durst, 1992; Makkee et al., 1985). Low molecular weight B-complexes with mannitol, sorbitol and fructose have been isolated and characterized (Hu et al., 1997; Penn et al., 1997). Besides, B complexes with rhamnogactoronan-II in the cell wall have been also characterized (Ishii and Matsunaga, 1996; Kobayashi et al., 1996; O'Neil et al., 1996). Martini and Thellier (1993) found that most of the absorbed B was associated with cell wall, with a small portion in the cytoplasm, and practically zero in the vacuole according to B stable isotope (^{10}B and ^{11}B) tracing studies. Future studies on B absorption should consider both short and long term uptake periods, realistic B concentrations, and the impact of B-complex formation.

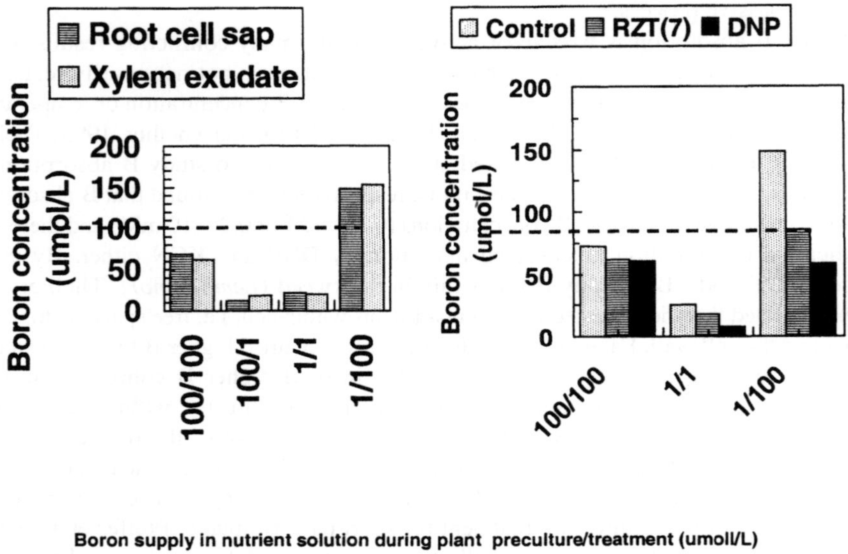

Figure 3. Effects of boron supply levels, low temperature, and metabolic inhibitors on boron concentrations in root cell sap and xylem sap of sunflower. (Adapted from Pfeffer et al., 1997, 1998)

3.1.2. Physiological Bases for B Efficiency.

There is good evidence to show that plant species or genotypes vary greatly in their responses to low B in soils (Rerkasem et al., 1988, 1993a; Rerkasem and Jamjod, 1997a). Dicotyledons are considered to require more B than monocotyledons, and vegetables of the *Cruciferous* and *Umbelliferous* families to have high B requirement (Martens and Westermann, 1991). Plants normally considered to be sensitive to low soil B include lucerne (*Medicago sativa*), rape (*Brassica spp.*), beet (*Beta vulgaris*), celery (*Apium graveolens*), grape (*Vitis vinifera*), apple (*Malus sylvestris*), pear (*Pyrus communis*), cotton (*Gossypium hirsutum*), and sunflower *(Helianthusa annuus*) (Jones, 1991). Among forage legumes, red clover (Trifolium pratense) is most tolerant (Sherrell, 1983). Large genotypic differences in B efficiency were noted under field trials. The range of sensitivity varies with crop species, such as, 9-71% for black gram Vignamungo), 34-100% for green gram (Vigna radiata)(Rerkasem and Jamjod, 1997a), 11-71% for peanut (Arachis hypogaea) (Keerati-Kasikorn et al., 1993), and 64-100% for wheat (Rerkasem and Jamjod, 1997a). The genotypic difference in B efficiency of wheat can be as large as 0-100% in terms of grain set index when 253 genotypes were compared in the field. At low external B level (e.g. below 0.1 mg kg^{-1} hot water soluble soil B), the responses of different genotypes to low B greatly varied, ranging from complete male sterility /grain set, through those partially affected (inefficient to moderately inefficient), and finally to no deficiency symptom (B-efficient)(Rerkasem and Jamjod, 1997a).

There are a number of mechanisms suggested for the genotypic differences in B efficiency. They include the ability to acquire B from the soil, the way that B is distributed, and its utilization within the plants. Boron efficiency mechanisms may differ among crop species. For a given species, the mechanisms for B efficiency may differ with genotypes and with the intensity of deficiency. The clarification of B efficiency mechanisms greatly depends on the understanding of physiological functions of B in plants.

Internal Requirement for B. Boron concentration in plant tissues varied largely among crop species and genotypes. Such as B sufficient level is 5-10 mg/kg for wheat, 6-12 mg/kg for ryegrass, while 35-80 mg/kg for alfalfa and 40-100 mg/kg for sugar beet (Bergman, 1992). So grasses are termed as plants species for low B requirement (Marschner, 1995). Low B requirement for grasses such as wheat was found to be associated with a low B absorption rate per unit root weight when compared to a dicotyledon crop species such as pea (Chapman et al., 1997). The distribution of B in different compartments may be another factor controlling the internal B requirement. Hu et al., (1996) found that B concentration in the cell wall of grasses was much lower than that of dicotyledons. The internal requirement for B in vegetative growth varies often from that for reproductive growth, so it may be necessary to consider vegetative and reproductive efficiency separately. In vegetative stage, B concentrations in the youngest opened leaf or youngest fully expanded leaf seemed to be better indication of genotypic deference's than total B concentrations in shoots, which has been demonstrated in rape, black and green gram, sunflower, and wheat. In wheat, higher functional B requirement for reproductive development was observed than that for vegetative growth, with higher B concentrations in anthers and the carpel than those in leaves (Rerkasem and Jamjod, 1997b). Thus grain set index has been used as a parameter for identifying B-efficient genotypes at reproductive growth stage for wheat, and also for other cereals. The effect of B deficiency on grain set failure has been reported also for barley (Ambak and Tadano, 1991), rice (Garg et al., 1979), and maize (Agrawala et < biblio >) besides wheat. Similarly, dicotyledons showed a higher B requirement for reproduction, mostly due to higher requirement of B for pollen germination and hence fertilization. For instance, maximum fertilization of grape flowers require 50-60 mg/kg B in the stigma (Marschner, 1995). In vitro pollen germination in mango (*Mangifera indica*) (de Wet and Robbertse, 1989), avocado (Smith et al., 1997a), and wheat (Cheng and Rerkasem, 1993) has been shown to respond to B supplied internally through the plant. However, it remains unclear whether B efficient genotypes simply require less B for anther and pollen development, or they are better in B re-translocation within the plant.

Boron retranslocation and Phloem mobility. The ability of a plant genotype to re-translocate B into reproductive non-transpiring organs where higher B is required for the development seems to be important for B efficiency (Brown and Hu, 1996). Boron efficient genotypes of broccoli were shown to be able to redistribute B into young leaves and florets (Shelp et al., 1992). In China, foliar B application was common field practice for correcting B deficiency in rape, and fertilizer responses differed with genotypes and years. The study by Luo (1998) with different genotypes showed that the ability of re-translocation B to younger leaves was closely associated with B efficiency in rape. Boron efficient genotypes have been identified with higher B concentration in newly developed leaves. In sunflower, efficient genotypes were those with higher B concentration in the uppermost mature leaves (Blamey et al., 1979). Similarly, in black gram and green gram, genotypes which were less severely affected in low B soils also had higher B concentration in their youngest fully expanded leaf (Rerkasem, 1990). In oil seed rape, B efficient genotypes contained higher B concentration in the youngest open leaf growing in a low B soil than the B -inefficient genotypes (Yang et al., 1993; Xue et al., 1998). These results indicated a role of B re-translocation and phloem mobility in B efficiency. Direct evidences are needed to prove the relationship of B efficiency with B re-translocation from vegetative parts to important reproductive organs like anthers and pollen tubes.

Interaction with environmental factors. Plant responses to B are influenced by environmental factors. Genotypic variations in B efficiency are likely associated with the ability to alleviate the effects of environmental factors. Boron efficiency ranking for wheat has been consistent over a wide range of environments (Rerkasem et al., 1993b), implying that efficient genotypes have better overcome some environmental stresses than inefficient genotypes. Low temperature at seedling stage, water deficit in soils and light intensity in flowering and pod setting stages seemed to be related to incidence of B deficiency in rape. We found that more tolerance to soil drought and low temperature at seedling growth stage after transplanting resulted in lower seedling mortality and higher yield for the B efficient genotypes than that for the B-inefficient genotypes (double low) (unpubl. data). The interaction of aluminum toxicity and B deficiency has been reported (LeNoble et al., 1996ab; Lukaszewski and Blevins, 1996), whereas whether B efficiency relates to Al tolerance is unknown. Root geometry and soil-roots interaction including rhizosphere effects may also play a role in B efficiency, but direct evidence is not available yet.

3.1.3. Genetics of B Efficiency

The genetic study of B efficiency has been reported in 1950s. Most of the researchers reported that responses to low B of plant species was under a single gene control, such as celery (Pope and Munger, 1953), tomato (Wall and Andrus, 1962) and red beet (Tehrani et al., 1971). However, in table beet the genetic control of B efficiency was found to be complex (Kelly and Gabelman (1960). Boron efficiency in sunflower was found to be highly heritable, with additive gene action predominant (Blamey et al., 1984). Similarly, both additive and dominant gene effects were involved in controlling B efficiency in bread wheat (Jamjod et al., 1992). However, complicated genetics of B efficiency is also suggested for wheat (Rerkasem and Jamjod, 1997b). Further genetic studies on B efficiency are needed to provide bases for transferring B-efficient traits into modern cultivars with better agronomic traits through genetic improvement.

Substantial genetic variation in response to high B has been identified in a wide ranges of plant species including wheat, barley, oats, field bean, and annual pasture medics (Jamjod et al., 1997; Nable et al., 1997; Paull, et al., 1992). The mechanism of tolerance for all species studied is related to a reduced accumulation of B by tolerant genotypes in both shoots and roots. The reduced accumulation of B in the tolerant genotypes is mainly due to a reduced B uptake and translocation, although the latter may be less important than the former (Kalayci et al., 1998). The response of wheat to high B supply seems under the control of several major additive genes, one of which has been located to Chromosome 4A

(Paull, et al., 1992). More genetic and molecular studies are needed to understand the nature of B uptake by plants.

3.2. CHLORINE AND MOLYBDENUM

3.2.1. Absorption Mechanisms

Chlorine. Chlorine is presented in aqueous solution as the monovalent ion chloride (Cl^-), and is taken up by plants mainly as anion (Cl^-). Chloride concentration in plants is in the range of 2-20 g kg^{-1} dry matter, but chlorine required for optimal plant growth is about 0.2-0.4 g/kg dry matter (Marschner, 1995). As chloride can be supplied to plants from different sources (soils, irrigation water, rain, fertilizers and air), chlorine toxicity is much more concerned than chlorine deficiency.

The absorption of Cl^- by higher plants was found to be by a proton-coupled Cl^- transport system i.e. via an anion-proton symport system. The driving force for the operation of the system is the electrochemical potential gradient developed by the H^+-translocating ATPase across the plasma membrane (Kochian, 1991). The Cl^--proton symport system allows the Cl^- accumulation against concentration gradients across the membrane (Tyerman, 1992). In a green alga, Cl^- uptake is cotransported with H^+, activated by blue light, suggesting that the mono-valent anion transporter is a plasma membrane-bound flavoprotein (Quinones et al., 1997).

Interaction between Cl^- and NO_3^- is related to some general aspect of ion absorption. However, addition of Cl^- decreased NO_3^- uptake is not the result of direct competition for functional plasma membrane binding sites in transport proteins (Robson and Pitman, 1983). The results from Liu and Shelp (1996) showed that nitrate absorption by broccoli (*Brassica oleracea* var. italica) was not inhibited by the presence of chloride in the growing medium and that the decrease in nitrate accumulation of the shoot resulted from the organic-nitrogen formation from absorbed nitrate. For leafy vegetable it is of importance to reduce NO_3^- accumulation through competition of Cl^-. More research is required to clarify the nature of Cl^- and NO_3^- interactions.

Molybdenum. Molybdenum is presented in aqueous solution mainly as MoO_4^{2-}, and is absorbed as the oxyanion MoO_4^{2-}. It is also assumed that MoO_4^{2-} is the form for long-distance translocation within plant (Marschner, 1995). The requirement of plants for Mo is lower than any other mineral nutrient except for Ni, but Mo deficiency is widespread in legumes and other plant species (e, g. cauliflower, maize, and wheat) grown in acid mineral soils or very sand soils. However, little work has been done to understand the mechanisms of MoO_4^{2-} uptake by plants, especially plant adaptation to Mo deficiency in soils.

The chemical properties of MoO_4^{2-} are similar to those of several other divalent inorganic anions, sulfate and phosphate in particular, so sulfate competitively inhibits MoO_4^{2-}.uptake by roots (Marschner, 1995). For SO_4^{2-} uptake, it was shown to be an active process via a plasma membrane transport protein, which is regulated by the additive and non-additive gene in the plant (Schmidt and Jager, 1992). However, evidence is lacking to show if MoO_4^{2-} may be controlled in a similar way as sulfate anion.

3.2.2. *Molybdenum Efficiency*

Molybdenum deficiency is frequently observed in both dicots (like legumes) and monocots (like Maize and wheat) grown in acid soils or sand soils. In acid soils Mo deficiency is mainly due to high adsorption of MoO_4^{2-} by iron oxide hydrate, decrease in deprotonation of molybdate (H_2MoO_4 - $HMoO_4^-$ - MoO_4^{2-}), and favor formation of polyanions (molybdate – tri- - hexa-molybdate) with decreasing pH (Marschner, 1995). In China, large area of Mo deficiency soils were mainly reported (Liu, 1996), mainly distributed in the Southeast and Northeast regions where most of Chinese population is concentrated. Effective response to Mo fertilizers were noted in legumes like soybean, some other dicotyledon species (Liu, 1996) and even in wheat grown in the Yellow Brown soil with extremely low Mo (Wei et al., 1998). There existed large genotypic differences in responses to Mo deficiency, although a little information is available so far. The critical values for Mo deficiency vary between 0.1 and 1.0 mg kg^{-1} leaf dry weight depending on nitrogen source and plant species (Bergmann, 1992). A high Mo in seeds is important for ensuring proper seedling growth and high grain yield of the plants grown in low Mo soils. The uptake rate of Mo by soybean is relatively lower at seedling stage. So the molybdenum requirement for growth has to be met mainly by retranslocation from the seed at early seedling stage (Ishizuka, 1982). These evidences suggested that the size of the seeds might be one of major factors for Mo efficiency in higher plants.

Brodrick and Biller (1991) found large genotypic differences of Mo translocation and Mo distribution in *Phaseolus*. Under low Mo, the efficient cutivar obtained higher seed yield, which is resulted from higher Mo distribution in seeds and retranslocation from shoots to seeds (Brodrick and Biller, 1991). When the concentrations of Zn, Cu, Mo and Ni in both shoots and seeds of legume were compared with those of grass plant species, the Mo and Ni concentrations in legume seeds were found to be 10-15 fold higher than those in grass (Hotak, 1985). These results implied that there might be existed large genotypic differences in translocation or retranslocation of Mo from vegetative organs to seeds during seed forming stage. More future research is needed to determine the physiology and genetic as well as molecular bases for Mo efficiency in higher plants.

4. Conclusions and Outlook

Micronutrient efficiency in higher plants is closely related to rhizosphere processes, uptake and retranslocation of the micronutrient in plants. The uptake is a decisive process in micronutrient acquisition. In the last decade, our knowledge on the mechanisms of Fe uptake has been advanced mostly. The adaptation of dicot and nongraminaceous plant species to Fe deficiency is mainly through increasing Fe reduction, while the adaptation of graminaceous plant species to Fe deficiency is mainly through enhanced synthesis, release of phytosiderophore and uptake of Fe(III)PS. However, the mechanisms of Zn, B and Mn uptake are not fully understood although researches on these elements have been increased in the last decade. More extensive and intensive studies are needed to understand the mechanisms of plant adaptation to Zn, B, Mn deficiency. Molybdenum deficiency is widespread and Cu deficiency is also a serious problem in some cases, however, a little information is available on physiological and genetic aspects of Cu and/or Mo uptake by higher plants. Future research could direct more on understanding the physiological bases and molecular backgrounds for Cu, especially Mo efficiency in higher plants. Extensive gene resources for micronutrient efficiency are available, however, up to now this potential has not been exploited to any great extent by breeders. Modern biotechnological methods such as gene transformation will result in a promising prospect for improving plant adaptation to micronutrient deficiency, thus increased considerably crop production and quality in future.

References

Agrawala, S.C., Sharma, P.N., Chestterjee, C. and Sharma, C.O. (1981) Development and enzymatic change during pollen development in boron deficiency maize plants, *J. Plant Nutr.* **3**, 329-336.

Ambak, K. and Tadano, T. (1991) Effect of micronutrient application on the growth and occurrence of sterility in barley and rice in a Malaysian deep peat soil. *Soil Sci. Plant Nutr.* **37**, 715-724.

Ambler, J.E. and Brown, J.C. (1969) Cause of differential susceptibility to zinc deficiency in varieties of navy beans (*Phaseolus vulgaris L.*). *Agron. J.* **61**, 41-43.

Asher, C.J. (1991) Beneficial elements , functional nutrients, and possible new essential elements. In: Morovedt, J., Cox, F.R., Shuman, L.M., and Welch, R.M., (Eds.) Micronutrients in Agriculture, Soil Science Society of America, Madison. WI 703-723.

Alloway, B.J. and Tills, A.R. (1984) Copper deficiency in world crops. *Outlook Agree.* **13**, 32-42.

Basso, B., Bagnaresi, P., Bracaale, M., and Soave, C. (1994) The yellow-stripe-1 and yellow-stripe-3 mutants of maize-nutritional and biochemical-studies, *Maydica* **39**, 97-105.

Bansal, R.L., Nayyar, V.K., and Takkar, P.N. (1991) Field screening of wheat cultivars for manganese efficiency, *Field Crop Res.* **29**,107.

Bergmann, W. (1992) Nutritional Disorders of Plants- Development, Visual and Analytical Diagnosis. *Fisher Verlag*, Jena

Blamey, F.P.C., Mould, D. and Chapman, J. (1979) Critical boron concentration in plant tissues of two sunflower cultivars. *Agron. J.* **71**, 243-247.

Blamey, F.P.C., Vermuelen, W.J., and Chapman, J. (1984) Inheritance of boron status in sunflower, *Crop Sci.* **24**, 43-46

Bienfait, H.F. (1988a) Proteins under the control of the gene for Fe efficiency in tomato *Plant Physiol.* **88**, 785-

Bienfait, F. and Luttge, U. (1988) On function off two systems that can transfer electrons across plasma membrane, *Olant Physiol. Biochem.* **26**, 665-671.

Bienfait, H.F. (1988b) Mechanisms in Fe-efficiency reactions of higher plants, *J. Plant Nutr* **11**, 605-629.

Bienfait, H.F. (1989) Prevention of stress in iron metabolism of plants, *Acta Bot. New Zl J.* **38**, 105-129.

Bingham, F.T., Elseewi, A. and Oertli, J.J. (1970) Characteristics of boron absorption by barley roots, *Soil Sci. Soc. Am. Proc* **34**, 613-618.

Brown, J.C., Chaney, R.L., and Ambler, J.E. (1971) A new tomato mutant inefficient in the transport of iron, *Physiol. Plant* **25**, 48-51.

Blamey, F.P.C., Vermuelen, W.J. and Chapman, J. (1979) Critical boron concentration in plant tissues of two sunflower cultivars. *Agron. J.* **71**, 243-247.

Bowen, J.E. and Nissen, P. (1976) Boron uptake by excised barley roots. I. Uptake into the free space, *Plant Physiol.* **57**, 353-357.

Brodrick, S.J. and Giller,K.E. (1991) Genotypic difference in molybdenum accumulation affects N_2-fixation in tropical *Phaseclus vulgaris* L, *J. Experimental Botany* **42**, 1339-1343.

Brown , P.H. and Hu, H. (1993) Boron uptake in sunflower, squash and cultured tobacco cells-studies with stable isotope and ICP-MS, *Plant Soil* **156**, 147-150.

Brown, P.H. and Hu, H. (1994) Boron uptake by sunflower, squash and cultured tobacco cells, *Physiol. Plant.* **91**, 435 - 441.

Brown , P.H. and Hu, H. (1996) Phloem mobility of boron are species dependent: evidence for phloem mobility in sorbitol-rich species. *Ann. Bot* **77**, 497-505.

Brown, J.C. and Jolley, V.D. (1989) Plant metabolic responses to iron deficiency stress, *BioScience* **39**, 546-550.

Brown, J.C. and McDaniel, M.E. (1978) Factors associated with differential response of two oat cultivars to zinc and copper stress, *Crop Sci* **18**, 817-820.

Brown, P.H., Welch, R.M., and Cary, E.E. (1987) Nickel: a micronutrient essential for all higher plants, *Plant Physiol* **84**, 801-803.

Brown, J.C., Chaney, R.L., and Ambler, J.E. (1971) A new tomato mutant inefficient in the transport of iron, *Physiol. Plant* **25**, 48-54.

Bruggemann, W., Moog, P.R., Nakagawa, H., Janiesch, P., and Kuiper, P.J.C. (1990) Plasma membrane-bound NADH: Fe^{3+}-EDTA reductase and iron deficiency in tomato (*Lycopersicon esculentum*). Is there a intro-reductase? *Physiol. Plant* **79**, 339-346.

Buckhout, T.J., Bell, P.F., Luster, D.G., and Chaney, R.L. (1989) Iron-stress induced redox activity in tomato (*Lycopersicum esculentum* Mill) is localized on the plasma membrane, *Plant Physiol* **90**, 151-156.

Cakmak, I. and Marschner, H. (1988) In crease in membrane permeability and exudation of roots of zinc deficient plants, *J. Plant Physiol* **132**, 356-361.

Cakmak, I., Gulut, K.Y., Marschner, H., and Graham, R.D. (1993) Effect of zinc and iron deficiency on phytosiderophore release in wheat genotypes differing in zinc efficiency, *J. Plant Nutr.*

Cakmak, I., Derici, R., Torun, B. Tolay, I., Braun, H.J., and Schlegel, R. (1997) Role of rye chromosomes in improvement of zinc efficiency in wheat and triticale. In T. Ando et al., (eds.) *Plant nutrition-for sustainable food production and environment*, pp 237-241. Kluwer Acad. Publ. Japan.

Carroll, M.D. and Loneragan, J.F. (1968) Response of plant species to concentrations of zinc in solution, I. Growth and zinc content of plants. Austr. *J. Agree. Res.* **19**, 859-868.

Chapman, V.J., Edwards, D.G., Blamey, F.P.C., and Asher, C.J. (1997) Challenging the dogma of a narrow supply ranges between deficiency and toxicity of boron. In: R.W. Bell and B. Rerkasem (Eds.) *Boron in Soils and Plants*, Proceedings, Kluwer Acad. Publ., Dordrecht, pp. 151-155

Chen, C. and Rerkasem, B. (1993) Effects of boron on pollen viability in wheat, *Plant and Soil* **155/156**, 313-315.

Cataldo, D.A., Garland, T.R., and Wildung, R.E. (1978) Nickel in plants. I. Uptake kinetics using intact soybean seedlings, *Plant Physiol* **62**, 563-565.

Chaney, R.L., Chen, Y., Green C.E., Holden, M.J., Bell, P.F., Luster, D.G., and Angle, J.S. (1992) Root hairs on chlorotic tomatoes are an effect of chlorosis rather than part of the adaptive Fe-stress response, *J. Plant Nutr.* **15**, 1857-1862.

Chaney, R.L., Brown, J.C., and Tiffin, L.T. (1972) Obligatory reduction of ferric chelates in iron uptake by soybean, *Plant Physiol* **50**, 208-213.

Chairidchai , P., and Ritchie, G.S.P. (1993) Zinc adsorption by sterilized and non-sterilized soil in the presence of citrate and catechol. Commun, *Soil Sci. Plant Anal* **24**, 261-275.

Clarkson, D.T. (1988) Movements of ions across roots. Pp.251-304. In D..A. Baker and J.L. Hall (ed.) *Solute transport in plant cells and tissues*, Longman Scientific and Technical, Essex, England

Cohen, C.K., Fox, T.C., Garvin, D.F., and Kochian, L.V. (1998) The role of iron-deficiency stress responses in stimulating heavy-metal transport in plants. *Plant Physiol* **116**, 1063-1072

Cohen, C.K., Norvell, W.A., and Kochian, L.V. (1997) Induction of the root cell plasma membrane ferric reductase-An exclusive role for Fe and Cu, *Plant Physiol* **114**, 1061-1069

Conner, R.L., MacDonald, M.D., and Whelan, E.D. (1988) Evaluation of take-all resistance in wheat-alien amphiploid and chromosome substitution lines, *Genome* **30**, 597-601

Crane, F. L., and Barr, R. (1989) Plasma membrane oxidoreductase, *Critical Reviews in Plant Sci.* **8**, 273-307

Crowley, D.E., Wang, Y.C., Reid, C.P.P., and Szaniszlo, P.J. (1991) Mechanisms of iron acquisition from siderophores by microorganisms and plants: *In* Y. Chen and Y. Hadar. (eds.) *Iron Nutrition and Interactions in Plants*, VCH Chemie, Weinheim, Germany, pp. 213-232

Dannel, F., Perffer H. and RÖmheld, V. (1997) Effect of pH and boron concentration in the nutrient solution on translocation of boron in the xylem of sunflower. In Bell, R.W., Rerkasem, B.(eds.) *Boron in Soils and Plants* ,.Kluwer Academic Publishers, pp.183- 187

Dancis, A., Klausner, R.D., Hinnebusch , A.G., and Barriocanal, J.G. (1990) Genetic evidence that ferric reductase is required for iron uptake in Saccharomyces cerevisiae, *Mol. Cell. Biol.* **10**, 2294-2301

Dancis, A., Roman, D.G., Anderson, G.J., Hinnebusch, A.G., and Klausner, R.D. (1992) Ferric reductase of Sacchromyces cerevisiae. Molecular characterization, role in iron uptake, and transcriptional control by iron, *Proc. Natl. Acad. Sci. U.S.A.* **89**, 3869-3872.

de Wet, E. and Robbertse, P.J. (1989) The influence of temperature and boron on pollen germination in *Mangifera indica* L. Sth. Africa, *J. Plant Soil* **6**, 228-234.

Eskew, D.L., Welch, R.M., and Norvell, W.A. (1984) Nickel: an essential micronutrient for legumes and

possibly all higher plants, *Science* **222**, 621-623.
Fehr, E.R. (1984) Current practices for correcting iron deficiency in plants with emphasis on genetics, *J. Plant Nutr* **7**, 347-351.
Faber, B.A., Zasazki, R.J., Burau, R.G., and Uriu, K. (1990) Zinc uptake by corn as affected by vescular-arbuscular mycorrhizae, *Plant Soil* **129**, 121-130.
Dannel, F., Pfeffer, H., and RÖmheld, V. (1997) Effect of pH and boron concentration in the nutrient solution on translocation of boron in the xylem of sunflower. In R.W. Bell and B. Rerkasem (eds) *Boron in Soils and Plants*, Kluwer Acad. Publ. Dordrecht, pp. 183-187.
Forgo, D A., Yoshida, S., and Asher, C.J. (1975) Zinc deficiency in rice. 1. Soil factors associated with the deficiency, *Plant Soil* **42**, 537-550.
Garg , O.K., Sharma, A.N., and Kona, G.R.S.S. (1979) Effect of boron on the pollen vitality and yield f rice plants (*Oryza sativa* L. var. Jaya), *Plant and Soil* **52**, 591-594.
George, E., RÖmheld, V., and Marschner, H. (1994) Contribution of mycorrhizal fungi to micronutrient uptake by plants. In J..A. Manthey et al., (eds.) *Biochemistry of Metal Micronutrients in the Rhizosphere*, Lewis Publishers, London. pp. 93-110
Gilmore, A.E. (1971) The influence of endotriphic mycorrhizae on the growth of peach seedlings, *J. Amer. Soc. Hort. Sci.* **96**, 35-38.
Graham, R.D. (1984) Breeding for nutritional characteristics in cereals, *Adv. Plant Nutr.* **1**, 57-102.
Graham, R.D. (1988) Genotypic differences in tolerance to manganese deficiency. In Graham, R.D., R.J. Hannam, and N.C. Uren (eds.) *Manganese in Soils and Plants*, Kluwer Acad. Publ., Dordrecht. pp. 261-284.
Graham, R.D., and Webb, M.J. (1991) Micronutrient and disease resistance and tolerance in plants. In Morovedt, J.J., Cox, F.R., Shuman, L.M., and Welch, R.M., (eds.) *Micronutrients in Agriculture,* Soil Science Society of America., Madison, WI, pp329-370.
Graham, R.D., Asher, J.S., and Hynes, S.C. (1992) Selecting zinc-efficient cereal genotypes for soils low in zinc status, *Plant Soil* **146**, 241-250.
Graham, R.D. and Rengel, Z. (1993). Genotypic variation in zinc uptake and utilization by plants. In Robson AD et al. (eds) *Zinc in Soils and Plants*, Kluwer Acad. Publ. Dordrecht, pp 107-118.
Graham, R.D. (1981) Absorption of copper by plant roots. In J..F. Loneragan et al., (ed.) *Copper in soils and plants*. Academic Press, New York, pp.141-163.
Grewal, H.S., Lu, Z.G., and Graham, R.D. (1997) Influence of subsoil zinc on dry matter production, seed yield and distribution of zinc in oilseed rape genotypes differing in zinc efficiency, *Plant Soil* **192**, 181-189.
Grotz, N., Ox, T., Connolly, E., Park, W., Guerinot, M.L., and Eide, D. (1998) Identification of a family of zinc transporter genes from Arabidopsis that respond to zinc deficiency, *Proc. Nat. Acad. Sci. of USA* **95**, 7220-7224.
Grusak, M.A., Welch, R.M., and Kochian, L.V. (1990a) Physiological characterization of a single-gene mutant of *Pisum sativum* exhibiting excess iron accumulation. I. Root iron reduction and iron uptake, *Plant Physiol.* **93**, 976-981.
Grusak, M.A., Welch, R.M., and Kochian, L.V. (1990b) Dose iron deficiency in Pisum sativum enhance the activity of the root plasmalemma iron transport protein? *Plant Physiol* **94**, 1353-1357.
Harrison, S.J., Lepp, N.W., and Phipps, D.A. (1978) Uptake of copper by excised roots. II. Copper desorption from the free space. *Z, Pflanzenphysiol* **94**, 27-34.
Hartwig, E.E., Jones, W.F., and Kilen, T.C. (1991) Identification and inheritance of inefficient zinc absorption in soybean, *Crop Sci* **31**, 61-63.
Hewitt, E. J. (1983) A perspective on mineral nutrition: essential and functional metals in plants: In Robb, D.A and Pierpoint, W.S. (eds.) *Metals and Micronutrients: Uptake and Utilization by Plants*, Academic Press, New York, pp. 277-323
Hewitt, E J. (1984) The essential and functional mineral elements. In C. Bould , E.J. Hewitt and P. Needham (eds.) *Diagnosis of Mineral Disorders in Plants Vol1*, Chemical Publishing, NY, pp 7-53.
Higuchi, K., Kanazwa, K., Nishizawa, N.K., and Mori, S. (1996) The role of nicotianamine synthase in response to Fe nutrition status in Gramineae, *Plant and Soil* **178**, 171-177.
Holden, M.J., Luster, D.G., Chaney, R.L., Buckhout, T.J., and Robinson, C. (1994) Fe^{3+}-chelate reductase activity of plasma membranes isolated from tomato (Lycopersicon escuentum Mill.) roots, *Plant Physiol* **97**, 537-544.
Holden, M.J., Luster, D.G., Chaney, R.L., and Buckhout, T.J. (1992) Enzymology of ferric chelate reduction at the root plasma membrane, *J. Plant Nutr* **15**, 1667-1678.
Horden, M.J., Luster, D.G., and Chaney, R.L. (1994) Enzymatic iron reduction at the root plasma membrane:

partial purification of the NADH-Fe chelate reductase. In Manthey, J.A., D.E. Crowley, and D.G. Luster (eds.) *Biochemistry of Metal Micronutrients in the Rhizosphere*. Lewis Publishers, Boca Raton, pp. 285-294

Pfeffer, H., Dannel, F., and Römheld, V. (1997) Compartmentation of boron in roots and its translocation to the shoot of sunflower as affected by short term changes in boron supply, In R.W. Bell and B.Rerkasem (eds.) *Boron in Soils and Plants*, Kluwer Academic Publishers, Dorsdrecht, pp 203-209.

Hu, H.N., and Brown, P.H. (1997) Absorption of boron by plant roots, *Plant Soil* **193**, 49-58.

Hu, H., Brown, P.H. and Labavitch, J.M. (1996) Species variability in boron requirement is correlated with cell wall pectin, *J. Exp. Bot* **47**, 227-232.

Hu ,H. , Penn, S.G., Lebtilla, C.B., and Brown, P.H. (1997) Isolation and characterization of soluble of B-complexes in higher plants ,The mechanism of phloem mobility of boron, *Plant Physiol.* **113**, 649-655.

Huang, C.Y., and Graham, R.D. (1997) Efficient Mn uptake in barley is a constitutive system. In. T. Ando et al., (eds.) *Plant nutrition - for sustainable food production and environment*, Kluwer Acad. Publ., Japan, pp.269-270.

Huang, C.Y., Webb, M.J., and Graham, R.D. (1994) Manganese efficiency is expressed in barley growing in soil system but not in the solution culture, J. Plant Nutr 17, 83-95.

Huang, C.Y., Webb, M.J. and Graham, R.D. (1993) Effect of pH on Mn absorption among barley genotypes in a chelate-buffered nutrient solution. In N.J. Barrow (ed.) *Plant nutrition -from genetic engineering to field practice*, Proc. XIIth Int. Plant Nutr. Colloq. Kluwer Acad. Publ. Dordrecht, pp. 653-657.

Ishii, T. and Matsunaga,T.,(1996)Isolation and characterization of a boron-rhamnogalacturoman compiex from cell walls of sugar beet pulp, *Carbohydr.Res* **284**, 1-9.

Ishii T and Matsunaga T (1996) Isolation and characterization of a borron-rhamnogalacturonan-II complex from cell walls of sugar beet pulp, *Carbohydr. Res.* **284**, 1-9.

Ishizuka, J. (1982) Characterization of molybdenum absorption and translocation in soybean plants. *Soil Sci. Plant Nutr.* (Tokyo) **28**, 63-78.

Jamjod, S., Mann, C.E., and Rerkasem, B. (1992a) Combining ability of the response to boron deficiency in wheat ,In.. P.J. Randall, E. Delhaize, R A Richards and R Munns (eds.) *Genetic Aspects of Plant Mineral Nutrition*. Dev. *Plant Soil Sci.* **50**, 359-362.

Jamjod, S., Paull, J. G., Brooks, B.J., and Rathjen, A.J. (1997) Genetic variation in the tolerance of durum wheat (Triticum turgidum L. var. durum) to high concentrations of boron.. In Bell RW and Rerkasem B (eds.) *Boron in Soils and Plants*, Kluwer Acad. Publ. Dordrecht, pp. 111-115.

Jolley, V.D., and Brown, J.C. (1991) Factors of iron-stress response mechanism enhanced by Zn-deficiency stress in Sanilac, but not Saginaw navy bean, *J Plant Nutr* **14**, 257-265.

Jones, J. B. Jr. (1991) Plant tissue analysis in micronutrients. In Morovedt, J.J., Cox, F.R., Shuman, L.M., and Welch, R.M. (eds.) *Micronutrients in Agriculture*, SSSA Book Series no. 4, SSSA , Madison. WI, pp.523-548.

Kabata-Pendias, A., and Pendias, H. (1992) *Trace Elements in Soils and Plants*. 2nd Ed., CRC Press, Boca Raton.

Kalayci, M., Alkan, A., Cakmak, I., Bayramouglu, O., Yilmaz, A., Aydin, M., Ozbek, V., and Ozberisoy, F. (1998) Studies on differential response of wheat cultivars to boron toxicity, *Euphytica* **100**, 123-129.

Kanazawa, K., Higuchi, K., Nishizawa, N.K., Fushiya, S., and Mori, S. (1995) Detection of two distinct isozymes of nicotiamine aminotransferases in Fe-deficient barley roots, *J. Exp. Botany* **46**, 1241-1244.

Kastrup. V., Steiger, S., Luttge U,, and Fischer Schliebs, E. (1996) Regulatory effects of zinc on corn root plasma membrane H^+-ATPase, *New Phytologist* **134**, 61-73.

Kaszewkski K.L. and Blevins D.G. (1996) Root growth inhibition in boron-deficient or aluminum stressed squash may be a result of impaired ascorbate metabolism, *Plant Physiol* **112**, 1135-1140.

Kaur, N.P., Takkar, P.N., Arora,C.L. and Nayyar, V.K. (1989) Relative Mn efficiency of wheat cultivar-HD 2009. Indian, *J. Plant Physiol.* **32**, 306-310.

Kawai, S., Ito,K., and S. Takagi, S. (1988) Biosynthesis pathway of mugineic acid in the roots of Fe-deficient barley, *Soil Sci. Plant Nutr.* **34**, 73-77.

Kastrup, V.S., Steiger, S., Luttge, U., and FischerSchliebs, E. (1996) Regulatory effects of zinc on corn root plasma membrane H^+-ATPase, *New Phytol* **134**, 61-73.

Keerati-Kasikorn , P., Bell, R.W., Panya, P., Gilmour, R.F., and Loneragan, J.F. (1993) Comparison of seed yield and quality of peanut (Arachis hypogaea L.) Cultivars in low fertility soils and their response to boron and complete fertilizer. In N.J. Barrow (ed.) *Plant Nutrition from Genetic Engineering to Field Practice*, Kluwer Acad. Publ., Dordrecht, pp. 409- 412.

Khan, H.R., McDonald, G.K., and Rengel, Z. (1998) Chickpea genotypes differ in their sensitivity to Zn deficiency, *Plant Soil* **198**, 11-18.

Kneen, B.E., LaRue, T.A., Welch, R.W., and Weeden, N.F. (1990) Pleiotropic effects of brz, a mutation in pisum sativum (L), CV "sparkle" conditioning decreased nodulation and increased iron uptake and leaf necrosis, *Plant Physiol.* **93**, 717-721.

Kobayshi, M., Motoh,T., and Azuma, J.L.(1996)Two chains of rhamnogalacturonan are cross-linked by orate-diol ester bonds in higher plant cell walls, *Plant Physiol.* **110**, 1017-1020

Kobayashi, M., Matoh, T. ,and Azuma J.L.(1996)Two chains of rhamnogalacturonan II are cross-linked by borate-diol ester bonds in higher plant cell walla, *Plant Physiol.* **110**, 1017-1020.

Kochian, L.V. (1991) Mechanisms of micronutrient uptake and translocation in plants. In Morovedt, J.J., F.R. Cox, L.M. Shuman, and R.M. Welch (eds.) *Micronutrients in Agricultur,* SSSA Book Series no. 4, SSSA , Madison. WI, pp 229-296.

Kochian, L.V. (1993) Zinc absorption from hydroponic solutions by plant roots. In Robson, A.D. (ed.) *Zinc in Soils and Plants.* Kluwer Acad. Publ., Dordrecht, pp. 15-58.

Kochian ,LV. (1995) Zinc absorption from hydroponic solutions by plant roots. In Robinson A.D. et al. (eds) *Zinc in Soils and Plants.* Kluwer Acad. Publ. Dordrecht, pp 45-57.

Kochian, L.V., Norvell, W.A., Shaff, J.E., and Chaney, R.L. (1991) Characterizing root iron uptake using a ferrous chelate to buffer free Fe^{2+} ion activity in solution, *Plant Physiol.* **96**, 5142-5146.

Kochian, L.V. and Lucas,W.J. (1991) Do plasmalemma oxidoreductases play a role in plant mineral ion transport? In Crane, F.L., D.J. Morre, and H.E. Low (eds.) *Oxidoreduction at the Plasma Membrane: Relation to Growth and Transport,* Vol. II, Plants CRC Press, Boca Raton, pp. 189-205.

Korcak, R.F. (1987) Iron deficiency chlorosis. *Hortic. Rev.* **9**, 133-138.

LeNoble, M.E.,Blevins, D.G., Sharp, E., and Cumbie, B.G. (1996a) Prevention of aluminum toxicity with supplemental boron I. Maintenance of root elongation and cellular structure, *Plant Cell Environ* **19**, 1132-1142.

LeNoble, M.E, Blevins, D.G., and Miles, R.J. (1996b) Prevention of aluminum toxicity with supplemental boron II. Stimulation of root growth in an acidic, high-aluminum subsoil, *Plant Cell Environ* **19**, 1143-1148.

Lesuisse, E., Crichton, R.R., and Labbe, P. (1990) Iron-reductase in the yeast Saccharomyces cerevisiae. *Biochim. Biophys. Acta.* **1038**, 253-261.

Liu, L.X. and Shelp,B.J. (1996) Impact of chloride on nitrate absorption and accumulation by broccoli (Brassica oleracea var. italica), *Can. J. Plant Sci* **76**, 367-377.

Liu, Z. (1996) Microelements in soil of China, *Jiasngsu Science and Technology Publishing*

Longnecker, N.E., Graham, R.D., McCarthy, K.W., Sparrow, D.H.B., and Egan, J.P.(1990) Screening for manganese efficiency in barley (*Hordeum valgare* L.). In Webb, M.J., R. Nable, R.D. Graham, and R.J. Hannam (eds.) Int. Symposium Manganese in Soils an Plants : *Contributed papers.* Adelaide.

Luo Y.Sh. (1998) Boron uptake by different oilseed rape (Brassica napus L.) genotypes as influenced by soil moisture and NPK fertilizers. *Ph.D Thesis.* Zhejiiang Agricultural University, Hangzhou, China

Loomis W.D. and Drust R.W. (1992) Chemistry and biology of boron, *Biofactors* **3**, 229-239.

Majumder, N.D., Rashit, S.C., and Borthakur D.N. (1990) Genetic effects on uptake of selected nutrients in some rice (Oryza sativa L.) Varieties in phosphorus-deficient soil, *Plant and Soil* **123**, 117-120.

Makkee, M., Kieboom, A P.G., and Van Bekku, H. (1995) Studies on borate esters III. Borate esters of D-mannitol, D-glucitol,D-fructose and D-glucose in water. *Recl.Trav, Chim. Pays-Bas* **104**, 230-235.

Marcar, N.E. and Graham, R.D. (1987) Genotypic variation for manganese efficiency in wheat, *J. Plant Nutr.* **10**, 20-49.

Manthey, J.A., Crowley, D.E., and Luster, D.G. (ed.) (1994) *Biochemistry of Metal Micronutrients in the Rhizosphere,* Lewis Publishers, Boca Raton, pp. 285-294.

Marschner, H. (1995) Mineral Nutrition of Higher Plants. 2nd Ed. Academic Press. London

Marschner, H., Römheld, V., and Kissel, M. (1986) Different strategies in higher plants in mobilization and uptake of iron, *J. Plant Nutr.* **9**, 695-697.

Marschner, H. (1993) Zinc uptake from soils. In Robinson A D et al. (eds). *Zinc in Soils and Plants,* Kluwer Acad. Publ. Netherlands, pp. 107-118.

Marschner, H., Treeby, H. M., and Römheld, V. (1989) Role of root induced changes in the rhizosphere for iron acquisition in higher plants. *Z. Pflanzenernahr Bodenk.* **152a**, 197-204.

Marschner, H. (1988) Mechanisms of manganese acquisition by roots from soilsIn R.D. Graham et al., (eds.) *Manganese in Soils and Plants,* Kluwer Acad. Publ.. Dordrecht, pp. 191-205.

Martens, D.C. and Westermann, D.T.(1991) Fertilizer applications for correcting micronutrient deficiencies. In Morovedt, J.J., F.R. Cox, L.M. Shuman, and R.M. Welch, (eds.) *Micronutrients in Agriculture,* SSSA Book Series no. 4, SSSA , Madison. WI, pp.5 49-592

Martini , F. and Thellier, M. (1993) Boron distribution in parenchyma cells of cloverleaves, *Plant Physiol.*

Biochem. **31**, 777- 786.
Mengel, K. (1994) Iron availability in plant tissues – iron chlorosis on calcareous soils, Plant and Soil 165, 275-283.
McCarthy, K.W., Longnecker, N.E., Sparrow, D.H.B., and Graham, R.D.(1988) Inheritance of manganese efficiency in barley (*Hordeum vulgare* L.). In Webb, M.J., R. Nable, R.D. Graham, and R.J. Hannam. (eds.) Int. Symposium Manganese in Soils and Plants, *Contributed papers*. Adelaide.
Miller, E.R., Lei, X., and Ullrey, D.E.(1991) Trace Elements in animal nutrition. In: Morovedt, J.J., F.R. Cox, L.M. Shuman, and R.M. Welch (eds.) *Micronutrients in Agriculture*, SSSA Book Series no. 4, SSSA, Madison. WI, pp. 593-662.
Mihashi, S., Mori, S., and Nishizawa, N.K. (1991) Enhancement of mugineic acid-Fe^{3+} uptake by Fe-deficient barley roots with coexistence of excess free mugineic acid in the medium, *Plant and Soil* **30**, 135-140.
Mihashi, S. and Mori, S. (1989) Characterization of mugineic acid-Fe transporter in Fe-deficient barley roots using the multicompartment transport box method, *Biol. Metals* **2**, 146-149.
Mori, S. and Nishizawa, N. (1987) Methionine as a dominant precursor of hytosiderophores in graminaceous plants, *Plant Cell Physiol* **28**, 1081.
Mori, S., Nishizawa, N., Kawai, S., Sato, Y., and Takagi, S. (1987) Dynamic state of mugineic acid and analogous phytosiderophores in Fe-deficient barley, *J. Plant Nutr* **10**, 1003.
Mori, S., Nishizawa, N., Hayashi, H., Chino, M., Yoshimura, E., and Ishihara, J. (1991) Why are young rice plants highly susceptible to iron deficiency? *Plant and Soil* **130**, 143-147.
Mori, S., and Nishizawa, N. (1989) Identification of barley chromosome no.4, possible encoder of genes of mugineic acid synthesis from 2_-deoxymugeneic acid using wheat-barley addition lines, *Plant Cell Physiol* **30**, 1057.
Mori, S., Nishizawa, N., and Fujigaki, J. (1990) Identification of rye chromosome 5R as a carrier of the genes for mugineic acid synthease and 3-hydroxymugineic acid synthease using wheat-rye addition lines. *Jpn. J. Genet.* **65**, 343.
Mori, S., Hachisuka, M., Kawai, S., Takagi, S., Nishizawa, N.K. (1988) Peptides related to phytosiderophore secretion Btu Fe-deficient barley roots.
Mori, S. (1997) Reevaluation of the genes induced by iron deficiency in barley roots. In: T. Ando et al., (ed.) *Plant Nutrition-for sustainable food production and environment*, Kluwer Acad. Publ. Japan, pp. 249-254.
Mori, S. (1994) Mechanisms of iron acquisition by graminaceous (Strategy II) plants. In: Manthey, J.A., D.E. Crowley, and D.G. Luster (eds.) *Biochemistry of metal micronutrient in the erhizosphere*, Lewis Publishers. London, pp. 225-250.
Mortvedt, J.J., Cox, F.R., Shuman, L.M., and Welch, R.M. (1991) Micronutrients in Agriculture. 2 nd Edition, Soil Science Society of America, Madison, WI
Mullins, G. L. and Sommers, L.E. (1986) Cadmium and zinc flux characteristics by intact corn (*Zea mays* L.) Seedlings, *Plant Soil* **96**, 153-164.
Nable, R.O., Banuelos, G.S., and Paull, J.G. (1997) Boron toxicity, *Plant Soil* **193**, 181-198.
Nakanishi, H., Okumura, N., Umehara,Y., Nishizawa ,N.K., Chino, M., and Mori, S. 1993 Expression of a gene specific for iron deficiency (ids3) in the roots of *Hordeum vulgare, Plant Cell Physiol.* 34, 401-410.
Nielsen, F.H. (1992) Nutritional requirements for boron, silicon, vanadium, nickel, and arsenic: current knowledge and speculation, *FASEB J* **5**, 2661-2667.
Nishizawa, N.K., Mori, S., Takahashi, T. Ueda (1989) Mugeneic acid secretion by cultured barley cells derived from anther, *Protoplasma* **148**, 264-270.
Nomoto, K., Yoshioka, H., Arima, M., Takemoto, T., Fushiya, S., and Takagim, S.(1981) Structure of 2'-deoxymugeic acid, a novel amino acid possessing an iron chelating activity, *Chimia* **35**, 249-254.
Nomoto, K., Sugiura, Y., and Takagi, S. (1987) Mugineic acid, studies on phytosiderophores. In Winkelmann, G., D. Van der Helm, J.B. Neilands (eds.) *Iron Transport in Microbes, plants and Animals*, **chap. 22**.. VCH Publishers, Weinheim, Germany.
Nobel, P.S. (1983) Biophysical plant physiology and ecology. W.H. Freeman and Co., New York
Nyborg, M. (1970) Sensitivity to manganese deficiency of different cultivars of wheat, oat and barley, *Can, J. Plant Sci.* **50**, 198-203.
Okumura, N., Nishizawa, N.K., Umehara,Y., Ohata, T., and Mori, S. (1992) Iron-deficiency specific cDNA with 2 homologous cysteine rich MT domains from the roots of barley, *J. Plant Nutr* **15**, 2157-2172.
Okumura, N., Nishizawa, N.K., Umehara,Y., and Mori, S. (1991) An iron deficiency specific cDNA from barley roots having two homologous cystein-rich MT domains, *Plant Mol. Biol.* **17**, 531-533.
Okumura, N., Nishizawa, N.K., Umehara,Y., Ohata, T., Nakanishi, H., Yamaguchi, T., Chino, M., and Mori, S. (1994) A dioxygenase gene (ids2) expressed under iron-deficiency conditions in the roots of *Hordeum*

valgare, Plant Mol. Biol. **25,** 705-719.

Oida, F., Ota, N., Mino, Y., Nomoto, K., and Sugiura, Y. (1989) Stereospecific iron uptake mediated by phytosiderophore in graminaeous plants, *J. Am. Chem. Soc.* **111**, 3436-3437.

O'Neill, M.A., Warrenfeltz, D., Kate, K., Pellerin, P., Doco, T., Darvill, A. G. and Albersherim, P. (1996) Rhamnogalacturonan-II, a pentic polysaccharide in the walls of growing plant cells, forms a dime that is covalently cross-linked by a borate ester, *.J. Biol. Chem.* **271**, 22923-2930.

Quinones, M.A., Giraldez, N., Witt, F.G., and Aparicio, P.J.(1997) Blue light-dependent monovalent anion uptake, *Physiologia Plantarum* **100**, 45-52.

Patzer, S.I., and Hantke, K. (1998) The ZnuABC high-affinity zinc uptake system and its regulator zur in Escherichia coli, *Mol.. Microbiol.* **28**, 1199-1210.

Parker, D.R., Aguilera, J.J., and Thomason, D.T. (1992) Zinc-phosphorus interactions in two cultivars of tomato *(Lycoperiscon esculentum* L.) Grown in chelator-buffered nutrient solutions, *Plant Soil* **143**, 263-177.

Paull, J.G., Nable, R.O., and Rathjen, A.J. (1992) Physiological and genetic-control of the tolerance of wheat to high- concentrations of boron and implications for plant-breeding, *Plant Soil* **146**, 251-260.

Pfeffer, H., Dannel, F., and RÖmheld, V. (1997) Compartmentation of boron in roots and its translocation to the shoot of sunflower as affected by short term changes in boron supply. In: Bell, R.W., and Rerkasem, B. (eds.) *Boron in Soils and Plants,*.Kluwer Acad. Publishers, Dordrecht, pp. 203-209.

Pich, A., Scholz, G., and Seifert, K. (1991) Effect of nicotianamine on iron uptake and citrate accumulation in two genotypes of tomato, Lycopersicon esculentum Mill, *J. Plant Physiol.* **137**, 323-327.

Pope, D.T., and Munger, H.M.(1953) The inheritance of susceptibility to boron deficiency in celery, *Amer. Soc. Hort. Sci.* **61**, 163-172.

Raven, J.A. (1980) Short- and long-distance transport of boric acid in plants, *New Phytol.* **84**, 231-249.

Rerkasem , B. and Jamjod, S. (1997a) Genotypic variation in plant response to low boron and implications for plant breeding, *Plant and Soil* **193**,168-180.

Rerkasem , B. and Jamjod, S. (1997b) Boron deficiency induced male sterility in wheat (Triticum aestivum L.) And implications for plant breeding *Euphytica*

Rerkasem, B., Bell, R.W., Lordkaew, S., and Loneragan, J.F. (1993a) Boron deficiency in soybean (Glycine max (L.) Merr.), peanut (Arachis hypogaea L.) And black gram (Vigna mungo (L.) Hepper.): Symptoms in seeds and differences among soybean cultivars in susceptibility to boron deficiency, *Plant and Soil* **150**, 289-294.

Rerkasem, B., Netsangtip, R., Lordkaew, S., and Cheng,, C.(1993b) Grain set failure in boron deficiency, *Plant and Soil* **155/156**, 309-312.

Rerkasem , B., Netsangtip, R., Bell, R.W., Loneragan ,J.F., and Hiranburana ,N. (1988) Comparative species responses to boron on a Typic Tropaqualf in Northern Thailand, *Plant and Soil* **106**, 15-21.

Rengel, Z. (1997) Root exudation and microflora populations in rhizosphere of crop genotypes differing in tolerance to micronutrient deficiency, In T. Ando et al., (eds.) *Plant nutrition-for sustainable food production and environment.* Kluwer Acad. Publ. Japan, pp. 243 -248.

Rengel, Z., Pedler, J.F., and Graham, R.D. (1994) Control of Mn status in plants and rhizosphere: Genetic aspects of host and pathogen effects in the wheat take-all interaction. In: J.A. Manthey et al. (eds.) *Biochenistry of Metal Micronutrients in the Rhizosphere*, Lewis Publishers, London, pp. 125-146.

Reid, R. J., Brookes, J.D., Tester, M.A., and Smith, F..A. (1996) The mechanism of zinc uptake in plants-characterization of the low-affinity system, Planta 198,39-45.

Rerkasem, B. (1990) Comparison of green gram (*Vigna radiata*) and black gram (*Vigna mungo*) in boron deficiency. In Thavarasook, C., P. Srinives, N. Bookerd, H. Imai, A. Pookpakdi, P. Laosuwan, and U. Pupipat (eds.) , *Proc. of the Mungbean Meeting 90*, Bangkok Office of Tropical Agree. Res. Center, Japan, pp. 167-174.

Rerkasem, B., and Jamjod. S.(1997) Genotypic variation in plant response to low boron and implications for plant breeding, *Plant Soil* **193**, 169-180.

Robb, D.A. and Pierpoint, W.S. (eds.) (1983*) Metals and Micronutrients: Uptake and Utilization by Plants*, Academic Press, New York.

Robson, A.D., and Pitman, M.G. (1983) Interactions between nutrients in higher plants. In: A. And Bieleski, R.L. (eds.) *Inorganic plant nutrition Lauchli*, Springer-Verlag, Berlin, pp. 147-180.

Robinson, N.J., Sadjuda and Groom ,Q.J.(1997) The froh gene family from Arabidopsis thaliana: Putative iron-chelate reductase. In T. Ando et al. (eds*) Plant Nutrition- fr sustainable food production and environment,* Kluwer Acad. Publ., Japan. , pp.191- 194.

Roman, D.G., Dancis, A., Anderson, G.J., and Klausner, R.D. (1993) The fission yeast ferric reductase gene *frp1+* is required for ferric iron uptake and encodes a protein that is homologous to the gp91-phox subunit of the human NADPH phagocyte oxidoreductase. *Mol. Cell. Biol.* **13**, 4342-4350.

Römheld, V. (1987) Different strategies for iron acquisition in higher plants, *Physiol. Plant* **70**, 231-234.

Römheld , V.(1991) The role of phytosiderophores in acquisition of iron and other micronutrients in graminaceous species: an ecological approach. In Chen Y. Amd Hadar, Y. (eds.) *Iron Nutrition and Interactions in Plants,* Kluwer Acad. Publ. Dodrecht, pp. 159-166.

Römheld, V. and Marschner, H. (1986) Evidence for a specific uptake system for iron phytosiderophore in roots of grasses, *Plant Physiol* **80**, 175-180.

Römheld, V., Muller, C., and Marschner, H. (1984) Localization and capacity of proton pumps in roots of intact sunflower plants, *Plant Physiol.* **76**, 603-606.

Römheld ,V. and Marschner, H. (1986) Evidence for a specific uptake system for iron phytosiderophores in roots of grasses, *Plant Physiol* **80**, 175.

Römheld, V. and Marschner, H.(1990) Genotypic differences among graminaceous species in release of phytosiderophores and uptake of iron phytosiderophores, *Plant and Soil* **123**, 147-153.

Runjin, L. (1989) Effect of vesicular arbuscular mycorrhizas and phosphorus on water status and growth of appl,. *J. Plant Nutr.* **12**, 997-1017.

Santa Maria, G.E., and Cogliatti, D.H. (1988) Bi-directional Zn-fluxes and compartmentation in wheat seedling roots, *J. Plant Physiol.* **132**, 312-315.

Saberi, H.K., Graham, R.D., and Rathjen A.J. (1997) Genotypic variation for Mn efficiency in durum wheat (*Triticum turgidium* L. Var. durum). In T. Ando et al. (eds). *Plant Nutrition- for sustainable food production and environment,* Kluwer Acad. Publ. Japan, pp. 289-290.

Schelegel, R., Kynast, R., Schwarzacher, T., Römheld ,V., and Walter, A. (1993) Mapping of genes for copper efficiency in rye and the relationship between copper and iron efficiency, *Plant Soil* **154**, 61-65.

Schlegal, R., Cakmak, I., Torun, B., Eker, S., Tolay, I., Ekiz, H., Kalaycy, M., and Braun, H.J. (1997) Screening for zinc efficiency among wheat relatives and their utilization for an alien gene transfer. In *Proc. of the 5th International Wheat Conference*. Kluwer Acad. Publ., Dordrecht.

Schmidt, W.E., Haag,, H.P., and Epstein, E. (1965) Absorption of zinc by excised barley roots, *Physiol. Plant.* **18**, 860-869.

Schmidt, W., Janiesch, P., and Bruggemann, W. (1990) Fe-EDTA reduction in roots of Plantago lanceolata by a NADH-dependent plasma membrane-bound redox system, *J. Plant Physiol* **136**, 51-55.

Schmidt, A. and Jager, K. (1992) Open questions about sulfur metabolism in plant. *Ann. Rev. Plant Physiol* **43**, 325-349.

Scholz, G., Schlesier, G., and Seifert, K. (1985) Effect of nicotanamine on iron uptake by the tomato mutant A*chloronerva, Physiol. Plant* **63**, 99.

Schlegel, R. and Cakmak, I. (1997) physical mapping of rye genes determing micronutrient efficiency in wheat. In T. Ando et. al., (eds.) *Plant Nutrition - for sustainable food production and environment,* Kluwer Acad. Publ., Japan, pp. 287-288.

Schlegel, R., Kynast, R., Schwarzacher, T., Römheld, V., and Walter, A. (1993) mapping of genes for copper efficiency in rye and the relationship between copper and iron efficiency, *Plant and Soil* **154**, 61-65.

Sharma, U.C. and Grewal. J.S. (1990) Potato response to zinc as influenced by genetic variability, *J. Indian Potato Assoc.* **17**, 1-5.

Sharma, A.K., Srivastava, P.C., and Johri, B.N. (1994) Contribution of VA mycorrhizal to zinc uptake in plants. In Manthey, J..A., D.E. Crowley, and D.C. Luster (eds) *Biochemistry of metal micronutrients in the rhizosphere*, Lewis Publisher, London, pp. 111-124.

Shelp, B.J., Marentes, E., Kitheka, A.M., and Vevekanandan, P. (1995) Boron mobility in plants, *Physiol. Plant* **94**, 356-361.

Shelp, B.J., Penner, R., and Zhu, Z. (1992) Broccoli (Brassica oleracea var italica) cultivar response to boron deficiency, *Can. J. Plant Sci* **72**, 883-888.

Sherrell, C.G. (1983) Effects of boron application on seed production of New Zealand herbage legumes, New Zealand, *J. Exp. Agree.* **11**, 113-117.

Shkonik, M.Y. (1984) Trace Elements in Plants. Elsevier, Amsterdam

Shojima, S., Nishizawa, N.K., and Mori, S.(1989) Establishment of a cell-free system for the biosynthesis of nicotianamine, *Plant Cell Physiol* **30**, 673.

Shojima, S., Nishizawa, N.K., Fushiya, S., Nozoe, S., Irifune, Y., and Mori, S. (1990) Biosynthesis of phytosiderophores - in vitro biosynthesis of 2'-deoxymugineic acid from l-Methionine and nicotianamine, *Plant Physiol* **93**, 1497.

Shojima, S., Nishizawa, N.K., Fushiya, S., Nozoe, S., Kumashiro, T. ,Nagata, T., Ohota, T., and Mori, S. (1989) Biosynthesis of nicotianamine in the suspension-cultured cells of tobacco (*Nicotiana megalosiphon*),

Biol. Metals **2**, 142.
Shukla ,U.C., and Raj (1974) Influence of genetic variability on zinc response in wheat (Triticum ssp.), *Soil Sci. Soc. Am. Proc* **38**, 477-479.
Shukla ,U.C., Arora, S.K., Singh, z., Prasad, K.G., and Safaya., N.M. (1973) Differential susceptibility of some sorghum (*Sorghum valgare*) genotypes to zinc deficiency in soil, *Plant Soil* **39**, 423-427.
Smith ,T.E., Stephenson, R..A., Asher, C.J., and Hetherington, E. (1997) Boron deficiency of avocado. 1.Feffects on pollen viability and fruit set .In: Bell, R.W., B. Rerkasem (eds.) *Boron in Soils and Plants,* Kluwer Academic Publishers, Dordrecht, pp. 131-134.
Spalding, E.P., Slayman, C.L., Goldsmith, M.H.M., Gradmann, D., and Berti, A. (1992) Ion channels in *Arabidopsis* plasma membrane. Transport characteristics and involvement in light-induced voltage changes, *Plant Physiol.* **99**, 96-102.
Stephen, U.W. and Grun, M. (1989) Physiological disorders of the nicotianamine auxotroph tomato mutant *chloronerva* at different levels of iron nutrition. II. Iron deficiency response and heavy metal metabolism, *Biochem. Physiol. Pflanz.* **185**, 189-205.
Swamvinathan, K., and Verma, B.C. (1979) Response of three crop species to vesicular-arbuscular mycorrhizal infections in zinc deficient Indian soils, *New Phytol* **82**, 481-487.
Takagi, S. (1984) Difference of iron absorption mechanisms among the plants. *Studies on Soil Sci. Plant Nutr.*, Tohoku, Japan.
Takagi, S. (1991) Mugineic acid. *Nutrition and Physiology of Metal Relating Compounds*. Hakuyusha Press, Tokyo.
Takkar, P.N., Bansal, R.L., Singh, S.P., and Nayyar, V.K. (1988) Response of pearl millet cultivars to zinc under field conditions, *Intern. J. Tropical Agree* **6**, 247-251.
Tehran, G., Munger, H.M., Robinson ,R.W., and Shannon, S. (1971) Inheritance and physiology of response to low boron in red beet (Beta vulgaris L.). *J. Amer. Soc. Hort. Sci.* **96**, 226-230.
Tester, M. (1990) Tansley Review No. 21. Plant ion channels: whole-cell and single-channel studies. *New Phytol*, **114**, 305-340.
Thellierr ,M., Duvaal, Y., and Demarty, M. (1997) Borate exchanges of Lemna minor L. as studied with the help of the enriched stable isotopes and of a (n,a) nuclear reaction, *Plant Physiol.* **63**, 283-288.
Treeby, M., Marschner, H., and RÖmheld ,V.(1989) Mobilization of iron and other micronutrients from a calcareous soil by plant-borne, microbial, and synthetic metal chelators, *Plant and Soil* **114**, 217-226.
Tyerman, S.D. (1992) Anion channels in plants, *Ann. Rev. Plant Physiol* **43**, 351-373.
Van Campen, D.R. (1991) Trace elements in human nutrition. In Morovedt, J.J., F.R. Cox, L.M. Shuman, and R.M. Welch (eds.) *Micronutrients in Agriculture*, Soil Science Society of America, Madison. WI, pp. 663-701.
Veltrup, W. (1978) Characteristics of zinc uptake by barley roots, *Physiol. Plant* **42**, 190-194.
Wall, J.R. and Andrus, C.F. (1962) The inheritance and physiology of boron response in tomato. *Am. J. Bot.* **49**, 758-762.
Wallwork, H. (1989) Screening for resistance to take-all in wheat, triticale and wheat triticale hybrid lines, *Euphytica* **40**, 103.
Webb, M.J. (1994) Recent aspects of Mn and Zn absorption and translocation in cereals. In Manthey, J.A. et al., (eds.) *Biochenistry of Metal Micronutrients in the Rhizosphere*. Lewis Publishers, London, pp. 183-198.
Webb. M.J., Norvell, W. A., Welch, R.M., and Graham R.D. (1993) Using a chelate-buffered nutrient solution to establish the critical solution activity of Mn^{2+} required by barley (*Hordeum vulgare* L.), *Plant and Soil* **153**, 195-199.
Weiss, M G. (1943) Inheritance and physiology of efficiency in iron utilization in soybeans, *Genetics* **28**, 253-261.
Wei, W.X., Tan, Q L., and Wang, Y.H. (1996) Effect of molybdenum nutrition on the balance of nutrients of winter wheat,. *J. Huazhong Agree. University* **15**, 437-441.
Welch, R.M. (1995) Micronutrient nutrition of plants, Critical Reviews, *Plant Sciences* **14(1)**, 49-82.
Welch, R.M. and LaRue, T.A. (1990) Physiological characteristics of Fe accumulation in the –bronze-mutant of *Pisum sativum* (L.), CV "sparkle" E107 (*brz brz*), *Plant Physiol.* **93**, 723-727.
Welch, R.M. (1981) The biological significance of nickel, *J. Plant Nutr.* **3**, 345-356.
Welch, R.M., Norvell, W.A., Schaefer, S.C., Shaff, J.E., and Kochian, L.V. (1993) Induction of iron (III) and copper (II) reduction in pea(Pisum sativa L.) by Fe and Cu status: Does the root-cell plasmalemma Fe(III)-chelate reductase perform a general role in regulating cation uptake? *Planta*:
Welch, R.M., Webb, M.J., and Loneragan., J.F. (1982) Zinc in membrane function and its role in phosphorous toxicity. In A Scife (ed.) *Plant Nutrition 1982*, Proceedings of the ninth International Plant nutrition

Colloquium Commonwealth Agricultural Bureau, UK, pp. 710-715.

Welkie, G.W. and Miller, G.W. (1993) Plant iron uptake by nonsiderophore systems. In Barton, L.L. and B.C. Hemming (eds.) *Iron Chelation in Plants and soil Microorganisms*, Academic Press, New York, pp. 345-355

Willaert, G., Verloo, M., and Sarrazyn, R. (1990) Influence of the cultivar on the accumulation of inorganic compounds in spinach. *Mededelingen van de Faculteit Landbouwwetenschappen Rijksunniversiteit Gen* **155**, 83-92.

Xue, J.M., Yang, Y., Bell, R.W., and Yang, X. (1998) Differential response of rape varieties to boron deficiency, *Plant and Soil* (in press).

Yang, X., Römheld, V., and Marschner, H. (1993) Effect of bicarbonate and root-zone temperature on uptake of Zn, Fe, Mn, and Cu by different rice cultivars (*Oryza Sativa L.*), *Plant and Soil* **156**, 441-444.

Yang, X., Römheld, V., and Marschner, H. (1994a) Uptake of iron, zinc, manganese, and copper by seedlings of hybrid and traditional rice cultivars from different soil types, *J. Plant Nutr.* **17**, 319-331.

Yang, X., Römheld, V., and Marschner, H. (1994b) Effect of bicarbonate on root-growth and accumulation of organic acids in Zn-inefficient and Zn-efficient rice cultivars (*Oryza Sativa L.*), *Plant and Soil* **164**, 1-7.

Yang, X., Baligar, V.C., Martens, D.C., and Clark R.B. (1996a) Plant tolerance to nickel toxicity 1. Influx, transport, and accumulation of nickel in four species, *J. Plant Nutr* **19**, 73-85.

Yang ,X., Baligar, V.C., Martens, D.C., and Clark, R.B. (1996b) Plant tolerance to nickel toxicity 2.Nickel effects on influx and transport of mineral nutrients in four plants species, *J. Plant Nutr* **19**, 265-279.

Yang ,X., Zhang, C.D., Wang, Y.L., and Römheld, V. (1997) Characteristics of Fe uptake by Ni-tolerant mutant and wild type of ryegrass. In *Proc. The Sixth Int. Iron Sym.*, Hohenheim, Germany.

Yang ,Y., Xue, J., Ye, Z., and Wang. K. (1993) Responses of rape genotypes to boron application, *Plant and Soil* **155/156**, 321-324.

Yehuda, Z., Shenker, M., Römheld, V., Marschner, H., Hadar, Y., and Chen, Y. (1996) The role of ligand exchange in the uptake of iron from microbial siderophores by gramineous plants, *Plant Physiol* **112**, 1273-1280.

Yi, Y., Jennifer, A.S., Guerinot, M.L. (1994) Iron uptake in *Arabidopsis thaliana*. In: Manthey, J.A., D.E. Crowley and D.C. Luster (eds.) *Biochemistry of metal micronutrients in the rhizosphere*, Lewis Publisher, London, pp. 295-308.

Zaharieva, T., Yamshita, K., and Matsumoto, H. (1997) Characterization of ferric chelate reductase in plasma membranes isolated from cucumber roots. In T. Ando et al., (eds.) *Plant Nutrition- for sustainable food production and environment,* Kluwer Acad. Publ, Japan, pp. 125-126.

Zhang, F., Römheld, V., and Marschner, H. (1989) Effect of zinc deficiency in wheat on the release of zinc and iron mobilizing root exudates. *Z. Pflanzen. Bodenk.***152**, 205-210.

Zhang, F., Römheld, V., and Marschner, H.(1991)Release of zinc mobilizing root exudates indifferent plant species as affected by zinc nutritional status, *J. Plant Nutr* **14**, 675-686.

EFFECT OF IRON DEFICIENCY ON TRANSLATABLE mRNA POPULATIONS IN ROOT OF *MEDICAGO SATIVA* (L.)

Y. ARAKAWA[1], Y. MASAOKA[1], K. HIGO[2]
[1]Kyushu National Agricultural Experiment Station
Suya Nishigoshi Kumamoto 861-1192, Japan
[2]National Institute of Agrobiological Resources
Kannondai Tsukuba Ibaraki 305-8602, Japan

1. Abstract

Messenger RNA was isolated from alfalfa (*Medicago sativa* L.) root tissue at various times after the start of iron deficiency treatment and translated *in vitro* in a massenger-dependent wheat germ extract. [^{35}S]-methionine-labelled translation products were then analyzed by 1- and 2-D gel electrophoresis. The changes in mRNA population were detectable in the plant subjected to iron deficiency prior to the occurrence of iron-deficient symptoms. A new translation product with pI=5.4 and M_rs=29kDa was found in day 10 and day 12 samples taken from iron-deficient root. On the other hand, a translation product with pI=5.3 and M_rs=42kDa were suppressed in all iron-deficient samples. Translation products with pI=5.6-6.3 and M_rs=43kDa gradually decreased in day 10 and day 12 samples taken from iron-deficient root. Whether these changes observed are specific to iron deficiency remains to be seen.

2. Introduction

Alfalfa (*Medicago sativa* L.) grows well in soils with a moderately high pH. Like any other dicot, alfalfa lowers its rhizosphere pH by releasing protons and activates extracellular ferric chelate reductase in response to iron deficiency[1]. Moreover, iron-deficient alfalfa excretes alfafuran, a benzofuran derivative, from its roots. This compound is very effective in dissolving ferric phosphate and is structurally different from organic acids or phytosiderophore-type amino acid derivatives[2]. Alfalfa developed multiple strategies against iron deficiency stress.

As the first step toward cloning genes related to iron efficiency in alfalfa, we here describe the changes in mRNA populations occurred in alfalfa root during the course of iron deficiency .

G. Gissel-Nielsen and A. Jensen (eds.), Plant Nutrition – Molecular Biology and Genetics, 187–192.

3. Materials and Methods

3.1 PLANT CULTURE

Seven seedlings of 20 days-old alfalfa (cv Natsuwakaba) with two or three expanded trifoliolate leaves were bundled together. Twenty-seven bundles were transplanted in 15-L plastic box covered with opaque plastic board having 27 holes, each 2 cm in diameter. The seedlings were cultured in a complete nutrient solution for four days, and then transferred to iron-free solution. Seedlings kept in iron sufficient solution were used as control. The nutrient composition of the medium was the same as that described previously [1]. Seedlings were grown in a growth chamber under 16hr light provided by fluorescent light (74-80μmol /cm^2 /s). Temperature was maintained at 23°C. Hydroponic solution was aerated continuously and adjusted to pH 6.0 daily with dilute NaOH. The solution was replaced every four day.

Control and iron-deficient plants were harvested at 0, 2, 8, 10 and 12 days after the start of iron deficiency treatment. Plants were cut into shoots and roots, and promptly immersed in liquid nitrogen. These samples were stored at –80°C until RNA preparation.

3.2 RNA PREPARATION AND *IN VITRO* TRANSLATION

Total RNA was extracted from the root tissue using an RNeasy plant mini kit (Qiagen). Poly (A)$^+$ RNAs were purified with Oligotex™-dT30 super (Nippon Roche Ltd.).

About 0.5μg of mRNA was translated *in vitro* in Wheat Germ IVT™ (Ambion) in the presence of [^{35}S]-methionine (>37TBq/mmol; Amersham) according to the manufacturer's instruction. The amount of [^{35}S]-methionine incorporated into translation product was determined as follows: To a 1μl aliquot of reaction mixture was added 0.5ml of 1M NaOH (containing 1.5% H_2O_2) and incubated at room temperature for 10min. Then 50μl of bovine serum albumin (1mg/ml) and 1ml of 25%(w/v) trichloroaceticacid were added. Insoluble materials were collected on Whatmann GF/C filters and counted with a liquid scintillation counter.

3.3 ELECTROPHORESIS OF TRANSLATION PRODUCTS

Products of *in vitro* translation were analyzed by sodium dodecyl sulfate-polyacrylamide gel electrophoresis (SDS-PAGE) by Laemmli[3]. The separation gel contained 12.5% (w/v) polyacrylamide. The two-dimensional polyacrylamide gel electrophoresis was performed as described by O'Farell[4] with slight modifications. Nonidet P-40 was replaced by 3-[(3-cholamidopropyl) dimethylammonio]–1–propanesulfonate.

The first dimension gel was 2.0 mm I.D and 11 cm in length. To cover the pH range from 5 to 8, the gel contained 2% preblended Ampholine (Pharmacia) . A portion of the *in vitro* translation product was subjected to isoelectric focusing (IEF). Electrophoresis was performed at 200V for 15 min, 300V for 15 min and 360V for15 hr. For the second dimension electrophoresis, the IEF gels were loaded onto slab gels. The acrylamide concentration of the separation gel was 12.5%(w/v). For autofluorography, gels were immersed in a mixture of ethanol, acetic acid and distilled water (3:1:6) and dried.

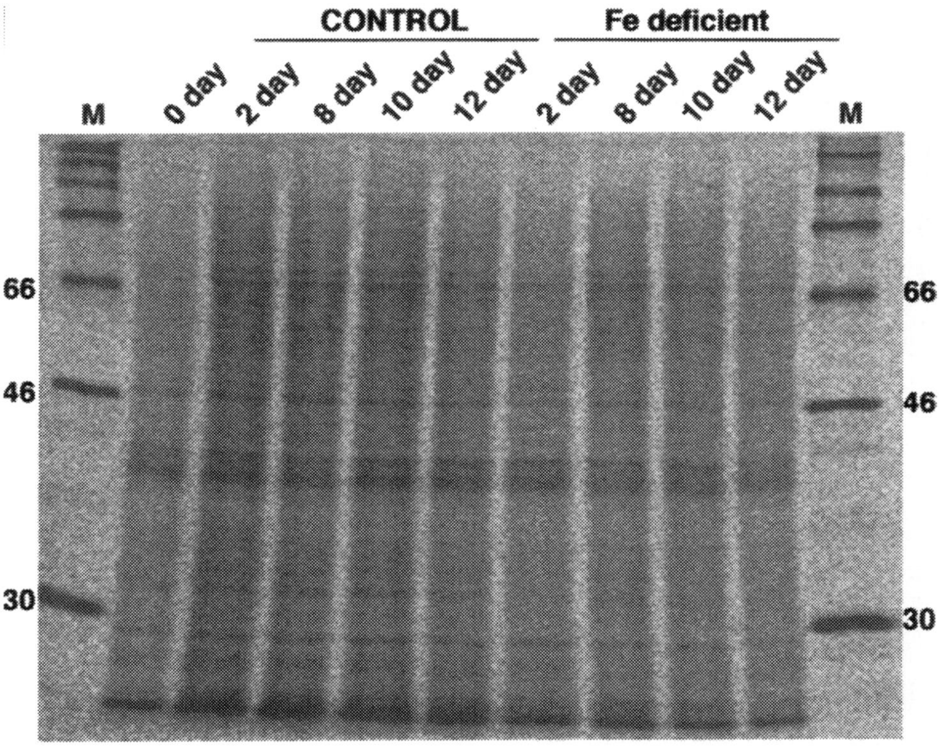

Figure 1. One dimensional SDS-PAGE analysis of *in vitro* translation products of mRNA isolated from control or iron-deficient alfalfa root at the times shown. Equal amounts of radioactivity (about 5×10^4 cpm) were applied to each lane. Lanes M: Molecular weight markers expressed in kDa.

Translation product spots were visualized using a Bio-imaging Analyzer BAS 2000 (Fuji Photo Film). The molecular weights of the resolved products were estimated from the positions of standard molecular weight markers included in the gel. The pH gradient in gels was determined as described before [4].

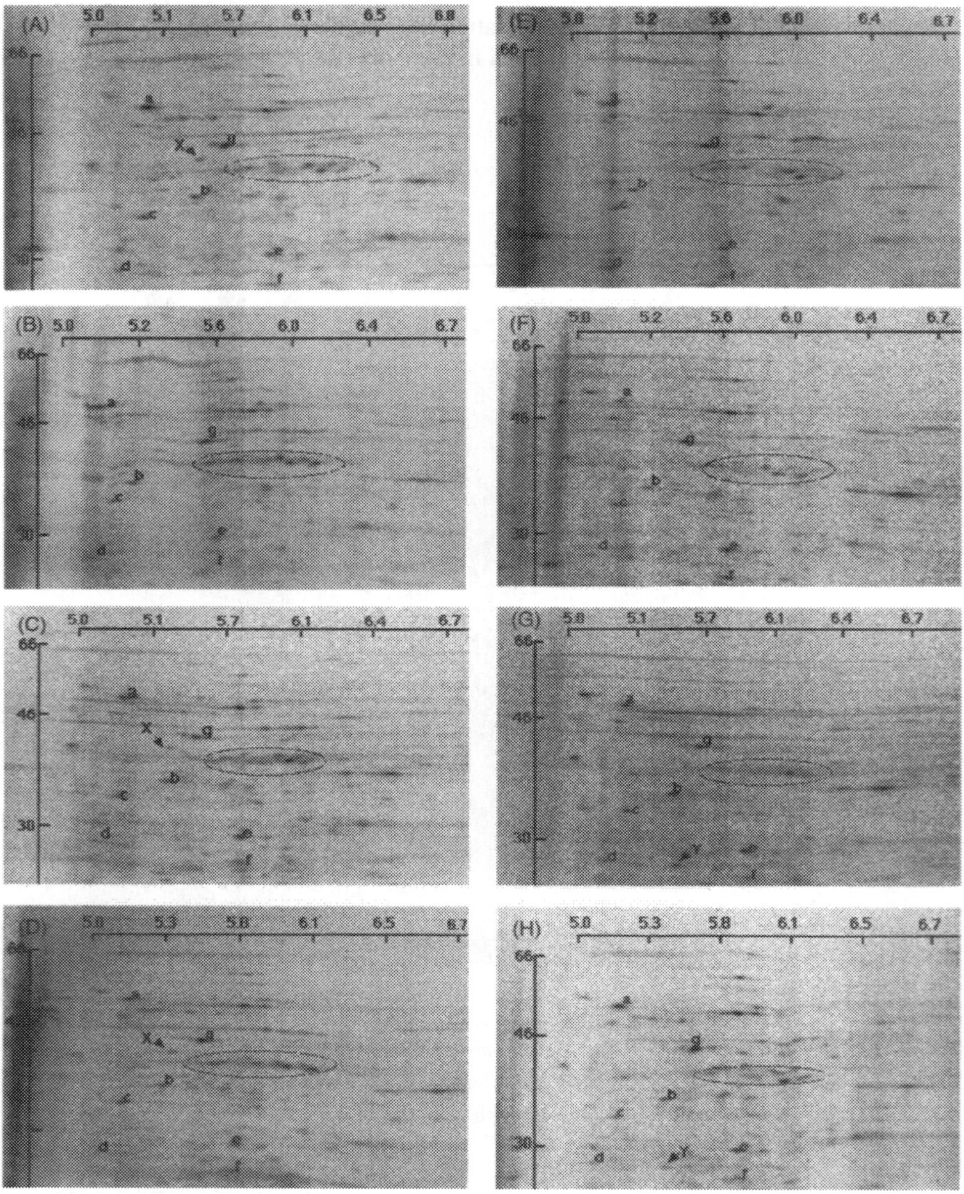

Figure 2. Two-dimensional IEF:SDS-PAGE analysis of proteins translated in vitro translation products of mRNA isolated from either control (+Fe) or iron-deficient (–Fe) alfalfa roots . A: +Fe day2; B: +Fe day8; C: +Fe day10; D: +Fe day12; E: –Fe day2; F: –Fe day8; G: –Fe day10; H: –Fe day12. The horizontal axis indicates pI values. The vertical axis indicates molecular weights expressed in kDa. X and Y arrows and open oval indicate the proteins affected by these treatment. Small letters, a to g, indicate reference products, which seemed not to be affected by these treatment.

4. Results

After 2 days of growth in iron-free solution, no symptoms of iron deficiency became apparent. After 5 days of growth in iron-free solution, symptoms of new leaf chlorosis, yellowing and subapical swelling of the lateral roots became apparent. The symptoms of leaf chlorosis became severe and growth rate decreased to zero by day 10 (data not shown).

As illustrated in Fig 1, no significant changes in the pattern of mRNA translation products were detected by one-dimensional SDS-PAGE between control and iron deficiency samples. The translation products were further analyzed by two-dimensional polyacrylamide gel electrophoresis. A new translation product of pI=5.4 and M_rs=29kDa (indicated by Y in Fig·2G, H) was detected in day 10 and day 12 samples taken from iron-deficient root. On the other hand, a translation product with pI=5.3 M_rs=42kDa (indicated by X in Fig·2A, C, D) was suppressed in all iron-deficient samples. Products of pI=5.6-6.3 M_rs=43kDa (indicated by open oval in Fig 2·G, H) were gradually decreased in day 10 and day 12 samples taken from iron-deficient root.

5. Discussion

In vitro translation analysis provides information on the relative abundance of mRNA species. Results showed that the response of alfalfa roots to iron deficiency is characterized by changes in the patterns of mRNA population. The changes were detectable in the plants subjected to iron deficiency prior to the occurrence of iron-deficient symptoms. In most cases, however, the products synthesized in iron-deficient plants were also synthesized in control plants. One product of pI=5.4 and M_rs=29kDa (indicated by Y in Fig·2G,H) was newly detected in day 10 and day 12 samples taken from iron-deficient root. It remains to be seen whether this product is related to adaptation mechanism to iron deficiency stress.

Recently, Schmidt and Buckhout [5] have investigated the effects of iron deficiency on protein synthesis in tomato roots. They identified seven translation products that were significantly stimulated under iron deficiency. The pI of the products ranged from 5.8 to 7.3. In the present study, translation products corresponding to their products were not observed. The difference in the results may be due to the plant species used or the experimental conditions.

Iron deficient alfalfa roots containe flavonoid compounds and their secretion increased with iron deficiency [6]. However, prolonged cultivation under iron deficiency resulted in a decrease in flavonoid secretion (unpublished data). The flavonoid biosynthesis is regulated by several genes including chalcone synthase (CHS). Thus, CHS genes are expected to be upregulated by iron deficiency. The apparent molecular weights, and isoelectric points of the translation products with pI=5.6-6.3 M_rs=43kDa (indicated by open oval in Fig 2·G, H) were similar to those of CHS [7]. Suppression of these genes suggested that root function was damaged extremely during prolonged iron-deficiency. We will examine this possibility by Northern blot analysis in future.

We are currently constructing subtraction cDNA libraries from which we aim to isolate clones encoding enzymes specific for iron efficiency. The kinetics of mRNA changes described in this study provides a basis for deciding the time to isolate mRNA for library construction or production of subtractive probes for cloning specific transcripts.

6. Acknowledgment

The authors wish to thank Dr. Hiromi Higo (National Institute of Agrobiological Resources) for her guidance and encouragement for this work.

7. References

1. Arakawa, Y. and Masaoka, Y. (1997) Characteristics of root Fe^{3+}-reduction and root exudate secretion under Fe-deficient conditions in alfalfa, In T. Ando et al.(Eds.) *Plant nutrition for sustainable food production and environment,* Kluwer Academic Publishers, Japan, pp 265-266.
2. Noguchi, A., Yoshihara, T., Ichihara, A., Sugihara, S., Koshino, M., Kojima, M. and Masaoka, Y. (1994) Ferric phosphate-dissolving compound, alfafuran, from alfalfa (*Medicago sativa L.*) in response to iron-deficiency stress, *Biosci. Biotech. Biochem.* **58**, 2312-2313.
3. Laemmli, U.K. (1970) Cleavage of structural proteins during the assembly of the head of bacteriophage T4, *Nature* **227**, 680-685.
4. O'Farrel, P.H. (1975) High resolution two-dimensional electrophoresis of proteins, *J. Biol. Chem* **250**, 4007-4021.
5. Schmidt, A., Buckhout, T.J. (1997) The responce of tomato roots (*Licopersicon esculentum* Mill.) to iron deficiency stress: alterations in the pattern of protein synthsis, *J. Exp. Botany* **48**, 1909-1918.
6. Masaoka, Y, Kojima, M., Sugihara, S., Yoshihara, T., Koshino, M. and Ichihara, A. (1993) Dissolution of ferric phosphate by alfalfa (*Medicago sativa* L.) root exudates, *Plant and Soil* **155/156**, 75-78.
7. Dalkin, K., Jorrin, J. and Dixon, A. (1990) Stress responses in alfalfa (*Medicago sativa* L.) VII. Induction of defence related mRNAs in elicitor-treated cell suspension cultures. *Physiol. Mol. Plant Pathol.* **37**, 293-307

ZINC STATUS OF ARID SOILS, ITS RESPONSE AND UPTAKE BY WHEAT

RAM DEO AND MATHUR, G.M.

*DEPARTMENT OF SOILS, AGRICULTURAL RESEARCH STATION,
SRI GANGANAGAR-335001(RAJ) INDIA*

Abstract :

The soils in the sandy arid plains of North-West part of India are deep, well drained, calcareous, sandy in nature and classified as Typic Torripsamments and Torrifluvent. Wheat is an important grain crop and occurance of widespread Zinc deficiency in soils focused attention on Zinc nutrition of Wheat crop.The available Zn content of soils varied from 0.20 to 2.80 mg kg^{-1} with mean value of 0.58mg kg^{-1} Zn. Correlation coefficient showed negative relationship between available Zn and $CaCo_3$ and pH of the soil. Wheat grain and straw yield increased significantly at 5 mg kg^{-1} Zn over control by Zinc application. Zn content of wheat grain and straw increased due to Zinc application.

Introduction :

Zinc (Zn) as a plant nutrient come to occupy an important position in Indian Agriculture and has become indispensable to the health and produtivity of plants and animals. Soils and foliar tests have revealed widespread deficiency of Zinc in many parts of India. Zinc as a plant nutrient now stands third in importance, next to nitrogen and phosphorus (Takkar and Randhawa, 1980). Zinc deficiency have resulted mainly due to nutrient imbalances caused by soil fertility depletion through intensive agriculture, introduction of high yielding varities, increased use of high analysis fertilizers and decreased recycling of crop residues and animal manures.
In India, the defeciency of Zinc have been observed in a variety of soils but more so in light textured calcareous soils. The soils in the sandy arid plains of north-west part of India are deep, excessively drained, calcareous, sandy in nature. Wheat is an important grain crop ranking second in respect of area under cultivation. This problem is likely to acquire greater significance in case of wheat, since the Zn

G. Gissel-Nielsen and A. Jensen (eds.), Plant Nutrition – Molecular Biology and Genetics, 193–197.

content of Wheat tract of India is mostly light to medium textured alluvials, appears to be rather low or marginal (Lal, 1989).

Plants absorbs Zinc mainly as Zn^{2+} ions, owing to their poor mobility and low Zn concentration of the soil solution, uptake is largly by direct contact between root and soil particles. Zinc mobility decreases as the pH value rises and its deficiency is therefore most common on soils with pH values of 6.5 to 8.0 and hence most often be expected in sandy soils (Shuman, 1975).

Zinc plays an important role in plant nutrition. Like other metallic micronutrients, Zinc owes its importance in the plant metabolism to its activity as a specific activator of several enzymes and its non specific activating or inhibitory action on various others. It has been reported that Zinc increases grain yields and improves quality of different crops (Bergmann, 1992).

The object of proposed investigation is to find out the distribution of available Zinc and its relationship with soil properties and response of wheat to Zinc application.

Material and Methods

The soils of irrigated North-Western part of Rajasthan State of India are mainly derived from alluvium. The mean annual temperature is 29°c, while the mean annual summer temperature approaches to 35.4° c. The annual rainfall ranges between 200-300 mm. The soils are classified as Typic Torripsamments of alluvial nature and Torrifluvent of flood plain region. Cotton-Wheat & Paddy- Wheat are the dominant rotation under these two group of soils, respectively. A total 150 surface soils (0-30 cm) samples were collected from both the arid soil group of irrigated North-West plain zone of Rajasthan. Soil samples were crushed with a wooden pestle and mortar and passed through a 2 m.m. stainless steel sieve. The soil samples were analysed for their physico-chemical characterstics as per satandard methods. The range and mean values of these properties have been reported in Table-3. The DTPA extractable Zinc was analysed with atomic absorption spectrophotometer (Lindsay and Norvell,1978)

A green house experiment was conducted with 4 Kg of soils taken in acid washed polythene lined earthen pots. Each soil received 3 levels of Zinc, viz., Zno (No Zn), Zn5 (5 mg kg^{-1} Zn) and Zn10 (10 mg kg^{-1} Zn) replicated thrice. A basal dose of 60:60:40 of N, P_2O_5, K_2O mg kg^{-1} soil was applied through urea, ammonium dihydrogen phosphate and sulphate of potash, respectively. Wheat variety Raj-3077, was sown. The two split doses each of 30 mg kg^{-1} were applied during crop growth. The crop was harvested at maturity and the data on yield of grain and straw were recorded after drying the samples at 60-65° c.

Result and Discussion :-

The results revealed that available Zinc status of Torrifluvent and Torripsamment soils group under arid condition varied from 0.20 to 2.80 mg kg^{-1} and 0.26 to 1.52

mg kg^{-1} Zn respectively. The mean values of available Zn content for these soil group are 0.85 and 0.58 mg kg^{-1} Zn and the percent soil defecient in available Zn observed was 56.9 and 78.0 percent, respectively (Table-1). Joshi et al. (1983) reported that DTPA extractable Zinc ranged from 0.27 to 2.36 ppm with a mean value of o.88 ppm in soils of Arid Rajasthan. Correlation coefficient of available Zn with pH, $CaCo_3$ and organic Carbon content of the soils was calculated. Calcium carbonate content showed significantly negative correlation with available Zn (table 2,3). Similar relationship between available Zn and $CaCo_3$ in redloam soils of Rajasthan was reported by Akbari et al. (1995).

The green house studies conducted to see the response of Zinc application revealed that yield of wheat grain as well as of straw increased significantly from 7.64 and 13.25 g/pot (at control) to 10.31 and 17.35 g/pot at 5 mg kg^{-1} Zn application to soil (Table 4 & Fig.1). The average increase in grain yield was 35.0% and that in straw was 30% on application of 5 mg kg^{-1} Zn over no Zinc. In comparision with control, increase in yield of wheat grain and straw with zinc regardless of its rate of application was significent.Similarly the Zn content in grain and straw increased significantly with the application of Zinc in comparision to control and consequently the uptake by crop. The increase in grain & straw yield of wheat with Zinc application was because of defeciency of plant available Zinc in the soils under studies. Maximum response to 5 mg kg^{-1} Zn indicate that this much Zinc was enough to raise the soil available Zinc to a level sufficient to meet the requirment of wheat crops. Similar results were also observed by Khamparia et.al. (1994)

In arid soils of North West India, a close monitoring is needed to regulate availability of soil Zinc and Zinc nutrition for sustainable Wheat production.

Table 1 Range, Mean and PSD values of DTPA Zn under Aridsoils

Arid Soil Group	Range	Mean	PSD*
Torrifluvent	0.20-2.80	0.85	56.9
Torripsamment	0.26-1.52	0.51	78.0

* Percent soil defecient

Table 2 Values of Correlation coefficients of soil parameters with DTPA Zn.

Soil parameters	DTPA Extractable Zn
pH	- 0.170
$CaCo_3$	- 0.379 **
Org. Carbon	0.0079

** Significent at 5% level

Table 3 Range and mean values of pH, EC, Org. Carbon and $CaCo_3$ under Aridsoils.

Porperties	Range	Mean
	Torrifluvent	
pH (1:2)	7.7 - 8.4	7.9
EC (1:2) (dsm^{-1})	0.2 - 2.8	0.4
Org. Carbon (%)	0.15 - 0.75	0.57
$CaCo_3$(%)	0.24 - 11.0	5.75
	Torripsamment	
pH (1:2)	8.0 - 8.6	8.1
EC (1:2) (dsm^{-1})	0.2 - 1.4	0.3
Org. Carbon (%)	0.15 - 0.51	0.25
$CaCo_3$(%)	1.5 - 3.0	2.00

Table 4 Average response of Wheat to Zinc levels

Levels of	Mean Yield (g/pot)		Zinc content (mg kg^{-1})	
Zn (mg kg^{-1})	Grain	Straw	Grain	Straw
0	7.64	13.25	20.5	6.5
5	10.31	17.35	30.5	12.5
10	9.18	15.85	31.2	13.5
C.D.at 5%	0.12	0.13	1.6	1.0

FIG. 1 :- RESPONSE OF WHEAT TO ZINC FERTILIZATION

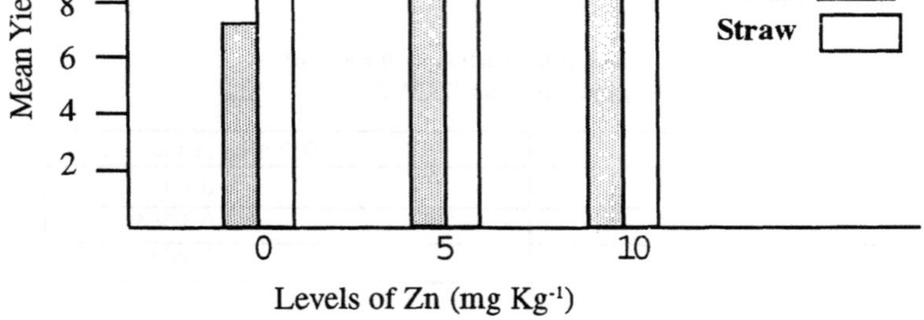

Reference

Akbari, K.N, Karan, F. and Pandya, H.R. (1995).Availlability of Zinc in redloam soils of Zone IV-A of Rajasthan as related to certain soil characteristics. J. Indian Soc. Soil sci. 43 (1), 131-132.

Bergmann, Werner (1992). Nutritional Disorders of plants. Development, visual and Analytical Diagnosis. VCH Publishers Inc. 303 N.W.12th Avenue, Florida. U.S.A.

Joshi, D.C. ,Dhir, R.P. and Gupta, S. B. (1983). Influnence of soil parameters on DTPA Extractable micronutrients in arid soils. Plant and soil 72:31-38.

Khamparia. R.S., Sharma, B.L. and Rathore, G.S. (1994). Suitability of soil test methods and response, in wheat to Zinc in some Alfisols and Vertisols. J. Indian Soc. Soil sci. 42 (1) :88-92.

Lal,F. (1989). Micronutrient status of sandy soils of Jaipur region and effect of their application on the yield of some major crops. Abstract part I. International symposium on managing sandy soils P. 385-387. CAZRI. Jodhpur.

Lindsay, W.L. and Norvell, W.A. (1978). Devlopment of a DTPA soil Test for Zinc, Iron, Manganese and Copper. Soil Sci. Soc. Am. J. 42:421-428.

Shuman, L.M. (1975). The effect of soil properties on Zinc adsorption by soils. Proc. Soil sci. Soc. Amer. 39:454-458.

Takkar, P.N. and Randhawa, N.S. (1980). Zinc defeciency in Indian soils and plants. Seminar on Zinc wastes and their utilization, 15-16 oct., 1980. Organised by Indian Lead Zinc Information Centre and Fertilizer Association of India.

Response of faba bean genotypes to low and high zinc levels

Mohamed M. El-Fouly, E. A. A. Abou El-Nour , M. R. Shabana* and Z. M. Moubarak
Botany Department ,National Research Centre, Cairo-Dokki, Egypt,
** Faculty of Agriculture, Cairo University, Giza, Egypt*

Key words: carbonic anhydrase, faba bean, genotypes, zinc, deficiency, toxicity

Abstract

Response of faba bean to Zn deficiency and toxicity was studied in 9 genotypes. Zn-deficiency induced reductions in shoot dry weight except in Giza 3. The negative effect of Zn deficiency on Zn concentration was more pronounced in young leaves than in old ones. Zn concentration in roots was increased and carbonic anhydrase activity (CAA) was decreased in Zn deficient plants. The decrease in CAA was correlated with Zn concentrations in young leaves. Response of CAA at before flowering stage in Giza 3 was not much affected. This genotype proved to be tolerant to Zn-deficiency. In case of Zn-excess, shoot dry weight was reduced. Zn-concentration in young and old leaves were increased. Decreases in CAA were recorded. CAA became unmeasurable at flowering stage, possibly due to protein denaturation.

Introduction

Nutrient deficiency and toxicity affect different physiological processes in plants and are reflected, in changes of growth parameters and yield. Zinc deficiency occurs at levels below 20 ppm in leaves. Values higher than 400 ppm are regarded as toxic (Ohnesorge and Wilhelm, 1991). Zn deficiency induced inhibition of photosynthesis which is coincident with a decrease in activity of key photosynthetic enzymes, (Brown et al, 1993). Under zinc deficiency or toxicity plants cannot give optimum growth. Zn concentration is not always a reliable measure of Zn nutritional status of plants (Bar-Akiva and Lavon, 1969; Gibson and Leece, 1981; Sinr, 1983 and Cayton et al, 1985). Activity of specific enzymes can be used to determine the status of a particular nutrient in the plant. CAA is used to determine Zn status. Bar-Akiva and Lavon, 1969 suggested CAA as an indicator for diagnosing Zn deficiency.

This paper was aiming to determine the effects of Zn stress conditions on growth, Zn content and CAA, as well as screen the ability of different genotypes to withstand Zn stress conditions

Materials and Methods

Nine genotypes of faba bean (*Vicia faba*) were used (Three Egyptian cultivars Giza 402, Giza 3 and Giza 2, together with 6 pure lines; 241/15, 348 / 7, 2062 / 4, 1495 / 2, 348 / 8 and 402/4). The plants grew in water cultures using half strength complete Hoagland

G. Gissel-Nielsen and A. Jensen (eds.), Plant Nutrition – Molecular Biology and Genetics, 199–204.

solution. under controlled conditions (16/8 hr. light/dark regime, 25/18°C day/night temperature and relative humidity 65-70%). Treatments were as follows:

1- 0.10 ppm Zn as control
2- 0.02 ppm Zn as deficient
3- 30.00 ppm Zn as toxic

Each treatment was replicated three times. The nutrient solutions were renewed weekly. At both before and at flowering stages, 40 and 55 days from sowing, respectively, the first fully mature leaf was taken for determining CAA (Waygood, 1955). At harvest time (58 days from sowing) shoot dry matter yield g/plant was recorded. Harvested plants were divided to roots, young leaves (the first five fully mature leaves) and old leaves (the rest leaves). The plant organs were prepared for Zn analysis according to Chapman and Pratt, 1979.

Results and Discussion

Table 1 Effect of Zn supply on shoot dry weight (g/plant)

Genotype	Zn supply		
	Low	Normal	High
Giza-2	5.7 ±0.1	8.3 ±0.4	2.8 ±0.1
Giza-3	7.0 ±0.1	6.8 ±0.2	1.6 ±0.1
Giza-402	6.0 ±0.2	7.2 ±0.1	2.0 ±0.1
402/4	6.4 ±0.1	7.0 ±0.3	1.2 ±0.1
348/8	5.2 ±0.0	6.5 ±0.2	2.6 ±0.1
348/7	5.4 ±0.5	7.2 ±0.4	1.4 ±0.2
241/5	4.6 ±0.1	5.0 ±0.3	0.7 ±0.0
2062/4	5.0 ±0.1	5.5 ±0.1	1.3 ±0.2
1495/2	2.2 ±0.1	6.3 ±0.0	1.4 ±0.1

Shoot dry weight :

Low Zn concentration reduced shoot dry weight/plant. Genotypes differed in their response (Table 1). Giza 3 was not affected by Zn deficiency in the media, while Giza 2 was most affected within the cultivars. The opposite trend was found in plants grown in solutions containing high Zn concentration (Figs. 1A/2A). Generally, least affected genotypes by low Zn was most affected by high Zn.

Zn concentration :
Old leaves:

Table 2. Effect of Zn supply on Zn concentration (ppm) in old leaves

Genotype	Zn supply		
	Low	Normal	High
Giza-2	18*	24	412
Giza-3	18	23	412
Giza-402	14	17	409
402/4	21	25	398
348/8	27	29	416
348/7	20	25	395
241/5	30	31	411
2062/4	28	32	415
1495/2	23	25	413

* Mean of two reblicates

Low Zn in the nutrient solution caused reductions in Zn concentration in old leaves (Table 2). The commercial cultivars were, generally, more affected than the pure lines (Fig. 1B). High Zn in the medium caused Zn increases in old leaves, which were more pronounced in the cultivars (Table 2-Fig. 2B).

Young leaves :
In general, low Zn in the nutrient solution was reflected in low Zn concentrations. Higher concentrations were found in plants grown in high Zn solutions (Table 3).

Table 3. Effect of Zn supply on Zn concentration (ppm) in young leaves

Genotype	Zn supply		
	Low	Normal	High
Giza-2	18*	25	314
Giza-3	24	29	294
Giza-402	9	21	398
402/4	12	24	387
348/8	34	35	307
348/7	8	26	395
241/5	8	30	346
2062/4	13	32	404
1495/2	14	28	386

* Mean of two replicates

The effect of low Zn in the solution on reducing Zn content in young leaves was at minimum in genotypes 348/8 and Giza-3 and much higher in the other genotypes. (Fig. 1C). High Zn in the nutrient solution led to increases in Zn concentration (Fig. 2C), which were less than the increase in the old leaves.

Roots :
Both low and high Zn concentration led to high Zn concentrations in the roots. However, to different extents (2-4 times in Zn deficient plant and 22 - 43 times in plants grown in high Zn solution (Table 4). Reaction of different genotypes to Zn deficiency and Toxicity showed more or less the same trend (Figs. 1D - 2D).

From these results, it could be concluded that Zn concentration of the different plant organs increased by increasing the availability of Zn in the growth medium. Rahimi and Schropp, 1984 and Ruano et al, 1988 reported similar finding. Genotypes of 348/8 and Giza-3 showed the less zinc reduction in the different plant organs and showed better Zn translocation when grown in Zn deficient medium. While, genotype of 241/5 which showed the greatest reduction in zinc measured in young leaves (73% reduction) as well as 61% more Zn accumulation in the roots.

Table 4. Effect of Zn supply on Zn concentration (ppm) in roots

Genotype	Zn supply		
	Low	Normal	High
Giza-2	23*	12	445
Giza-3	24	15	437
Giza-402	28	10	449
402/4	33	8	408
348/8	32	20	454
348/7	26	11	430
241/5	33	13	423
2062/4	30	16	421
1495/2	27	10	432

- Mean of two reblicates

Carbonic anhydrase activity (CAA) :
CAA determined before flowering was reduced due to low and high Zn supply (Table 5). The reduction in Zn deficient plants was much less than in plants grown under toxicity conditions (Figs. 3 and 4). Giza 3 and 402/4 were not affected at before flowering stage (Fig. 3). At flowering stage, CAA was much more affected than when measured at flowering stage (Fig. 4). Here also Giza 3 was the least affected among the cultivars and 348/8 among the pure lines. Mengel, 1974 reported that under zinc deficiency conditions the normal photosynthetic process may be affected and also disturbances in plant enzymes system may be occurred. The results are in agreement

with the findings of Randall and Bouma, 1973; Snir, 1983; Dell and Wilson, 1985 and 1989.

Conclusion

Low Zn concentration in the nutrient solution led to low Zn concentration in both old and young leaves and high Zn accumulation in the roots. CAA was reduced. These changes were negatively reflected on dry weight/plant. Zn efficient genotypes showed low reduction in dry weight/plant correlated with reduction in Zn concentration in old and young leaves and lowest increase in root Zn concentration as well as low decrease in CAA.

CAA measured at flowering stage is better correlated with Zn status in plant than at before flowering stage. This led to recommend that samples taken for this assay should be taken at flowering stage to reflect the true variation of genotypes.

Giza 3 was less affected by Zn deficiency among the cultivars and can be considered as Zn-effecient genotype, while 348/8 showed more or less similar trend among the pure lines.

Table 5. Effect of Zn supply on carbonic anhydrase activity (EU)*

Genotype	Before flowering			At flowering		
	Zn supply			Zn supply		
	Low	Normal	High	Low	Normal	High
Giza-2	11.2±	13.1±	4.6±	4.1 ±	12.2±	-
Giza-3	12.5±	12.4±	0.9±	8.1 ±	12.6±	-
Giza-402	8.2±	10.4±	0.3±	1.0 ±	15.4±	-
402/4	11.5±	10.6±	0.8±	1.5 ±	12.8±	-
348/8	8.5±	11.2±	6.8±	4.0 ±	13.8±	-
348/7	11.8±	13.1±	2.2±	0.7 ±	14.5±	-
241/5	11.8±	14.7±	0.5±	0.9 ±	15.4±	-
2062/4	6.9±	12.7±	0.8±	1.6 ±	8.7±	-
1495/2	10.0±	12.5±	1.0±	3.2 ±	13.5±	-

* EU = Rate of enzymic hydration of CO_2 moles/ Liter/ Second

Acknowledgement

E.A.A. Abou El-Nour wishes to express his appreciation to The German Academic Exchange Service (DAAD) for awarding a scholarship to carry out a part of this work.

References

Bar-Akiva, A. and R. Lavon, 1969. Carbonic anhydrase activity as an indicator of zinc deficiency in citrus leaves. J. Hort. Sci., 44: 359- 362.

Borown, P. H.; I. Cakmak and Q. Zhang, 1993. Form and function of zinc plants. Proc. of the International Symposium on "Zinc in Soil and Plants" 27-28 Sept. Australia, Ed. Robson, A. D., 93-106.

Cayton, M.T.C.; E.D. Reyes and H. U. Neue, 1985. Effect of zinc fertilization on the mineral nutrition of rice differing in tolerance to zinc deficiency. Plant and Soil, 87:319-327.

Chapman, H. D. and P. F. Pratl, 1978. Methods of analysis for soils, plants and water. University of California, Dept. of Agric. Sci., USA, pp 1-309.

Dell, B. and S. A. Wilson, 1985. Effect of zinc supply on growth of three species of *Eucalyptus maculate seedling and wheat.* Plant and Soil, 88, : 377-388.

Dell, B. and S. A. Wilson, 1989. Zinc nutrition and leaf carbonic anhydrase activity of *Eucalyptus maculate seedlings and Trifolium subterraneuum.* Plan. And Soil, 133,: 287-290.

Ghoneim, M. F. and W. Bussler, 1980. Diagnosis of zinc deficiency in cotton. Z. Pflanzeneraehr. Bodenkd., 134: 377-384.

Gibson, T. S. and D. R. Leece, 1981. Estimation of physiologically active zinc in maize by biochemical assay. Plant and Soil, 63: 395-406.

Hoagland, D. R. and D. I. Aron, 1950. The water culture method for growing plant without soils Cric. 347. Calif. Agric. Expt. Stn. Berkley, CA.

Mengel, K., 1974. Plant ionic status. In: THE PLANT ROOT AND ITS ENVIRONMENT. ED. Carson, E. W., Univ. Press of Virginia, Charlottesville : 63-81.

Ohnesorge, F.K. and M. Wilhelm, 1991. In: Merian, E. (ed.) Metals and Their Compounds in the Environment, pp 1307-1342. VCH, Weinheim, New York, Cambridge.

Rahimi, I. and A. Schropp, 1984. Carboanhydrase activity and extractable zinc as indicator of zinc supply of plants. Z. Pflanzenernehhr. Bodenkd.,147 : 572-583.

Randall, P. J. and B. Bouma, 1973. Zinc deficiency, carbonic anhydrase and photosynthesis in leaves of spinach. Plant Physiol. 52, : 229-232.

Sinr, I. 1983. Carbonic anhydrase activity as an indicator of zinc deficiency in peacan leaves. Plant and Soil, 74: 287-289.

GENOTYPIC VARIATION IN THE RESPONSE OF BARLEY TO ZINC DEFICIENCY

Y. GENC, G. K. McDONALD, Z. RENGEL*, R. D. GRAHAM
Department of Plant Science, University of Adelaide, 5064 Glen Osmond, Australia
**Soil Science and Plant Nutrition, University of Western Australia, 6907 Nedlands, Australia*

Key words: deficiency, efficiency, genotype, Hordeum vulgare, variation, Zinc

1. Abstract

Two experiments were conducted to evaluate the response of 48 genotypes of barley (*Hordeum vulgare* L.) to Zinc (Zn) deficiency using a soil-based screening method. Plants were grown in a Zn-deficient soil fertilized with 0, 0.02 and 0.8 mg Zn kg^{-1} soil and sensitivity to Zn deficiency was assessed by visual deficiency symptoms and by measuring the ratio of shoot growth with 0.02 mg Zn kg^{-1} soil to that with 0.8 mg Zn kg^{-1} soil (defined as Zn efficiency) at tillering (Feekes growth stage, 5). Genotypes exhibited a wide range in visual symptoms of Zn deficiency when grown with no applied Zn. Those which did not develop severe deficiency symptoms included Schooner, WI-2875 (mild symptoms) and Sahara, Stein, Ellice, Mona and Kinukei 21 (no symptoms). Severe symptoms were observed in Forrest, Kinukei 15, Blenheim, Franklin and Haruna Nijo.

Zn efficiency ranged from 28% (Forrest) to 55% (Skiff), and from 26% (Manley) to 91% (Kinukei 21) in the two experiments respectively. However, seed Zn content in both experiments showed a considerable range and it is likely that a proportion of variation in expression of Zn deficiency symptoms and Zn efficiency may be due to differences in seed Zn content. Following analysis of covariance using seed Zn content as a covariate, genotypic variation in tolerance to Zn deficiency was still

G. Gissel-Nielsen and A. Jensen (eds.), Plant Nutrition – Molecular Biology and Genetics, 205–221.

evident among barley genotypes. The results suggest that a soil-based pot screening method based on visual scores and ratio of the shoot growth with low Zn to the growth with adequate Zn can be used to assess differential responses of barley genotypes to Zn deficiency.

2. Introduction

Zinc (Zn) deficiency occurs on millions of hectares of the world's cereal-growing soils, especially in the arid and semi-arid regions (Graham *et al.*, 1992; Singh, 1982; Singh, 1983; Takkar and Walker, 1993). In barley, Zn deficiency has been recognized for many years and yield responses to both soil and foliar applications of Zn have been reported in many different regions (Graham *et al.*, 1992; MacNaeidhe and Fleming, 1990; Pathak *et al.*, 1978; Takkar *et al.*, 1983; Yilmaz *et al.*, 1995). Responses to Zn fertilizer vary considerably and are influenced by soil available Zn. For example, in studies with ten barley cultivars and two Zn fertilization levels (0 and 23 kg Zn ha^{-1}) on various soils, Yilmaz *et al.* (1995) observed increases in grain yield ranging from 15% (DTPA extractable Zn in soil=0.15 mg kg^{-1}) to 202% (DTPA extractable Zn=0.11 mg kg^{-1}). Another study reported the grain yield of barley increased up to 60% after foliar application of Zn on six soils with ETDA-extractable Zn of 0.6-2.0 mg kg^{-1} soil (MacNaeidhe and Fleming, 1990).

Apart from the marked affect of soil type on Zn response, there is considerable variation in the response to Zn within cereals (Cakmak *et al.*, 1997a; Cakmak *et al.*, 1997b; Graham *et al.*, 1992). A differential response to Zn application indicates that genotypic variation could be exploited in a breeding programme to produce genotypes with higher Zn efficiency (i.e. the ability of a genotype to grow and yield better on a soil deficient in Zn compared to the standard genotype; Graham, 1984). The first step in breeding for Zn efficiency is to establish a reliable system for screening a large number of genotypes. So far, field screening has been employed extensively (Graham *et al.*, 1992; MacNaeidhe and Fleming, 1990; Takkar *et al.*, 1983; Yilmaz *et al.*, 1995), however, the results of field tests are often variable across sites and years, due to the effects of other growth limiting factors (eg. drought, diseases and other nutrient deficiencies and/or toxicities), soil variability and seasonal variations. Reliable alternative methods are therefore required. In the study presented here, a soil-based pot screening method, under controlled conditions, was employed as an alternative to field trials (i) to overcome problems encountered in the field screening method and (ii) to determine genetic variation in the response of barley to Zn deficiency.

3. Materials and methods

A Zn-deficient, sandy, surface soil, collected from uncleared land near Lancelin, Western Australia, (DTPA-extractable Zn=0.2 mg kg^{-1} soil), was sieved through a 2 mm stainless steel sieve and one kg of soil was placed into cardboard cartons (7x7x17.5 cm) of 600 ml volume following the addition of calcium carbonate powder (0.5 % w/w). Basal nutrients (in mg kg^{-1} dry soil) of NH_4NO_3, 95; K_2PO_4, 90; K_2SO_4, 140; $MgSO_4.7H_2O$, 20; $CuSO_4.5H_2O$, 2; $MnSO_4.4H_2O$, 15; H_3BO_3, 0.7; $CoSO_4.7H_2O$, 0.4; $Na_2MoO_4.2H_2O$, 0.2, together with zinc treatments (0.0, 0.02, 0.8 mg Zn kg^{-1} soil applied as $ZnSO_4.7H_2O$ and designated hereafter as Zn_0, $Zn_{0.02}$ and $Zn_{0.8}$) were applied to the surface of the soil and allowed to dry and mixed again throughout the soil. A preliminary experiment with two genotypes (Amagi Nijo and Tantangara) and seven Zn fertilization rates (0, 0.04, 0.08, 0.2, 0.8, 3.2 and 12.8 mg Zn kg^{-1} soil) had established that 0 (very severe Zn stress), 0.02 (severe Zn stress) and 0.8 mg Zn kg^{-1} soil (no Zn stress) were appropriate rates for screening studies.

To determine the extent of genotypic variation in the response of barley to Zn deficiency, 48 genotypes of barley (*Hordeum vulgare* L.) were tested in the two experiments: Experiment 1 (16 genotypes, Table 1) and Experiment 2 (39 genotypes, Table 2). Most of the seed in Experiment 1 was obtained from field plots from Lameroo, South Australia in which genotypes were grown without applied Zn. The seed in Experiment 2 was obtained from the barley breeding programme of University of Western Australia.

The seed in both experiments was hand sorted to a uniform size, surface-sterilized, and imbibed on moist filter paper in petri dishes for 24 h at room temperature.

Four pregerminated seeds of each genotype were sown in each pot and seedlings were thinned to two per pot after emergence.

TABLE 1. Zn concentration and content in the seeds of barley genotypes used in Experiment 1. Data were means of three replicates containing 10 seeds each; standard errors are given in parentheses

Genotype	Weight ($g\ seed^{-1}$)	Zn concentration ($ug\ g^{-1}$)	Zn content ($ug\ seed^{-1}$)
CI3576	0.041	10.7 (0.4)	0.44 (0.02)
WI-2868	0.048	10.9 (1.2)	0.53 (0.06)
WI-2597	0.047	11.6 (0.6)	0.55 (0.03)
Yagan	0.050	11.0 (0.2)	0.55 (0.01)
Clipper	0.047	12.2 (0.6)	0.57 (0.03)
Schooner	0.045	12.9 (0.5)	0.58 (0.02)
Chebec	0.046	12.8 (0.1)	0.59 (0.01)
Harrington	0.045	13.6 (0.3)	0.62 (0.01)
Galleon	0.047	13.5 (0.5)	0.63 (0.02)
Forrest	0.050	13.0 (0.3)	0.65 (0.02)
Amagi Nijo	0.045	14.7 (0.4)	0.65 (0.04)
Tantangara	0.047	13.9 (0.8)	0.66 (0.04)
Skiff	0.045	15.0 (0.3)	0.68 (0.01)
WI-2875	0.046	15.2 (0.8)	0.70 (0.04)
Haruna Nijo	0.048	16.6 (1.3)	0.80 (0.06)
Sahara	0.029	41.6 (2.5)	1.21 (0.07)

Plants were grown in a glasshouse at approximately 20/10 °C day/night temperature and 12 h photoperiod. Plants were watered with double deionized water (DD) (18 MOhms cm^{-1} resistivity) daily by weight, keeping water content at 12 % (w/w) and harvested four weeks after sowing (tillering stage, FS 5, Large, 1954). At harvest, the adhering soil was washed off roots under running tap water then dipped into dionized water. Following a quick rinse in DD water, plants were separated into roots and shoots. Plant samples were oven dried at 80 °C for 48 h followed by digestion in 70% HNO_3 as described elsewhere (Zarcinas, 1984).

Zinc efficiency was calculated as the ratio of shoot dry matter at 0 or 0.02 to 0.8 mg Zn kg^{-1}, but comparisons were based on the 0.02 to 0.8 ratio since there was

TABLE 2. Zinc concentration and content in seeds of barley genotypes used in Experiment 2.

Genotype	Weight (g seed^{-1})	Zn concentration (ug g^{-1})	Zn content (ug seed^{-1})
Cheri	0.053	11.4	0.61
Blenheim	0.051	13.0	0.66
Kinukei 15	0.042	17.3	0.72
Molloy	0.056	13.0	0.73
Natasha	0.049	15.9	0.79
Manley	0.045	18.1	0.81
Europa	0.049	17.1	0.83
Kinukei 19	0.052	16.4	0.85
Waverney	0.048	20.0	0.97
Franklin	0.061	16.3	0.99
Chebec	0.056	18.5	1.04
Bearpaw	0.052	20.1	1.05
Mundan	0.064	17.6	1.12
Onslow	0.043	26.1	1.12
Fitzgerald	0.054	21.5	1.16
Haruna Nijo	0.056	21.7	1.21
Stirling	0.062	20.0	1.24
83SM522	0.057	21.8	1.24
Galleon	0.057	22.2	1.27
Igri	0.056	25.3	1.43
Gairdner	0.053	30.2	1.60
Yagan	0.074	22.9	1.68
Skiff	0.060	28.7	1.73
Steptoe	0.058	30.8	1.79
Nudinka	0.050	38.8	1.92
Chariot	0.059	32.7	1.94
Proctor	0.048	40.3	1.95
Morex	0.053	39.0	2.07
Dicktoo	0.035	60.1	2.09
Prisma	0.041	54.3	2.24
Harrington	0.052	45.6	2.36
Capulet	0.053	59.4	3.16
Mona	0.056	58.4	3.26
O'Connor	0.063	55.0	3.44
Ellice	0.057	63.0	3.61
TR-306	0.058	68.3	3.98
Stein	0.053	80.9	4.30
Clipper	0.061	86.3	5.23
Kinukei 21	0.058	122.0	7.07

Due to limited availability of seed, seed Zn concentration values were based on one replication containing 10 seeds per genotype better discrimination between genotypes at this ratio than the 0.0 to 0.8 ratio because the level of Zn deficiency stress was less severe and there was less reliance on seed Zn content.

The experiment was set up in a completely randomized block design with three replicates. Results were analyzed by the GENSTAT statistical package (GENSTAT 5, 1988). Tukey's Honestly Significant Differences (HSD) at α=0.05 was employed in pairwise comparisons (Steel and Torrie, 1960). In Experiment 2, there was considerable variation in Zn concentration of seeds (Table 2) which can affect interpretation of the results (Genc et al, submitted for publ.). Therefore, analysis of covariance was performed using seed Zn as the covariate to eliminate that proportion of experimental error for shoot and root growth which can be attributed to differences in seed Zn content.

4. Results

4.1. EXPERIMENT 1

4.1.1. *Visual symptoms*

At harvest, most of genotypes developed very severe Zn deficiency symptoms when no Zn was applied, except for Schooner, WI-2875, Skiff (mild symptoms) and Sahara (no symptoms). Zinc fertilization of 0.02 mg kg^{-1} soil resulted in less severe symptoms. Symptoms were absent in Sahara and were mild in all other genotypes. No symptoms of Zn deficiency occurred at $Zn_{0.8}$.

4.1.2. *Dry matter accumulation*

Shoot dry matter production was depressed by Zn deficiency in all genotypes, but relative depression was less in Sahara, Skiff, WI-2875 and Yagan than in other genotypes (Table 3). Zinc efficiency varied from 28 % for Forrest to 55 % for Skiff (Figure 1). Root dry matter was also depressed in all genotypes due to Zn deficiency but the reduction was more marked in Forrest and Clipper (Table 3). Skiff and Sahara produced higher root dry matter than other genotypes tested in this experiment under Zn deficiency (Zn≤0.02 mg kg^{-1} soil).

TABLE 3. Shoot and root dry weight of barley genotypes grown at 3 levels of Zn 4 weeks after sowing in Experiment 1

Genotype	Shoot, g (2 plants)$^{-1}$			Root, g (2 plants)$^{-1}$		
	Zn fertilization (mg kg^{-1} soil)			Zn fertilization (mg kg^{-1} soil)		
	0	0.02	0.80	0	0.02	0.8
CI3576	0.11	0.23	0.72	0.10	0.13	0.47
WI-2868	0.17	0.32	0.74	0.15	0.24	0.53
WI-2597	0.12	0.28	0.68	0.13	0.21	0.49
Yagan	0.18	0.34	0.80	0.11	0.20	0.38
Clipper	0.11	0.25	0.64	0.10	0.19	0.53
Schooner	0.14	0.23	0.67	0.14	0.17	0.50
Chebec	0.14	0.25	0.66	0.10	0.17	0.49
Harrington	0.15	0.26	0.76	0.15	0.23	0.54
Galleon	0.13	0.27	0.67	0.11	0.21	0.50
Forrest	0.12	0.22	0.77	0.09	0.14	0.58
Amaji Nijo	0.14	0.28	0.82	0.12	0.20	0.57
Tantangara	0.13	0.31	0.77	0.13	0.24	0.49
Skiff	0.16	0.37	0.66	0.18	0.27	0.47
WI-2875	0.14	0.30	0.60	0.12	0.22	0.52
Haruna Nijo	0.17	0.4,	0.90	0.15	0.29	0.52
Sahara	0.21	0.36	0.69	0.21	0.27	0.40
Tukey's $HSD_{0.05}$						
Genotype (G) x Zn fertilization (Zn)		0.14			0.14	

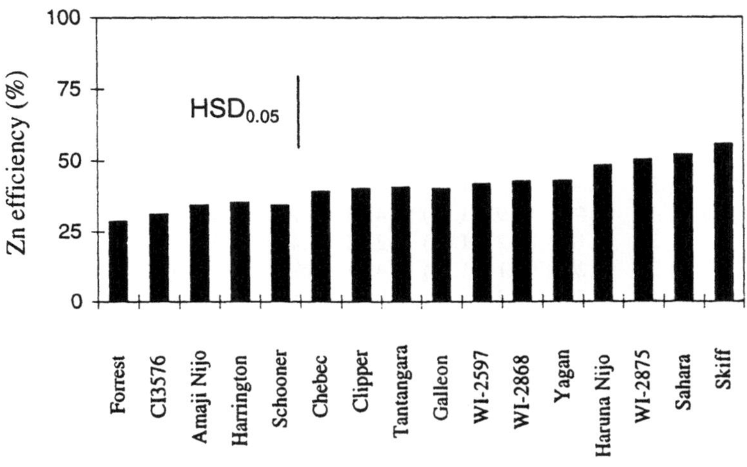

Figure 1 . Zn efficiency of barley genotypes in Experiment 1. Vertical bar represents Tukey's $HSD_{0.05}$ value for genotype effect.

4.1.3 *Concentration and content of Zn in shoots*

In all genotypes, concentrations of Zn were much higher in plants supplied with adequate Zn ($Zn_{0.8}$) than plants with inadequate Zn ($Zn \leq 0.02$ mg kg^{-1} soil) (Table 4). Zinc concentrations were similar among genotypes at each level of Zn except for Sahara at $Zn_{0.8}$ and Forrest at Zn_0 that had high Zn concentration at these levels respectively. Zinc contents in shoots were also greater in plants supplied with adequate than inadequate Zn (Table 4). Under Zn deficiency, genotypes did not differ significantly in Zn content. With Zn fertilization of 0.02 mg kg^{-1} soil, some variation occurred; Haruna Nijo had higher Zn content than most of the genotypes; WI-2875 and Sahara had higher Zn content than some other genotypes. When supplied with adequate Zn ($Zn_{0.8}$), Sahara had the greatest and Skiff the least Zn content in shoot.

TABLE 4. Zinc concentration and content in the shoots of barley genotypes grown at 3 levels of Zn weeks after sowing in Experiment 1

Genotype	Zn concentration (ug g^{-1}) Zn fertilization (mg kg^{-1} soil) 0		0.02		0.8		Zn content, ug (2 plants)$^{-1}$ Zn fertilization (mg kg^{-1} soil) 0		0.02		0.8	
CI3576	4.6	(1.52)[a]	5.5	(1.71)	43.6	(3.77)	0.5	(0.40)	1.2	(0.81)	31.2	(
WI-2868	3.2	(1.17)	4.5	(1.50)	41.1	(3.70)	0.5	(0.43)	1.4	(0.88)	30.7	(
WI-2597	4.0	(1.39)	4.7	(1.54)	43.8	(3.77)	0.5	(0.39)	1.4	(0.84)	30.0	(
Eagan	4.4	(1.45)	5.3	(1.67)	36.1	(3.59)	0.8	(0.56)	1.8	(1.04)	28.9	(
Clipper	4.2	(1.44)	4.9	(1.60)	49.9	(3.90)	0.5	(0.39)	1.2	(0.81)	31.6	(
Schooner	3.8	(1.31)	4.7	(1.55)	46.3	(3.82)	0.5	(0.43)	1.1	(0.73)	32.6	(
Chebec	5.1	(1.60)	5.3	(1.64)	46.5	(3.82)	0.7	(0.51)	1.3	(0.83)	29.7	(
Harrington	3.3	(1.17)	4.8	(1.57)	33.3	(3.50)	0.5	(0.40)	1.3	(0.82)	25.7	(
Galleon	3.3	(1.19)	5.1	(1.62)	41.2	(3.71)	0.4	(0.35)	1.4	(0.86)	28.2	(
Forrest	6.7	(1.91)	5.9	(1.77)	40.4	(3.68)	0.8	(0.58)	1.3	(0.83)	31.5	(
Amaji Nijo	3.7	(1.30)	3.9	(1.35)	36.9	(3.59)	0.5	(0.43)	1.0	(0.71)	30.2	(
Tantangara	4.1	(1.41)	4.9	(1.58)	42.5	(3.75)	0.6	(0.44)	1.5	(0.92)	32.7	(
Skiff	3.2	(1.18)	4.2	(1.44)	31.5	(3.43)	0.5	(0.41)	1.5	(0.93)	21.0	(
WI-2875	3.4	(1.20)	6.1	(1.81)	49.7	(3.90)	0.5	(0.38)	1.9	(1.05)	30.0	(
Haruna Nijo	4.9	(1.58)	6.3	(1.82)	35.8	(3.57)	0.8	(0.59)	2.7	(1.30)	32.4	(
Sahara	4.0	(1.38)	5.4	(1.70)	67.3	(4.21)	0.9	(0.61)	2.0	(1.09)	46.0	(
Tukey's $HSD_{0.05}$												
G x Zn			(0.48)[b]						(0.44)[b]			

[a] Numbers in parentheses refer to averages obtained from the analysis of variance of log-transforme

[b] The HSD values are applicable to log-transformed data (in parentheses)

4.2. EXPERIMENT 2

4.2.1. *Visual symptoms*

In this experiment, there were large differences in the severity of Zn deficiency symptoms among genotypes when they were grown without Zn fertilization. Deficiency symptoms became severe in Kinukei 15, Blenheim, Franklin and Haruna Nijo, while they were not visible in Stein, Ellice, Mona and Kinukei 21.

4.2.2. *Dry matter accumulation*

Shoot dry matter production was reduced by Zn deficiency in all genotypes but the relative reduction was more marked in Kinukei 19, Waverney, Blenheim, Manley, Kinukei 15, Onslow, O'Connor and Franklin. In contrast, Kinukei 21, Stein, Mona and Ellice performed well (Table 5). Zinc efficiency ranged from 26 % for Manley to 85 % for Stein (Figure 2). Root dry matter production was also significantly decreased in almost all genotypes. However, Steptoe, Fitzgerald, Gairdner and Skiff had higher root dry matter production at Zn_0 than other genotypes (Table 5). Interestingly, Kinukei 21, a superior genotype in shoot dry matter production under Zn deficiency, had the lowest root dry matter accumulation at Zn_0.

4.3.3. *Concentration and content of Zn in shoots*

The concentration of Zn in the shoot was significantly increased only when plants were supplied with sufficient Zn ($Zn_{0.8}$). Under Zn deficiency ($\leq Zn_{0.02}$), shoot concentrations of Zn were similar in almost all genotypes with the exception of Harrington which had highest value at Zn_0. The total amount of Zn in shoots was also significantly enhanced by Zn fertilization (Table 6). Similar to Zn concentration, genotypes differed in their capacity to accumulate Zn only when grown in Zn-sufficient conditions. Kinukei 21 had the highest and Nudinka the lowest Zn accumulation in shoots.

5. Discussion

Genotypes differed in appearance and severity of Zn deficiency symptoms. Based on visual symptoms in Experiment 1, Forrest the most sensitive, Schooner, Skiff and WI-

TABLE 5. Shoot and root dry weight of barley genotypes grown at 3 levels of Zn 4 weeks after sowing in Experiment 2

Genotype	Shoot, g (2 plants)$^{-1}$			Root, g (2 plants)$^{-1}$		
	Zn fertilization (mg kg^{-1} soil)			Zn fertilization (mg kg^{-1} soil)		
	0	0.02	0.8	0	0.02	0.8
Cheri	0.29*	0.50	0.93	0.32	0.38	0.61
Blenheim	0.22	0.31	0.94	0.26	0.28	0.50
Kinukei 15	0.22	0.41	0.94	0.24	0.29	0.55
Molloy	0.25	0.40	0.80	0.24	0.34	0.48
Natasha	0.32	0.49	1.07	0.32	0.31	0.59
Manley	0.23	0.29	0.94	0.22	0.28	0.62
Europa	0.23	0.43	0.97	0.25	0.28	0.53
Kinukei 19	0.23	0.55	1.14	0.22	0.33	0.60
Waverney	0.20	0.45	0.89	0.23	0.28	0.47
Franklin	0.23	0.41	0.91	0.24	0.29	0.57
Chebec	0.23	0.32	0.86	0.29	0.27	0.50
Bearpaw	0.25	0.45	0.94	0.26	0.38	0.61
Mundah	0.28	0.43	0.89	0.23	0.25	0.46
Onslow	0.22	0.40	0.91	0.23	0.31	0.59
Fitzgerald	0.32	0.49	0.89	0.35	0.37	0.68
Haruna Nijo	0.32	0.49	1.02	0.27	0.32	0.53
Stirling	0.26	0.45	0.81	0.24	0.31	0.56
83SM522	0.30	0.40	0.78	0.25	0.27	0.49
Galleon	0.30	0.50	0.85	0.33	0.40	0.68
Igri	0.34	0.49	0.78	0.28	0.25	0.33
Gairdner	0.32	0.50	0.87	0.34	0.35	0.60
Yagan	0.36	0.62	1.02	0.34	0.40	0.62
Skiff	0.33	0.54	0.96	0.33	0.36	0.54
Steptoe	0.40	0.57	0.81	0.43	0.45	0.53
Nudinka	0.35	0.47	0.65	0.30	0.30	0.33
Chariot	0.36	0.49	0.79	0.27	0.30	0.55
Proctor	0.31	0.53	0.74	0.31	0.35	0.35
Morex	0.29	0.48	0.88	0.23	0.36	0.49
Dicktoo	0.35	0.47	0.75	0.31	0.29	0.45
Prisma	0.35	0.50	0.90	0.29	0.32	0.38
Harrington	0.36	0.48	0.81	0.32	0.35	0.48
Capulet	0.36	0.41	0.68	0.26	0.24	0.40
Mona	0.56	0.69	0.92	0.29	0.26	0.32
O'Connor	0.19	0.36	0.76	0.13	0.18	0.40
Ellice	0.47	0.59	0.72	0.30	0.28	0.40
TR-306	0.32	0.50	0.77	0.11	0.23	0.30
Stein	0.50	0.63	0.83	0.30	0.37	0.44
Clipper	0.40	0.45	0.60	0.12	0.13	0.29
Kinukei 21	0.50	0.68	0.79	0.05	0.13	0.27
Tukey's $HSD_{0.05}$						
G x Zn		0.19			0.17	

* Values were adjusted for differences in seed Zn content

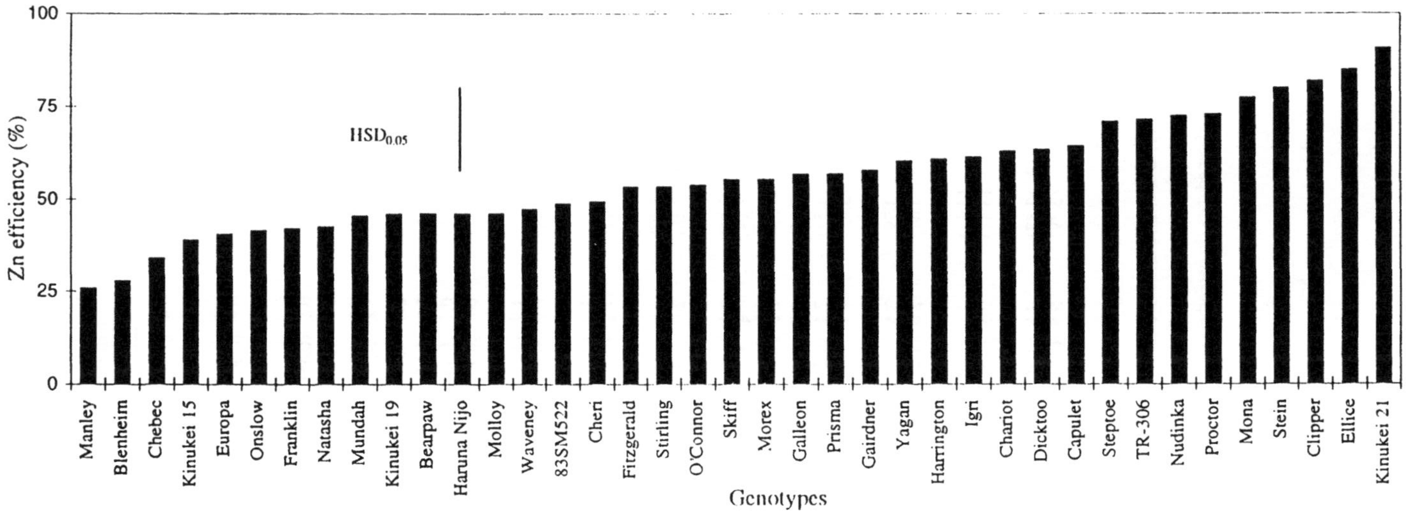

Figure 2. Zn efficiency of barley genotypes in Experiment 2. Values adjusted by the analysis of covariance were used to compute the ratio. Vertical bar represents the Tukey's HSD$_{0.05}$ value for genotype effect

TABLE 6. Zn concentration and content in the shoots of barley genotypes grown at 3 levels of Zn 4 weeks after sowing in Experiment 2

Genotype	Zn concentration (ug g^{-1})			Zn content, ug (2 plants)$^{-1}$		
	Zn fertilization (mg kg^{-1} soil)			Zn fertilization (mg kg^{-1} soil)		
	0.0	0.02	0.8	0	0.02	0.8
Cheri	3.9 (1.37)[a]	4.8 (1.57)	32.2 (3.47)	0.9 (0.62)	2.0 (1.10)	27.3 (3.34)
Blenheim	5.9 (1.77)	3.6 (1.21)	36.3 (3.59)	0.9 (0.62)	0.9 (0.61)	31.4 (3.48)
Kinukei 15	5.3 (1.67)	6.9 (1.93)	42.2 (3.74)	0.8 (0.59)	2.3 (1.19)	36.6 (3.63)
Molloy	4.2 (1.44)	6.9 (1.92)	37.2 (3.62)	0.8 (0.59)	2.3 (1.18)	27.4 (3.33)
Natasha	4.1 (1.40)	5.2 (1.64)	39.8 (3.68)	1.0 (0.69)	2.2 (1.15)	39.8 (3.70)
Manley	4.4 (1.47)	5.6 (1.70)	39.7 (3.68)	0.7 (0.53)	1.3 (0.81)	34.5 (3.57)
Europa	3.8 (1.23)	5.9 (1.77)	38.9 (3.66)	0.6 (0.45)	2.2 (1.15)	35.1 (3.59)
Kinukei 19	6.5 (1.87)	5.8 (1.75)	39.4 (3.68)	1.1 (0.74)	2.8 (1.33)	42.3 (3.77)
Waveney	3.9 (1.36)	6.1 (1.80)	31.0 (3.43)	0.6 (0.44)	2.4 (1.22)	25.8 (3.29)
Franklin	3.7 (1.32)	5.8 (1.76)	53.4 (3.98)	0.6 (0.47)	2.1 (1.13)	45.5 (3.84)
Chebec	4.5 (1.50)	4.9 (1.57)	40.9 (3.71)	0.8 (0.59)	1.4 (0.85)	33.1 (3.53)
Bearpaw	4.0 (1.37)	5.5 (1.71)	31.7 (3.45)	0.8 (0.56)	2.2 (1.16)	28.0 (3.36)
Mundah	3.2 (1.16)	5.6 (1.72)	43.0 (3.76)	0.7 (0.53)	2.2 (1.14)	36.2 (3.62)
Onslow	4.5 (1.50)	5.7 (1.74)	32.4 (3.47)	0.8 (0.56)	2.1 (1.11)	27.8 (3.35)
Fitzgerald	4.4 (1.48)	6.4 (1.86)	36.5 (3.60)	1.2 (0.76)	2.9 (1.35)	30.8 (3.46)
Haruna Nijo	7.0 (1.94)	6.0 (1.79)	43.6 (3.77)	1.9 (1.06)	2.7 (1.29)	42.5 (3.77)
Stirling	4.0 (1.38)	5.8 (1.76)	32.6 (3.48)	0.9 (0.62)	2.4 (1.22)	25.2 (3.27)
83SM522	4.8 (1.57)	6.6 (1.89)	52.8 (3.97)	1.3 (0.81)	2.4 (1.22)	39.2 (3.69)
Galleon	4.2 (1.41)	6.8 (1.91)	46.6 (3.84)	1.2 (0.76)	3.1 (1.40)	37.8 (3.65)
Igri	3.9 (1.35)	6.0 (1.78)	30.9 (3.43)	1.3 (0.81)	2.8 (1.32)	23.2 (3.18)
Gairdner	4.6 (1.52)	5.9 (1.78)	42.5 (3.78)	1.4 (0.85)	2.9 (1.35)	36.2 (3.61)
Yagan	5.3 (1.63)	4.9 (1.58)	33.2 (3.50)	1.9 (1.04)	2.9 (1.36)	33.4 (3.54)
Skiff	6.1 (1.78)	5.7 (1.74)	39.8 (3.68)	2.0 (1.06)	3.0 (1.39)	37.7 (3.67)
Steptoe	5.4 (1.69)	5.9 (1.77)	43.3 (3.77)	2.2 (1.14)	3.3 (1.45)	34.8 (3.58)
Nudinka	5.9 (1.77)	6.0 (1.78)	31.2 (3.44)	2.1 (1.11)	2.8 (1.33)	20.3 (3.06)
Chariot	6.9 (1.93)	6.5 (1.88)	37.3 (3.62)	2.5 (1.24)	3.2 (1.43)	29.4 (3.41)
Proctor	7.4 (1.95)	6.3 (1.84)	34.2 (3.53)	2.3 (1.17)	3.4 (1.48)	25.0 (3.26)
Morex	6.1 (1.81)	6.0 (1.77)	31.3 (3.44)	1.9 (1.05)	2.8 (1.34)	28.0 (3.35)
Dicktoo	6.4 (1.85)	5.6 (1.71)	35.5 (3.57)	2.4 (1.21)	2.7 (1.29)	26.8 (3.33)
Prisma	5.1 (1.62)	5.9 (1.78)	31.4 (3.45)	1.9 (1.05)	3.1 (1.41)	28.8 (3.39)
Harrington	11.4 (2.43)	5.8 (1.76)	31.7 (3.46)	4.4 (1.68)	2.9 (1.36)	26.5 (3.31)
Capulet	5.8 (1.75)	7.0 (1.94)	36.5 (3.59)	2.5 (1.25)	3.4 (1.47)	27.5 (3.34)
Monar	4.1 (1.40)	6.0 (1.80)	38.5 (3.64)	2.6 (1.28)	4.7 (1.73)	38.5 (3.67)
O'Connor	4.4 (1.47)	6.8 (1.91)	41.5 (3.72)	1.3 (0.81)	3.1 (1.40)	35.0 (3.58)
Ellice	5.5 (1.71)	6.7 (1.91)	31.1 (3.44)	3.1 (1.41)	4.7 (1.73)	25.5 (3.28)
TR-306	5.5 (1.70)	5.9 (1.78)	47.9 (3.87)	2.5 (1.24)	3.8 (1.55)	43.0 (3.77)
Stein	5.3 (1.66)	6.6 (1.87)	36.7 (3.60)	3.4 (1.48)	5.1 (1.79)	35.5 (3.59)
Clipper	5.8 (1.76)	7.5 (2.00)	46.3 (3.83)	3.5 (1.50)	4.9 (1.77)	36.8 (3.63)
Kinukei 21	6.6 (1.87)	6.4 (1.86)	44.3 (3.79)	5.5 (1.83)	6.4 (2.00)	48.8 (3.91)
Tukey's $HSD_{0.05}$						
G x Zn		(0.55)[b]			(0.43)[b]	

[a]Numbers in parentheses refer to averages obtained from the analysis of covariance of log-transformed data

[b]The $HSD_{0.05}$ values are applicable to log-transfromed data (in parentheses)

2875 were moderately sensitive and Sahara the least sensitive to Zn deficiency. The tolerance of Sahara to low supplies of Zn in these experiments may be associated with its higher Zn content in seed (Table 1). This result is in accordance with the conclusion of Genc et al (submitted for publication) that deficiency symptoms were absent or less evident in barley plants grown in Zn-deficient conditions from seed with high Zn content. However, the differences in visual symptoms among the other genotypes were not related to their seed Zn contents because they did not show much variation.

Genotypic differences in Zn deficiency symptoms were much greater in Experiment 2 than those in Experiment 1. One reason for this may be the larger differences in seed Zn content than Experiment 1. For instance, Kinukei 21 had a 10 fold higher seed Zn content than Kinukei 15 which may have contributed to absence of symptoms and the higher shoot growth of Kinukei 21 over Kinukei 15.

Genotypes differed in their response to Zn application in Fxperiment 1, but not markedly. The increases in shoot and root dry matter production were similar among all the genotypes, except for Sahara and Skiff. In contrast, there was greater genotypic variation among barley genotypes in Experiment 2. As has been observed in this study, there are a number of reports on differential responses of barley genotypes to Zn deficiency in the literature. For example, in their field study with six barley genotypes, Takkar et al. (1983) found that all the genotypes were tolerant to Zn deficiency stress. In contrast, Pathak et al. (1978) in a glasshouse study demonstrated that barley genotypes differed considerably in their response, and Yilmaz et al. (1995) observed considerable genetic differences in response to Zn deficiency among ten genotypes grown in the field. The differential responses reported in the literature may be due to inherent differences among barley genotypes or to differences in experimental conditions. None of these studies reported the seed Zn content, but it has been reported that higher seed Zn content improves both vegetative growth and grain yield of wheat (Rengel and Graham, 1995a,b). A similar result has been found for vegetative growth in barley (Genc et al., submitted for publication) and it is likely that the seed Zn effect persists to grain yield. It is, therefore, possible that the seed used in other experiments may have differed in Zn content, which in turn would have confounded the published results on Zn efficiency/ responsiveness.

Genotypic variation in Zn efficiency was much greater in Experiment 2 than that in Experiment 1, which may be partly related to differences in the Zn content or concentration in the seeds. However, the variation remained after comparisons were based on values for shoot dry weight adjusted for differences in seed Zn content after analysis of covariance. The analysis of covariance may have assessed the relative differences among seed Zn contents and accordingly down-scaled relative growth of plants but it is unlikely that it took into account the boosting effect that greater seed Zn content gives to one genotype over the other. Greater root growth, as a result of boosting effect, may result in higher Zn uptake and consequently compound in further increases in root and shoot growth. Seeds used in Experiment 2 differed considerably in Zn content or concentration. The positive correlation between seed Zn content and Zn efficiency (Figure 3) indicates that seed Zn content plays an important part in screening studies. Nevertheless, based on adjusted values, a few conclusions can be drawn from Experiment 2 in terms of differential variation in Zn efficiency. For example, Manley, Blenheim, Chebec, Kinukei 15 and Europa can be classified as Zn-inefficient; Mona, Stein, Clipper, Ellice, Kinukei 21 as Zn-efficient (Figure 2).

Apart from the analysis of covariance, genotypes with similar seed Zn contents can be compared to examine genotypic variation independently of seed Zn content. Again, this indicates considerable variation in Zn efficiency. For instance, at a similar seed Zn content, O'Connor was 50 % less Zn efficient than Mona and Ellice. O'Connor also had greater Zn content in seed but was less Zn efficient than Steptoe, Nudinka and Proctor (see Table 2 and Figure 2). Therefore, the differences in Zn efficiency among these genotypes may result from inherent genotypic variation, independent of seed Zn content.

Differences in Zn efficiency of genotypes used in Experiments 1 and 2 (Chebec, Haruna Nijo, Skiff, Clipper, Galleon, Harrington and Yagan) further suggest that not only seed Zn content but also level of Zn deficiency stress influence growth and, accordingly, Zn efficiency of genotypes (Figures 1 and 2). Generally, genotypes achieved higher Zn efficiency when grown from high-Zn seed (Experiment 2) than low-Zn seed (Experiment 1), but ranking of genotypes varied with Zn fertilization. At Zn fertilization of 0.02 mg kg^{-1} soil, Clipper, Galleon, Harrington and Yagan grown from seed with high Zn content had higher Zn efficiency than when grown from seed with low Zn content but Chebec, Haruna Nijo and Skiff remained unaffected. This has implications for breeding programmes aimed at selecting genotypes with higher Zn efficiency. Seeds of genotypes should be generated under identical conditions to ensure similar Zn content or concentration in seed, and the levels of Zn deficiency stress should be chosen in such a way that genotypes are to be under equal stress. However, if a large number of genotypes are to be tested, then the task is very difficult (if not practically impossible) to accomplish. Therefore, research is needed to develop a screening technique independent of measuring growth.

The results suggest that barley genotypes tested differed considerably in their response to Zn deficiency. However, from earlier studies and as well as high correlation between seed Zn content and Zn efficiency in the present study (Figure 3), the differential Zn efficiency of barley genotypes observed in this study can be partly attributed to differences in seed Zn content.

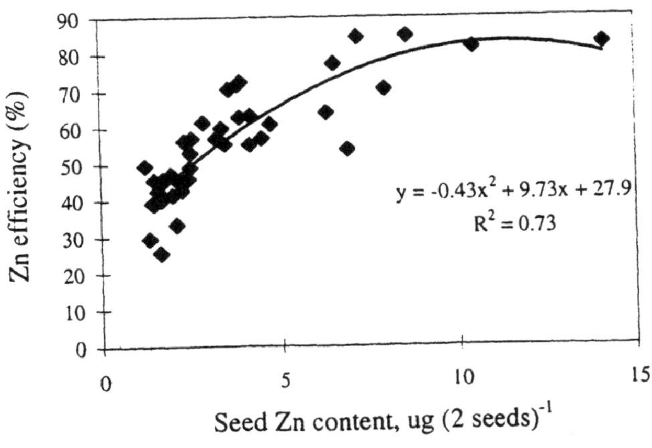

Figure 3 . The relationship between seed Zn content and Zn efficiency in Experiment 2. Values for shoot dry matter were raw and not adjusted for differences in shoot Zn content.

The results also clearly demonstrate that breeding for Zn efficiency can not rely on the current method, soil-based pot screening, to differentiate between genotypes by relying on low-Zn seed in Zn-deficient environments. Instead, development of a more reliable screening method such as biochemical/ molecular markers is needed. However, given difficulties of selection for Zn efficiency in field, the current method, seems to be useful to further our knowledge of mechanisms of Zn efficiency in order to develop biochemical/molecular markers that may facilitate screening process and speed up breeding for Zn efficiency in the near future.

6. Acknowledgments

The first author acknowledges Adnan Menderes University for the award of a post-graduate scholarship which enabled him to complete this study. Technical assistance with ICP analyses from Mrs T O Fowles is greatly appreciated. This study was partly supported by the Soil Science and Plant Nutrition, University of Western Australia. The authors wish to thank Dr. R Boyd and Mrs L Osborne (University of Western Australia) for their valuable assistance throughout this study.

7. References

Cakmak, I., Yilmaz, A., Kalayci, M., Ekiz, H., Torun, B., Erenoglu, B., and Braun, H.J. (1996) Zn deficiency as a critical problem in wheat production in Central Anatolia, *Plant and Soil* **180**, 165-172.

Cakmak, I., Ekiz, H., Yilmaz, A., Torun , B., Koleli, N., Gultekin, I., Alkan, A., and Eker, S. (1997a) Differential response of rye, triticale, bread and durum wheats to Zn deficiency in calcareous soils, *Plant and Soil* **188**, 1-10.

Cakmak, I., Derici, R., Torun, B., Tolay, I., Braun, H.J., and Schlegel, R. (1997b) Role of rye chromosomes in improvement of Zn efficiency in wheat and triticale, in T. Ando et al. (eds), *Plant Nutrition for Sustainable Food Production and Environment*, Kluwer Academic Publishers, Dordrecht, pp. 33-35.

GENSTAT 5 (1988) *Reference Manual*, Genstat 5 Committee, Rothamsted Experimental station, Clarendon Press, Oxford.

Graham, R.D., (1984) Breeding for nutritional characteristics in cereals, *Adv. Plant Nutr* **1**, 57-102.

Graham, R.D.,. Ascher, J.S., and Hynes, S.C. (1992) Selecting Zn-efficient cereal genotypes for soils of low zinc status, *Plant and Soil* **146**, 241-250.

Large, E.C. (1954) Growth stages in cereals, *Plant Pathol* **3**, 128.

MacNaeidhe, F.S., and Fleming, G.A. (1990) The effect of zinc and magnesium application on ear density, grain development and grain yield of winter and spring barley (Hordeum vulgare L.) crops on some Irish soils, in M. L. van (ed.), *Plant Nutrition, Physiology and Applications*, Kluwer Academic Publishers, Dordrecht, pp.251-255.

Pathak, A.N., Tiwari, K.N.,. Upadhyay, R.L, and. Jha, A.K. (1978) Zinc response of barley as influenced by genetic variability, *Indian J. Agric. Sci*, **49**, 25-29.

Rengel, Z., and Graham, R.D. (1995a). Importance of seed Zn content for wheat growth on Zn-deficient soil. I. Vegetative growth, *Plant and Soil*, **173**, 259-266.

Rengel, Z., and Graham, R.D. (1995b) Importance of seed Zn content for wheat growth on Zn-deficient soil. II. Grain yield, *Plant and Soil*, **173**, 267-274.

Singh, T.N. (1982) Zn requirement of wheat in alkali soils of U.P, *Indian Farming*, **31**, 22.

Singh, T. N. (1983) When Zn is vital, *Intensive Agric*, **20**, 12-13.

Steel, R.G.D., and Torrie, J.H. (1960) *Principles and Procedures of Statistics*, McGraw-Hill Book Company Inc, New York.

Takkar, P.N., Singh, S.P., Bansal, R.L., and Nayyar, V.K. (1983) Tolerance of barley varieties to Zn deficiency, *Indian J. Agric. Sci*, **53**, 971-979.

Takkar, P. N., and Walker, C.D. (1993) The distribution and correction of Zn deficiency, in A.D. Robson. (Ed.), *Zinc in Soils and Plants*,. Kluwer Academic Publishers, Dordrecht, pp.151-166.

Yilmaz, A., Kalayci, M., Ekiz, H., Torun, B., Ozturk, L., and Cakmak, I. (1995) Tolerance of barley genotypes to Zn deficiency, III. Barley-Malt symposium, 5-7 Sep'rmber 1995, Konya, Turkey (Orta Anadolu bolgesinde yetistirilen bazi arpa genotiplerinin cinko noksanligina tepkileri. III. Arpa-Malt simpozyumu, 5-7 Eylul 1995, Konya Turkiye)

Zarcinas, B.A. (1984) Analysis of soil and plant material by inductively coupled plasma-optical spectrometry: comparison of digestion procedures for major and trace constituents in plant material, Aus. CSIRO Div. of Soils, Div. Rep. No. 70.

COMPARTMENTATION OF MANGANESE IN THE VACUOLES AND IN THE APOPLAST OF LEAVES IN RELATION TO GENOTYPIC MANGANESE LEAF-TISSUE TOLERANCE IN *VIGNA UNGUICULATA* (L.) WALP.

WALTER J. HORST and PETER MAIER

Institute of Plant Nutrition, University of Hannover,
30419 Hannover, Germany

Abstract

In cowpea (*Vigna unguiculata* (L.) Walp.) resistance to manganese (Mn) excess is due to a higher Mn leaf-tissue tolerance of the Mn-tolerant genotypes. The physiological mechanisms for improved Mn leaf-tissue tolerance are still poorly understood. In the present study the effect of an increased Mn supply in hydroponics on the occurrence of Mn toxicity symptoms and the compartmentation of Mn on the cellular level within the leaf tissue was evaluated under controlled environmental conditions.
Plants were precultured for 15 days with low Mn supply and treated with 50 μM Mn for 2 to 8 days. On the basis of the density of brown spots on the leaves which are typical Mn toxicity symptoms and even more sensitively of the callose content, it was confirmed that cultivar TVu 1987 was more Mn-tolerant than cultivar TVu 91. Vacuolar sap was extracted from vacuoles isolated from the bulk leaf-tissue and the Mn and organic anion concentrations determined and compared to the bulk-tissue Mn content. With increasing duration of the Mn treatment Mn concentration of the vacuolar sap increased. A close positive relationship existed between the bulk Mn contents of the leaves and the vacuolar Mn concentrations from the same leaves. All organic anions except succinate in the vacuolar sap increased with increasing vacuolar Mn concentrations, with a much steeper increase in the Mn-sensitive cultivar TVu 91. Computations of Mn complexation by organic anions in the vacuoles (GEOCHEM-PC) indicated that about 30% of the Mn was complexed by malate, the most abundant organic anion in the vacuoles. The results do not suggest that accumulation of Mn in the vacuoles and complexation through organic anions plays a fundamental role in Mn leaf-tissue tolerance in cowpea.
Therefore, Mn and organic anions were also determined in the intercellular washing fluid (IWF) extracted from whole leaves by an infiltration/centrifugation technique. Above a bulk-tissue Mn content of 1 mmol (kg fw)$^{-1}$ a near linear relationship could be found between Mn contents and concentrations of free (H_2O-soluble) and exchangeable-bound ($BaCl_2$-exchangeable) Mn in the IWF. With increasing Mn contents of the leaves, concentrations of organic anions in the H_2O-IWF also increased malate being the most abundant organic anion. Computation of the Mn speciation in the IWF (GEOCHEM-

G. Gissel-Nielsen and A. Jensen (eds.), Plant Nutrition – Molecular Biology and Genetics, 223–234.

PC) resulted in an increasing percentage of complexed Mn in the leaf apoplast (IWF) with increasing bulk-leaf Mn content. Up to 70% of Mn was calculated to be in a complexed form, with citrate as the dominant complexor. The higher Mn tolerance of TVu 1987 compared to TVu 91 was consistent with the lower absolute concentration and relative proportion of free Mn^{2+} in the leaf apoplast of this cultivar.
We interpret our results as evidence that the key for a better understanding of the physiology of Mn toxicity and tolerance lies in the Mn^{2+}-mediated oxidation/reduction reactions in the leaf apoplast.

Introduction

Manganese (Mn) toxicity is a major factor limiting plant growth and crop yields on acidified and waterlogged soils (Schlichting and Sparrow, 1988). Plant species, and genotypes within plant species may differ considerably in resistance to excess Mn (Foy et al., 1988). Large differences between cultivars of cowpea (*Vigna unguiculata* L.) have been reported (Horst, 1983). In cowpea, genotypic Mn resistance is primarily due to a higher Mn tolerance of the shoot tissue (Horst, 1988). Genotypic differences in Mn tolerance could be easily assessed using the appearance of Mn toxicity symptoms which are brown speckles on older leaves (Horst, 1982, Wissemeier and Horst, 1991). The induction of callose formation by Mn proved to be an even more sensitive parameter for Mn sensitivity (Wissemeier and Horst, 1987). The brown speckles represent local accumulations of oxidized Mn and phenolics mainly in the cell walls (Wissemeier and Horst, 1992). In Mn-tolerant leaf tissue local accumulation of Mn is prevented by a more homogenous distribution (Horst, 1988). The reason for this more homogenous distribution is still unknown. Results of Wissemeier and Horst (1990) indicate that compartmentation of Mn is of great significance in Mn tolerance of cowpea genotypes. Accumulation of Mn in the vacuoles could lead to a more homogenous distribution of Mn on the tissue and cellular level and at the same time detoxify Mn by sequestration preferentially by organic anions such as citrate and oxalate which form strong complexes with heavy metals. Such a detoxification mechanism has been proposed for Zn and Cd (Krotz et al., 1989, Wang et al., 1992, Godbold et al., 1983) and Mn in a Mn hyperaccumulator (Memon and Yatazawa, 1984). Since the Mn toxicity symptoms indicate a special role of Mn oxidation in the cell wall, sequestration of Mn could also be important for Mn leaf-tissue tolerance. In this paper we address the possible role of Mn accumulation and sequestration of Mn by organic acids in the vacuoles and the apoplast for the differences in Mn tolerance of two cowpea cultivars.

Materials and Methods

Cowpea (*Vigna unguiculata* (L) Walp.) cultivars TVu 91 and TVu 1987 were grown in a growth chamber under controlled environmental conditions at 30/25 °C day/night temperatures, 75% relative air humidity and a photon flux-density of 270 μmol m^{-2} s^{-1} photosynthetic radiation in mid-plant height during 16 h. After germination between filter paper moistened with 1 m*M* $CaSO_4$ solution, plant seedlings were transferred to constantly aerated nutrient solution. The composition of the nutrient solution was [µ*M*]: $Ca(NO_3)_2$ 1000, K_2SO_4 700, KH_2PO_4 100, $MgSO_4$ 325, FeEDDHA 20, NaCl 10, H_3BO_3 8, $MnSO_4$ 0.2, $CuSO_4$ 0.2, $ZnSO_4$ 0.2, and Na_2MoO_4 0.05. The pH of the solutions which were renewed every three days was 5.8.

Assessment of Mn toxicity was made by estimating the density of brown spots on the upper side on 1 cm^2 size areas of the base, middle and tip of the second oldest trifoliate leaf prior to each harvest and determining leaf callose-content according to Köhle et al. (1985).

Vacuoles were isolated from one leaflet of the second oldest trifoliate leaf discarding the middle rib. For the isolation of the vacuoles quadratic pieces of 5 mm side-length were cut and incubated in a 25 m*M* MES-Tris and 0.7 *M* mannitol buffer medium (pH 6.5) containing 0.5% cellulase Onozuka R-10 (EC 3.2.1.4) from *Trichoderma viride* (1 U mg^{-1}), 0.15% Macerozyme R-10 (EC 3.2.1.15) from *Rhizopus sp. lyophil.* (pectinase: 0.55 U mg^{-1}), and 1.0% BSA. After shaking at 85 rpm for 16 h at 28 °C on a rotary shaker the suspension was filtered through a 40 µm gaze-cloth for the separation of released vacuoles from other tissue and cell fragments. The vacuoles were purified by centrifugation at 100 x g for 5 min, discarding the supernatant, and resuspension of the pelleted vacuoles in buffer solution four times. The isolated vacuoles were finally resuspended in 1 ml buffer solution and counted under a light microscope using a Fuchs-Rosenthal hemocytometer. The average yield of vacuoles was $9.5*10^5$ per g fresh weight, the average diameter of the vacuoles was 23 µm with no difference between the two cowpea cultivars. Vacuolar sap volume was calculated from the diameter of the vacuoles assuming complete spherical shape by multiplication with the total number of isolated vacuoles. The vacuolar sap was released from the vacuoles by resuspension of the purified vacuoles in bidistilled water leading to the bursting of the vacuoles and separated from few adhering chloroplasts by immediate centrifugation at 500 *x g* for 10 min.

Intercellular washing fluid (IWF) was extracted from the middle leaflet of the second oldest trifoliate leaf. After cleaning of the leaf surface by dipping the leaflet in 50 m*M* EDTA and twice in bidistilled water fresh weight was recorded. Leaflets were then infiltrated with either bidistilled water or in 50 m*M* $BaCl_2$ with 150 m*M* MES (pH 3.5) in a vacuum dessicator by 2 times reducing the pressure to 35 hPa followed by slow relaxation to atmospheric pressure within 2 min. Afterwards fresh weight was recorded for estimating the intercellular volume. The intercellular volume was 365 ± 0.08 µL (g $fw)^{-1}$. Intercellular washing fluid was recovered by centrifugation at 7°C at 150 *x g* for 15 min of the intact leaflet mounted into a centrifugation tube. The IWF was immediately frozen in liquid nitrogen. Malate dehydrogenase (MDH) activity in the IWF was lower than 1% of MDH activity of the leaf homogenate.

Manganese in the bulk-leaf tissue was determined after dry ashing and dissolving the ash in 6 *M* HCl with 1.5% (w/v) hydroxylammonium chloride in 1/10 diluted solution by flame-AAS (Hitachi Z 8000) in the presence of Cs/La-chloride. In the IWF Mn was determined by GF-AAS (Hitachi Z 8000) after filtration through a 0.45 µm nylon membrane. In this filtrate also organic anions were determined by isocratic HPLC (Kroma System 2000, Kontron Instruments) equipped with an Aminex HPX-87H column (BioRad) supplemented with a cation H^+ micro guard cartridge, using 5 m*M* H_2SO_4 as eluant.

Speciation of Mn in the vacuolar sap and IWF was computed using GEOCHEM-PC (Parker *et al.*, 1995) with measured concentrations of Mn, K, Ca, Mg, NO_3^-, Cl^- PO_4^{3-}, and organic anions as input. The pH was set to 5.5 and SO_4^{2-} was estimated from the cation/anion balance. Speciation of Mn in the IWF was computed on the basis of measured concentrations in the H_2O extract of Mn, Ca, and organic anions at a pH of 6.5.
Statistical analysis was carried out using SAS Release 6.10. Coefficients of determination from regression analysis are given according to their level of significance as ***, **, * for $p < 0.001$, 0.01, and 0.05, respectively.

Results

First visible Mn toxicity symptoms in cowpea appear as distinct dark brown spots on old leaves. The density of these spots per unit of leaf area in relation to the bulk-leaf Mn content is an indicator of the Mn tolerance of the leaf tissue.

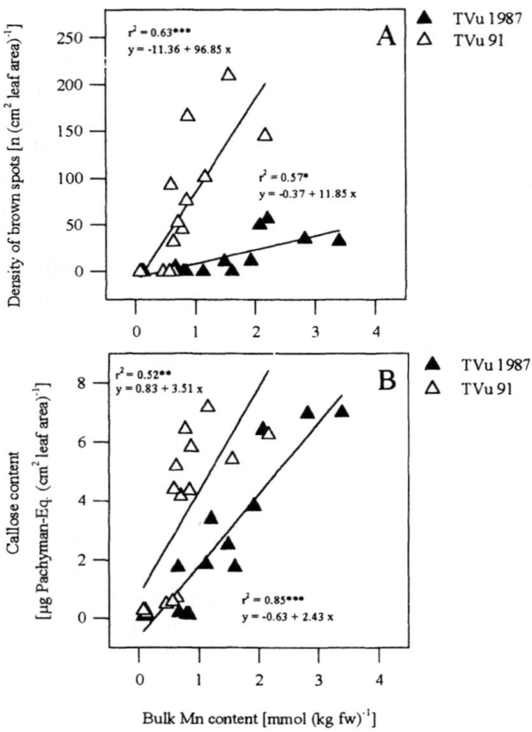

Figure 1. Relationship between bulk Mn content and density of (A) brown spots (Mn toxicity symptoms) and (B) callose content of the second oldest trifoliate leaf of two cowpea cultivars. Plants were precultured for 15 days at low Mn supply and than treated with 50 μM Mn for 2 up to 8 days.

Based on this positive relationship cowpea cultivar TVu 1987 proved to be more Mn-tolerant than cultivar TVu 91 (Fig 1A). The genotypic difference in Mn tolerance is also apparent when Mn-induced callose formation is used as an indicator for Mn injury (Fig. 1B). This parameter appeared to be more sensitive than the density of brown spots because enhanced callose formation could be measured well before the appearance of brown spots in the Mn-tolerant cultivar TVu 1987.

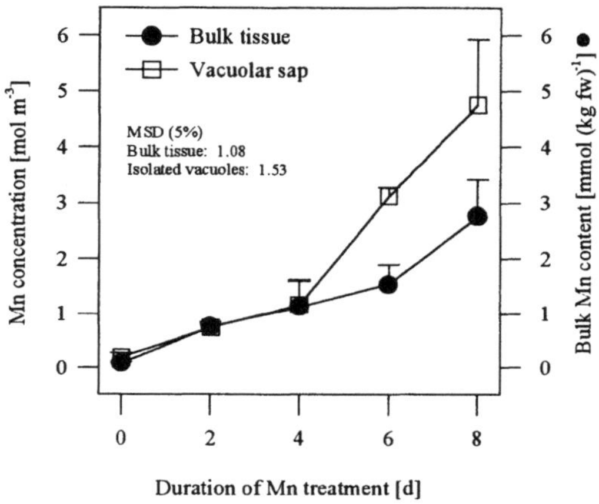

Figure 2. Manganese concentration of the vacuolar sap extracted from vacuoles isolated from the second oldest trifoliate leaf and bulk Mn content as affected by the duration of treatment at elevated Mn supply (50 μM). Plants of cultivar TVu 1987 were precultured for 15 days at low Mn supply.

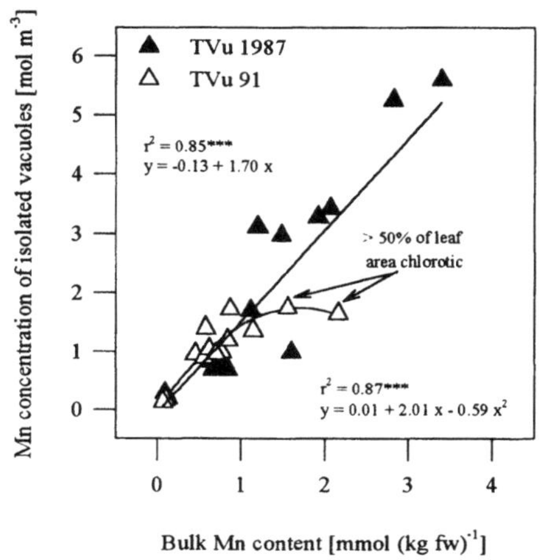

Figure 3. Relationship between bulk Mn content and Mn concentration of the vacuolar sap extracted from vacuoles isolated from the second oldest trifoliate leaf of two cowpea cultivars differing in Mn tolerance. Plants were precultured for 15 days at low Mn supply and than treated with 50 μM Mn for 2 up to 8 days.

Manganese concentration in the vacuolar sap increased with the duration of the treatment at elevated Mn supply (Fig. 2). This increase is faster than the increase of the bulk-leaf Mn content with longer treatment duration. When all data from the different harvests after 2, 4, 6, and 8 days of Mn treatment are pooled, vacuolar Mn concentrations in the Mn-tolerant cultivar TVu 1987 increased linearly with increasing bulk-leaf Mn contents (Fig 3). The slope of the regression is significantly greater than 1. In the Mn-sensitive cultivar TVu 91 vacuolar Mn concentrations leveled off at higher Mn contents probably due to severe leaf injury by Mn toxicity.

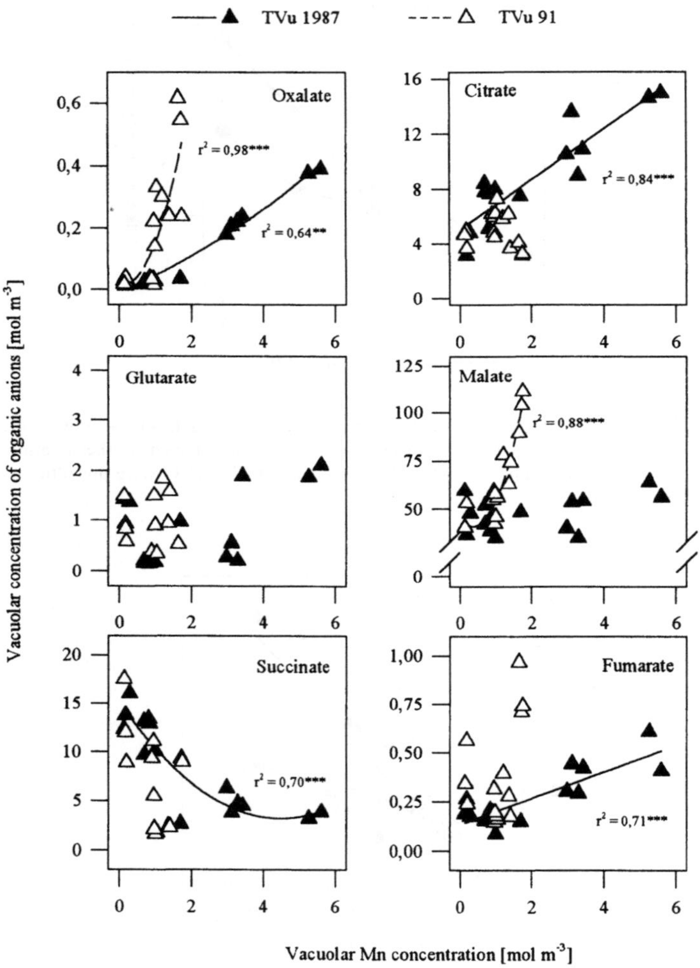

Figure 4. Relationships between vacuolar concentrations of Mn and organic anions extracted from vacuoles isolated from the second oldest trifoliate leaf of two cowpea cultivars differing in Mn tolerance. Plants were precultured for 15 days at low Mn supply and than treated with 50 μM Mn for 2 up to 8 days. Only significant regression lines are shown.

Binding of Mn to organic anions could be responsible for the accumulation of Mn in the vacuoles at elevated Mn supply. Six organic anions could be identified in the vacuolar sap: malate >> succinate > citrate > glutarate > fumarate > oxalate (Fig. 4). Oxalate, citrate, malate, and fumarate concentrations generally increased with the vacuolar Mn concentration, whereas glutarate remained unchanged and succinate decreased. In the Mn-sensitive cultivar TVu 91 the concentrations of oxalate, malate, and fumarate were much higher especially at elevated vacuolar Mn concentrations than in cultivar TVu 1987. In this cultivar, only the citrate concentration linearly increased with the vacuolar Mn concentration to much higher levels than in cultivar TVu 91. Computation of the Mn speciation in the vacuolar sap of plants grown at elevated Mn supply for 6 days (Tab. 1) revealed that most of the Mn was present as free Mn^{2+}. Among the Mn complexes Mn-malate was the most important followed by Mn-citrate.

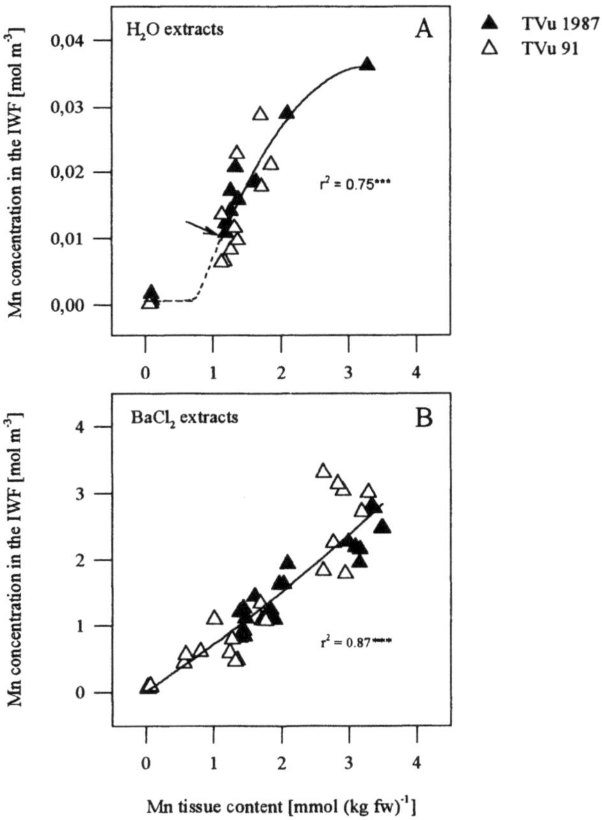

Figure 5. Relationship between bulk Mn content and concentration of (A) free, H_2O-extractable and (B) adsorbed, $BaCl_2$-extractable Mn in the intercellular washing fluid of the second oldest trifoliate leaf of two cowpea cultivars differing in Mn tolerance. Plants were precultured for 15 days at low Mn supply and than treated with 10 or 50 μM Mn for 8 days, while control plants received 0.2 μM Mn continuously.

Table 1. Simulated Mn speciation (GEOCHEM-PC) in the vacuolar sap extracted from vacuoles isolated from the second oldest trifoliate leaf of two cowpea cultivars differing in Mn tolerance. Plants were precultured for 15 days at low Mn supply and than treated with 50 μ*M* Mn for 6 days.

	Mn species in the vacuole						
	as	complexed with					
Cultivar	free Mn^{2+}	Cl	SO_4	Oxalate	Citrate	Malate	Succinate
				[% of total Mn]			
TVu 1987	60.85	4.02	0.19	0.11	7.91	24.86	2.06
TVu 91	64.35	4.13	0.09	0.06	2.24	28.76	0.36

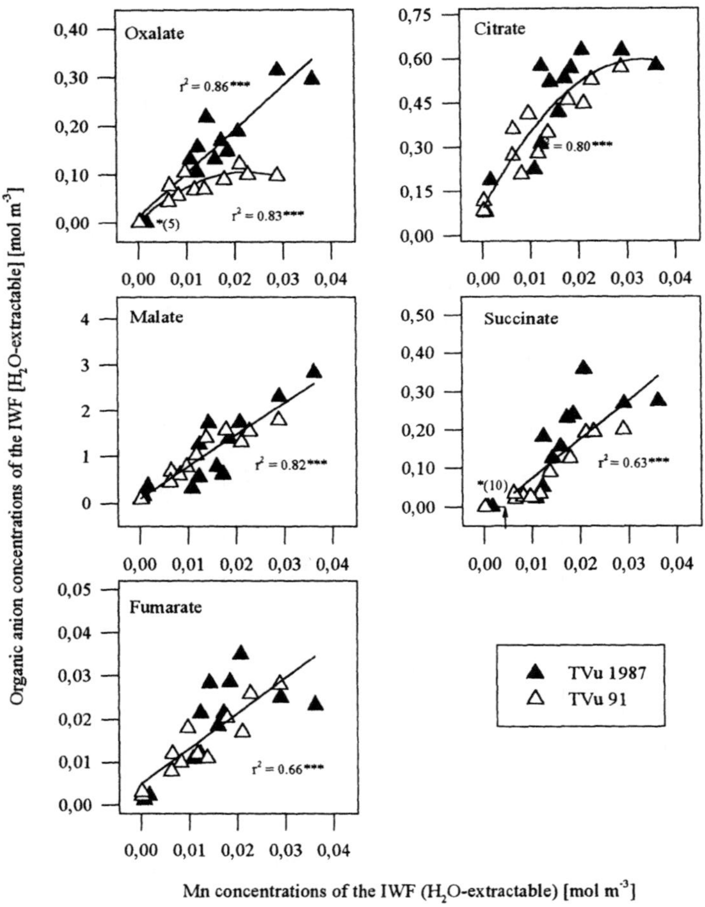

Figure 6. Relationships between concentrations of H_2O-extractable Mn and organic anions in the intercellular washing fluid of the second oldest trifoliate leaf of two cowpea cultivars differing in Mn tolerance. Plants were precultured for 15 days at low Mn supply and than treated with 10 or 50 μ*M* Mn for 8 days, while control plants received 0.2 μ*M* Mn continuously.

There were no clear differences between the cultivars. The results shown are representative also for all other Mn treatment durations (not shown).
Not only in the vacuolar sap but also in the IWF, Mn concentrations increased with the bulk-leaf Mn contents (Fig. 5). This was not only true for the free, water-soluble Mn (Fig. 5A) but also for the adsorbed, $BaCl_2$-exchangeable Mn (Fig 5B) which was higher by a factor of 100. There was no difference between the cultivars in this relationship. Malate was the most abundant organic anion also in the IWF followed by citrate > succinate > oxalate > fumarate (Fig 6). Concentrations of all organic anions increased with the Mn concentrations. Especially at higher apoplastic Mn concentrations, organic anion concentrations tended to be higher in Mn-tolerant cultivar TVu 1987. However, this was only significant for oxalate.
Computation of Mn speciation in the IWF revealed a difference between the cultivars (Tab. 2). The Mn-tolerant cultivar TVu 1987 appeared to be able to maintain the free Mn concentration in the IWF at a lower level than the Mn-sensitive cultivar TVu 91 because a higher percentage of Mn was complexed mainly with citrate.

Table 2. Total Mn concentration, free Mn^{2+} concentration and simulated speciation of Mn (GEOCHEM-PC) in the intercellular washing fluid of the leaves with high bulk Mn content of two cowpea cultivars. Plants were precultured for 15 days at low Mn supply and than treated with 50 *μM* Mn for 8 days.

Cultivar	Mn concentration Mn_t [*μM*]	Mn concentration Mn^{2+} [*μM*]	Mn species in the IWF: free Mn^{2+}	Complexed with Citrate	Complexed with Malate	Complexed with Succinate	Complexed with Oxalate
			[% of total Mn]				
TVu 1987	23	6.3	27.3	51.9	14.3	0.5	6.0
TVu 91	20	8.4	41.3	35.1	18.6	0.5	4.5

Discussion

The results presented confirm previous reports (Horst 1983, Wissemeier and Horst, 1991) that Mn resistance in cowpea is due to differences in Mn leaf-tissue tolerance. Based on the two indicators of Mn toxicity, density of brown spots and callose content, cultivar TVu 1987 was more Mn-tolerant than cultivar TVu 91 (Fig. 1). Tissue tolerance of heavy metals may be the consequence of avoiding mechanisms operating at the subcellular level (Verkleij and Schat, 1990). In this regard the sequestration of the metals in metabolic less active cell compartments such as the cell wall (Memon *et al.*, 1981, Lytle *et al.*, 1996) and the vacuole (Vögeli-Lange and Wagner, 1990, Brune *et al.*, 1994) are of special importance. The vacuole can be regarded as the most appropriate compartment because it represents the largest storage volume of the leaf (Martinoia, 1992) making up for 67-99% of the total cell volume depending on plant species and leaf tissue (Woodrow et al., 1984, Fricke *et al.*, 1994).
In this study we estimated vacuolar concentrations of Mn and organic anions using isolated vacuoles. The integrity of the vacuole preparations was excellent based on their capability to exclude Evans-Blue and accumulate Neutral-Red (Maier, 1997). The vacuolar sap recovered from the isolated vacuoles showed only slight cytosolic contamination based on the cytosolic marker enzyme MDH. Calculated Mn

concentrations agreed well with concentrations of vacuolar sap extracted from single cells using a modified cell pressure-probe according to Malone *et al.*, 1989 (Maier and Marienfeld, personal communication). A special role of the vacuoles in the storage of Mn and Mn tolerance is indicated by the fact that after longer term Mn treatment the Mn concentrations in the vacuolar sap increased steeper than in the bulk tissue (Fig. 2). Also the vacuolar Mn concentrations continued to increase to much higher levels in the Mn-tolerant cultivar TVu 1997 whereas they leveled off at a lower level in the Mn-sensitive cultivar TVu 91 (Fig. 3). However, assuming that the isolated vacuoles are mainly from the leaf mesophyll and the vacuoles of the mesophyll represent 79% of the leaf water, similar contributions of the mesophyll vacuoles to the total leaf Mn content of 54% and 52% for cultivar TVu 1987 and TVu 91, respectively, were calculated. Furthermore, accumulation of Mn in the bulk-leaf tissue and in the vacuolar sap seemed to be impeded owing to severe damage by Mn toxicity in the Mn-sensitive cultivar TVu 91 (Fig.3). It therefore appears that accumulation of high concentrations of Mn in the vacuoles is the consequence rather than the cause of Mn tolerance.

Sequestration of metals in the vacuoles by organic acids has been suggested as a key element of plant strategies of heavy metal tolerance (Wang and Evangelou, 1995). However, with regard to Mn unequivocal experimental evidence is still lacking. Memon and Yatazawa (1984) concluded from their study of the Mn hyperaccumulator *Acanthopanax sciadophylloides* that oxalate acts as terminal acceptor of Mn in the vacuoles. The presented results (Fig. 4) do not support a role of vacuolar concentrations of organic anions in genotypic Mn tolerance in cowpea: (i) although organic anions accumulated in the vacuolar sap with increasing Mn concentration (the only exception was succinate) this increase was much steeper in the Mn-sensitive cultivar TVu 91 especially for oxalate and malate, (ii) computation of the speciation of Mn in the vacuolar sap revealed that Mn will be present mainly in the free ionic form in both genotypes (Tab. 1). This can be explained by the fact that most organic anions, especially oxalate and citrate are bound to Ca which strongly accumulated in the vacuolar sap in response to increasing bulk-leaf Mn contents, particularly in cultivar TVu 91 (not shown).

Inactivation of metals in the cell wall by binding to cell-wall constituents has also been suggested as tolerance mechanism (Wang and Evangelou, 1995). Accumulation of Mn in leaf cell-walls has been demonstrated using X-ray microanalysis by Memon *et al.* (1980, 1981) and Lytle *et al.* (1996). Using the infiltration/centrifugation method for the extraction of IWF we demonstrated that, also in cowpea, the Mn content in the apoplast is positively related to the bulk-leaf Mn content (Fig. 5). Most of the Mn was bound to the exchange sites of the cell wall ($BaCl_2$-exchageable), only 1% was present as water-soluble Mn. There was no difference between the cowpea cultivars in the accumulation of Mn in the apoplastic fraction. This suggests that the storage of Mn in the apoplast thus reducing the accumulation in the symplast, is not an important tolerance mechanism in cowpea.

Oxidation of Mn(II) in the apoplast has been proposed as the key process leading to Mn toxicity (Horst, 1988), because Mn(III) may act as a powerful oxidant of proteins and lipids (Archibald and Fridovich, 1982) leading to damage of biological membranes. Organic anions may protect Mn(II) from oxidation through complexation and acting as reductants for Mn(III, IV). The Mn-tolerant cultivar TVu 1987 appears to be able to

maintain higher concentrations of organic anions in the apoplast (Fig. 6) thus reducing the concentration of free Mn^{2+} (Tab. 2). Enhanced complexation of apoplastic Mn by organic anions may also explain the more homogeneous distribution of Mn in the leaf tissue and less deposition in the cell walls of Mn oxides and phenolics in this cultivar compared to the Mn-sensitive cultivar TVu 91 (Horst, 1988).
The results clearly indicate that the sequestration of Mn in the vacuoles cannot explain genotypic differences in Mn tolerance in cowpea. The Mn-tolerant cultivar seems to be capable to reduce the Mn^{2+} activity in the apoplast through complexation by organic anions. The characterization of Mn^{2+}-mediated oxidation/reduction reactions in the leaf apoplast appears to be necessary for a better understanding of the physiology of Mn toxicity.

Acknowledgments

We thank the German Science Foundation (DFG) for financial support.

References

Archibald, F.S. and Fridovich, I. (1982) The scavenging of superoxide radical by manganous complexes: *In vitro*, *Arch. Biochem. Biophys.* **214**, 452-463.

Brune, A., Urbach, W., and Dietz, K.-J. (1994) Compartmentation and transport of zinc in barley primary leaves as basic mechanisms involved in zinc tolerance, *Plant Cell Environ.* **17**, 153-162.

Foy, C.D., Scott, B.J., and Fisher, J.A. (1988) Genetic differences in plant tolerance to manganese toxicity, in M.J. Webb, R.O. Nable, R.D. Graham, and R.J. Hannam (eds.), *Manganese in Soils and Plants*, Kluwer Academic Publishers, Dordrecht/Boston/London, ISBN 90-247-3758-3, pp. 293-307.

Fricke, W., Leigh, R.A., and Tomos, A.D. (1994) Concentrations of inorganic and organic solutes in extracts from individual epidermal, mesophyll and bundle-sheath cells of barley leaves, *Planta* **192**, 310-316.

Godbold, D.L., Horst, W.J., Marschner, H., Collins, J.C., and Thurman, D.A. (1983) Root growth and zinc uptake by two ecotypes of *Deschampsia caespitosa* as affected by high Zn concentrations, *Z. Pflanzenphysiol.* **112**, 315-324.

Horst, W.J. (1982) Quick screening of cowpea genotypes for manganese tolerance during vegetative and reproductive growth, *Z. Pflanzenernähr. Bodenkd.* **145**, 423-435.

Horst, W.J. (1983) Factors responsible for genotypic manganese tolerance in cowpea (*Vigna unguiculata*), *Plant Soil* **72**, 213-218.

Horst, W.J. (1988): The physiology of manganese toxicity, in M.J. Webb, R.O. Nable, R.D. Graham, and R.J. Hannam (eds.), *Manganese in Soils and Plants*, Kluwer Academic Publishers, Dordrecht/Boston/ London, ISBN 90-247-3758-3, pp. 175-188.

Köhle, H., Jeblick, W., Poten, F., Blaschek, W., and Kauss, H. (1985) Chitosan-elicited callose synthesis in soybean cells as a Ca^{2+}-dependent process, *Plant Physiol.* **77**, 544-551.

Krotz, R.M., Evangelou, B.P., and Wagner, G.J. (1989) Relationship between cadmium, zinc, Cd-peptide, and organic acid in tobacco suspension cells, Plant Physiol. **91**, 780-787.

Lytle, C.M., Lytle, F.W., and Smith, B.N. (1996) Use of XAS to determine the chemical speciation of bio-accumulated manganese in *Potamogeton pectinatus*, *J. Environ. Qual.* **25**, 311-316.

Maier, P. (1997) Bedeutung der Kompartimentierung von Mangan und organischen Säuren für die Mangan-toleranz von Cowpea (*Vigna unguiculata* (L.) Walp.), Verlag Ulrich Grauer, Stuttgart, ISBN 3-86186-170-4.

Malone, M., Leigh, R.A., and Tomos, A.D. (1989) Extraction and analysis of sap from individual wheat leaf cells: the effect of sampling speed on the osmotic pressure of extracted sap, *Plant Cell Environ.* **12**, 919-926.

Memon, A.R. and Yatazawa, M. (1984) Nature of manganese complexes in manganese accumulator plant - *Acanthopanax sciadophylloides*, *J. Plant Nutr.* **7**, 961-974.

Memon, A.R., Chino, M., Takeoka, Y., Hara, K., and Yatazawa, M. (1980) Distribution of manganese in leaf tissues of manganese accumulator: *Acanthopanax sciadophylloides* as revealed by electronprobe X-ray microanalyzer, *J. Plant Nutr.* **2**, 457-476.

Memon, A.R., Chino, M., Hidaka, H., Hara, K., and Yatazawa, M. (1981) Manganese toxicity in field grown tea plants and the microdistribution of manganese in the leaf tissue as revealed by electron probe X-ray micrography, *Soil Sci. Plant Nutr.* **27**, 317-328.

Martinoia, E. (1992) Transport processes in vacuoles of higher plants, *Bot. Act.* **105**, 232-245.

Parker, D.R., Norvell, W.A., and Chaney, R.L. (1995) GEOCHEM-PC: A chemical speciation program for IBM and compatible personal computers, in: R.H. Loepppert, A.P. Schwab, and S. Goldberg (eds.), *Soil Chemical Equilibrium and Reaction Models* , ASA and SSSA, Madison, Wis. SSSA Spec. Publ. 4,. pp. 253-269.

Schlichting, E. and Sparrow, L.A. (1988) Distribution and amelioration of manganese toxic soils, in M.J. Webb, R.O. Nable, R.D. Graham, and R.J. Hannam (eds.), *Manganese in Soils and Plants*, Kluwer Academic Publishers, Dordrecht/Boston/London, ISBN 90-247-3758-3, pp. 277-292.

Verkleij, J.A.C. and Schat, H. (1990) Mechanisms of metal tolerance in higher plants, in J. Shaw (ed.) , *Evolutionary Aspects of Heavy Metal Tolerance in Plants*), CRC Press, Boca Raton, USA., pp. 179-193.

Wang, J. and Evangelou, V.P. (1995) Metal tolerance aspects of plant cell wall and vacuole, in M. Pessarakli (ed.), *Handbook of Plant and Crop Physiology*, Marcel Dekker, Inc., New York, pp. 695-717.

Wang, J., Evangelou, B.P., Nielsen, M.T., and Wagner, G.J. (1992) Computer-simulated evaluation of possible mechanisms for sequestering metal ion activity in plant vacuoles. II. Zinc, *Plant Physiol.* **89**, 621-626.

Wissemeier, A.H. and Horst, W.J. (1987) Callose deposition in leaves of cowpea (*Vigna unguiculata* [L.]. Walp.) as a sensitive marker to high Mn supply, *Plant Soil* **102**, 283-286.

Wissemeier, A.H. and Horst, W.J. (1990) Manganese oxidation capacity of cowpea [*Vigna unguiculata* (L.) Walp.] leaves differing in manganese tolerance, *J. Plant Physiol.* **136**, 103-109.

Wissemeier, A.H. and Horst, W.J. (1991) Simplified methods for screening cowpea cultivars for manganese leaf-tissue tolerance, *Crop Sci.* **31**, 435-439.

Wissemeier, A.H. and Horst, W.J. (1992) Effect of light intensity on manganese toxicity symptoms and callose formation in cowpea (*Vigna unguiculata* (L.) Walp.), *Plant Soil* **143**, 299-309.

Woodrow, I., Ellis, J.R., Jellings, A., and Foyer, Ch.H. (1984) Compartmentation and fluxes of inorganic phosphate in photosynthetic cells, *Planta* **161**, 525-530.

TRITICUM TAUSCHII - A POTENTIAL SOURCE OF GENES FOR THE IMPROVEMENT OF ZINC EFFICIENCY IN BREAD WHEAT

L.J. MERRY[1], R.D. GRAHAM[2] and D.P. SCHACHTMAN[1]
Departments of Botany[1] and Plant Science[2], The University of Adelaide, South Australia 5005, AUSTRALIA

1. Introduction

Zinc is an essential mineral for animals and plants. Marginal zinc deficiency in humans or in animals can lead to conditions such as dermatitis, a reduced growth rate and prolonged healing of wounds (Welch, 1995), and humans can also experience a diminished learning ability and impaired dark adaptation (Welch & Norvell, 1993). Increasing the dietary intake of zinc for livestock such as cattle and sheep reportedly reduces the incidence of skin lesions and improves growth rates (White, 1993). Therefore, maintaining a high zinc concentration in pastures, straw, grains, and other foodstuff is an important objective (White, 1993). In plants, zinc is important as a cofactor or activator of various enzymes that participate in cellular processes such as photosynthesis, RNA formation (Brown et al., 1993; Römheld & Marschner, 1991) and the dissipation of oxygen free radicals (Marschner, 1995). Plants deficient in zinc tend to have low resistance to fungal diseases (Sparrow & Graham, 1988; Thongbai et al., 1993; Grewal et al., 1996), and this is largely attributed to the fact that zinc is important for maintaining the integrity of the plasma membrane (Brown et al., 1993).

Zinc deficiency in plants is a widespread problem geographically (Takkar & Walker, 1993) due to the many different biotic and abiotic factors which affect soil Zn availability (Graham et al., 1992). Regions with Mediterranean climates, (e.g. southern Australia and Central Anatolia in Turkey), with dry, alkaline subsoils and low DTPA-extractable zinc levels (Cakmak et al., 1996c), are particularly prone to zinc deficiency, and it is a major objective for plant breeders in these regions to improve the zinc efficiency of crop varieties. Increasing the diversity of germplasm available to wheat breeding programs will raise the likelihood of the production of highly Zn-efficient wheat varieties suited to growth in such regions. A high level of variation has been demonstrated for various agronomic traits in wheat relatives such as *Triticum tauschii* (Gill & Gill, 1994), and work is currently underway to determine whether *T. tauschii* (also known as *Aegilops squarossa* and *Aegilops tauschii*) is also a good source of Zn efficiency germplasm.

G. Gissel-Nielsen and A. Jensen (eds.), Plant Nutrition – Molecular Biology and Genetics, 235–244.

2. Factors affecting the availability of zinc to plants

Plants may develop Zn deficiency when grown in regions where the soil is inherently low in Zn, or where certain soil characteristics lower the level of zinc available for uptake by the plant root. For example, some soils are low in zinc content, whereas highly calcareous soils can reduce the effectiveness of Zn fertilizers by making Zn partially unavailable (Welch et al., 1991). Zinc is relatively immobile in the soil, and therefore, the rate of zinc uptake by the roots of plants is generally limited by the rate of transport of zinc to the soil solution. Zinc availability from the soil solution is modified by a variety of abiotic and biotic soil characteristics (Marschner, 1995).

Soil pH is the factor with the most marked effect on Zn availability in the soil solution (Marschner, 1993; Harter, 1991). The equilibrium concentration of zinc in the soil solution can be reduced by a factor of 45 as the soil pH increases from 5.5-7.0 (Moraghan & Mascagni, 1991). Soils with a high calcium carbonate content also have a low levels of plant-available Zn (Cakmak et al., 1996c), and farming practices such as soil liming can decrease the amount of zinc available to plant roots from the soil solution (Marschner, 1993). Products of the breakdown of soil organic and mineral matter can complex Zn ions and either increase or decrease their mobility (Stevenson, 1991). Zinc deficiency in plants is also commonly associated with flooded soils, due to precipitations with sulfides or sesquioxides in anoxic conditions (Takkar & Walker, 1993).

There are numerous biological factors that also contribute to the rate of zinc acquisition in higher plants. Firstly, root growth and root surface area are important for overcoming the depletion of Zn from the soil solution in the rhizosphere (Marschner, 1993). Vesicular-arbuscular mycorrhizae assist plants with the uptake of Zn in soils by effectively increasing the surface area of soil available for Zn uptake (Moraghan & Mascagni, 1991). High P fertilization rates or the rotation of crops with non-mycorrhizal species can result in Zn deficiency due to a lowered rate of mycorrhizal infection (Marschner, 1993). Plants may also influence the availability of Zn in the rhizosphere more directly, by the excretion of mucilage (Nambiar, 1976), organic acids, reducing substances, and the activation of reductases in response to abiotic stresses (Marschner, 1993). For example, the release of phytosiderophores as a response to iron deficiency in grasses (Guerinot & Yi, 1994) also aids zinc and copper uptake (von Wirén et al,, 1996).

3. Zinc efficiency, a desirable trait for crops grown in soils with low zinc availability

Zinc efficiency in agricultural crops is a form of tolerance to zinc-deficient soils. Zinc-efficient plant genotypes are those which are able to extract more Zn from soils and yield more dry matter and grain compared to other genotypes when grown under zinc deficient conditions (Graham et al., 1992). The development of zinc-efficient crops is desirable in regions where low soil zinc availability limits the yield. The application of effective zinc fertilizers can be costly, especially if there is a need to re-apply frequently (Mortvedt & Gilkes, 1993). Their effectiveness is also dependent on the method of application, and the mixing of Zn fertilizers with other types can cause a reduced availability of Zn (Mortvedt, 1991). Another problem with zinc fertilizers is that most application methods only supply Zn to the top 10 cm of the soil. Zinc deficiency commonly occurs throughout the soil profile (Nable & Webb, 1993), and Zn may not be readily mobilized from roots exposed to Zn into the actively growing root regions where it is required (Webb & Loneragan, 1990). Zinc-efficient plants also have increased disease resistance, growth rate and yield (Grewal et al., 1996) compared with less efficient varieties. Zinc efficient plants may also allow a higher exploitation of the water resources available to plants in micronutrient-poor subsoils where root growth would usually be impaired (Nable & Webb, 1993).

4. Important factors affecting the zinc efficiency of plants

At present, the number of mechanisms involved with zinc efficiency is not clear. It is likely that different crop species vary in the number and type of zinc efficiency mechanisms employed, and that more than one zinc efficiency mechanism is responsible for the level of zinc efficiency in each genotype (Graham & Rengel, 1993). Many investigations into the physiological nature of zinc efficiency have revealed factors that contribute to the zinc efficiency of some genotypes or species.

Seed size and nutrient content is a factor contributing to the successful emergence and yield of wheat seedlings in soil (Mian and Nafziger, 1992). It follows that a high micronutrient content in the seed contributes to the establishment of plants when grown in soil conditions where micronutrient deficiencies typically occur, particularly when the root system has not yet been established. Seed zinc content has also been shown to affect the growth, yield and grain zinc concentration of wheat plants (Yilmaz et al., 1997; Rengel & Graham, 1995a & b). Although the effect of seed Zn is small compared to that of fertilization rate, (Yilmaz et al., 1997; Rengel & Graham, 1995b), a higher seed Zn content may still be useful in overcoming problems in the field such as temporal and spatial variation in Zn availability (Yilmaz et al., 1997; Rengel & Graham, 1995a & b).

The morphology and density of wheat roots is determined more by genotype than by zinc supply, and this suggests that physiological traits associated with zinc efficiency, such as zinc uptake capacity and zinc usage within plants, may be more important than root growth characteristics (Dong et al., 1995). Some zinc-efficient cultivars take up more zinc from the soil or nutrient solution than their inefficient counterparts (Grewal et al., 1996; Rengel & Hawkesford, 1997; Rengel & Graham, 1996). This could be due to a higher density of Zn uptake sites on root cell plasma membranes, however this has yet to be proven, and Zn uptake efficiency is likely to be more complex. For example, Zn deficiency caused by low Zn availability or low Zn content in the soil must require different mechanisms (Graham & Welch, 1996). The Zn-efficient bread wheat cultivar, Excalibur, was the most zinc efficient cultivar when tested in a calcareous soil, but the least efficient genotype in when compared with the same cultivars in chelate buffered nutrient solution studies (Graham & Rengel, 1993). Excalibur is better adapted for extracting Zn from calcareous conditions, but not for the conditions provided by a chelate buffered nutrient solution. The fact that different cultivars have different mechanisms of Zn efficiency has important implications for the use of appropriate screening techniques to select Zn-efficient genotypes.

Phytosiderophore release is important in the adaptation of wild grasses to Zn-deficient calcareous soils (Cakmak et al., 1996a). Bread wheats are generally more Zn-efficient than durum wheats (Cakmak et al., 1997), and also have a higher rate of phytosiderophore release under low-Zn conditions (Cakmak et al., 1996b). Zinc-efficient bread wheat genotypes increase the rate of phytosiderophore release ten-fold in zinc-deficient conditions compared with fertilized conditions, whereas in most durum wheats, the increase is only three-fold (Cakmak et al., 1996a). However, a comparison of bread wheat genotypes differing in Zn-efficiency showed that the cultivar most susceptible to zinc deficiency had a similar rate of phytosiderophore release to the most Zn-efficient cultivar (Erenolglu et al., 1996). This suggests that while a higher rate of phytosiderophore release may contribute to the higher zinc efficiency of bread wheats compared to durum wheats, other Zn efficiency mechanisms contribute to the differential efficiency of bread wheats.

5. Genetic variability of zinc efficiency in cereals

5.1 INTERSPECIFIC VARIATION

Some crop species exhibit a greater ability to grow in Zn-limiting conditions (Graham & Rengel, 1993). Cereal species differ greatly in their ability to tolerate soils with low Zn availability , and the zinc efficiency of different cereal species tends to increase, from the Zn-inefficient durum wheats, as follows (Cakmak et al., 1997):

Durum wheats < oats ≤ bread wheats < barley < triticales < rye

Results of a field trial comparing zinc-efficient *T. aestivum* (bread wheat) cultivars (Excalibur, Schomburgk, Warigal) with a rye addition line (Warigal 5RL), a Zn-inefficient bread wheat (Kite), and two durum wheats (Durati & Kamilaroi) are shown below (Table 1). Durum wheats perform poorly in soils with low Zn availability (Graham & Rengel, 1993). When comparing the Zn efficiency of bread wheats, durum wheats are often used to provide the most Zn-inefficient end of the scale.

5.2 INTRASPECIFIC VARIATION

A large survey of Zn deficient soils in India revealed that 16 different crops showed intraspecific variation in the degree of response to Zn fertilization (Takkar, 1992). Significant differences between the response to Zn fertilization in Zn efficient and inefficient lines of soybean suggest that there is the potential to breed varieties more

Table 1. Grain yields and grain Zn contents of wheat genotypes grown on a low rainfall, Zn-responsive sand-over-clay site in 1988 (Lameroo, South Australia). Zinc was applied at 25kg/ha as Zn oxysulfate granules with the seed. Adapted from Graham et al., 1992.

Genotype	Yield (t/ha)			Zn content (g/ha)		
	-Zn	+Zn	Efficiency	-Zn	+Zn	Efficiency
Excalibur	1.43	1.76	81	10.8	45	24
Schomburgk	1.11	1.32	84	8.3	31	27
Warigal 5RL	1.04	1.14	91	10.9	33	33
Warigal	1.00	1.22	82	13.0	31	41
Kite	0.95	1.37	69	9.0	33	27
Durati	0.45	1.12	40	4.3	28	15
Kamilaroi	0.45	1.29	35	4.0	30	13

suited for Zn deficient soils (Hartwig et al., 1991). The zinc efficiency ratio of *Brassica napus* ranged from 32 to 74% in a controlled temperature, pot experiment where 25 canola and mustard seed varieties were screened for Zn efficiency. When replicated in the field, in Zn deficient soil, the zinc efficiency ranged from 62 to 76%, and the genotypes were ranked in a different order to that of the pot experiments (Grewal et al., 1997). The results obtained in controlled growth conditions are not easily replicated in the field, and a genotype may show different degrees of Zn efficiency over consecutive years (Graham et al., 1992). Studies comparing the extent of Zn efficiency of bread and durum wheat species frequently compare only a few genotypes, and are carried out in pots or with nutrient solution culture in controlled temperature conditions (Erenoglu et al., 1996; Cakmak et al., 1996; Grewal et al., 1996; Rengel & Hawkesford, 1997; Rengel & Graham, 1995a & b).

6. *Triticum tauschii* as a potential source of zinc efficiency genes for bread wheat

6.1 THE ROLE OF *TRITICUM TAUSCHII* IN BREAD WHEAT EVOLUTION

The Asian goat grass, *Triticum tauschii* (also known as *Aegilops squarrosa* and *Aegilops tauschii*) is endemic to a wide geographical area across Asia. The species has been divided into three sub-species – *strangulata*, *eusquarrosa* and *anathera* - and two varieties have been classified within the sub-species *eusquarrosa* (meyeri and typica) (Eastwood, 1991). *T. tauschii* is widely accepted as being the contributor of the D genome to modern day bread wheat, *T. aestivum* L. ($6n = 42$; AABBDD) (Lagudah et al., 1991 a; Eastwood et al., 1991; Cox et al., 1994), and there is considerable evidence, from rDNA studies that the subspecies *strangulata* was the original D genome donor (Lagudah et al., 1991b). The bread wheat species (*T. aestivum*) is believed to be the result of a hybridization event between a tetraploid wheat species (*T. turgidum)* and the diploid *T. tauschii* (Fig. 1). This hybridization may have only occurred between a small number of plants, suggesting that the amount of genetic variation in present day bread wheat could be limited from having originated from a narrow range of *T. tauschii* genes. Therefore, *T. tauschii* may prove to be a useful source of genes for the improvement of micronutrient efficiency in bread wheat.

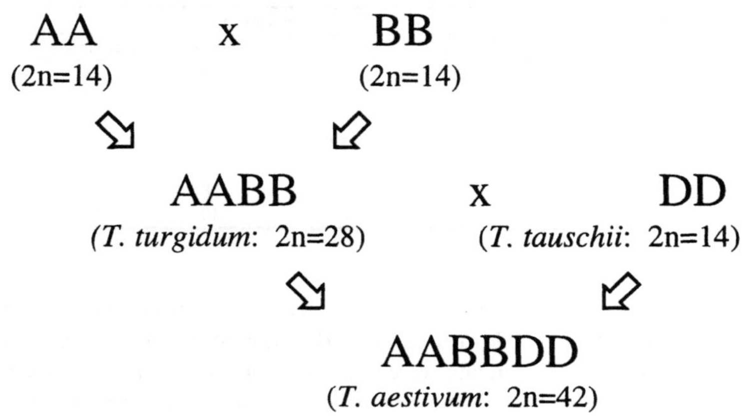

Fig. 1. The evolution of hexaploid wheat, whereby a hybridization event between *T. tauschii* (diploid) and *T. turgidum* (tetraploid) is believed to have resulted in the formation of *T. aestivum* (Kihara, 1965).

6.2 THE USE OF TRITICEAE GENES FOR WHEAT IMPROVEMENT

Hexaploid wheat has a low level of genetic polymorphism (Gill & Gill, 1994), and so for breeding purposes, wheat relatives are sometimes used to introduce additional genetic diversity for wheat improvement. These can be sources of resistance to wheat diseases, tolerances to abiotic stresses (Graham et al., 1987), or genes to improve grain quality (Ren et al., 1990). Relatives of wheat were first used as a source of genes for wheat breeding in late nineteenth century, with crosses between rye (*Secale cereale*) and *T. aestivum* (Appels & Lagudah, 1990). There are difficulties associated with the crossing of species that are not closely related, and the transfer of genes between two such species requires induced or spontaneous recombination (Appels & Lagudah 1990). Comparative mapping studies between different species within the *Poaceae* has revealed a high level of synteny (Devos et al., 1995), however of all wheat relatives, the genome of *T. tauschii* has the most similarity to a wheat genome (Gill & Gill, 1994).

6.3 THE USE OF *T. TAUSCHII* GENES FOR WHEAT IMPROVEMENT

Triticum tauschii has been used as a source of novel genes for bread wheat resistance to a wide range of plant diseases and pests, such as leaf rust (Cox *et al.*, 1994) and cereal cyst nematode (Eastwood, 1991). These resistance traits provide useful alternatives to those bred from hexaploid wheats alone, and in the case of leaf rust-resistance, at least 4 different genes controlling resistance have been identified (Cox *et al.*, 1994). A considerable amount of variation in salt tolerance traits also exists within the species, with some accessions performing better under high Na^+ conditions than the more salt tolerant bread wheats (Schachtman et al., 1992b).

Before *T. tauschii* genes can be used, a collection must be screened for the agronomic trait of interest. Growing a wide range of *T. tauschii* lines has a few problems, for example: (a) some of the lines have a winter growth habit, (b) conventional seed (thrashing) techniques result in a large proportion of shattered seed – glumes must be removed manually to ensure maximum seed recovery, and (c) *T. tauschii* is a potential, weed (Villereal & Mujeeb-Kazi, 1996). However, the similarity between the D genome of hexaploid wheat (*T. aestivum*) and that of diploid *T. tauschii* permits the two species to be crossed with relative ease. The genome of *T. tauschii* can be induced to recombine with the D genome of wheat by two methods: the production of synthetic hexaploids, and direct crossing (Appels & Lagudah, 1990; Eastwood, 1996).

Synthetic hexaploids provide an indirect means by which to incorporate *Triticum tauschii* genes with bread wheat genomes, and are produced by directly crossing *T. turgidum* (AABB) with a chosen *T. tauschii* (DD) line followed by a doubling of chromosomes with colchicine (Appels & Lagudah, 1990). An advantage of using synthetic hexaploids is that they provide a fixed breeding line that can be used directly (Eastwood, 1996) for back crossing to commercial breeding varieties (Appels & Lagudah, 1990). Besides being a useful means of *T. tauschii* gene transfer, synthetic hexaploids provide an indication of how *T. tauschii* genes might interact with those of the A and B genome (Schachtman *et al*, 1992a). As an alternative to synthetic hexaploids, directly crossing *T. tauschii* with bread wheat has some advantages. For example, the first back cross (BC_1) contains 92% of the hexaploid wheat parentage (compared with synthetic hexaploids, which require around 6 back-crosses to incorporate 94% of the hexaploid parent genome); there is no need to use colchicine (a carcinogen); and the A and B genomes are not rearranged in the process (Eastwood, 1996). Direct crossing to and elite cultivar reduces the amount of time needed for production of elite cultivars (Appels and Lagudah, 1990).

6.4 THE POTENTIAL USE OF *T. TAUSCHII* FOR THE IMPROVEMENT OF ZINC EFFICIENCY IN WHEAT

Zinc deficiency limits the yield of bread wheat in a range of soil types, and the development of zinc-efficient varieties is a high priority in regions where the soils are prone to severe zinc deficiency. *Triticum tauschii* is a geographically and genetically diverse taxa, and some accessions were originally collected from regions where zinc deficiency is common. The large amount of variation in the genome of *T. tauschii* has been used widely as a source of disease and pest resistance for bread wheat. This variation has also been used to improve the abiotic stress tolerance on bread wheat (Schachtman *et al*, 1992a) and therefore, may also prove to be an important source of zinc efficiency germplasm that could be used for the further improvement of this economically important crop.

7. References

Appels, R. & Lagudah, E.S. (1990) Manipulation of chromosomal segments from wild wheat for the improvement of bread wheat, *Australian Journal of Plant Physiology* **17,** 253-266.

Brown, P.H., Cakmak, I. & Zhang, Q. (1993) Form and function of zinc in plants, in A. D. Robson (ed.), *Zinc in Soils and Plants*, Kluwer Academic Publishers, Dordrecht, pp. 93-106.

Cakmak, I., Derici, R., Torun, B., Tolay, I., Braun, H.J. & Schlegel, R. (1997) Role of rye chromosomes in improvement of zinc efficiency in wheat and triticale, in T. Ando, K. Fujita, T. Mae, H. Matsumoto, S. Mori and J. Sekiya (eds.), *Plant Nutrition for Sustainable Food Production and Environment*, Kluwer Academic Publishers, Dordrecht, pp. 237-241.

Cakmak, I., Öztürk, L., Karanlik, S., Marschner, H. & Ekiz, H. (1996a) Zinc-efficient wild grasses enhance release of phytosiderophores under zinc efficiency, *Journal of Plant Nutrition* **19,** 551-563.

Cakmak, I., Sari, N., Marschner, H., Ekiz, H., Kalayci, M., Yilmaz, A. & Braun, H.J. (1996b) Phytosiderophore release in bread and durum wheat genotypes differing in zinc efficiency, *Plant and Soil* **180,** 183-189.

Cakmak, I., Yilmaz, A., Kalayci, M., Ekiz, H., Torun, B., Erenolglu, B. & Braun, H.J. (1996c) Zinc deficiency as a critical problem in wheat production in Central Anatolia, *Plant and Soil* **180,** 165-172.

Cox, T.S., Raupp, W.J. & Gill, B.S. (1994) Leaf rust-resistance genes *Lr41*, *Lr42* and *Lr43* transferred from *Triticum tauschii* to common wheat, *Crop Science* **34,** 339-343.

Devos, K.M., Moore, G. & Gale, M.D. (1995) Conservation of marker synteny during evolution, *Euphytica* **85,** 367-372.

Dong, B., Rengel, Z. & Graham, R.D. (1995) Root morphology of wheat genotypes differing in zinc efficiency, *Journal of Plant Nutrition* **18,** 2761-2773.

Eastwood, R.F. (1991) *Triticum tauschii*: a novel source of resistance to cereal cyst nematode (*Heterodera avanae*), *Australian Journal of Agricultural Research* **42,** 69-77.

Eastwood, R.F. (1996) Using *Triticum tauschii* for wheat improvement. In *Wheat Breeding Society of Australia; Eighth Assembly*, Canberra, A.C.T., pp. O189-93.

Erenolglu, B., Cakmak, I., Marschner, H., Römheld, V., Eker, S., Daghan, H., Kalayci, M. & Ekiz, H. (1996) Phytosiderophore release does not relate well with zinc efficiency in different bread wheat genotypes, *Journal of Plant Nutrition* **19,** 1569-1580.

Gill, K.S. & Gill, B.S. (1994) Mapping in the realm of polyploidy: the wheat model, *Bioessays* **16,** 841-846.

Graham, R.D. (1984) Breeding for nutritional characteristics in cereals, *Advances in Plant Nutrition* **1,** 57-102.

Graham, R.D., Ascher, J.S., Ellis, P.A.E. & Shepherd, K.W. (1987) Transfer to wheat of the copper efficiency factor carried on rye chromosome arm 5RL, *Plant and Soil* **99,** 107-114.

Graham, R.D., Ascher, J.S. & Hynes, S.C. (1992) Selecting zinc-efficient cereal genotypes for soils of low zinc status, *Plant and Soil* **146,** 241-250.

Graham, R.D. & Rengel, Z. (1993) Genotypic variation in zinc uptake and utilization by plants, in A.D. Robson (ed.), *Zinc in Soils and Plants*, Kluwer Academic Publishers, Dordrecht, pp. 107-118.

Graham, R. D. & Welch, R. M. (1996), *Breeding for staple-food crops with high micronutrient density.* International Food Policy Research Institute, Washington D.C.

Grewal, H.S., Graham, R.D. & Rengel, Z. (1996) Genotypic variation in zinc efficiency and resistance to crown rot disease (*Fusarium graminearum* Schw. Group 1) in wheat, *Plant and Soil* **186,** 219-226.

Grewal, H.S., Stangoulis, J.C.R., Potter, T.D. & Graham, R.D. (1997) Zinc efficiency of oilseed rape (*Brassica napus* and *B. juncea*) genotypes, *Plant and Soil* **191,** 123-132.

Guerinot, M.L. & Yi, Y. (1994) Iron: nutritious, noxious, and not readily available, *Plant Physiology* **104,** 815-820.

Harter, R.D. (1991) Micronutrient adsorption-desorption reactions in soils, in J.J. Mortvedt, F.R. Cox, L.M. Shuman & R.M. Welch (eds.), *Micronutrients in Agriculture*, Soil Science Society of America, Madison, pp. 59-88.

Hartwig, E.E., Jones, W.F. & Kilen, T.C. (1991) Identification and inheritance of inefficient zinc absorption in soybean, *Crop Science* **31,** 61-63.

Kihara, H., Yamashita,K., & Tanaka, M. (1965) Morphological, physiological, genetical and cytological studies in *Aegilops* and *Triticum* collected form Pakistan, Afghanistan and Iran, in K. Yamashita (ed), *Cultivated plants and their relatives*, KOEI Printing, Japan, pp. 1-118.

Lagudah, E.S., Appels, R., Brown, A.D.H. & McNeil, D. (1991) The molecular-genetic analysis of *Triticum tauschii*, the D-genome donor to hexaploid wheat, *Genome* **34**, 375-386.

Lagudah, E.S., Appels, R. & McNeil, D. (1991) The Nor-D3 locus of *Triticum tauschii*: natural variation and genetic linkage to markers in chromosome 5, *Genome* **34**, 387-395.

Marschner, H. (1993) Zinc uptake from soils, in A. D. Robson (ed.), *Zinc in Soils and Plants*, Kluwer Academic Publishers, Dordrecht pp. 59-77.

Marschner, H. (1995) *Mineral Nutrition of Higher Plants*, Academic Press, London.

Mian, A.R. & Nafziger, E.D. (1992) Seed size effects on emergence, head number, and grain yield of winter wheat, *Journal of Production Agriculture* **5**, 265-268.

Moraghan, J.T. & Mascagni, H.J. (1991) Environmental and soil factors affecting micronutrient deficiencies and toxicities, in J.J. Mortvedt, F.R. Cox, L.M. Shuman & R.M. Welch (eds.), *Micronutrients in Agriculture*, Soil Science Society of America, Madison, pp. 371-426.

Mortvedt, J.J. (1991) Micronutrient fertilizer technology, in J.J. Mortvedt, F.R. Cox, L.M. Shuman & R.M. Welch (eds.), *Micronutrients in Agriculture*, Soil Science Society of America, Madison, pp. 523-548.

Mortvedt, J.J. & Gilkes, R.J. (1993) Zinc fertilizers, in A.D. Robson (ed.), *Zinc in soils and plants*, Kluwer Academic, Dordrecht, pp. 33-44.

Nable, R.O. & Webb, M.J. (1993) Further evidence that zinc is required throughout the root zone for optimal plant growth and development, *Plant and Soil* **150**, 247-253.

Nambiar, E.K.S. (1976) The uptake of zinc-65 by oats in relation to soil water content and root growth, *Australian Journal of Soil Research* **14**, 67-74.

Ren, Z.L., Lelley, T. & Röbbelen, G. (1990) The use of monosomic rye addition lines for transferring rye chromatin into bread wheat, *Plant Breeding* **105**, 257-264.

Rengel, Z. & Graham, R.D. (1995a) Importance of seed Zn content for wheat growth on Zn-deficient soil, 1: Vegetative growth, *Plant and soil* **171**, 259-266.

Rengel, Z. & Graham, R.D. (1995b) Importance of seed Zn content for wheat growth on Zn-deficient soil, 2: Grain yield, *Plant and Soil* **171**, 267-274.

Rengel, Z. & Graham, R.D. (1996) Uptake of zinc from chelate-buffered nutrient solutions by wheat genotypes differing in zinc efficiency, *Journal of Experimental Botany* **47**, 217-226.

Rengel, Z. & Hawkesford, M.J. (1997) Biosynthesis of a 34-kDa poly peptide in the root-cell plasma membrane of Zn-efficient wheat genotype increases upon Zn deficiency, *Australian Journal of Plant Physiology* **24**, 307-315.

Römheld, V. & Marschner, H. (1991) Function of Micronutrients, in Plants in J.J. Mortvedt, F.R. Cox, L.M. Shuman & R.M. Welch (eds.), *Micronutrients in Agriculture*, Soil Science Society of America, Madison, pp. 297-328.

Schachtman, D.P., Lagudah, E.S., & Munns, R. (1992a) The expression of salt tolerance from *Triticum tauschii* in hexaploid wheat, *Theoretical and Applied Genetics* **84**, 714-719.

Schachtman, D.P., Munns, R., & Whitecross, M.I. (1992b) Variation in sodium exclusion and salt tolerance in *Triticum tauschii*, *Crop Science* **31**, 992-997.

Sparrow, D.H. & Graham, R.D. (1988) Susceptibility of zinc-deficient wheat plants to colonization by *Fusarium graminarium* Schw. Group 1, *Plant and Soil* **112**, 261-266.

Stevenson, F.J. (1991) Organic matter-micronutrient reactions in soil, in J.J. Mortvedt, F.R. Cox, L.M. Shuman & R.M. Welch (eds.), *Micronutrients in Agriculture*, Soil Science Society of America, Madison, pp. 145-186.

Takkar, P.N. & Walker, C.D. (1993) The distribution and correction of zinc deficiency, in A.D. Robson (ed.) *Zinc in Soils and Plants*, Kluwer Academic, Dordrecht, pp. 151-165.

Thongbai, P., Graham, R.D., Neate, S.M. & Webb, M.J. (1993) Interaction between zinc nutritional status of cereals and *Rhizoctonia* root rot severity. II: Effect of Zn on disease severity of wheat under controlled conditions, *Plant and Soil* **153**, 215-222.

Villereal, R. L. & Mujeeb-Kazi, A. (1996) Exploitation of synthetic hexaploids (*Triticum turgidum* x *T. tauschii*) for some biotic resistances in wheat, *Wheat Breeding Society of Australia: Proceedings of the eighth assembly, Canberra,* pp.185-188.

von Wirén, N., Marschner, H. & Römheld, V. (1996) Roots of iron-efficient maize also absorb phytosiderophore-chelated zinc, *Plant Physiology* **111**, 1119-1125.

Wee, M.J., & Loneragan, J.F. (1990) Zinc translocation to wheat roots and its implications for a phosphorus/zinc interaction in wheat plants, *Journal of Plant Nutrition* **13**, 1499-1512.

Welch, R. M. (1995) Micronutrient nutrition of plants, *Critical Reviews in Plant Sciences* **14**, 49-82.

Welch, R.M., Allaway, W.H., House, W.A. & Kubota, J. (1991) Geographic distribution of trace element problems, in J.J. Mortvedt, F.R. Cox, L.M. Shuman & R.M. Welch (eds.), *Micronutrients in Agriculture*, Soil Science Society of America, Madison, pp. 31-58.

Welch, R.M. & Norvell, W.A. (1993) Growth and nutrient uptake by barley (Hordeum vulgare L. cv Herta): studies using an N-(2-Hydroxyethyl)ethylenedinitrilotriacetic acid-buffered nutrient solution technique ii: Role of zinc in the uptake and root leakage of mineral nutrients *Plant Physiology* **101,** 627-631.

White, C. L. (1993) The zinc requirements of grazing ruminants, in A.D. Robson (ed.), *Zinc in Soils and Plants,* Kluwer Academic, Dordrecht, pp.197-206.

Yilmaz, A., Ekiz, H., Gültekin, I., Torun, B., Karanlik, S. & Cakmak, I. (1997) Effect of seed zinc content on grain yield and zinc concentration of wheat grown in zinc-deficient calcareous soils, in T. Ando, K. Fujita, T. Mae, H. Matsumoto, S. Mori & J. Sekiya (eds.), *Plant nutrition - for sustainable food production and environment,* Kluwer Academic Publishers, pp. 283-284.

TOLERANCE OF GROUNDNUT CULTIVARS TO MANGANESE TOXICITY

K.B. REDDY[1]., G. VENUGOPLA RAO[1]., C. MASTAN REDDY[2] and
B. RAVINDRANATH REDDY[3]

[1]Department of Plant physiology, [2]Department of Soil Science and Agrl. Chemistry, [3]Department of Agronomy, S.V Agricultural colloge, Titupati, A.P., India.

Key words: Manganese Toxicity, relative leaf weight, differential response.

Abstract

To investigate the differential response of groundnut genotypes to manganese toxicity, twenty groundnut (*Arachis hypogaea* L.) genotypes were grown at toxic level of Mn (44.0 ppm) in solution culture. The susceptible genotypes (TCGS 150, TCGS 476 etc) showed severe toxicity symptoms such as inword curling of leaves, isolated rusty brown spots with severe necrosis while the tolerant genotype (TCGS 1709/1518) showed reduced growth but not marked toxicity symptoms. Leaf area, chlorophyll content and shoot and root weight decreased significantly with increase in Mn concentration in the nutrient medium. Relative leaf dry weight (Wt. with toxic level of Mn/Wt with control level of Mn x 100) was 95% for TCGS 1709 and only 62.6% for TCGS 476. Leaf area and total chlorophyll content were found to be the most sensitive parameters to Mn toxicity. Plant analytical results indicated that the TCGS 1709 genotype tolerate a higher Mn concentration due to reduced translocation of Mn to shoots.

Introduction

Aluminium and manganese toxicities are major factors limiting crop growth on acidic and poorly drained soils (Foy, 1984). Plant physiological disorders associated with excess Mn are crinkle leaf of cotton, stem streak necrosis in potato and growth retardation and leaf tip burning in carnation. Manganese toxicity induced by various soil conditions is not always economically prevented or corrected by conventional soil management practices. But plant species and genotypes within species, differ widely in tolerance to excess Mn (Foy and Campbell, 1984). Hence, an alternative approach is to select or breed cultivars having greater resistance to Mn toxicity. This approach is being used successfully in different countries.

G. Gissel-Nielsen and A. Jensen (eds.), Plant Nutrition – Molecular Biology and Genetics, 245–252.

Helyar (1978) suggests that the plant damage under high tissue levels of Mn may be due to loss of control of Mn-activated catalytic systems. Physiological and biochemical responses to toxic level of Mn have generally been used as screening techniques in determining Mn tolerance in a variety of crop plants (Scott and Fischer, 1989). Tolerance may operate by Mn exclusion, by restriction of Mn transport to plant shoots and by tolerance of shoots to higher internal Mn concentration (Foy, 1984).

The availability of such information is scanty in groundnut. Keeping these points in view, the experiment was conducted to study the differential response of groundnut genotypes to Mn toxicity and to examine the reasons for differential response.

Materials and methods

Seeds of twenty groundnut genotypes were germinated in acid washed quartz sand in plastic trays. Seven days after sowing, ten uniform seedlings were selected and transferred to a plastic pot containing 3 liters of aerated, modified Hoagland solution (Johnson *et al.*, 1957). The basal solution consisted (mg L^{-1}): NO_3-N: 196, NH_4-N:28, P: 62, K: 235, Ca: 160, Mg: 24, S: 32, Fe: 5.0, B: 0.27, Zn: 0.019, Cu: 0.03, MO: 0.05, Cl: 1.77. Half strength nutrient solution was used for first seven days. A randomized block design was used with 20 genotypes and 2 treatments (T_1: control – 0.11 mg Mn L^{-1} and T_2 – toxic level of Mn, 44.0 mg Mn L^{-1}). Manganese was added as $MnSO_4.7H_2O$. The nutrient contents were renewed weekly so that plants received the same elemental exposure for the duration of the experiment. The experiment was conducted inside a glass house where the mean day and night temperatures were 30 and 25°C respectively with three replications. Aeration was given regularly by bubbling air into the nutrient solution with an aquarium air pump. Variations in tolerance to Mn toxicity were recorded based on the visual toxicity symptoms by using the scale:

1: interveinal chlorosis, 2: interveinal chlorosis with inward curling of leaves.
3: marginal necrosis, 4: marginal necrosis with brown spots.
5: brown spots wih severe necrosis.

Total chlorophyll content was estimated by using third leaf from top as per Arnon (1949). Leaf area was estimated by using LI-3100 area meter (LI-COR Inc, USA). Twenty five days after transplantating, plants were harvested and separated into roots and shoots. Tissues were dried at 80°C, weighed and analysed for Mn by atomic absorption spectrophotometry (Jackson, 1958).

Results

To conserve space and simplify the presentation, only data for five genotypes are presented. The visual toxicity score on manganese toxicity (Table 1) reveals that there is variation in tolerance to Mn toxicity among twenty groundnut genotypes. The susceptible genotypes (TCGS 150, TCGS 476 etc) with high toxicity score showed severe toxicity symptoms such as a inward curling of leaves, isolated rusty brown spots with severe necrosis, while the plants of TCGS 1709/1518 showed reduced growth but not marked toxicity symptoms. Chlorophyll 'a + b' content expressed on a FWT basis decreased significantly in all the genotypes and the decrease was only 6.2% in TCGS 1709 while in susceptible genotypes it has reduced by more than 25.0% (Fig.1). The total chlorophyll content has significant negative correlation (r = -0.528) with toxic level of Mn. Chlorophyll 'a' and chlorophyll 'b' content showed a similar reduction thus the chlorophyll a/b ratio did not differ among the genotypes (data not given).

Table1. Toxicity score, Relative leaf dry weight (RLW) and leaf area of groundnut genotypes under manganese toxicity.

Genotype	Visual toxicity score	RLW %	Leaf area (cm^2 $plant^{-1}$)		
			T_1	T_2	%ROC
TCGS 150	4	66.0	85.2a	46.3e	46
TCGS 476	5	62.5	78.5b	47.1d	40
TCGS 83	3	82.4	76.6c	53.5a	24
TCGS 37	3	72.7	65.7d	50.8c	23
TCGS1709/1518	1	95.0	57.8e	52.0b	10

* Means in a column not followed by the same letter are significantly different at the 0.05 probability level.

T_1: Control (0.11 ppm Mn)

T_2 :Toxic level of Mn (44.0 ppm Mn)

% ROC: Percent reduction over control

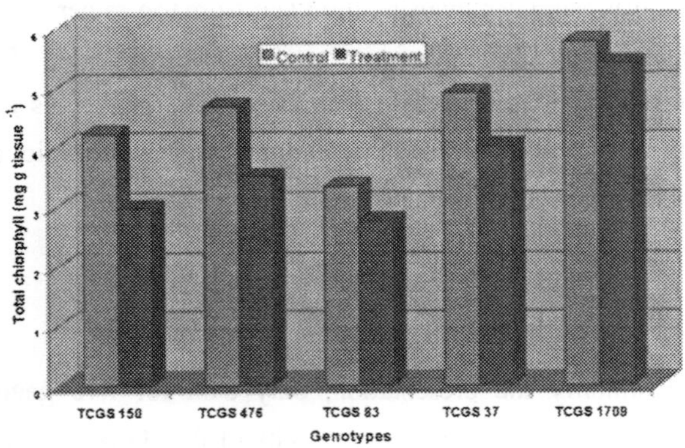

Figure 1. Genotypic differences in total chlorophyll content (mg g tissue $^{-1}$) under manganese toxicity.

The total dry matter production differed significantly among groundnut genotypes. The reduction in dry matter yields at toxic level (44.0 ppm)of Mn indicates a range of susceptibility with TCGS 1709 being the most tolerant and TCGS 476 the most sensitive of the twenty genotypes (Fig.2). The reduction in total dry matter due to Mn toxicity was more than 25.0% in susceptible genotypes while it was less than 20.0% in moderately tolerant and tolerant genotypes.

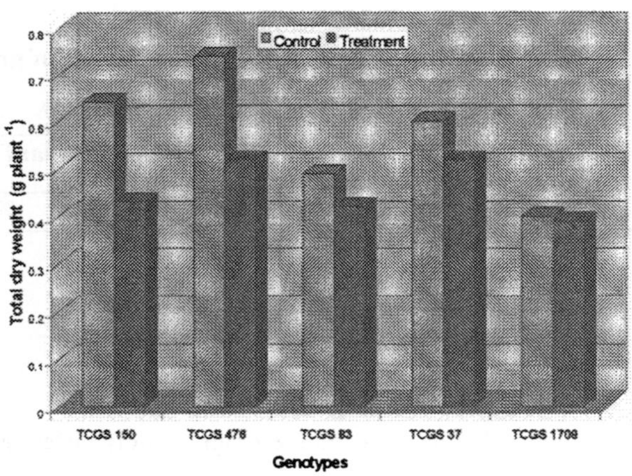

Figure 2. Genotypic differences in total dry weight (g plant^{-1}) under manganess toxicity

The correlation between Mn at toxic level and the total dry matter (r = + 0.058) was not significant at the 5% level. The detrimental effects of excess Mn were more pronounced in roots than shoots (Table 3).

Relative leaf dry weight (RLW) (wt. with toxic level of Mn/wt. with control level of Mn x100) decreased in all susceptible genotypes. The RLW was 95.0% for TCGS 1709 and only 62.6% for TCGS 476 (Table 1). Excess Mn also reduced the leaf area significantly in all the cultivars. The reduction was more (40.0%) in susceptible genotypes (Table 1) while it was less (10.0%) in case of tolerant genotypes.

Groundnut genotypes differed significantly in Mn concentration of root and shoot and the correlation coefficients between Mn at toxic level and Mn in root (r = + 0.144) was not significant at 5% level but Mn in shoot (r = + 0.85) was significant at the 5% level. The root contain higher concentration of Mn compared to shoot (Table 2). The strain showing tolerance to excess Mn (TCGS 1709) contain 140.0% increased concentration of Mn in the root than shoot. While the increase was less than 50.0% in susceptible genotypes.

Table 2. Genotypic differences in manganese concentration in shoot and root dry matter at toxic level (44.0 ppm) of manganese.

Genotype	Manganese concentration (ppm)		
	Shoot	**Root**	**% Increase in Root**
TCGS 150	1201b	1801e	50
TCGS 476	1249a	1861c	49
TCGS 83	1001d	1840d	84
TCGS 37	1141c	1931	69
TCGS1709/1518	840e	2015a	140

* Means in a column not followed by the same letter are significantly different at the 0.05 probability level.

Discussion

Results of this study have shown that groundnut genotypes differ widely in their tolerance to Mn toxicity. Toxic level of Mn reduced total chlorophyll content and the reduction was more (25.0%) in susceptible genotypes and was less (6.0%) in tolerant genotypes. It has been reported that high Mn concentration inhibit an iron requiring step in chlorophyll synthesis in tobacco callus (Clairmont *et al.*, 1986). Excess Mn reduced total dry matter production and the reduction was more in susceptible genotypes. Mn toxicity in plants has been associated with auxin destruction in cotton (Morgan *et al.*, 1966) and reduced number of cells $leaf^{-1}$ and cell volume in sugar beets (Terri *et al.*, 1975). These Mn toxicity effects may be adversely affect groundnut growth and development.

Shoot and root dry weights in the toxic level of Mn treatment were reduced to 84.0 and 80.0% respectively to those of controls indicating that roots were more sensitive to Mn toxicity stress than shoot. These results were in agrement with those of Ohki (1984). How ever Foy *et al* (1984) reported that excess Mn generally affected plant tops more severly than roots.

Plant analytical results indicated that the roots contained higher concentration of Mn compared to shoot. The tolerant cv TCGS 1709 contain high Mn concentration in roots than shoot at toxic level of Mn. This suggests a reduced ability of the roots to translocate absorbed Mn to the shoots which is presumed to be the main site of action for Mn. This reduced translocation of Mn to tops can be attributed for tolerance of TCGS 1709 to toxic level of Mn. Similar observations were reported by Heenan and Carter (1976) in soyabeen and Foy (1973) in cotton genotypes. Andrew and Hegarty (1969) suggested that plants tolerant to excess Mn retain a greater portion of absorbed Mn in the roots than sensitive plants.

To identify, which of the growth parameter is more sensitive to Mn toxicity, decrease in plant height, dry weight of root and shoot, total dry matter, leaf area and total chlorophyll content of the control plants were compared to that of plants under Mn toxicity (Table 3). It is found that the leaf area was the most sensitive parameter for identifying Mn toxicity. Total chlorophyll and root dry matter were the second and third most sensitive parameters. Therefore measurement of leaf area followed by chlorophyll content of Mn-stressed seedlings apper to be suitable techniques for screening groundnut genotypes for their tolerance to Mn toxicity.

Based on visual toxicity score (Table 4), the genotypes were classified into three categories, tolerant (toxicity score below 2.0) moderately tolerant (toxicity score above 2.0 and below 3.0) and susceptible (toxicity score above 3.0). Among the 20 genotypes, one was tolerant, five were moderately tolerant and fourteen were susceptible.

Table 3. Influence of manganese toxicity on growth parameters of twenty groundnut genotypes.

Character	T_1	T2	% ROC
Plant height (cm)	26.9 ± 1.7	23.3 ± 1.4	13
Shoot dry matter (g plant^{-1})	0.44 ± 0.03	0.37 ± 0.02	16
Root dry mattr (g plant^{-1})	0.15 ± 0.01	0.12 ± 0.007	20
Total dry matter (g plant^{-1})	0.59 ± 0.04	0.49 ± 0.02	17
Leaf area (cm^2 plant^{-1})	74.8 ± 2.1	48.1 ± 1.2	36
Total chlorophyll (mg total chll g f.wt)	4.8 ± 0.28	3.7 ± 0.17	23

T_1: 0.11 ppm manganese (Control)
T_2: 44.0 ppm manganese (Toxic level)
% ROC: Per cent reduction over control

Table 4. Relative tolerance of groundnut genotypes to manganese toxicity based on toxicity score.

Tolerant	Moderately tolerant	Susceptable
	TCGS 83	TCGS 150, TCGS 30
	TCGS 40	TCGS 91, TCGS 334
	TCGS 37	TCGS 29, TCGS 88
TCGS 1709/1518	ICGV 92113	TCGS 320, TCGS 1706
	JL 2220	ICGV 9218, ICGV 86031,
		JL 24

Results of this study have shown that the differential tolerance of groundnut genotypes to Mn toxicity was due to reduced translocation of Mn to shoots. Based on visual toxicity score, total chlorophyll content, leaf area and root and shoot dry weights cv TCGS 1709/1518 was significantly more tolerant than other genotypes.

References

Andrew, C.S., and Hegarty, M.P. (1969) Comparitive responses to manganese excess of eight tropical and four temperate pasture legume species, Aust. J.Agric.Res. 20, 687-696.

Arnon, D.I. (1949) Copper enzyme in isolated chloroplast polyphenol oxidase in *Beta vulgaris*, Plant physiol. 24, 1-13.

Clairmont, K.B., Hagar, W.G., and Davis, E.A. (1990) Manganese toxicity to chlorophyll synthésis in tabaum callus, Plant physiol. 80, 291-293.

Foy, C.D. (1973) Manganese and Plants, Manganese, National academy science, National research council, Washington, D.C pp57-76.

Foy, C.D. (1984) Physiological effects of hydrogen, aluminium and manganese toxicities in acid soils, In F.Adams (ed), Soil acidity and liming, Second Edition, Agronomy, 12, 57-97.

Foy, C.D., and Campbell, T.A. (1984) Differential manganese tolerance of cotton genotypes in nutrient solution, J. Plant Nutr. 7, 1365-1388.

Foy, C.D., Chaney, R.L, and White, M.C. (1978) The physiology of metal toxicity in plants, Ann. Rev. Plant Physiol. 29, 511-566.

Heenan, D.P., and Carter, O.G. (1975) Response of two soybean cultivars to manganese toxicity as affected by pH and calcium levels, Aust. J. Agric. Res. 26, 967-974.

Helyar, K.R. (1978) Effects of aluminium and manganese toxicity on legume growth, In C.S. Andrw and E.J.Kamprath (eds.) Mineral nutrition of legumes in tropical and subtropical soils, CSIRO, Melbourne, Australia. pp.207-231

Jackson, M.L. (1958) Soil Chemical analysis. Prentice-Hall Inc., Eagle Wood Wlis, NJ.

Jhonson, C.M., Stout, P.R., Broyer, J.C, and Carltont, A.B (1957) Comparitive chlorine requirements of different plant species, Plant and Soil 8, 337-386.

Morgan, P.W., Joham, H.E., and Amin, J.V. (1966) Effects of manganese toxicity on the indole acetic acid oxidase system of cotton, Plant Physiol. 41, 718-724.

Ohki, K. (1984) Mn deficiency and toxicity effects on growth, development and nutrient composition in wheat, Agron, J. 76, 213-218.

Scott, B.J., and Fisher, J.A. (1989) Selection of genotypes tolerant to aluminium and manganese, In Soil acidity and Plant growth. (ed) A.D.Robson, Academic Press, Sydney, Australia, pp 167-203.

Terry, N., Evans, P.S., and Thomas, D.F. (1975) Manganese toxicity effects on leaf cell multiplication expansion and on dry matter yield of sugar beets, Crop Sci. 15, 205-208.

RHIZOSPHERE ACQUISITION TRAITS FOR PHENOTYPING Zn EFFICIENT RICES (*Oryza sativa L.*)

P. THONGBAI, G. J. D. KIRK, D. SENADHIRA

Project Scientist, Soil Chemist and Plant Breeder
International Rice Research Institute. PO. Box 933, 1099 Manila. Philippines

Introduction

Zinc (Zn) deficiency is the most widespread micronutritional disorder in wetland rice crop [16, 27], contributing as high as 3 t/ha of yield loss [23]. Although total content of Zn in the soil may be sufficiently high, Zn availability is always reduced when the soil is calcareous, saline and/or high in organic matter [11, 35]; and most importantly when the soil is flooded regardless of its original pH; precipitation of $Zn(OH)_2$ occurs in acid soils as the result of pH increase after flooding [32], ZnS and bicarbonate precipitate in sodic or calcareous soil [27, 19], and there can be formation of $ZnFe_2O_4$ in the normal soil [28]. Using Zn efficient rice varieties is considered more appropriate because, for soil application, at least 10 kgZn/ha is needed in the field while plants take up less than 1 kgZn/ha, leaving the remaining Zn accumulated in the soil which might lead to toxicity if the soil conditions change [10]. Since 1971, IRRI tested more than 23,000 lines in both experimental and farmers' fields and found 9% efficient. The disadvantage of field selection is poor repeatability due to environmental variation, while greenhouse screening technique using buffer chelate solution does not differentiate traits for rhizosphere acquisition and therefore marker-assisted selection techniques seem to be a most effective measure [9, 15]. However, as the segregation pattern suggests at least 3 independent incompletely dominant genes involved in this character [1, 18], quantitative trait loci (QTL) analysis is needed [33]. At present, lack of phynotypic characterization due to poor understanding of the physiology of tolerance and/or efficiency mechanism(s) is the main limitation [9, 24, 15].

Differential responses to Zn deficiency among varieties might be attributed either to greater physiological efficiency due to low requirement and/or better distribution (internal efficiency), or higher absorption from a deficient medium (external effciency) [26]. Internal efficiency could be easily examined by analyzing tissue Zn content and the already-identified Zn transporter gene, ZRT1 and ZRT2 in yeast [36] may make it possible to map and study Zn transport and distribution. External efficiency, on the

G. Gissel-Nielsen and A. Jensen (eds.), Plant Nutrition – Molecular Biology and Genetics, 253–260.

other, is more complicated. For other cereals, it has been proved to be a differential ability of the root to induce changes in rhizosphere oxidation state and pH or release Zn-mobilizing root exudates such as phytosiderophores [21]; but there is no such information in rice. Using soil-sandwich technique, Kirk & Bajita [12] showed that Zn efficient IR36 could acidify its rhizosphere by H^+ generated from the oxidation of Fe(II) by root-released O_2 and, thereby, release Zn from highly insoluble fractions to the acid-soluble forms which are accessible to the plant. However, varietal differences have not yet been compared. Moreover, these various mechanisms are likely to be additive and operating together at different levels in the plant [20]. Clearly, an understanding of the type of efficiency and the responsible mechanism(s), especially rhizosphere acquisitions, is needed to facilitate development of screening tools, which, therefore, is the main objective of this research.

Performance of different Zn efficient rice lines in a Zn deficient soil

To start with, a group of rice lines were tested for internal and/or external efficiency, using incorporation of nine quantitative nutrient efficiency indices suggested by Blair [5]. Ten lines of rice thought by breeders to differ in their Zn efficiency were grown in a Zn deficient soil (a fine montmorillonitic calcareous Typic Hydraquent from Tiaong, Quezon, Philippines. DTPA extractable Zn 0.04 mg/kg, pH 7.8) with 10 mg Zn/kg soil (Zn10) as $ZnSO_4$ or without Zn (Zn0). Three kg soil in a plastic pot was flooded with water for 3 weeks before sowing with pre-germinated rice seeds. The plants were then grown for 3 weeks in a glasshouse under typical humid tropic wet season conditions, then visually scored for Zn deficiency according to IRRI Standard Evaluation System. The number of leaves with Zn deficiency symptom were counted, and plants were harvested for tiller number, height, dry wt and Zn content of shoots. To categorize varieties for their Zn efficiency, all nine indices were calculated, ranked and used to distinguish different types of efficiency (Table 1A, 1B). These indices are all quantitative

Table 1. Indices ranking (A) and contribution to characterize all rice varieties for Zn efficiency (B).

TABLE 1A.

Efficiency indices	score	Ranking									
		IR26	IR58	IR65	CSR10	Ketumbar	KDML 105	IR9764	FR13A	IR8192	Madhukar
1. Visual score of Zn0 plants	1=best	4	3	2	3	2	3	1	1	1	2
2. No leaves with def symptom in Zn0	1=less	4	3	2	2	3	6	1	5	1	6
3. Shoot dry wt./Zn conc. of Zn0	1=highest	6	4	6	5	4	3	4	2	1	2
4. Shoot dry wt./ uptake in Zn0 shoot	1=highest	4	2	5	4	3	2	3	2	1	2
5. Concentration of Zn in Zn0 shoot	1=highest	2	4	1	2	5	4	3	4	5	4
6. Amount of Zn uptake in Zn0 shoot	1=highest	5	5	6	4	4	3	3	2	1	2
7. Average of % growth reduction	1=less	3	3	2	3	2	1	2	1	1	1
8. DWt Zn0 / DWt Zn10	1=highest	5	5	5	6	5	2	5	3	1	4
9. (DWtZn10 -DWtZn0)/ Zn added	1=highest	6	4	6	2	3	5	3	2	1	2

TABLE 1B.

Efficiency Category	indices & feature	IR26	IR58	IR65	CSR10	Ketumbar	KDML 105	IR9764	FR13A	IR8192	Madhukar
1. **Zn deficiency tolerance**	L 1, 2, 7	S	S	MS	S	MT	MT	T	MT	T	MT
2. **Zn efficiency**	L 1, 2, 7	I	I	MME	I	ME	ME	E	E	E	E
2.1. ~**internal** (low requirment; efficient transport & metabolisms)	H 3, 4, 6 ; L5			-		+	+	-	+	+	+
2.2. ~**external** (rhizosphere mobilization and absorption)	L 3, 4, 6; H 5			+		+		+	-	-	-
3. **Responsive** (increase growth in +Zn)	8,9	N	M	N	MH	M	N	M	H	VH	MH
4. **Accumulator** (high Zn in tissue without growth increase)	H 5, L 6	H	M	VH	H	N	N	M	N	N	N
5. **Efficient transport** (lower Zn but well distributed)	L 5, H 6	N	MM	N	MM	N	M	M	MH	H	MH

Note S = sensitive, T = tolerant, I = Inefficient, E = efficient, H = high, L = low, M = moderate, V = very, N = non, none

and Zn-specific, and therefore could be used directly for QTL analysis. Some varieties were then picked up for more detailed study of mechanisms.

Quantification of Zn associated with Fe plaque for an accurate index to calculate root internal Zn.

Differential compartmentalization and/or utilization of Zn within plant parts is considered a key mechanism in internal efficiency; Zn content in the root is, therefore, necessary for study uptake kinetics and rate of transport from root to shoot [6, 9, 17, 22]

However, on the root surface of submerged aquatic plants, Zn was usually found precipitated with Fe oxyhydroxides, known as 'iron plaque' [3, 26]. Although Fe oxides on root surface will adsorb or otherwise immobilize Zn leading to decreased plant uptake, oxidation of Fe & accompanying acidification in the rhizosphere will solubilize Zn leading to increased plant uptake [12]. Campbell & St. Cyr [7] suggested that 'iron plaque' should play a role in regulating Zn bioavailability. In any case, when analyzing plant for Zn content, Zn absorbed on Fe oxides on the roots will be included in the assessment of total Zn unless somehow excluded. Using dithionite-citrate-bicarbonate technique (DCB) to extract iron plaque from the root [30] may provide the better value of the real Zn content inside the rice root.

To measure amount of Zn precipitate at the root surface by DCB extraction and to assess its effect on total Zn content in the root of rice lines differing in their Zn efficiency, the stagnant culture with 0.1% agar [34] was used. This technique prevents any O_2 convection within the solution, with 5M K-MES buffer to maintain pH at 6.5 for 10 days without daily adjustment; Fe oxidation is then minimized from any source other than rice root itself. The nutrient composition was based on Yoshida et al (1976) with Zn level fixed at 10 uM. Five levels of Fe (15, 150, 300, 600, 1200μM) from $FeSO_4$ were used to test 2 rice varieties, IR26 (Zn inefficient) and IR8192 (Zn efficient). After 21 days, the plants were harvested and, after weighing for fresh weight, the root was divided half for DCB extraction [30] and the remaining half dried for dry wt and total Zn analysis.

Results in Fig. 1 show that 50% of Zn content in the rice root was associated with Fe plaque at the root surface.

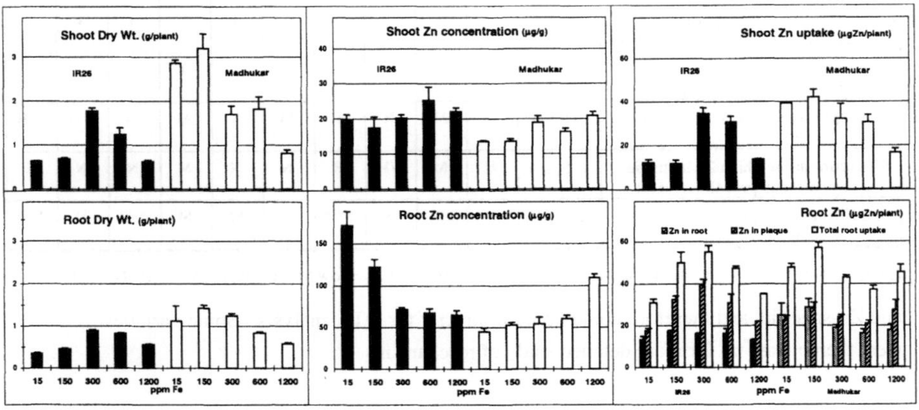

Figure 1. Dry weight, Zn concentration, Zn uptake in shoot and root, Zn content in the plaque and in the root after DCB washing from IR26 and IR8192.

Although concentration of Zn in the solution is constant in all treatments, Zn that associated with Fe plaque and Zn content in the shoot and root varied according to the concentration of Fe in the solution and rice variety, with the highest at 150 ppm Fe for IR8192 and 300 ppm Fe for IR26. The pattern of growth response differed between varieties. In IR26, growth started to increase from very low in the lowest Fe treatment, reached its peak at Fe 300 ppm and declined to the lowest at Fe 1200 ppm due to Fe toxicity. In IR8192, growth was generally higher than for IR26 at all Fe levels, especially at Fe 15 ppm, reaching a peak at Fe 150 ppm, then declining to the lowest at Fe1200 ppm. Zn concentration in the shoot of both varieties varied with the Fe treatment in the same manner as growth, much lower than in the root, and generally lower in IR8192 than in IR26. In IR26 root Zn concentration declined with increasing Fe treatments, while those of in IR8192 increased according to the Fe treatments.

Zn uptake in root and shoot responded to Fe treatments in the same pattern as the growth response for both varieties. However, at each level of Fe, uptake of Zn in shoot and root of IR 8192 was relatively higher than those of IR26, confirming that IR8192 is more efficient than IR26 due to lower requirement of Zn, the so-called internal efficiency.

We are now in a process of extracting root from the soil-grown plants to confirm measurements of 'internal' and 'external' Zn efficiency based on uptake redistribution.

Measurement of oxygen release from roots of Zn efficient and Zn inefficient rice lines

Armstrong [2], using oxygen electrode, showed that rice cultivars tolerant to Akagare, a disease caused by Zn deficiency, released more O_2 from their roots than the susceptible ones. Therefore the ability of rice root to oxidize its rhizosphere has been suggested as a major mechanism of Zn efficiency in rice [20, 21]; and molecular oxygen release may be a good indicator of Zn uptake efficiency [26]. With the hypothesis that Zn-efficient lines may release more O_2 to the rhizoshere than inefficient ones, a special experiment using deoxygenated nutrient solution [4, 14] was set up to estimate rate of O_2 released from Zn deficiency rice root and to compare O_2 releasing rate between Zn inefficient IR26 and Zn efficient, IR54 (external) and Madhukar (internal), grown in solution with or without Zn. Dissolved O_2 in the solution bathing the rhizosphere was measured every other day using oxygen electrode (Orion, model 97-08). The rate of O_2 release from the root in the pot at a particular time depends on its rate of inflow with the solution from the reservoir, plus its rate of release from the roots, less its rate of outflow through the overflow port. The concentration will change over the interval between samplings to an extent depending on the balance between removal in the outflow and addition from the roots and inflow. The instantaneous rate of change could be calculated from the equation:

$$\frac{dC}{dt} = \frac{v^*C^* + U - vC}{V} \tag{1}$$

where C = O_2 concentration in the solution in the pots, mol dm^{-3},
C^* = O_2 concentration in the solution in the reservoir, mol dm^{-3},
v = rate of solution flow out of the pot, dm^3 s^{-1},
v^* = rate of solution flow into the pot (=v + rate of transpiration), dm^{3}s^{-1},
U = rate of O_2 release from the roots, mol s^{-1},
V = volume of solution in the pot, dm^3

After rearranging and integrating, and substitute C with C_1 and C_2 which are the concentration at the beginning and end of the sampling interval Δt, eqn (1) then becomes:

$$U = \frac{vC_2 - v^*C^* - (vC_1 - V^*C^*)\exp(-v\Delta t/V)}{1 - \exp(-v\Delta t/V)} \tag{2}$$

Apparent O_2 released per pot per day is shown in Fig. 2. Rate of O_2 release from the plus Zn treatments was

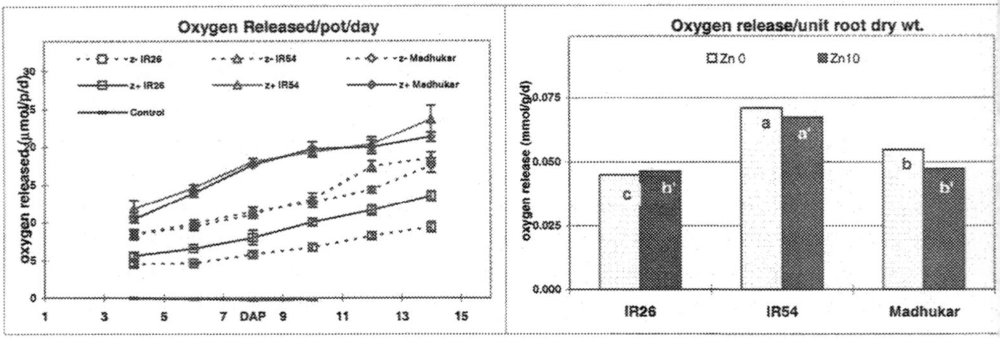

Figure 2. Rate of O_2 release, per pot and per unit root dry weight, from 3 rice lines differed in Zn efficiency.

higher than the minus Zn, thus explaining how this mechanism operated to adsorb Zn present in the solution. Zn inefficient IR26 released significantly less O_2 compared to the efficient IR54 and Madhukar, which are not significantly different from each other. However, the rate of O_2 release per g root dry wt was significantly higher in IR54 (external efficient) than Madhukar (internal efficient) and both are higher than inefficient IR26, but not between Zn treatments of the same variety. It could be concluded that this character is variety specific and therefore could be used as a good screening index for Zn efficiency. The technique is now being developed and refined for more precise and specific measurement as well as capacity to screen big populations simultaneously.

Conclusion

This study uses an aggregate of various scores to give an overall index for Zn deficiency tolerance, Zn efficiency and responsiveness to Zn where possible quantitative scores have been used. To use for screening, these techniques need more refinement before mapping the meaningful QTLs. However, additional mechanisms may yet be usefully added to the index. Organic acid root exudation in rice is yet to be explored, since this has been proved possible [31]. Also possible is the screening for free radical generation or activity of the enzyme superoxide dismutase, Zn metalloenzyme detoxifying free radicals and prevents membrane peroxidation, for which varietal differences exist in other cereal species [21]. It is necessary for a breeding program to have simple, fast and inexpensive techniques permitting the assessment of hundreds or thousands of segregants from Zn-efficient x Zn-inefficient parents without using expensive field testing. This cannot be achieved unless the physiology of Zn efficiency mechanisms are fully understood so proper phenotyping of recombinant populations can be carried out.

Acknowledgement

We would like to thank Ms. C. Quijano for help providing facilities, and Prof. Dr. R. D. Graham of Waite Institute, Adelaide, for kind comments on the manuscript. This paper is dedicated to (the late) Dr. D. Senadhira who was killed in a road accident in Bangladesh on July 7th, 1998.

References

1. Afzal, M. (1980) Studies on the inheritance of resistance to zinc deficiency in rice (Oryza sativa L.). *J. Agric. Res.* **18**, 49-54.
2. Armstrong ,W. (1969) Rhizosphere oxidation in rice: an analysis of intervarietal differences in oxygen flux from the roots. *Phyiol. Plant.* **22**, 296-303.
3. Bacha, R.E. and Hossner, L.R. (1977) Characteristics of coatings formed on rice roots as affected by iron and manganese additions. *SSSAJ* **41**:, 931-5.
4. Bedford, B.L., Bouldin, D.R., and Beliveau, B.D. (1991) Net oxygen and carbon dioxide balances in solutions bathing roots of wetland plants. *J. Ecol.* **79**:, 943-959.
5. Blair, G. (1993) Nutrient efficiency - what do we really mean? In Randall, P.J., Delhaize, E., Richards, R.A. and Munns, R. (eds) *Genetic Aspects of Plant Mineral Nutrition*, Kluwer Academic Publishers. The Natherlands. pp. 205-213.
6. Bowen, J.E. (1986) Kinetics of zinc uptake by two rice culivars. *Plant Soil* **94**, 99-107.
7. Cambell, P.G.C. and St Cyr, L. (1996) Metals (Fe, Mn, Zn) in the root plaque of submerged aquatic plants collected in situ: relations with metal concentrations in the adjacent sediments and in the root tissue. *Biogeochem.* **33**, 45-76.
8. Fischer, K.S. (1996) Research approaches for variable rainfed systems-thinking globally, acting locally. In Cooper, M. and Hammer, G. (eds.) *Plant Adaptation and Crop Improvement,*. CAB International, Wallingford, Oxon, UK, pp. 25-35.
9. Graham, R.D. and Rengel, Z. (1993) Genotypic variation in zinc uptake and utilization by plants. In A. D. Robson (ed.) *Zinc in Soil and Plants*. Kluwer Academic Publishers, Australia, pp. 107-118.
10. IAEA, (1981) Zinc Fertilization of Flooded Rice. IAEA-TECDOC-242. International Atomic Energy Agency, Vienna.
11. IRRI, (1972) Annual Report for 1971. IRRI, Los Banos, Philippines.
12. Kirk, G.J.D. and Bajita, J.B. (1995) Root-induced iron oxidation, pH changes and zinc solubility in the rhizosphere of lowland rice. *New Phytol.* **131**, 129-137.
13. Kirk, G.J.D. and Le Van Du. (1997) Changes in rice root architecture, porosity, and oxygen and proton release under phosphorus deficiency. *New Phytol.* **135**, 191-200.
14. Lafitte, H.R. (1998) Research opportunities to improve nutrient-use efficiency in rice cropping systems. *Field Crop Res.* **56**, 223-236.
15. Lindsay, W.L. (1972) Zinc in soils and plant nutrition. *Adv. Agron.* **24**, 147-186.
16. Lopes, A.S. (1980) Micronutrients in soils of the tropics as constraints to food production. In *Soil Related constraints to food production in the tropics*. IRRI. Los Baños, Philippines. Pp. 277-298.
17. Loneragan, J.F., Kirk, G.J., and Webb, M.J. (1987) Translocation and function of zinc in roots. *J. Plant Nutr.* **10**, 1247-54.
18. Mahadevappa, M., Ikehashi, H., Aurin, P. (1981) Screening rice genotypes for tolerance to alkalinity and zinc deficiency. *Euphytica* **30**, 253-257.
19. Majumder, N.D., Rakshit, S.C., and Borthakur, D.N. (1990) Genetics effects on uptake of selected nutrients in some rice (*Oryza sativa* L.) varieties in phosphorus-deficient soil. *Plant Soil* **123**, 117-120.
20. Marschner, H. (1993) Zinc uptake from soils. In A. D. Robson (ed.) *Zinc in Soil and Plants*. Kluwer Academic, Australia. pp. 59-77.
21. Marschner, H. (1995) *Mineral Nutrition of Higher Plant*. 2nd. Academic Press.
22. Marschner, H. (1998) Role of root growth, arbuscular mycorrhiza, and root exudates for the efficiency in nutrient acquisition. *Field Crop Res.* **56**, 203-8.
23. Neue, H.U., Lantin, R. S., Cayton, M.T.C., and Autor, N.U. (1990) Screening of rices for adverse soil tolerance. In El Bassam, N., Dambroth, M., Loughman, B.C. (eds) *Genetic Aspects of Plant Nutrition.* Kluwer Academic Publisher, Dodrecht, the Natherlands. pp. 523-531.
24. Neue, H.U. and Lantin, R.S. (1994) Micronutrient toxicity and deficiency in rices. In Yeo ,A.R. and Flower, T.J. (eds) *Monographs on Theoretical and Applied Genetics*, Vol. 21. Springer-Verlag Berlin Heidelberg. pp. 175-200.
25. Neue, H.U., Quijano, C., Senadhira, D. and Setter, T.L. (1998) Strategies for dealing with micronutrient disorders and salinity in lowland rice systems. *Field Crop Res.* **56**, 139-155.
26. Otto, M.L., Rozema, J., Koster, L., Haarsma, M.S., and Broekman ,R.A. (1989) Iron plaque on roots of *Aster tripolium* L.: interaction with zinc uptake. *New Phytol.* **111**, 309-17.
27. Ponnamperuma, F.N. (1972) The chemistry of submerged soils. *Adv. Agron.* **24**, 29-96.
28. Sajwan, K.S. and Lindsay, W.L. (1986) Effect of redox on zinc deficiency in paddy rice. *SSSAJ* **50**, 1264-69.
29. Scharpenseel, H.W., Eichwald, E., Haupenthal, Ch., and Neue, H.U. (1983) Zinc deficiency in a soil toposequence grown to rice at Tiaong, Quezon Province, Philippines.,*Cetena* **10**: 115-132.
30. Taylor, G.J. and Crowder, A.A. (1983) Use of the DCB technique for extraction of hydrous iron oxides from roots of wetland plants. *Amer. J. Bot.* **70**, 1254-7.

31. Thongbai, P., Neuemann, G., Roemheld, V. and Marchner, H. 1(997) Quantitative study of pH and root exudates of Zn efficient and inefficient rice lines. Report Project No. A/95/24286 submitted to Deutscher Akademischer Austauschdienst (DAAD), Bonn, Germany. (Unpublished).
32. Trierweiler, J.F. and Lindsay, W.L. (1969) EDTA ammonium carbonate soil test for zinc. *SSSAP* **33**, 49.
33. Tsunematsu, H., Yoshimura, A., Yano, M., and Iwata, N. (1996) Quantitative trait locus analysis using recombinant inbred lines and restriction fragment length polymorphism markers in rice. In Khush, G. S. (ed). *Rice Genetics III.* IRRI, Philippines. pp. 619-623.
34. Wiengweera, A, Greenway, H., and Thomson, C.J. (1997) The use of agar nutrient solution to stimulatelack of convection in waterlogged soil. *Ann. Bot.* **80**, 115-123.
35. Yoshida, S., Ahn, J. S., and Forno, D. H. (1973) Occurrence, diagnosis and correction of zinc deficiency of lowland rice. *Soil Sci. and Plant Nutr.* **16**, 147-149.
36. Zhao, H and Eide, D. (1996) The yeast ZRT1 gene encodes the zinc transporter protein of a high-affinity uptake system induced by zinc limitation. *Proc. Natl. Acad. Sci. USA* **93**, 2454-8.

ENHANCING BEANS (*PHASEOLUS VULGARIS* L.) AS SOURCES OF BIOAVAILABLE IRON THROUGH GENETIC SELECTION

R.M. WELCH[1], W.A. HOUSE[1], S. BEEBE[2], Z. CHENG[1]
[1]*U.S. Depart. Agr., Agr. Res. Serv. and Cornell University;* [2]*Centro Internacional de Agricultura Tropical*
[1]*U.S. Plant, Soil and Nutrition Laboratory, Tower Road, Ithaca, NY, 14853, USA;* [2]*Cali, Columbia*

Abstract

Iron (Fe) deficiency is the most prevalent nutrient deficiency in humans. Globally, this disorder adversely affecting the health, well being, and livelihood of over 2 billion people, especially those living in developing countries. Significantly increasing the bioavailable amount of Fe in staple plant foods could contribute greatly to reducing the incidence of Fe deficiency, particularly among those at greatest risk (i.e., poor women, infants and children). Fe-enriched genotypes of beans could be selected for in plant breeding programs to improve human nutrition and thereby enhance nutritional health in a sustainable manner.

Studies were conducted to determine the effects of Fe, phytate (*myo*-inositolhexaphosphate), and tannin concentrations in 24 select genotypes of bean seeds on iron bioavailability to rats. Beans selected from field trials in Cali, Columbia for their variations in Fe, phytate and tannin seed concentrations were grown in a green house in nutrient solutions radiolabeled with ^{59}Fe. Samples of radiolabeled mature seeds were autoclaved and lyophilized. The dried bean homogenate was used to prepare test meals fed to rats. Test meals contained 1 g of dried bean, 0.5 g sucrose and 1 g of basal Fe-deficient rat diet. The meals were fed to marginally Fe-deficient weanling rats over a 3-h period. The rats were radioassayed in a "whole body" gamma spectrometer immediately after feeding the test meals and daily thereafter for the next 10 d. "Whole body" radioiron retention data were used to calculate percent Fe absorption (i.e., Fe bioavailability) from the meals.

Concentrations of Fe in mature bean seeds harvested from the 24 accessions grown at the same time in a greenhouse varied with genotype and ranged from 51 to 157 $\mu g\ g^{-1}$ dry weight. The bioavailability to rats of Fe in the bean seeds depended on the genotype and varied from 53% to 76% of the Fe in the seed. There was no correlation between Fe concentration in the seeds and Fe bioavailability to rats fed the seeds even though the "antinutrients", phytate and tannins, varied widely in seeds of the different genotypes studied. Our results support the contention that breeding for enhanced concentrations of

G. Gissel-Nielsen and A. Jensen (eds.), Plant Nutrition – Molecular Biology and Genetics, 261–267.

Fe in bean seeds can result in significant increases in bioavailable Fe for humans eating the beans as staple food. Results of the effects of phytate and tannin levels on Fe bioavailability are also presented.

1. Introduction

Globally, there are approximately 2 billion people that are suffering from iron deficiency (U.S. Department of Health and Human Services, 1990; Mason and Garcia, 1993). Iron deficiency effects mostly women, infants and children resulting in increased mortality and morbidity rates, decreased labor productivity, and impaired mental development which diminishes the capacity of people to live healthy and productive lives and for nations to sustain development (Welch et al., 1997). Historically, the nutrition community has turned to intervention programs that employ iron supplementation or food fortification programs to alleviate iron deficiency. Unfortunately, many of these programs have not been sustainable in many regions of the world (Yip, 1997). Additionally, agriculture has never made improving human nutrition an explicit goal (Combs, Jr. et al., 1998). Currently, most policy makers agree that sustainable programs addressing iron deficiency must include food-based interventions that closely link agricultural production to this immense problem (Combs, Jr. et al., 1996). Increasing the concentration of Fe in staple foods could contribute greatly to decreasing the number of people suffering from iron deficiency. Furthermore, plant breeding is a tool that can be used to improve the nutritional quality of staple plant foods (Graham and Welch, 1996).

The common bean (*Phaseolus vulgaris* L.) is the most important grain legume for direct human consumption in the world, with global production reaching some 11 million MT annually. Beans supply the human diet with important quantities of protein and calories, and are also an excellent source of minerals, especially iron (Pennington and Young, 1990). The common bean is native to the New World and was domesticated in both Mesoamerica and the Andean zone, resulting in the formation of two major gene pools that differ in morphology and essential physiological processes (Singh et al., 1991). About 25,000 accessions of common bean are held in the gene bank of CIAT (International Center of Tropical Agriculture), Cali, Colombia. This collection serves as a source of useful genes for the improvement of the crop for both agronomic traits and nutritional quality. Due to the large size of the collection, it is difficult to fully evaluate the collection for many traits. Therefore, a core collection was formed to facilitate study and use of bean germplasm (Tohme et al., 1998). The core collection is a systematic sample of 1,441 accessions selected to represent the full range of agroecological environments and morphological variability of the species. An evaluation of the core offers a good approximation of the potential of the species for a given trait. We studied the effect of increasing bean seed-Fe concentration on the bioavailability of Fe to rats fed seed meals prepared from selected bean accessions obtained from the core collection of common beans maintained at CIAT.

2. Materials and Methods

2.1. PLANT MATERIAL AND GROWTH

Seeds from the CIAT core bean collection were initially screened for seed-iron concentrations using inductively-coupled, argon-plasma, emission spectrometry (ICPES), and 24 contrasting accessions were selected for this study. These accessions represented a range in seed types differing not only in their iron concentrations, but also in their seed colors, sulfur (S) and phosphorus (P) concentrations (Table 1).

TABLE 1. Some characteristics of the 24 bean (*Phaseolus vulgaris* L.) accessions selected for this study (received from CIAT).

CIAT Identification	Common name	Origin	Gene pool	Seed color	Seed size (g/100 seeds)	Growth habit
G87	De Seda	ELS	M	Cream	20	2
G734	Otz k'al tsaik	GTA	M	Cream	32	3
G1678	Baetao manteiga	BZL	A	Purple mottled	51	2
G1844	-	CRA	M	Cream	22	3
G2572	Rosado	ECD	A	Cream	40	3
G2774	-	MEX	M	Cream striped	34	3
G3096	Blanco Mono Ligero	GTA	M	Black mottled	26	3
G4825	Carioca	BZL	M	Cream striped	23	2
G5706	Jalpatagua 72	GTA	M	Black	20	2
G8465	Guatemala 0370	GTA	M-A	Red	17	3
G11350	Chiapas 36-3	MEX	M	Red	19	3
G11419	Puebla 44-2	MEX	M	Cream striped	40	4
G12610	Bayo mediano	PER	A	Cream	48	4
G13220	Tolima	CLB	A	Yellow	44	1
G14519	Hickman pole bean	USA	M	Brown	38	4
G15137	-	CMR	A	Red	58	1
G16267	Chimbolo	CRA	M	Pink	16	3
G18372	Coscorron No. 4	CLE	A	White	55	1
G18811	Alubia de Asturia	SPN	M	White	27	3
G19022	Chiapas 115	MEX	M	Cream & black	32	4
G21078	-	ARG	A	Cream	47	4
G21242	-	CLB	A	Cream mottled	43	4
G21725	IPA 6	BZL	M	Cream	19	4
G23063	Line 227	MWI	M-A	Purple	31	3

Seed color is somewhat correlated to seed-tannin levels which may interfere with iron bioavailability to humans. White-seeded beans are known to contain very low tannin levels; therefore two accessions of white-colored bean were included to assure variability in seed-tannin content among accessions selected for further study. Sulfur concentration is an indicator of S-containing amino acids in seeds that may serve as bioavailability promoters for Fe in plant foods. Phosphorus is an indicator of seed-phytate concentrations in seeds. Phytate is a known "antinutrient" that can inhibit Fe bioavailability from plant foods to humans. Variability was likewise sought in seed-S and seed-P concentrations among accessions included in the study.

Seeds were germinated between layers of water-saturated filter paper in glass-covered petrie dishes at 25°C in the dark. After hypocotyl and radical emergence, the plantlets were transferred to nutrient solutions of the following composition: (in mM) N,

16; K, 6; P, 2; Mg, 1; S, 1; Ca, 4; (in μM) Cl, 50; B, 12.5; Mn, 2; Zn, 2; Cu, 0.5; Mo, 0.1; Ni, 0.1; Fe, 50 (as the Fe(III)-EDDHA chelate). At flowering, Fe was supplied as radiolabeled Fe(III)-EDDHA (740 MBq ^{59}Fe mmole^{-1} Fe). Seeds were harvested at maturity and subsamples were autoclaved under steam pressure for 15 min to cook the beans. The autoclaved beans were homogenized in a Polytron Homogenizer and the homogenate lyophilized to dryness.

2.2. ANALYSES

Sub-samples (0.2 -.5 g) of the dried homogenates were wet digested (using nitric-perchloric acids) and the acid digestates were analyzed via ICPES for mineral concentrations. Phytate was determined in extracts (Lehrfeld, 1994) of sub-samples of dried bean meal via a liquid ion chromatography method (Dionex, 1990). Tannins and polyphenolics were determined in sub-samples of a methanol extraction of dried bean meal as catechin equivalents using the method of Deshpande and Cheryan (1987).

2.3. BIOAVAILABILITY STUDIES

Autoclaved and lyophilized homogenates of intrinsically ^{59}Fe-labeled mature beans were used to prepare single meals. The meals were fed to male weanling rats weighing between 103 to 120 g (mean, 111g) and maintained on a marginally Fe-deficient diet (modified AIN93-G containing 20 μg Fe g^{-1} dry wt. diet) for 7 d (blood-hemoglobin averaged 10.6±0.2 g dL^{-1}). The meal contained 1.0 g dried bean homogenate, 1.0 g of basal diet and 0.5 g of sucrose. The meal was fed *ad libitum* for 3 h. Immediately after being fed the meals, the rats were radioassayed in a "whole body" gamma spectrometer for radioactivity and daily thereafter for the next ten days (Welch et al., 1978). Rats were then provided *ad libitum* the basal diet for the remainder of the experiment. Plots of ^{59}Fe-label retained were constructed and percent absorption values were calculated from the "whole body" retention data obtained (Welch et al., 1974). Bioavailability was defined as the percent of radioiron absorbed from the bean meal. Statistical analyses were performed using SAS software (SAS Institute Inc., Cary, NC) [1].

3. Results and Discussion

Depending on the genotype, bean seeds, labeled intrinsically with ^{59}Fe contained a wide range of Fe, total inositolphosphate (*myo*-inositolpentaphosphoric acid + phytic acid), and total tannin concentrations. However, there were no significant correlations between Fe, inositolphosphates and tannin concentrations, and bioavailable Fe in the

---

[1]Mention of trademark, proprietary products or vendors does not constitute a guarantee or warranty of the product by the U.S. Department of Agriculture and does not imply its approval to the exclusion of other products or vendors that may also be suitable.

TABLE 2. Concentration of Fe, inositol phosphates, and tannins, and bioavailable Fe in mature bean seeds from 24 genotypes of beans grown in ^{59}Fe-radiolabeled nutrient solutions.

Bean Genotype	Fe ($\mu g\ g^{-1}$)	IP5+IP6[2] ($\mu mole\ g^{-1}$))	Tannin[3] ($mg\ g^{-1}$)	Bioavailable Fe (% absorbed)
G12610	51.63	19.57	2.65	64.09
G8465	55.67	25.00	1.64	68.24
G21725	60.25	26.45	1.18	60.55
G21242	69.16	27.77	1.56	53.81
G21078	74.53	23.68	1.88	62.88
G2774	75.27	20.66	0.96	67.94
G19022	81.45	22.99	1.28	53.08
G15137	82.07	27.48	1.55	56.29
G4825	82.26	23.23	1.85	67.08
G14519	85.53	33.53	0.90	53.56
G18372	88.01	19.60	0.35	72.34
G23063	88.88	21.94	2.60	61.84
G1678	89.74	25.68	1.21	56.13
G11419	91.96	23.58	0.91	60.72
G1844	92.53	29.16	1.15	58.67
G11350	98.71	25.91	1.32	56.87
G5706	100.51	24.27	0.65	60.13
G2572	103.55	23.06	2.22	65.99
G16267	110.59	25.41	1.32	60.05
G13220	122.70	23.74	2.49	59.37
G18811	130.70	22.94	0.54	76.24
G3096	139.91	25.27	0.89	58.81
G87	149.86	26.15	1.97	72.09
G734	156.91	24.09	1.41	59.23
LSD[1]	10.18	2.75	0.04	8.98

[1]LSD = Least significant difference ($P \leq 0.05$).
[2]IP5+IP6 = concentration of *myo*-inositolpentaphosphoric acid + phytic acid.
[3]Determined as catechin equivalences in bean homogenates.

intrinsically radiolabeled bean seeds as determined using the rat model (see Table 2). Figure 1 shows the relationship between bioavailable Fe in the bean meals and the concentration of Fe in the beans. Clearly, increasing the Fe concentrations in beans by selecting for high-seed-Fe genotypes resulted in increased bioavailable amounts of Fe to rats fed the bean meals irrespective of "antinutrient" levels in the beans.

Our results demonstrate that breeding beans for increased Fe concentrations will result in significant increases in bioavailable Fe to monogastric animals fed bean meals. This is true even though some of the bean genotypes contained very high levels of the "antinutrient" substances, phytic acid and tannins. Thus, our results support the contention that breeding for higher Fe density in common beans could contribute significantly to improving the Fe status of individuals dependent on beans as a staple food.

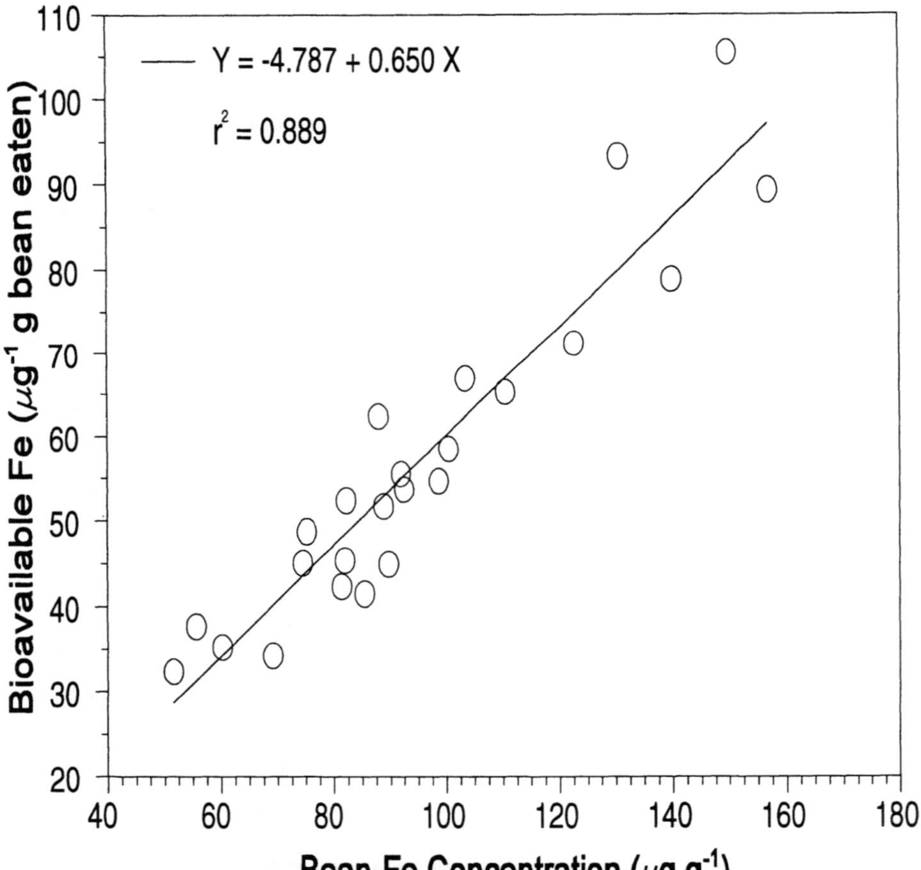

Figure 1. Relationship between bean-Fe concentration and amount of bioavailable Fe (Fe absorbed) from intrinsically radiolabeled bean meal fed to rats. Data points represent means for each of the 24 bean genotypes. The equation inserted represents the calculated regression line using a linear regression model. The r^2 value represents the correlation coefficient.

References

Combs, G.F., Jr., Welch, R.M., Duxbury, J.M., Uphoff, N.T. & Nesheim, M.C. (1996) Food-Based Approaches to Preventing Micronutrient Malnutrition: an International Research Agenda, pp. 1-68. Cornell International Institute for Food, Agriculture, and Development, Cornell University, Ithaca, NY.

Combs, J.F., Jr., Welch, R.M. & Duxbury, J.M. (1998) Fighting hidden hunger. *The World & I* April 1998: 174-181.

Deshpande, S.S. & Cheryan, M. (1987) Determination of phenolic compounds of dry beans using vanillin, redox and precipitation assays. *J Food Sci* 52: 332-334.

Dionex (1990) Application Note 65 - analysis of inositol phosphates. Dionex Corporation Application Notes, Dionex Corporation, Sunnyvale, CA.

Graham, R.D. & Welch, R.M. (1996) Breeding for staple-food crops with high micronutrient density, Ed.1, pp. 1-72. International Food Policy Research Institute, Washington, D.C.

Lehrfeld, J. (1994) HPLC separation and quantification of phytic acid and some inositol phosphates in foods: Problems and solutions. *J Agric Food Chem* 42: 2726-2731.

Mason, J.B. & Garcia, M. (1993) Micronutrient deficiency - the global situation. *SCN News* 9: 11-16.

Pennington, J.A.T. & Young, B. (1990) Iron, zinc, copper, manganese, selenium and iodine in foods from the United States total diet study. *J Food Comp Anal* 3: 166-184.

Singh, S.P., Gepts, P. & Debouck, D.G. (1991) Races of common bean (*Phaseolus vulgaris*, Fabaceae). *Econ Bot* 45: 379-396.

Tohme, J., Jones, P., Beebe, S. & Iwanaga, M. (1998) The combined use of agroecological and characterization data to establish the CIAT *Phaseolus vulgaris* core collection. In: Core Collections of Plant Genetic Resources (Hodgkin, T., Brown, A.H.C., van Hintum, Th.J.L. & Morales, E.A.V., eds.), pp. 95-107. J. Wiley and Sons, Chichester, UK.

U.S. Department of Health and Human Services (1990) Healthy People 2000. *Nutr Today* November/December: 29-39.

Welch, R.M., Combs, G.F., Jr. & Duxbury, J.M. (1997) Toward a "Greener" revolution. *Issues in Science and Technology* 14: 50-58.

Welch, R.M., House, W.A. & Allaway, W.H. (1974) Availability of zinc from pea seeds to rats. *J Nutr* 104: 733-740.

Welch, R.M., House, W.A. & Van Campen, D.R. (1978) Availability of cadmium from lettuce leaves and cadmium sulfate to rats. *Nutr Rep Int* 17: 35-42.

Yip, R. (1997) The challenge of improving iron nutrition: limitations and potentials of major intervention approaches. *Eur J Clin Nutr* 51: S16-S24

ISOLATION AND PHYSIOLOGICAL ANALYSIS OF A NOVEL *ARABIDOPSIS THALIANA* MUTANT THAT REQUIRES A HIGH LEVEL OF BORON

Miho YASUMORI, Kyotaro NOGUCHI, Mitsuo CHINO, Hiroaki HAYASHI, Satoshi NAITO* and Toru FUJIWARA** *Department of Applied Biological Chemistry, Graduate School of Agricultural and Life Sciences, The University of Tokyo, Tokyo 113-8657, Japan; *Department of Applied Bioscience, Faculty of Agriculture, Hokkaido University, Sapporo 060-8589, Japan* (**corresponding author, email atorufu@hongo.ecc.u-tokyo.ac.jp; FAX 81-3-5689-7226)

1. Abstract

A newly isolated *Arabidopsis thaliana* mutant, *bor 1-2* (high boron requirement) is described. The mutant was screened from 20,000 ethylmethane sulfonate (EMS)-mutagenized M2 lines looking for lines with poor seed sets under low B. After three rounds of screening, a line was isolated that clearly exhibited normal seed sets at high B (150 µM) but poor seed sets at low B (3 µM). Growth retardation at low B was evident in this line. The mutant carried a single recessive mutation responsible for the phenotype. The mutation was allelic to *bor1-1* (Noguchi et al., 1997) and the phenotypes were very similar to *bor1-1*.

2. Introduction

Boron has long been established as an essential element in higher plants (Loomis and Durst, 1992, for review), though its mechanism of action is still poorly understood. It is mostly deduced from disorders of B-deficient plants and biochemical characterization of

G. Gissel-Nielsen and A. Jensen (eds.), Plant Nutrition – Molecular Biology and Genetics, 269–275.

B associated materials. Studies on physiological functions of B were summarized previously (Loomis and Drust, 1992). Boron is predominantly localized to cell walls in association with rhamnogalacturonan II (Hu and Brown, 1994; Kobayashi et al., 1996; Ishii and Matsunaga, 1996; O'Neill et al., 1996). Positive correlation between pectin contents and B requirements has been reported in a number of plant species (Hu et al., 1996).

A number of studies have also been carried out on mechanisms of B transport and translocation. Studies at the cellular level established that the uptake of B mainly occurs by passive diffusion, although other mechanisms may also be involved (Raven, 1980; Brown and Hu, 1994). Recent study on sunflower suggests an active mechanism(s) of B absorption and translocation under low B (Pfeffer et al., 1997). Translocation of B is also known to occur through transpiration streams, although several observations suggest that B can also be mobilized through phloem (Brown and Hu, 1996).

One possible approach to gain insights into a molecular component(s) required for the action of B in plants is to use genetics. We previously described isolation and characterization of an *Arabidopsis thaliana* mutant, *bor1-1* that requires a high level of B for normal growth (Noguchi et al., 1997). Here, we present isolation and characterization of a novel allele, *bor1-2*.

3. Materials and Methods

3.1. PLANT MATERIALS AND GROWTH CONDITIONS

Arabidopsis thaliana (L.) Heynh. strains Col-0 (ecotype Columbia wild type), Col *gl1-1* and Col *bor1-1* were from our laboratory stock. EtMS M2 seeds of Col *gl1-1* was purchased from Lehle Seeds (Tucson, Arizona, USA). The *bor1-2* mutant line was isolated in this study (see below).

For the screening, plants were grown on rockwool and vermiculite with hydroponic media (Fujiwara et al., 1992) in a greenhouse at 22 °C under natural light supplemented with fluorescence lamps. B concentration was adjusted by altering H_3BO_3 concentration in the media. For physiological studies, plants were grown hydroponically (Hirai et al., 1995).

3.2. DETERMINATION OF B CONCENTRATION

Concentration of B was measured according to Noguchi et al. (1997) Briefly, plant materials were heat-dried, digested with conc. HNO_3 and subjected to ICP-MS (Model SII SPQ 9000, Seiko Instruments, Chiba, Japan) analysis.

4. Results

4.1. ISOLATION OF A MUTANT LINE

The new mutant was isolated from about 20,000 EMS-mutagenized M2 plants by looking for lines that set seeds poorly on low B (3 μM) but normally on high B (150 μM). Outline of the screening procedure is shown in Figure 1. After three rounds of screening, a line was isolated that clearly exhibited growth retardation on low B (Bor^- phenotype), i.e., very poor seed sets, reduced expansion of rosette leaves and loss of apical dominance at 3 μM B. F2 plants from the backcross with Col-0 wild type segregated Bor^+: Bor^- at 71:18, suggesting that the line carries a single recessive mutation responsible for the phenotype (χ^2=0.30). This mutant line was established after two rounds of backcrosses and used for subsequent studies.

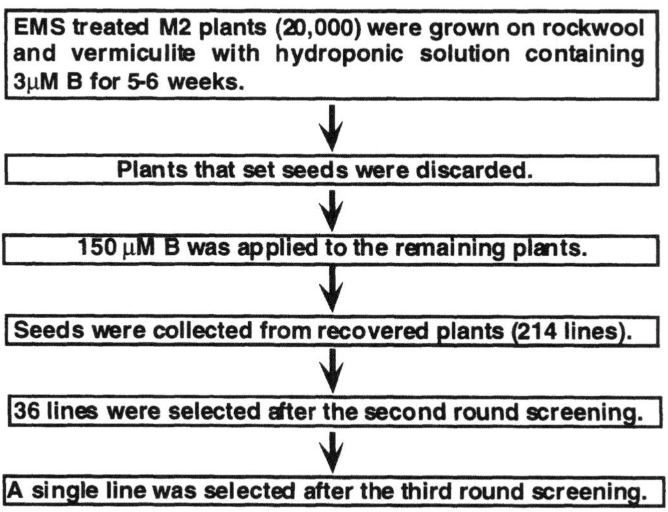

Figure 1 The scheme of the muatant isolation

4.2. ALLELISM TEST OF THE MUTANT

The newly isolated mutant was crossed with *bor1-1* to test allelism. All F1 (13 plants) and F2 (30 plants) from the crosses exhibited Bor$^-$ phenotype, suggesting the new mutation is allelic to *bor1-1*. Thus, the new mutant allele was named *bor1-2*.

4.3. GROWTH UNDER VARIOUS LEVELS OF B SUPPLY

Growth patterns of *bor1-2* plants at various concentrations of B were compared with the Col-0 and *bor1-1* plants by hydroponic cultures. Expansion of rosette leaves of *bor1-1* and *bor1-2* plants was severely affected when the B concentration in the medium was reduced to 3 µM (data not shown). Apical dominance of the mutants was also lost at this concentration (data not shown). At 10-30 µM B, expansion of rosette leaves was indistinguishable from the wild type plants but the mutants were mostly sterile (data not shown). Fertilities were recovered at 100 µM B. These features are very similar to *bor1-1* (Noguchi et al., 1997).

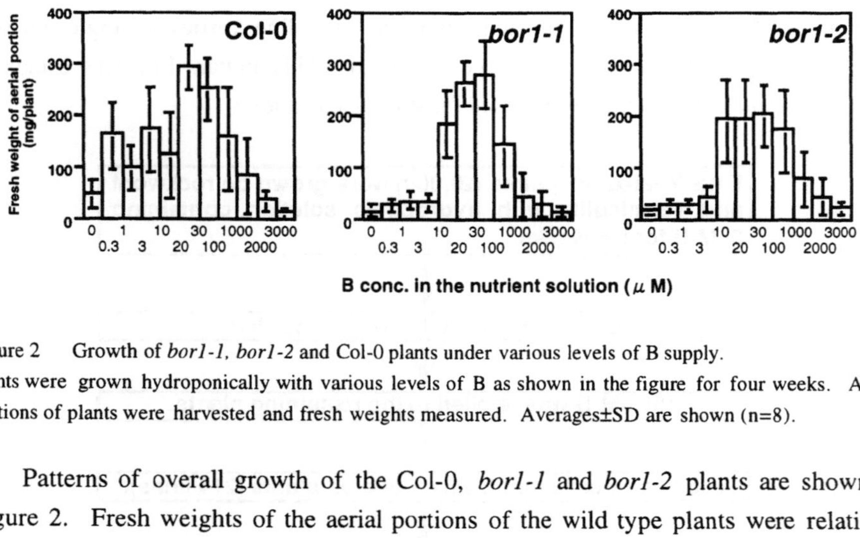

Figure 2 Growth of *bor1-1*, *bor1-2* and Col-0 plants under various levels of B supply.
Plants were grown hydroponically with various levels of B as shown in the figure for four weeks. Aerial portions of plants were harvested and fresh weights measured. Averages±SD are shown (n=8).

Patterns of overall growth of the Col-0, *bor1-1* and *bor1-2* plants are shown in Figure 2. Fresh weights of the aerial portions of the wild type plants were relatively constant within a range of 0.3 to 300 µM B, but omission of B from the nutrient solution reduced the growth of aerial portions severely. Wild type plants failed to set seeds at 0.3 µM B and root growth was reduced markedly (data not shown). Growth at 0 µM B presumably represents a minor contamination of B in our experiments. Growth of the aerial portion of the mutant plants was severely diminished when B supply was reduced to 0.3 μM. Growth of the mutants and wild type plants was similar when B supply was at or higher than 10 μM. No statistically significant difference (by the students' t-test) was observed between *bor1-1* and *bor1-2* plants at all the B concentrations tested. These observations establish that *bor1-2* plants are sensitive to B-deficiency, in a similar way as *bor1-1* plants.

4.4. BORON CONCENTRATION IN THE MUTANTS

We also measured concentrations of B in rosette leaves grown at various B levels (Figure 3). It was found that B concentrations in *bor1-1* and *bor1-2* were lower than that of the wild type within a range of 3 to 30 μM B. There was no significant difference (by the students' t-test) in B concentrations between *bor1-1* and *bor1-2* plants at all the B concentrations tested.

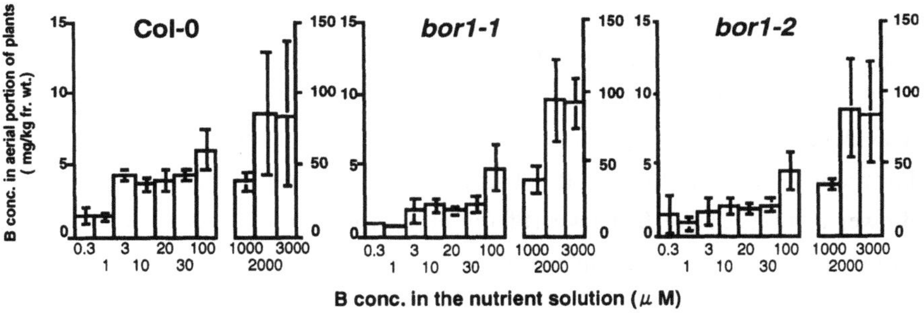

Figure 3 Boron concentrations in *bor1-1, bor1-2* and Col-0 plants.

Plants were grown hydroponically with various levels of B as shown in the figure for four weeks. Aerial portion of plants were harvested, dried and digested with conc. HNO_3 (61%) . Boron concentrations were determined by ICP-MS. Averages±SD are shown (n=5). Note the difference in scale in the right and left sides of each graph.

5. Discussion

Here we described isolation and characterization of *bor1-2*, a novel allele of *bor1*. The phenotypes of *bor1-2* were very similar to *bor1-1*. It was suggested that the *bor1-1* mutant has a reduced capacity of B uptake and/or translocation (Noguchi et al,. 1997). The *bor1-2* mutant plants also had lower B concentrations than the wild type plants in aerial portions under low levels (3-30 μM) of B supply (Figure 3). B concentrations in aerial portions of the *bor1-2* mutant plants were similar to the *bor1-1* at all the B concentrations tested. These findings suggest that the *bor1-2* mutant has a defect(s) in B uptake and/or translocation when B supply is limited, similar to the case of the *bor1-1*.

Although the *bor1-1* and *bor1-2* mutants exhibited similarly reduced growth and B concentrations at low B, it was intriguing to note that the range of B concentrations in the media that affected growth were different from that affected the *in vivo* B concentrations in the aerial portions. At 30 μM B, expansion of rosette leaves of the mutants was similar to the wild type (Figure 2) but B concentrations in leaves of the *bor1-1* and *bor1-2* mutants were lower (Figure 3). Growth of the mutant plants were much poorer than the wild type even when the *in vivo* concentrations of B were equivalent (e.g., the wild type at 3 μM compared with the mutants at 30 μM). These findings suggest the possibility that the *BOR1* gene product may not only affect B uptake/translocation but also B utilization. Alternatively, *in vivo* B concentrations of the mutants after four weeks from germination may not correlate with B requirements for normal growth. It is likely that B continues to accumulate after growth retardation become evident in early stages of leaf development in the mutants.

The fact that the difference in B concentration in aerial portions of the plants are only evident at low B in hydroponic cultures suggests that the mutation affects B uptake and/or translocation only at low B. Together with the recent report that low-B grown sunflower exhibits active B uptake (Pfeffer et al. 1997), our findings suggest that the mutation affects an active uptake system of B that operates only under low concentrations of B.

The novel allele described here will be helpful in the cloning process of *BOR1* gene and understanding of the *BOR1* function in B related phenomena in higher plants.

6. Acknowledgements

We thank N. Nakayama for technical assistance. This work was supported in part by Grant-in-Aid from Ministry of Education, Science, Sports and Culture of Japan to KN (No. 4247) and TF (No. 09760053).

7. References

Brown, P.H., and Hu, H. (1994) Boron uptake by sunflower, squash and cultured tobacco cells. *Physiol*

Plant **91**, 435-441

Brown, P.H,, and Hu, H. (1996) Phloem mobility of boron is species dependent: evidence for phloem mobility in sorbitol-rich species. *Ann. Bot.* **77**, 497-505

Fujiwara, T., Hirai, Y.M., Chino, M., Komeda, Y. and Naito, S. (1992) Effects of sulfur nutrition on expression of the soybean storage protein genes in transgenic petunia. *Plant Physiol.* **99**, 263-268

Hirai, M.Y., Fujiwara, T., Chino, M. and Naito, S. (1995) Effects of sulfate concentrations on the expression of a soybean seed storage protein gene and its reversibility in Transgenic *Arabidopsis thaliana. Plant Cell Physiol.* **36**, 1331-1339

Hu, H. and Brown, P.H. (1994) Localization of boron in cell walls of squash and tobacco and its association with pectin. *Plant Physiol.* **105**, 681-689

Hu, H., Brown, P.H. and Labavitch, J.M. (1996) Species variability in boron requirement is correlated with cell wall pectin. *J. Exp. Bot*. **47**, 227-232

Ishii, T. and Matsunaga, T. (1996) Isolation and characterization of a boron-rhamnogalacturonan-II complex from cell walls of sugar beet pulp. *Carbohydr. Res.* **286**, 1-9

Kobayashi, M., Matoh, T. and Azuma, J. (1996) Two chains of rhamnogalacturonan II are cross-linked by borate-diol ester bonds in higher plant cell walls. *Plant Physiol.* **110**, 1017-1020

Loomis, W.D. and Durst, R.W. (1992) Chemistry and biology of boron. *BioFactors* **3**, 229-239

Noguchi, K., Yasumori, M., Imai, T., Naito, S., Matsunaga, T., Oda, H., Hayashi, H., Chino, M. and Fujiwara, T. (1997) *bor1-1*, an *Arabidopsis thaliana* mutant that requires a high level of boron. *Plant Physiol.* **115**, 901-906

O'Neill, M.A., Warrenfeltz, D., Kates, K., Pellerin, P., Doco, T., Darvill, A.G. and Albersheim, P. (1996) Rhamnogalacturonan-II, a pectic polysaccharide in the walls of growing plant cell, forms a dimer that is covalently cross-linked by a borate ester. *J. Biol. Chem.* **271**, 22923-22930

Pfeffer, H., Dannel, F. and Roëmheld, V. (1997) Compartmentation of boron in roots and its translocation to the shoot of sunflower as affected by short term changes in boron supply. *in Boron in Soils and Plants*, Bell and Rerkasem eds., Kluwer Academic Publishers, Dordrecht pp203-207

Raven, J.A. (1980) Short- and Long-distance transport of boric acid in plants. *New Phytol.* **84**, 231-249

THE ROLE OF MYCORRHIZAS IN PLANT P NUTRITION: FUNGAL UPTAKE KINETICS AND GENOTYPE VARIATION

P.F. SCHWEIGER AND I. JAKOBSEN
Centre for Plant Microbe Symbioses, Plant Biology and Biogeochemistry Department, Risø National Laboratory, DK-4000 Roskilde, Denmark

1. Introduction

Because of the widespread occurrence of arbuscular mycorrhizas (AM) across plant families and across habitats and because of their mutualistic nature, nutrient uptake by soil-grown plants can be fully understood only if the contribution of the mycorrhizal fungi is considered. It is necessary to understand in detail the mechanisms of fungal nutrient transport from soil to plant and how this transport varies with characteristics of soil, fungus and plant. Arbuscular mycorrhizas are particularly important for plant uptake of phosphate (P), which is poorly soluble and poorly mobile in soil. Increased plant P uptake following root colonization by AM fungi has been attributed to P uptake by external hyphae outside the P depletion zone forming around an actively absorbing root, followed by translocation of P within the hyphae to the plant root.

The development of P efficient cropping systems via plant breeding or soil management will rely on an identification of the crucial rate-limiting steps in plant P acquisition. In this paper we derive estimates for kinetic parameters of P uptake by non-mycorrhizal and mycorrhizal root systems in soil including their associated external hyphae. We show that kinetic parameters have less importance for root P acquisition than for P acquisition by mycorrhizal hyphae. We also discuss the possible influence of interfungal variation in P transport on plant P acquisition and show that the intraspecific variation is rather small as compared to previously reported variations at the interspecific level (Jakobsen, 1995).

2. Phosphate uptake kinetics of mycorrhizas

So far no published reports have convincingly shown AM fungal access to additional pools of soil P other than soil P also available to plant roots. This pool of P is the P in the soil solution, which is in equilibrium with P sorbed onto clay particles and organic matter. Plant roots take up P from the soil solution at the root surface, and this disturbance in the equilibrium between sorbed and solution P causes some P to desorb and some to diffuse towards the root. Continuous root P uptake will eventually lead to the formation of a P depletion zone around the root. Uptake of P by the root and the diffusive supply of P from the soil to the root can both be mathematically described,

G. Gissel-Nielsen and A. Jensen (eds.), Plant Nutrition – Molecular Biology and Genetics, 277–289.

and these processes have been incorporated into mechanistic models of plant P uptake (Barber, 1984). A sensitivity analysis was performed with one of theses models (Silberbush and Barber, 1983) in an attempt to examine which soil and plant parameters had the greatest effect on P uptake. Root diameter and rate of root elongation were found to have the greatest effect on P uptake, while changes in the kinetic parameters of uptake only had a very small effect. Accordingly, the kinetic parameters of hyphal P uptake have received little attention so far. P uptake into roots and fungi follows a saturation curve as it is described by the Michaelis-Menten equation (Nielsen, 1976).

$$I = \frac{I_{max} * (C_s - C_{min})}{K_m + C_s - C_{min}}, \qquad (1)$$

where I is rate of uptake, I_{max} the maximum rate of uptake, C_s is the solution concentration of P, C_{min} is the solution concentration of P below which no uptake takes place and K_m is the solution concentration of P at which the rate of uptake equals $I_{max}/2$. To our knowledge there are only two studies dealing with uptake kinetics of arbuscular mycorrhizal root systems. Two uptake systems were detected in both mycorrhizal and non-mycorrhizal tomato root systems (Cress *et al.*, 1979). The high-affinity uptake system of the mycorrhizal plants had a significantly lower K_m value than non-mycorrhizal plants, while maximum uptake rates were similar. The kinetic parameters of the two low-affinity uptake systems were similar. A similar approach was used to examine P uptake by mycorrhizal and non-mycorrhizal soybeans where the highest I_{max} but the lowest affinity (= highest K_m) was found in the mycorrhizal plant (Karunaratne *et al.*, 1986). Apart from this high value, K_m values were similar in the two studies (1.5 to 4 μM) and also similar to the values obtained for isolated germ tubes of the AM fungus *Gigaspora margarita* (Thomson *et al.*, 1990). Maximum uptake rates were much higher in soybeans (19-58 nmol s^{-1} m^{-2}) than in tomatos (1.5-4.8 nmol s^{-1} m^{-2}) possibly in parts due to the high age of the tomato plants. Plants were pre-grown in sand in both studies and this probably caused external hyphae to be lost during preparation of the plants for absorption studies in nutrient solution.

While kinetic parameters of uptake have been shown only to have a small effect on P uptake by roots (Silberbush and Barber, 1983), they may play a more important role in determining effective P uptake by external hyphae. Due to the very small diameter of external hyphae, the solution at the hyphal surface will be much less depleted of P than the solution at the root surface (Barber, 1984), and in contrast to uptake by roots, will not be diffusion-limited. An increase in I_{max} will therefore result in a much greater increase in P uptake by a hypha than an equivalent increase would affect root P uptake. In a number of studies, mycorrhizal plants have been observed to absorb P at soil P concentrations at which there was no P uptake by non-mycorrhizal plants (Bolan, 1991). This could be explained by a lower C_{min} for mycorrhizal root systems, but P concentration profiles away from non-mycorrhizal and mycorrhizal roots and external hyphae were very similar (Li *et al.*, 1991), which does not support that possibility. Hyphae are however expected to be able to absorb P from lower soil solution P concentrations than roots just based on their small diameter and the lack of development of a depletion zone (Barber, 1984). Hyphal P uptake from within the P depletion zone of roots is therefore possible and may play an important role under conditions of low soil P and very high rooting densities (George *et al.*, 1995). An increase in the affinity for P (lower K_m) by the hyphal P uptake system would enable hyphae to take up P even more effectively under these very P-deficient soil conditions. The mycorrhizal growth response observed in ryegrass at very high rooting densities (Parfitt, 1979) may be a result of hyphal P uptake from within P depletion zones formed around roots. However, the frequently observed decrease in the growth-enhancing mycorrhizal effect following a reduction in the available soil volume per plant (Bååth and Hayman, 1984) supports the view that the undepleted bulk soil is the main source for additional P taken up by external mycorrhizal hyphae.

3. Estimated P uptake parameters for mycorrhizal and non-mycorrhizal subclover

In the following we estimate the kinetic parameters of P uptake of non-mycorrhizal and mycorrhizal root systems including associated external hyphae on the basis of results of an unpublished study (Schweiger, 1993) and a number of assumptions. We also examine whether root and hyphal P uptake are differently affected by changes in the kinetic parameters.

Our calculations are based on data for shoot P content and root length of mycorrhizal and non-mycorrhizal subterranean clover plants grown in a high P-fixing soil at 10 levels of added P (Schweiger, 1993). Plants were harvested at three and five weeks from sowing. Roots of mycorrhizal plants were heavily colonized by the AM fungus *Glomus invermaium* WUM 10, while roots of non-mycorrhizal plants remained uncolonized. Amounts of added P were converted to soil solution concentrations of P by using the P sorption isotherm and moisture content of the soil. Curves were fitted to shoot P content and root length data as functions of soil solution P concentrations. Curves describing shoot P content and root length had fits better than R^2=0.99 and R^2=0.86, respectively. Phosphorus inflow (V) into shoots was calculated using the following formula (Williams, 1948):

$$V = \frac{(P_2 - P_1) * (\log RL_2 - \log RL_1)}{(RL_2 - RL_1) * (t_2 - t_1)} \tag{2}$$

where P is shoot P content, RL is root length, t is time and subscripts 1 and 2 refer to harvests 1 and 2. Fitted values for P content and root length were used. Calculated inflow rates were used to estimate a soil solution P concentration at the root surface (C_{la}) necessary to sustain the calculated inflows using the following equation (Barraclough, 1986):

$$C_{la} = C_s - \frac{V}{4\pi D_l \Theta f} * \left(1 - \frac{1}{1 - \pi r_o^2 L_v} * \ln \frac{1}{\pi r_o^2 L_v}\right). \tag{3}$$

D_l is the diffusion coefficient of phosphate in water ($0.89\ 10^{-5}\ cm^2\ s^{-1}$; (Edwards and Huffman, 1959), Θ is the water content of the soil (0.23), f is a factor that describes the tortuosity of the path of a diffusing molecule (assumed to equal $\sqrt{\Theta}$; (Nye and Tinker, 1977)), r_o is the root radius (r_o = 0.15 mm) and L_v is the root length density ($cm\ cm^{-3}$). This equation is based on a number of simplifying assumptions. Roots are assumed to be smooth cylinders (no root hairs and no external mycorrhizal hyphae) of uniform diameter, distributed uniformly in the soil, with constant uptake properties that do not change during the experimental period. Each root exploits a concentric cylinder of soil with a radius equal to the half-distance between neighbouring roots and with a steady-state P concentration profile. The soil is assumed to be homogeneous, and transport of the nutrient through soil is by diffusion only.

A Hofstee plot of inflow against inflow divided by the P solution concentration at the root surface was used to estimate kinetic parameters of root P uptake (Seeling and Claassen, 1990); a value for C_{min} was not included. Two uptake systems were found for both mycorrhizal (M) and non-mycorrhizal (NM) root systems. The low-affinity uptake systems were very similar with a maximum inflow of about $1*10^{-14}$ mol $cm^{-1}\ s^{-1}$. The K_m value of the mycorrhizal root system (19 µM) was slightly lower than the K_m estimated for the non-mycorrhizal root system (28 µM). The high-affinity uptake systems were, however, very different. While maximum inflow of NM and M roots were $6*10^{-16}$ and $1*10^{-15}$ mol $cm^{-1}\ s^{-1}$, respectively, K_m was 0.14 µM for the NM roots in contrast to $0.66*10^{-4}$ µM for the M roots. It is important to keep in mind that this extremely low K_m was calculated without taking account of external hyphae, which need to be considered for a realistic description of mycorrhizal P uptake.

4. Estimated P uptake parameters for hyphae of *Glomus invermaium*

In an attempt to estimate kinetic parameters of P uptake by the external hyphae, a number of additional assumptions were made. Mycorrhizal roots were assumed to possess the very same uptake properties as NM roots. This assumption is supported by measurements of the shape of the depletion gradient next to M and NM roots (Li *et al.*, 1991) and our own measurements on the time-course of P uptake of M and NM roots (Jakobsen *et al.*, 1998). A negative feedback mechanism of hyphal P uptake on root P uptake has, however, been suggested in one study (Pearson and Jakobsen, 1993). The mechanism of this feedback is still unknown but may be equivalent to the negative feedback on root P uptake following pretreatment of the roots with solutions of high P concentration (Jungk *et al.*, 1990).

Hypothetical values for shoot P content were calculated using root uptake properties of the NM root system and measured lengths of the M root systems. The difference between these hypothetical and measured data for shoot P content of mycorrhizal plants was considered as the hyphal contribution. Hyphal lengths had not been determined, but values for hyphal length formed by the same AM fungus colonizing the same plant host have been published (Abbott *et al.*, 1984; Abbott and Robson, 1985). Based on their results hyphal lengths were assumed to increase from 7.7

m g^{-1} soil to 11 m g^{-1} soil at the second harvest irrespective of soil solution P concentration. External hyphae were treated as a separate cluster of very thin roots (10 µm diameter) (Darrah, 1993), and were assumed to absorb nutrients along their entire length. These assumptions were chosen to result in fairly conservative estimates for inflow since the mean diameter of external hyphae of *Glomus invermaium* is probably around 5 µm (Abbott and Robson, 1985). Observations made on the internal pH gradient along external hyphae also suggest that they absorb nutrients mainly along the first 150 µm behind the growing tip (Jolicoeur *et al.*, 1998).

Hypothetical hyphal P inflows were calculated, and as was done for the roots, P solution concentrations at the surface of the external hyphae necessary to sustain those inflows were calculated using equation 3. An Hofstee plot for hyphal P uptake gave an estimated K_m of 0.17 µM, which is very similar to the 0.14 µM obtained for the roots. These K_m values are much lower than the one obtained for a phosphate transporter isolated from an AM fungus (K_m=18 µM, (Harrison and van Buuren, 1995)) and also lower than values measured for plants and fungi in nutrient solutions (Clipson *et al.*, 1987; Jungk *et al.*, 1990). The calculations are based on data for plants grown in a soil with a very high P-fixing capacity, which will undoubtedly have affected our estimates. Maximum inflow was $2*10^{-17}$ mole cm^{-1} s^{-1}, which is of the same order as previous estimates (Jakobsen *et al.*, 1992b). This value is lower than the value for NM roots. Converting these inflow values into influx as expressed on the basis of surface area, maximum influx into hyphae ($2.55*10^{-11}$ mol cm^{-2} s^{-1}) is however much higher than maximum influx into roots ($8.97*10^{-13}$ mol cm^{-2} s^{-1}). Estimates for maximum influx rates are within the range measured for solution-grown plants (Karunaratne *et al.*, 1986).

A much higher maximum influx into external hyphae than into roots as suggested by the above calculations can be explained in various ways. Firstly, the number of P uptake sites per surface area could be much higher on external hyphae. Secondly, the hyphae may have a much more efficient biochemical pathway of incorporating inorganic ortho-phosphate followed by transfer into the vacuole and transport towards the root, thereby avoiding a negative feedback of increased internal P status on P uptake (Mimura *et al.*, 1998). Thirdly, hyphae may proliferate in patches enriched in nutrients (Cui and Caldwell, 1996) and as observed for organic patches (St John *et al.*, 1983). Finally, the increase may also be explained by the much smaller diameter of the hyphae relative to roots. Hyphae may thereby enter smaller soil pores than roots, which would add to the volume of soil effectively exploited. Smaller pores are also less likely to dry out and therefore provide a much more constant environment conducive to P uptake (O'Keefe and Sylvia, 1992). The assumption that soil is a homogeneous substrate does not cause major problems when modelling P uptake by the relatively coarse roots, but the situation may be different for the much thinner external hyphae. It would therefore be interesting to examine whether the very high estimate for maximum P influx into hyphae is due to physiological differences between hyphae and roots, or whether it reflects proliferation of hyphae in parts of the soil matrix relatively enriched in nutrient availability.

5. Importance of uptake parameters in relation to soil P level

The effect of changes in the kinetic parameters of hyphal or root P uptake on P content of plants grown at various levels of soil P was examined while keeping all other parameters constant (Fig. 1). For this exercise we considered the non-mycorrhizal plants

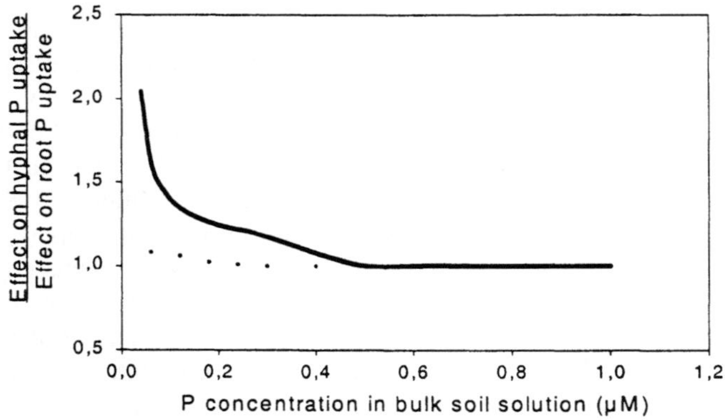

Figure 1. The effect of a change in K_m (dotted line) or V_{max} (solid line) on P uptake by very thin roots (diameter equivalent to hyphae) relative to P uptake by roots at various soil solution P concentrations. K_m was reduced to one fifth and V_{max} was increased to 10 times of their respective original values. Total volume of roots and hyphae were equal. All other parameters were held constant.

(root diameter = 0.3 mm) grown at the five soil solution P concentrations for which a growth-enhancing effect of mycorrhizal root colonization had been observed (Schweiger, 1993). Each of these plants was paired with a fictitious plant of the very same P content, but with roots of a diameter similar to external hyphae (10 μm). Root volume was held constant for both plants. P inflow was calculated for both sets of plants and solution concentrations at the root surfaces were calculated as described above. Kinetic parameters of uptake were again determined using a Hofstee plot. The effect of a change in V_{max} and K_m on plant P content was then estimated for both sets of plants. This was done by inserting the value for the altered inflow, caused by a change in either K_m or V_{max}, and the solution P concentration at the root surface into equation 3. A bulk soil solution concentration could thus be calculated, and inflow could be plotted against it. Fig. 1 shows the difference between the two sets of data expressed as the ratio between the effect on P content of plants with very thin roots and that of plants with roots of normal diameter. An increase in the affinity of the uptake system was simulated by reducing K_m to 20 % of its estimated value. The P contents of plants with the two different root systems were similarly affected by this change in K_m. On the other hand, an increase in V_{max} to 10 times its original value resulted in a much higher P content in plants with very thin roots than in plants with roots of normal diameter (Fig. 1) at solution P concentrations below 0.4 μM.

The calculations confirmed our original hypothesis that the kinetic parameters of P uptake have a greater effect on hyphal P uptake than on root P uptake. The view that root uptake characteristics have an only marginal effect on plant P uptake stems from analysing the effect of changes in the input parameters for a mechanistic model of plant P uptake on predicted plant P uptake (Silberbush and Barber, 1983). The data used in this sensitivity analysis were for plants grown on a fairly fertile soil (soil solution P concentration = 13.6 μM) with a relatively low buffer power and a high rate of P diffusion. Root uptake kinetics are however predicted to increase in importance when plants are grown on a soil with a higher P sorption capacity or when grown in a relatively poor soil with fertile patches (Caldwell *et al.*, 1992). This last point is especially relevant in regards to P uptake by external hyphae, because of the possibility that hyphae proliferate in organic soil patches and are able to enter pores from which roots are excluded (see above). These pores are water-filled at field capacity and can therefore be considered as fertile 'micropatches' with a higher nutrient availability than the larger pores accessible to roots.

6. Hyphal uptake from within P depletion zones

Root densities were high in the example taken for our calculations (>20 cm cm^{-3}), so we examined whether the additional P taken up via external hyphae was likely to have been taken up in soil not affected by root P uptake and therefore left undepleted. The progress of the depletion zone around the roots was calculated (Syring and Claassen, 1995):

$$r = r_o * e^{\pi/(2a)} \tag{4}$$

where r is the radius of the depletion zone from the root axis, r_o is the root radius and a is defined as:

$$a = \arctan\left(\pi * \left(\ln\left(1.26 * \frac{D * t}{r_o^2}\right)\right)^{-1}\right). \tag{5}$$

D is the effective diffusion coefficient of P in soil calculated as $D = D_l * \Theta * f / b$ (Nye and Tinker, 1977), where b is the buffer power. Values for D ranged from $1 * 10^{-11}$ to $2 * 10^{-10}$ cm^2 s^{-1} at the levels of soil P at which mycorrhizal colonization increased plant P uptake. During the two-week inter-harvest period, a zone of 0.86 mm diameter (corresponding to about 15 % of the total soil volume) is calculated to have been affected by root P uptake. This was calculated for the highest level of soil P at which mycorrhizal colonization increased plant P uptake and includes the 0.3 mm diameter root. Hyphal P uptake from within root P depletion zones apparantly did not contribute significantly to the total amount of P taken up, considering that external hyphae of *Glomus invermaium* can spread several cm away from root surfaces (Jakobsen *et al.*,

1992b). However, this assumption may not be valid for soils with a much lower P sorption capacity or for all mycorrhizal fungi.

7. Genotype variation in fungal P transport

Interfungal variation in plant growth responses to colonisation by AM fungi was demonstrated already over two decades ago (Gilmore, 1971; Mosse, 1972). Such variation is most often caused by differences in the mycorrhizal enhancement of P uptake and may be related to differences in colonised root length (Abbott and Robson, 1981; Sanders *et al.*, 1977) or not (Jensen, 1982). Direct measurements of fungal P transport are required in order to understand the nature of the genotype-determined variation. Growth systems with an inserted mesh barrier are suitable for this purpose: mycorrhizal plants are grown in soil on one side of the mesh which can be penetrated by AM fungal hyphae but not by roots. The fungal P uptake from the root-free soil may be measured either as depletion of the soil P (Li *et al.*, 1991) or as transfer of tracer isotopes from the soil to the plant (Jakobsen *et al.*, 1992a).

The fungal uptake of P described in the previous sections represents only the first step in the overall fungal P transport from soil to plant. Absorbed P must be translocated to the root through the root-external hyphae and subsequently be transferred to the plant cytosol via the fungus-plant interface. Each of these processes will contribute to the AM fungal effectiveness in P transport. The spread of the extraradical mycelium into undepleted soil volumes represents an important determinant of fungal effectiveness: The P uptake from root-free soil varied markedly between three fungi, such that the highest P uptake was observed with the fungus showing the furthest spread (Jakobsen *et al.*, 1992b). The effectiveness also depends on other genotype-determined factors, because fungal P transport expressed on the basis of a hyphal length unit varied markedly between different mycorrhizal fungi in experiments with *Trifolium subterraneum* and cucumber (Jakobsen *et al.*, 1992a; Pearson and Jakobsen, 1993). The background for this variation is unclear but could be related to fungal uptake kinetics as well as to fungal translocation of P and transfer of P across the symbiotic interface (Jakobsen, 1995). The picture may be even more complicated as the ranking of fungi in P transport capacity also appears to vary with the host genotype (Ravnskov and Jakobsen, 1995).

The interspecific differences in P uptake by AM fungi might be expected from the observation of genetic differences among these fungi (Lloyd-MacGilp *et al.*, 1996). Some intraspecific genetic variation has also been observed (Bago *et al.*, 1998; Dodd *et al.*, 1996; Lloyd-MacGilp *et al.*, 1996) which poses the question whether AM fungi also exhibit intraspecific variation at the functional level. One study compared 12 isolates of 5 different *Glomus* species and 6 isolates belonging to 3 *Gigaspora* species, all isolates being of different geographic origin (Carling and Brown, 1980). Both inter- and intraspecific variation in growth response to inoculation was observed for the *Glomus* isolates, whereas the *Gigaspora* isolates showed interspecific variation only. The background for the observed variation is obscure, as neither root colonisation nor P uptake were reported. In another study three isolates of *Glomus mosseae* differed in terms of improvement of plant growth and shoot P concentration and this variation was unrelated to root colonisation (Stahl *et al.*, 1990).

We investigated the intraspecific variation in terms of fungal P transport in 15 isolates of *G. fistulosum* which were originally collected from agricultural soils in Finland and propagated with *Plantago lanceolata* by Mauritz Vestberg. The isolates were later repropagated with *Trifolium subterraneum* before being used in the P transport experiment described in Fig. 2. A weak positive correlation was observed

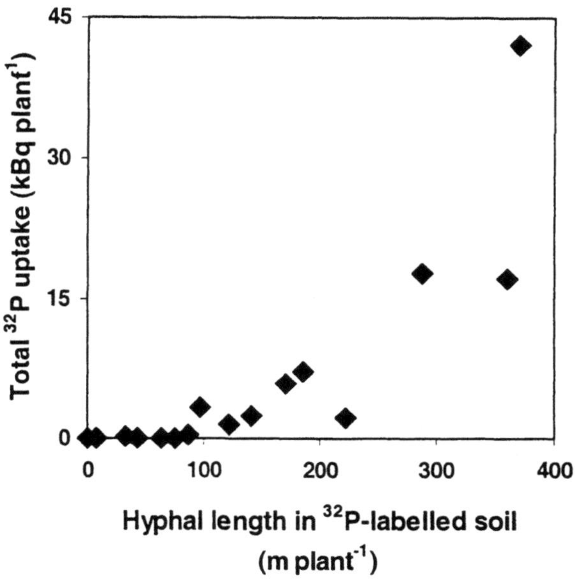

Figure 2. Fungal transport of ^{32}P into mycorrhizal cucumber plants growing in a two- compartment system made from PVC tubing (Pearson and Jakobsen, 1993). Hyphae from colonised roots in the main compartment grew into the side compartment containing a 10 mm buffer layer of unlabelled soil and a 30 mm layer of soil labelled with 90 kBq $H_3{}^{32}PO_4$.. Plants were harvested after 4 weeks and their dry weight, ^{32}P content, colonised root length and length density of mycorrhizal hyphae in the labelled soil were measured.

between plant growth and colonised root length (R^2=0.39). The length of external hyphae in the ^{32}P-labelled soil varied markedly between isolates, but this corresponded to a variation in root colonisation (R^2=0.81) caused by differences in inoculum strength between isolates. Still, the 15 isolates were rather similar in their ability to transport P as indicated by the close correlation (R^2=0.76) between ^{32}P uptake into the plant and hyphal length in the labelled soil (Fig. 2). The average ^{32}P uptake on a hyphal length basis was not constant, but this is to be expected as the hyphae would have grown into the labelled soil volume at different times.

Although these results demonstrate the existance of some variation between the 15 *G. fistulosum* isolates, this intraspecific variation is much smaller than the interspecific variation described above and also smaller than the intraspecific variation observed in *G. mosseae* (Stahl *et al.*, 1990). Our finding is in accordance with an apparantly limited genetic variation among the Finnish *G. fistulosum* isolates studied (M. Vestberg and S. Rosendahl, pers. comm.). Plants inoculated with several 'generations' of single spores of *G. etunicatum* varied considerably in root colonisation and growth response (Feldmann, 1998) and this suggests that some intraspecific functional variation may exist. Feldmann suggested that the multinucleate and probably heterocaryotic nature of AMF spores might be the key to understanding this variation in effectiveness at the single spore level. DNA sequences do actually differ between nuclei of a single spore (Trouvelot *et al.*, 1998) and it therefore seems feasible that the symbiotic effectiveness would depend on which nuclei were involved in the first contact between host and fungus (Feldmann, 1998). It is possible that a single isolate might differentiate with time into functionally rather different subpopulations. Colonisation under natural conditions would usually originate from a population or a community of spores and the overall mycorrhizal effectiveness would result from the average effect of colonisation originating from a number of single spores. The low variation observed in the present study with *G. fistulosum* was obtained by using isolates which were each added as a population of spores instead of a population of nuclei; the latter situation should be considered as artificial, compared to colonisation events in natural soil systems. It remains to be investigated whether a limited intraspecific variation in P uptake capacity is common among arbuscular mycorrhizal fungi or whether the genetic variation in some species is large enough to affect effectiveness in P nutrition. The small variation observed so far may reflect the clonal nature of arbuscular mycorrhizal fungi (Rosendahl and Taylor, 1997) such that spread of the fungus during evolutionary time would have occurred without a major influence on fungal effectiveness in nutrient uptake.

8. Conclusions

Models may accurately predict plant P uptake from rather fertile soil (Barber, 1984), while they drastically underestimate plant P uptake under P-deficient soil conditions (Ernani *et al.*, 1994). There have so far only been a few attempts to incorporate the mycorrhizal contribution to plant P uptake into models of plant P uptake (Ernani *et al.*, 1994; Fitter, 1985; Grant and Robertson, 1997; Sanders and Tinker, 1973) probably due to a number of so far unresolved questions concerning the physiology and morphology of the external mycelium. Based on increased information on the functioning of the external mycelium gained over the last years we consider it worthwhile to try and incorporate mycorrhizas in plant P uptake models. Our work has shown that it is possible to estimate the kinetic parameters of P uptake by external hyphae. The magnitude and variation of these parameters should be further investigated in experiments with sequential harvests including quantification of external hyphae and the data collected woud be most useful for the proposed modelling.

Inoculation with effective isolates of arbuscular mycorrhizal fungi is already feasible in micropropagated or pot grown plants in general and inoculation may also be feasible at the field scale when methods become available for the production of sufficient amounts of clean inoculum. Standard methods are available for selecting the most P-efficient fungi and the required source of variation for this selection appears to exist more between than within species. The length and spread of hyphae in soil are important determinants of fungal effectiveness and our estimates for the kinetic parameters of P uptake show that maximum influx can be considerably higher in hyphae than in roots.

An improved plant P acquisition will depend on the development of cropping systems which allow for the shortest possible distance of P transport from soil P sources to plant uptake systems. This includes breeding for higher root lengths and many long root hairs and high susceptibility to the formation of mycorrhizas. Soil conditions should be conducive to the development of an extensive mycelium of mycorrhizal fungi and the fungi involved should have the highest possible V_{max}. In contrast, changing the plant kinetic parameters of P uptake, either by plant breeding or by engineering of root P transporters, can be expected to have only relatively small effects on P uptake from soil. Future research at the physiological as well as molecular level should clarify the mechanisms whereby the mycorrhizal contribution to plant P nutrition is influenced by the environment and by the host genotype.

9. References

Abbott, L. and Robson, A.D.(1981). Infectivity and effectiveness of vesicular arbuscular mycorrhizal fungi: Effect of inoculum type. *Aust.J.Agric.Res.* **32**, 631-639.

Abbott, L.K. and Robson, A.D.(1985). Formation of external hyphae in soil by four species of vesicular-arbuscular mycorrhizal fungi. *New Phytol.* **99**, 245-255.

Abbott, L.K., Robson, A.D. and Boer, G.D.(1984). The effect of phosphorus on the formation of hyphae in soil by the vesicular-arbuscular mycorrhizal fungus, *Glomus fasciculatum*. *New Phytol.* **97**, 437-446.

Bago, B., Bentivenga, S.P., Brenac, V., Dodd, J.C., Piche, Y. and Simon, L.(1998). Molecular analysis of *Gigaspora* (Glomales, Gigasporaceae). *New Phytol.* **139**, 581-588.

Barber, S.A.(1984). *Soil Nutrient Bioavailability - a mechanistic approach*, Wiley-Interscience, New York, 398pp.

Barraclough, P.B.(1986). The growth and activity of winter wheat roots in the field: nutrient inflows of high yielding crops. *J. Agric. Sci* **106**, 53-59.

Bolan, N.S.(1991). A critical review on the role of mycorrhizal fungi in the uptake of phosphorus by plants. *Plant Soil* **134**, 189-207.

Bååth, E. and Hayman, D.S.(1984). Effect of soil volume and plant density on mycorrhizal infection and growth response. *Plant Soil* **77**, 373-376.

Caldwell, M.M., Dudley, L.M. and Lilieholm, B.(1992). Soil solution phosphate, root uptake kinetics and nutrient acquisition: implications for a patchy environment. *Oecologia* **89**, 305-309.

Carling, D.E. and Brown, M.F.(1980). Relative effect of vesicular-arbuscular mycorrhizal fungi on the growth and yield of soybean. *Soil Sci.Soc.Am.J.* **44**, 528-532.

Clipson, N.J.W., Cairney, J.W.G. and Jennings, D.H.(1987). The physiology of basidiomycete linear organs I. Phosphate uptake by cords and mycelium in the laboratory and the field. *New Phytol.* **105**, 449-457.

Cress, W.A., Throneberry, G.O. and Lindsey, D.L.(1979). Kinetics of phosphorus absorption by mycorrhizal and nonmycorrhizal tomato roots. *Plant Physiol.* **64**, 484-487.

Cui, M.Y. and Caldwell, M.M.(1996). Facilitation of plant phosphate acquisition by arbuscular mycorrhizas from enriched soil patches .2. Hyphae exploiting root- free soil. *New Phytol.* **133**, 461-467.

Darrah, P.R.(1993). The rhizosphere and plant nutrition - a quantitative approach. *Plant Soil* **156**, 1-20.

Dodd, J.C., Rosendahl, S., Giovannetti, M., Broome, A., Lanfranco, L. and Walker, C.(1996). Inter- and intraspecific variation within the morphologically similar arbuscular mycorrhizal fungi *Glomus mosseae* and *Glomus coronatum*. *New Phytol.* **133**, 113-122.

Edwards, O.W. and Huffman, E.O.(1959). Diffusion of aqueous solutions of phosphoric acid at 25°C. *J.Phys.Chem.* **63**, 1830-1833.

Ernani, P.R., Santos, J.C.P., Kaminski, J. and Rheinheimer, D.S.(1994). Prediction of phosphorus uptake by a mechanistic model in a low phosphorus highly weathered soil as affected by mycorrhizae inoculation. *J. Plant Nutr.* **17**, 1067-1078.

Feldmann, F.(1998). The strain-inherent variability of arbuscular mycorrhizal effectiveness: II. Effectiveness of single spores. *Symbiosis* **25**, 131-143.

Fitter, A.H.(1985). Functioning of vesicular-arbuscular mycorrhizas under field conditions. *New Phytol.* **99**, 257-265.

George, E., Marschner, H. and Jakobsen, I.(1995). Role of arbuscular mycorrhizal fungi in uptake of phosphorus and nitrogen from soil. *Critic. Rev. Biotech.* **15**, 257-270.

Gilmore, A.E.(1971). The influence of endotrophic mycorrhizae on the growth of peach seedlings. *J. Am. Soc. Hort. Sci.* **96**, 35-38.

Grant, R.F. and Robertson, J.A.(1997). Phosphorus uptake by root systems: Mathematical modelling in ecosys. *Plant Soil* **188**, 279-297.

Harrison, M.J. and van Buuren, M.L.(1995). A phosphate transporter from the mycorrhizal fungus *Glomus versiforme*. *Nature* **378**, 626-629.

Jakobsen, I.(1995). Transport of phosphorus and carbon in VA mycorrhizas. In *Mycorrhiza: Structure, Function, Molecular Biology and Biotechnology* (Varma, A. and Hock, B., Eds.) Springer-Verlag, Berlin. 297-324.

Jakobsen, I., Abbott, L.K. and Robson, A.D.(1992b). External hyphae of vesicular-arbuscular mycorrhizal fungi associated with *Trifolium subterraneum*. L. 1: Spread of hyphae and phosphorus inflow into roots. *New Phytol.* **120**, 371-380.

Jakobsen, I., Abbott, L.K. and Robson, A.D.(1992a). External hyphae of vesicular-arbuscular mycorrhizal fungi associated with *Trifolium subterraneum* L. 2. Hyphal transport of ^{32}P over defined distances. *New Phytol.* **120**, 509-516.

Jakobsen, I., Smith, F. A., Schweiger, P. F. and Smith, S. E. (1998). The influence of arbuscular mycorrhizal fungi on phosphate uptake by roots. *Noble Foundation Plant Biology – 10 year Symposium* Abstract

Jensen, A.(1982). Influence of four vesicular arbuscular mycorrhizal fungi on nutrient uptake and growth in barley *(Hordeum vulgare)*. *New Phytol.* **90**, 45-50.

Jolicoeur, M., Germette, S., Gaudette, M., Perrier, M. and Becard, G.(1998). Intracellular pH in arbuscular mycorrhizal fungi - A symbiotic physiological marker. *Plant Physiol.* **116**, 1279-1288.

Jungk, A., Asher, C.J., Edwards, D.G. and Meyer, D.(1990). Influence of phosphate status on phosphate uptake kinetics of maize and soybean. *Plant Soil* **124**, 175-182.

Karunaratne, R.S., Baker, J.H. and Barker, A.V.(1986). Phosphorus uptake by mycorrhizal and nonmycorrhizal roots of soybean. *J. Plant Nutr.* **9**, 1303-1313.

Li, X.L., George, E. and Marschner, H.(1991). Phosphorus depletion and pH decrease at the root-soil and hyphae-soil interfaces of VA mycorrhizal white clover fertilised with ammonium. *New Phytol.* **119**, 397-404.

Lloyd-MacGilp, S.A., Chambers, S.M., Dodd, J.C., Fitter, A.H., Walker, C. and Young, J.P.W.(1996). Diversity of the ribosomal internal transcribed spacers within and among isolates of *Glomus mosseae* and related mycorrhizal fungi. *New Phytol.* **133**, 103-111.

Mimura, T., Reid, R.J. and Smith, F.A.(1998). Control of phosphate transport across the plasma membrane of *Chara corallina*. *J. Exp. Bot.* **49** , 13-19.

Mosse, B.(1972). The influence of soil type and *Endogone* strain on the growth of mycorrhizal plants in phosphate deficient soils. *Rev. Ecol. Biol. Sol.* **9**, 529-537.

Nielsen, N.E.(1976). A transport kinetic concept for ion uptake by plants III. Test of the concept by results from water culture and pot experiments. *Plant Soil* **45**, 659-677.

Nye, P.H. and Tinker, P.B.(1977). *Solute movement in the soil-root system*, First edition. Blackwell, Oxford.

O'Keefe, D.M. and Sylvia, D.M.(1992). Chronology and mechanisms of P uptake by mycorrhizal sweet potato plants. *New Phytol.* **122**, 651-659.

Parfitt, R.L.(1979). The availability of P from phosphate-goethite bridging complexes. Desorption and uptake by ryegrass. *Plant Soil* **53**, 55-65.

Pearson, J.N. and Jakobsen, I.(1993). The relative contribution of hyphae and roots to phosphorus uptake by arbuscular mycorrhizal plants, measured by dual labelling with ^{32}P and ^{33}P. *New Phytol.* **124**, 489-494.

Ravnskov, S. and Jakobsen, I.(1995). Functional compatibility in arbuscular mycorrhizas measured as hyphal P transport to the plant. *New Phytol.* **129**, 611-618.

Rosendahl, S. and Taylor, J.W.(1997). Development of multiple genetic markers for studies of genetic variation in arbuscular mycorrhizal fungi using AFLP™. *Mol. Ecol.* **6**, 821-829.

Sanders, F.E. and Tinker, P.B.(1973). Phosphate flow into mycorrhizal roots. *Pestic. Sci.* **4**, 385-395.

Sanders, F.E., Tinker, P.B., Black, R.L.B. and Palmerley, S.M.(1977). The development of endomycorrhizal root systems: I. Spread of infection and growth-promoting effects with four species of vesicular-arbuscular endophyte. *New Phytol.* **78**, 257-268.

Schweiger, P. F. (1993). *Factors affecting VA mycorrhizal uptake of phosphorus*.Thesis, 127pp.

Seeling, B. and Claassen, N.(1990). A method for determining Michaelis-Menten kinetic parameters of nutrient uptake for plants growing in soil. *Z.Planzenernahr.Bodenk.* **153**, 301-303.

Silberbush, M. and Barber, S.A.(1983). Sensitivity of simulated phosphorus uptake to parameters used by a mechanistic-mathematical model. *Plant Soil* **74**, 93-100.

Stahl, P.D., Christensen, M. and Williams, S.E.(1990). Population variation in the mycorrhizal fungus *Glomus mosseae:* uniform garden experiments. *Mycol. Res.* **94**, 1070-1076.

St John, T.V., Coleman, D.C. and Reid, C.P.P.(1983). Association of vesicular-arbuscular mycorrhizal hyphae with soil organic particles. *Ecology* **64**, 957-959.

Syring, K.M. and Claassen, N.(1995). Estimation of the influx and the radius of the depletion zone developing around a root during nutrient uptake. *Plant Soil* **175**, 115-123.

Thomson, B.D., Clarkson, D.T. and Brain, P.(1990). Kinetics of phosphorus uptake by the germ-tubes of the vesicular-arbuscular mycorrhizal fungus, *Gigaspora margarita. New Phytol.* **116**, 647-653.

Trouvelot, S., Hijri, M., Van Tuinen, D., Dulieu, H. and Gianinazzi-Pearson, V.(1998). Characterization of internuclear variability in glomalean fungi by fluorescent *in situ* hybridization of ribosomal DNA loci. *Second International Conference on Mycorrhizas (ICOM2), Uppsala, Sweden* **Abstracts**, 169

Williams, R.F.(1948). The effect of phosphorus supply on the rates of intake of phosphorus and nitrogen and upon certain aspects of phosphorus metabolism in gramineous plants. *Aust.J.Sci.Res.(B)* **1**, 333-361.

MEMBRANES AND NUTRITION: OPPORTUNITIES FOR INTEGRATION AND PROGRESS

F.A. SMITH[1], S.E. SMITH[2] AND R.J. REID[3]
Centre for Plant Membrane Biology and
Departments of [1]Botany, [2]Soil Science and [3]Plant Science
The University of Adelaide, SA 5005, Australia

1. Introduction

Over the last 100 years massive effort in agronomy and plant breeding has been put into developing high-yielding crops and their products. Much of the improvement has resulted from breeding to divert photosynthate from sites within the plant that are (apparently) less useful agriculturally to more useful sites. Improved grain yield is an obvious example. In analysing these developments and future prospects, Lester Brown (1997) has emphasized that: *'once plant breeders have pushed genetic yield potential close to the physiological limit, then further advances rely on the expanded use of basic inputs such as... fertilizer.'* Certainly, there has been - at least in developed countries - massive investment in fertilizer production and increased understanding of how fertilizer use can be better managed (Loneragan, 1997). Nevertheless, it seems rather depressing to those working in the field of plant nutrition that it may be necessary to rely primarily on fertilizers further to increase plant productivity (or grain quality), even in developed countries. In developing countries where fertilizer use is minimal, to rely on high input of fertilizers to improve plant productivity seems even more depressing.

It is important to consider whether in fact it is possible to exploit better or improve the efficiency of nutrient acquisition and use by plants (i.e. efficiency in terms of external nutrient levels and/or utilization of photosynthate). In this context, important questions that relate to membrane biology are as follows.

- Can differences in mineral nutrition of plants (including cultivars or mutants) be explained in terms of presence, absence, or differences in regulation of individual membrane transport processes or processes closely linked to transport?
- Can membrane properties of plants be modified (e.g. by genetic engineering) to change usefully the mineral nutrition and productivity of the plant?

Further - and this is important:

- Can aspects of root morphology that relate to nutrient acquisition from soil also be exploited?

Our aim in considering these questions is to provide a bridge between detailed consideration of individual membrane transport processes and broader aspects of plant nutrition and productivity. We will discuss some of the problems and focus on areas where there is increasing scope for integration of experimental approaches.

G. Gissel-Nielsen and A. Jensen (eds.), Plant Nutrition – Molecular Biology and Genetics, 291–301.

2. Membrane Transport and Plant Nutrition

2.1. OVERVIEW

Concentrations of mineral nutrients in a plant are determined by a large range of processes that involve cell membranes. Key processes are of course uptake into roots and transport to the shoot. Compartmentation of nutrients within cells occurs throughout the plant. There is some loss from shoots by leaching and (in relatively few plants) through specialized glands (e.g. for Na^+ and Cl^-). In even fewer cases there are glands in shoots for uptake of nutrients (as in some epiphytes), but some absorption can occur through unspecialized shoot epidermes if nutrients are applied in appropriate form. Some nutrients are redistributed within the shoot or transported back down to the roots, involving loading and unloading processes in conducting tissues. Most plants have mycorrhizal associations of one sort or another that are very important for uptake of nutrients such as P, Zn, or sometimes N in return for photosynthate provided to the fungal partner. Others - but relatively few - have symbiotic N_2-fixing associations involving roots or sometimes shoots. In all of these cases there is movement of the nutrients through cell membranes at rates that the plant controls - except possibly with leaching or other loss from damaged tissues. In addition, mobilization and acquisition of nutrients from the bulk soil can also involve plant membrane activity, including excretion of H^+, organic acids, siderophores, enzymes, etc. Excretion of organic material provides substrates for free-living microorganisms in the rhizosphere that increase nutrient availability to plants.

It follows that to understand differences in mineral nutrition between plants, whether genetically related or otherwise, there are very many processes to be unravelled that relate directly or indirectly to membranes. Further, the full range of structural molecules must be considered. There are different types of phospholipids that can influence membrane fluidity (e.g. as a function of temperature) and the permeability to small molecules moving in or out of cells. Genotypic differences in uptake of boron by barley that are associated with different tolerances to boron deficiency or toxicity are very relevant in this context (Nable *et al.*, 1990). Boron is thought to enter the root mainly by simple diffusion of boric acid and not by a catalyzed transport process. As far as we are aware, it is not yet clear whether it is simply different structural properties of plasma membranes of the barley varieties that result in apparently very different permeabilities to boron - or whether there are more subtle physiological features.

Sterols, also known to have a role in determining membrane fluidity, may regulate some membrane enzymes (Hartmann, 1998). Carbohydrates occur as components of glycoproteins, which are part of a complex system of antenna molecules that sense both external and internal signals. These are an active field of research beyond the scope of this paper (see Hahn, 1996). Membrane-associated protein kinases are also involved in trans-membrane signalling (Harmon *et al.*, 1996). Other important proteins include of course the various types of transport proteins on which much research focuses.

2.2. CLASSES OF TRANSPORT PROTEINS

Transport proteins fall into three main classes: membrane ATPases, co-transporters and channels. The best known ATPases in terms of structure and function are the H^+-

ATPases. Families composed of several genes, with different specialized locations and functions, are known from some plants (Michelet and Boutry, 1995). H^+-ATPases help control both intracellular and extracellular (e.g. apoplastic) pH and generate trans-membrane electric potential differences. Thus, they have an important role in energizing membranes for other transport processes. Ca^{2+}-ATPases help to regulate cytoplasmic Ca^{2+}, which is in turn involved in many developmental processes. Broad-range 'ABC'-ATPases are a newly discovered class involved in transport of complex solutes to vacuoles. Co-transporters in plants predominantly use H^+ to drive active transport of other solutes. (The use of Na^+ as the driver in exceptional cases is mentioned below). There are many examples involved in uptake of nutrients such as phosphate (Pi), NO_3^- and K^+. Within the plant there are co-transporters for amino-acids and sugars. Channels allow rapid uptake of ions such as Ca^{2+}, and (again) K^+ and efflux of the latter. Channels will allow uptake of cations such as NH_4^+ and of course Na^+ at relatively high concentrations externally. A channel that allows efflux of Pi from roots has been found in ryegrass and white clover (Dunlop and Thai Phung, this conference).

Measurements of nutrient transport with isotopic tracers and more recently, patch-clamping and molecular studies, have provided a great deal of information about some transporters. However, very little is known about others, such as those for most micronutrients: it is not known whether there are specific transporters or whether they can 'slip through' other transporters, such as channels for Ca^{2+}. Comparisons with transporters in other systems, and especially yeast, are proving very informative (e.g. for Mg^{2+}: see MacDiarmid *et al.*, this conference). It is wise to bear in mind that nature might produce some surprises in terms of mechanisms that do not fall neatly within the three individual categories. Thus, the freshwater charophyte *Chara corallina* has both H^+-linked and Na^+-linked co-transporters (see Walker *et al.*, 1989). One of these, a Na^+-Pi co-transporter that is induced by P-starvation, apparently requires more energy than can be provided by the Na^+ influx alone. The form in which this energy is provided is not known (Reid *et al.*, 1997 and unpublished results). Also in *Chara*, the NH_4^+ transporter (a high-affinity uniporter that also transports $CH_3NH_3^+$) shows properties of both a channel and a carrier. In the absence of detailed measurements of stoichiometry of amine transported to electric charge transported it might have been wrongly concluded that the mechanism was a H^+ co-transporter (Walker *et al.*, 1979*a*, 1979*b*). The mechanism of the high-affinity NH_4^+ transporters in higher plants remains an open question despite the great advances made with molecular biology (W. B. Frommer, personal communication). Lastly, a possible role for water channels (aquaporins) in uptake of B as boric acid and Si as silicic acid (see Rafi and Epstein, this conference) remains a distinct possibility that might help explain the large differences in rates of uptake between different plants.

The different classes of transport proteins and their properties have relevance to many practical aspects of plant nutrition. In thinking about improving efficiency of uptake of micronutrients and perhaps overcoming soil deficiencies we must think in terms of high-affinity systems that cope with very low external concentrations but may be vulnerable to toxic ions such as Pb^{2+}, Hg^{2+} and Ni^{2+}, also at low concentrations. In thinking about overcoming sensitivity of plants to high salinity we have to bear in mind a large range of transporters - both high-affinity and low-affinity ones - involved in Na^+ uptake in competition with other cations such as K^+ and involved also in movement of

Na^+ within the plant. It is therefore hardly surprising that the molecular basis of Na^+ sensitivity versus tolerance is so difficult to unravel.

2.3. THE MOLECULAR DIMENSION: MULTIPLE TRANSPORT SYSTEMS

Since there are only 16 or so nutrients that are necessary for plant growth but hundreds if not thousands of genes for transport proteins, we can expect that plants have multiple systems for transport of many nutrients. This is turning out to be the case with major nutrients such as Pi (see Schachtman *et al.*, 1998), various N compounds, both inorganic and organic (von Wirén *et al.*, 1997), and K^+ (see below). Molecular studies of such transporters are rapidly extending the many kinetic studies (i.e. tracer fluxes and patch-clamping) that have been carried out. The latter - despite problems of interpretation - indicated duality (at least) of transport systems for many nutrients, i.e. the occurrence of high- and low-affinity transporters. It is essential for the different approaches to be integrated if the physiological significance of individual transport genes is to be resolved. For example, Dunlop *et al.* (1997) have shown that the operation of individual Pi transporters is blurred when Pi influx into wild-type *Arabidopsis* is measured over a range of external concentrations. This type of study gives a good foundation for measurements of uptake of Pi and other nutrients into *Arabidopsis* mutants.

Work with K^+ is showing that there is danger in ascribing too much significance to individual transport genes or even gene families. The role of the high affinity transporter for K^+ (HKT1) that was shown to be a K^+-Na^+ co-transporter (Schachtman and Schroeder, 1994; Rubio *et al.*, 1995) is still obscure. It does not seem to be the major transporter for K^+ uptake even into the plant from which the gene was obtained (Maathuis *et al.*, 1996; Walker *et al.*, 1996). Other genes for high-affinity K^+ transport are now known, as well as genes for low affinity K^+ transport (Santa-Maria *et al.*, 1997; Kim *et al.*, 1998; Fu and Luan, 1998). The existence of multiple transporters is likely to provide functional buffering for the plant that may complicate or frustrate attempts to change plant nutrition by manipulating gene expression experimentally.

2.4. KINETIC CHARACTERISTICS OF TRANSPORT: SOME COMPLICATIONS

Much work on the efficiency of nutrient uptake has focussed on the kinetic characteristics of uptake and especially the relative importance of parameters such as K_m and V_{max} (see Marschner, 1995 for a general review). As we have already shown, physiological concepts of high- and low-affinity transport are well entrenched despite uncertainty about what they tell us about the molecular nature of the transporters involved. The concept of 'C_{min}' is also important, i.e. the external level of nutrient below which there is no net uptake (Marschner, 1995). There are, however, many problems in analyzing uptake in these terms and then applying them to the plant growing in soil. One is that the concentration of a nutrient immediately adjacent to the transport protein ('substrate concentration' - C_o) may not be the same as that in the external medium. There may be restricted access - for example due to the presence of an apoplastic phase that produces diffusion limitations. Ionizable groups in the cell wall are relevant here. These adsorb cationic nutrients and give cation exchange capacity (CEC), in addition to which there can be specific chemical binding that can affect mobility of ions within the apoplast and hence uptake into cells. The CEC is affected by the local pH - itself a property that can be modified by membrane activity via H^+ fluxes. In addition, membrane molecules have ionizable head groups and hence surface charge. This charge will also determine the levels of ions immediately adjacent to the transport proteins. Surface charge can be affected by pH and presence of ions such as Ca^{2+}. This is an aspect often ignored in work with plants as opposed to work with animals (but see Clarkson, 1991).

The transporter being investigated will be particularly inaccessible to the bulk solution if it is located in cortical cells rather than the root hair or epidermis. Free access of nutrients to surfaces inside the epidermis may increase if external concentrations are high but the presence of a hypodermis may be an additional barrier (Peterson, 1988). All of the features that restrict access of nutrients to transport sites on membranes will lower the apparent affinity of transport and hence the measured K_m. Thus the simple approach of defining transporters by the K_m of the nutrient transported as measured using the concentration in a bulk solution outside the root can be very unrealistic.

Another confounding factor in measuring nutrient influx into intact roots is of course that there is variation along the root surface. As shown by Clarkson *et al.* (1978), high influx of Pi extends further along the roots of P-deficient barley than with P-sufficient plants. Results such as these reflect a range of issues that relate to location and regulation of transporters. It is in resolving these types of physiological results, which are obviously important in understanding plant nutrition, that molecular biology is becoming increasingly useful, e.g. with immunolocalization and *in situ* hybridization.

2.5. REGULATION OF MEMBRANE TRANSPORT

It has been known for many years that uptake of nutrients into roots is controlled directly or indirectly by the nutrient status of the plant. Marschner (1995) gave a classical view of what can loosely be called internal feedback mechanisms. Relevant processes (see Marschner's Fig. 2.23) that contribute to the levels of nutrients in roots and might be involved in feedback can be listed as follows:
(i) influx - affected via changes in affinity, numbers or activity of transporter molecules;
(ii) efflux - affected via changes (as above) in transporter(s) for efflux;
(iii) transport to or from the vacuole;
(iv) transformation and incorporation within the root (as with N, P and S);
(v) transport to the shoot;
(vi) feedback on (v) from 'shoot demand', possibly involving recycling from the shoot;
(vii) deficiency-enhanced excretion of solutes that can increase influx: see (i).
To this list we can add:
(viii) supply of nutrients from symbioses (mycorrhizas or N_2 fixers).

Until quite recently, feedback was thought to involve allosteric control of pre-existing transport proteins by cytoplasmic nutrients or (with P, N or S) by immediate organic derivatives: see Marschner (1995)[1]. In fact, changes in transport rates are

[1]This outstanding book was revised in 1993, at the end of the era 'BC' - 'before cloning'. It is a tragedy that Horst Marschner is not still with us to witness and contribute to the exciting new developments in the field of plant nutrition - and to produce a new edition!

increasingly being shown to result from changes in expression of genes for transport, as with induction of high affinity transporters in nutrient-stressed roots (see below). However, it is not known whether the cytoplasmic concentration of the nutrient within the root is an important regulatory factor - as is often assumed - or whether there is independent signalling that somehow relates to levels or demand elsewhere. For example, the cytoplasmic concentration of Pi is held within very narrow limits and it has been proposed that vacuolar Pi in roots can sometimes determine influx across the plasma membrane (see Marschner, 1995 and Schachtman *et al.*, 1998 for discussion and references). In individual *Chara* cells, neither cytoplasmic or vacuolar Pi changes during P starvation but Pi influx is much higher than in non-starved cells. The results suggested that there is a signalling mechanism on the outside of the plasma membrane that allows rapid influx following release from P starvation, overriding any internal controls (Mimura *et al.*, 1998).

More and more, there is emphasis on control of xylem loading by signals from shoot to root, not surprisingly as the shoot is the sink for many of the nutrients and is of course the supplier of energy for uptake. An excellent example of shoot/root interactions is the study by Liu *et al.* (1998) of differential regulation by plant P of transporter genes for Pi in tomato roots (see also Raghothama, this conference). Gene expression was markedly induced by Pi starvation. A split-root system demonstrated similar expression of transport genes in both parts of the root system, irrespective of local Pi supply, via internal signals from shoots. These results are in accord with earlier measurements of Pi influx (e.g. Lefebvre and Glass, 1982; Drew *et al.*, 1984).

The same principles that apply to regulation of Pi uptake can be seen to apply to other nutrients, though evidence is more fragmentary. For example, K^+ uptake into excised barley roots is strongly influenced by the internal content of the roots (Glass and Dunlop, 1979). In contrast, K^+ uptake into intact maize plants is related to the K^+ concentration in the shoots and not that in roots (Barber, 1979). This was demonstrated with a split-root system in which K^+ uptake was the same into two parts of the root system with very different K^+ concentrations, depending on whether or not they had previously been supplied with K^+ (see Table 1). The inference is again that some signal reflecting shoot demand is transmitted to roots, overriding K^+ in the roots. It would be nice to see this experimental system applied to the study of K^+ transport gene expression, as has been done with Pi (Liu *et al.*, 1998).

TABLE 1. Effects of growth in a split-root system on K^+ content of maize: different proportions of roots grown in solution with (+) or without (-) K^+ Plants were grown 17 days before measurement of Vmax (Barber, 1979).

Roots grown + or -K^+ (%)	Shoot K^+ (% dry wt)	Root K^+ (% dry wt)	Vmax (pmol cm^{-1} s^{-1})
100+	8.0	5.9	15.8
50+	6.5	5.6	28.0
50-		1.9	30.8
15+	4.1	5.5	36.8
85-		1.2	38.0

It is feasible that nutrients which are relatively mobile within the plant, such as Pi and K^+, might contribute to shoot-to-root signalling via recycling in phloem followed by a sensing mechanism that allows feedback to the transporters for uptake into the root. The same might apply for N, with amino-N moving down the phloem. However, this type of long distance signalling does not seem feasible for nutrients that are immobile within shoots. A possible role for signalling via proteins (see Lucas, 1997) or other macromolecules in phloem has to be considered seriously for all nutrients, whether mobile or otherwise. In any case, to emphasize xylem loading as a key site in the regulation (e.g. Drew *et al.*, 1984; Wegner and De Boer, 1997) does not clarify how uptake into the root is regulated.

2.6. IMPROVING PLANT MEMBRANE FUNCTION

We can return now to the general questions that we asked in the Introduction. First:

- can differences in mineral nutrition of plants (including cultivars or mutants) be explained in terms of presence, absence, or differences in regulation of individual membrane transport processes or processes closely linked to transport?

Recent molecular studies are indeed starting to explain differences in mineral nutrition of plants in terms of presence, absence or up- or-down regulation of individual membrane transport processes - especially with Pi. While the number of plants studied in detail is limited (e.g. *Arabidopsis*, barley and tomato) we can expect some rapid advances in other plants.

Second:

- can membrane properties of plants be modified (e.g. by genetic engineering) to change usefully the mineral nutrition of the plant?

At present, membrane components (e.g. transport proteins) can only be modified to a very limited extent but this situation too may soon change. However, the interrelations between multiple transporters at the same location (e.g. root epidermis) and transporters at different locations (e.g. root epidermis, cortex and xylem parenchyma) will have to be taken into account. Understanding the nature and role of regulatory factors (transcriptional, translational and post-translational) will be essential. Until these interactions are understood, *useful* manipulation of transport genes (in terms of improving plant nutrition and productivity) seems an unrealistic goal. Interestingly, enhanced efficiency of boron use in transgenic tobacco has been achieved not via a membrane transport process but by increasing B mobility within phloem (see Brown *et al.*, this conference). The issue of regulation of membrane transport is also important in studies involving transport gene function in heterologous systems such as yeast mutants or oocytes, because absence of regulatory proteins or protein sub-units from the plant can give incomplete or misleading information about the properties of the transporter in the plant.

3. Larger Dimensions: Roots and Mycorrhizas

The last general question that we asked was:

- can aspects of root morphology that relate to nutrient acquisition from soil also be exploited?

The answer has to be a strong 'yes'. These aspects include growth rates of roots, root architecture including thickness and length, numbers and length of root-hairs, and distribution patterns of roots in soil. They all relate to membrane activity in that the amount of root material produced affects the amount of transport activity that is possible per plant. Also, we must remember the molecules excreted from the root to help overcome P deficiency, Al toxicity, etc. Plasticity in response to environmental factors (e.g. different levels of soil nutrients) includes changes in branching patterns and shoot/root ratios and occurs more in some plants than in others. There is sometimes a strong genetic influence even at the cultivar level (see Lynch, 1995; Delhaize and Ryan, 1995). Hence there is great opportunity for genetic manipulation to improve plant nutrition and productivity. But there is also a danger in that inherent plasticity may confound attempts to improve mineral nutrition by manipulating transport genes alone - as would be the case if the plant with more efficient transporters responded by growing shorter roots. A relevant example of plasticity in action is the response of *Arabidopsis* transformed with an anti-sense construct that decreased the formation of water channels (Kaldenhoff *et al.*, 1998). Decreased permeability to water (about 33% of wild-type) led to no change in size of shoots but a 500% increase in size of roots. It can be deduced that uptake of nutrients per unit length of root was probably decreased compared with the wild-type, but transport to the shoot was unchanged (because shoot growth was unaffected) and that transfer of organic C to roots was greatly increased (to produce the much larger roots).

The relative importance of transport parameters (e.g. K_m or V_{max}) versus other factors associated with root architecture will depend on the individual nutrient - a point well made in the modelling done by Barber and associates (e.g. Claassen and Barber, 1976; Silberbush and Barber, 1983). Thus, uptake of P and other immobile nutrients may be influenced more by root architectural properties than by K_m or V_{max} of the transporter(s). With more mobile nutrients such as K^+ and NO_3^-, the transport parameters will be relatively more important (Silberbush and Barber, 1984). This is something to be borne in mind if manipulation of genes to benefit the nutrition of plants growing in soil is to be a worthwhile exercise. Rather than attempting to engineer transporters that might lower C_{min} (e.g. via a lower K_m) it may be more profitable to focus on increasing the number or length of root-hairs.

We must remember here the extra dimension of mycorrhizas that can help overcome some of the limitations of immobility of nutrients such as P or Zn. This is done via the extensive external fungal hyphae with their own suites of transporters (and hence their own K_m and V_{max}), and other suites of transporters at the various fungal-plant interfaces (see Smith and Read, 1997). Mycorrhizas have a cost to the plant in terms of organic C provided to the fungus. Whether this is a serious cost depends to some extent again on plasticity of root development because some plants reduce root growth in response to the mycorrhizal colonization that allows more efficient exploitation of the soil. Additional plasticity in response comes from the ability of some plants to reduce mycorrhizal colonization when external nutrient levels are high, i.e. when the mycorrhizal fungus is 'not needed' by the plant. This subject goes far beyond the scope of this review but it is worth noting that there are strong genetic elements in growth responses of plants and their cultivars to colonization by individual mycorrhizal fungi. Agricultural plant breeding programs ignore the mycorrhizal dimension and have generally been done in the high fertilizer environment in which mycorrhizal benefits are minimized (Smith *et al.* 1992; see also Smith *et al.*, this conference). This is almost certainly to the detriment of improved plant productivity in developing countries. Intriguingly, many tropical plants of economic importance show particularly high growth responses to mycorrhizal colonization. In the context of root architecture, as well as of membrane biology, symbiotic N_2 fixation deserves separate consideration which - due to lack of space and expertise - we have not attempted.

4. Conclusions

First, it is worth noting how developments in techniques within the last 40 years have enabled the investigation of membrane transport to evolve from the whole plant or organ level (using chemical analyses to measure net uptake and distribution within organs) progressively to the cellular and sub-cellular level with the intensive use of radio-isotopes and electrophysiology dating from the 1960's and 70's, advanced biochemistry with subcellular preparations (e.g. protoplasts and vesicles) and more recently patch-clamping and molecular techniques. Very importantly, the downward trend - in terms of the level of organisation - is now being reversed as molecular techniques are combined with structure and anatomy (e.g. immunolocalization and *in situ* hybridization). Laser scanning confocal microscopy is playing a very useful part in linking function to structure at the cellular level and beyond. Thus we do now have vastly improved opportunities for integrating approaches to this field and hence to understanding the molecular and physiological bases of plant nutrition.

Lastly, returning to Lester Brown's belief in the necessity for greater input of fertilizer to improve plant productivity in future, we do hope that useful advances in plant mineral nutrition can be made that will reduce this need and that greater emphasis will be put on maximizing nutrient efficiency in its various forms for individual nutrients. However, we will always have to bear in mind the realities of biodiversity in function - i.e. the diversity of strategies for nutritional acquisition in different plants and the various regulatory processes about which we know so little. In these respects, the extent to which *Arabidopsis* is a useful model for maize or mango remains to be seen!

Acknowledgements

Our own experimental work is supported by the Australian Research Council.

References

Barber, S.A. (1979) Growth requirements for nutrients in relation to demand at the root surface, in J.L. Harley and R. Scott-Russell (eds.), *The Soil-Root Interface*, Academic Press, London, pp. 5-20.

Brown, L.R. (1997). Facing the challenge of food scarcity, in T. Ando, K. Fujita, T. Mae, H. Matsumoto, S. Mori and J. Sekija (eds.), *Plant Nutrition for Sustainable Food Production and Environment*, Kluwer Academic Publishers, Dordrecht, pp. 15-24.

Claassen, N and Barber, S.A. (1976) Simulation models for nutrient uptake from soil by a rapidly growing plant root system. *Agronomy Journal* **68**, 961-964.

Clarkson, D.T. (1991) Root structure and sites of ion uptake, in W.Y. Waisel, A. Eshel and Y. U. Kafkafi (eds.), *Plant Roots: The Hidden Half*, M. Dekker, New York, pp. 417-453.

Clarkson, D.T., Sanderson, J., and Scattergood, C.B. (1978) Influence of phosphate stress on phosphate absorption and translocation by various parts of the root system of *Hordeum vulgare* L. (Barley). *Planta* **139**, 47-53.

Delhaize, E. and Ryan, P.R. (1995) Aluminium toxicity and tolerance in plants. *Plant Physiol*ogy **107**, 315-321.

Drew, M.C., Saker, L.R., Barber, S.A. and Jenkins, W. (1984) Changes in the kinetics of phosphate and potassium absorption in nutrient-deficient barley roots measured by a solution-depletion technique. *Planta* **160**, 490-499.

Dunlop, J., Phung, H.T., Meeking, R. and White, D.W.R. (1997) The kinetics associated with phosphate absorption and its regulation by phosphorus status. *Australian Journal of Plant Physiology* **24**, 623-629.

Fu, H.-H. and Luan, S. (1998) A dual-affinity K^+ transporter from Arabidopsis. *Plant Cell* **10**, 63-73.

Glass, A.D.M. and Dunlop, J. (1979) The regulation of K^+ influx in excised barley roots. Relationship between K^+ influx and electrochemical potential differences. *Planta* **145**, 395-397.

Hahn, M.G. (1996) Signal perception at the plasma membrane: binding proteins and receptors, in M. Smallwood, J.P Knox and D.J. Bowles (eds.), Membranes: Specialized Functions in Plants, BIOS Scientific Publishers, Oxford, pp. 103-136.

Harmon, A.C., Lee, J.-Y., Yoo, B.C. and Shao, J. (1996) Plant membrane-associated protein kinases: great communicators, in M. Smallwood, J.P Knox and D.J. Bowles (eds.), Membranes: Specialized Functions in Plants, BIOS Scientific Publishers, Oxford, pp. 147-150.

Hartmann, M.-A. (1998) Plant sterols and the membrane environment. *Trends in Plant Science* **3**, 170-175.

Kaldenhoff, R., Grote, K., Zhu, J.J and Zimmermann, U. (1998) Significance of plasmalemma aquaporins for water transport in *Arabidopsis*. *Plant Journal* **14**, 121-128.

Kim, E.J., Kwak, J.M., Uozumi, N. and Schroeder, J M. (1998) *AtKUP1*: an Arabidopsis gene encoding high-affinity potassium transport activity. *Plant Cell* **10**, 51-62.

Lefebvre, D.D. and Glass, A.D.M. (1982) Regulation of phosphate influx in barley roots: effects of phosphate deprivation and reduction of influx with provision of orthophosphate. *Physiologia Plantarum* **54**, 199-206.

Liu, C., Muchhal, U.S., Uthappa, M., Kononowicz and Raghothama, K. (1998). Tomato phosphate transporter genes are differentially regulated in plant tissues by phosphorus. *Plant Physiology* **116**, 91-99.

Loneragan, J.F. (1997) Plant nutrition in the 20th and perspectives for the 21st century, in T. Ando, K. Fujita, T. Mae, H. Matsumoto, S. Mori and J. Sekija (eds.), *Plant Nutrition for Sustainable Food Production and Environment*, Kluwer Academic Publishers, Dordrecht, pp. 3-14.

Lucas, W.J. (1997) Application of microinjection techniques to plant nutrition, in T. Ando, K. Fujita, T. Mae, H. Matsumoto, S. Mori and J. Sekija (eds.), *Plant Nutrition for Sustainable Food Production and Environment*, Kluwer Academic Publishers, Dordrecht, pp. 25-40.

Lynch, J. (1995) Root architecture and plant productivity. *Plant Physiology* **109**, 7-13.

Maathuis, F.J.M., Verlin, D., Smith, F.A., Sanders, D., Fernandéz, J.A. and Walker, N.A. (1996) The physiological relevance of Na^+-coupled K^+ transport. *Plant Physiology* **112**, 1609-1616.

Marschner, H. (1995) *Mineral Nutrition of Higher Plants, 2nd edition,* Academic Press, London.

Michelet, B. and Boutry M (1995) The plasma membrane H^+-ATPase. *Plant Physiology* **108**, 1-6.

Mimura, T., Reid, R.J. and Smith, F.A. (1998) Control of phosphate transport across the plasma membrane of *Chara corallina*. *Journal of Experimental Botany* **49**, 13-19.

Nable, R.O., Lance, R.C.M. and Cartwright, B. (1990) Uptake of boron and silicon by barley genotypes with differing susceptibilities to boron toxicity. *Annals of Botany* **66**, 83-90.

Peterson, C.A. (1988) Exodermal Casparian bands: their significance for ion uptake by roots. *Physiologia Plantarum* **72**, 204-208.

Reid, R.J., Mimura, T. and Smith, F.A. (1997) A high-affinity Na-dependent phosphate transporter induced by P starvation in *Chara*, in T. Ando, K. Fujita, T. Mae, H. Matsumoto, S. Mori and J. Sekija (eds.), *Plant Nutrition for Sustainable Food Production and Environment*, Kluwer Academic Publishers, Dordrecht, pp. 137-138.

Rubio, F., Gassmann, W. and Schroeder, J.I. (1995) Sodium-driven potassium uptake by the plant potassium transporter HKT1 and mutations conferring salt tolerance. *Science* **270**, 1660-1663.

Santa-Maria, G., Rubio, F., Dubcovsky, J and Rodriguez-Navarro, A. (1997) The *HAK1* gene of barley is a member of a large gene family and encodes a high-affinity potassium transporter. *Plant Cell* **9**, 2281-2289.

Schachtman, D.P., Reid, R.J. and Ayling, S.M. (1998) Phosphorus uptake by plants: from soil to cell. *Plant Physiology* **116**, 447-453.

Schachtman, D.P. and Schroeder, J.I. (1994) Structure and transport mechanism of a high-affinity potassium uptake transporter from higher plants. *Nature* **370**, 655-658.

Silberbush, M. and Barber, S.A. (1983) Sensitivity of simulated phosphate uptake parameters used by a mechanistic mathematical model. *Plant and Soil* **74**, 93-100.

Silberbush, M. and Barber, S.A. (1984) Phosphorus and potassium uptake of field-grown soybean cultivars predicted by a simulation model. *Soil Science of America Journal* **48**, 592-596.

Smith, S.E., Robson, A.D. and Abbott, L.K. (1992) Involvement of mycorrhizas in assessment of genetically dependent efficiency of nutrient uptake and use. *Plant and Soil* **146**, 169-179.

Smith, S.E. and Read, D.J. (1997) *Mycorrhizal Symbiosis, 2nd edition*, Academic Press, London.

Walker, N.A., Beilby, M.J. and Smith, F.A. (1979*a*) Amine uniport at the plasmalemma of charophyte cells. I. Current-voltage curves, saturation kinetics, and effects of unstirred layers. *Journal of Membrane Biology* **49**, 29-56.

Walker, N.A., Sanders, D. and Maathuis, F.J.M. (1996) High-affinity potassium uptake in plants. *Science* **273**, 977-978.

Walker, N.A., Smith, F.A. and Beilby, M.J. (1979*b*) Amine uniport at the plasmalemma of charophyte cells. II. Ratio of matter to charge transported and permeability of free base. *Journal of Membrane Biology* **49**, 283-296.

Walker, N.A., Smith, F.A. and McCullough, S.R. (1989) Plasmalemma transport in the life of the cell: *Chara* uses both sodium and protons as chemiosmotic drivers, in J. Dainty, M.I. deMichelis, E. Marre and F Rasi-Caldogno (eds.), *Plant Membrane Transport: the Current Position*, Elsevier, Amsterdam, pp. 19-24.

Wegner, L.H. and De Boer, A.H. (1997) Properties of two outward-rectifying channels in root xylem parenchyma cells suggest a role in K^+ homeostasis and long-distance signaling. *Plant Physiology* **115**, 1707-1719.

von Wiren, N., Gazzarrini, S. and Frommer, W.B. (1997) Regulation of mineral nitrogen uptake in plants, in T. Ando, K. Fujita, T. Mae, H. Matsumoto, S. Mori and J. Sekija (eds), *Plant Nutrition for Sustainable Food Production and Environment*, Kluwer Academic Publishers, Dordrecht, pp. 41-49.

MYCORRHIZAL INVOLVEMENT IN PLANT MINERAL NUTRITION: A MOLECULAR AND CELL BIOLOGY PERSPECTIVE.

S.E. SMITH[1], G. ROSEWARNE[2], S.M. AYLING[1], S.DICKSON[1], D.P. SCHACHTMAN[2], S.J. BARKER[3*] AND F.A. SMITH[2]
The Centre for Plant Root Symbioses and the Departments of Soil Science[1], Botany[2] and Plant Science[3], The University of Adelaide Adelaide 5005, South Australia.

1. Introduction

The importance of vesicular-arbuscular (VA) mycorrhizas as one of a number of plant strategies ensuring effective absorption of phosphate (P) from soil is now widely recognised, but few practical efforts to ensure that their potential is realised in sustainable agricultural production have been made. On the contrary, high use of fertilizers, coupled with plant breeding approaches which have largely ignored mechanisms of nutrient efficiency (including mycorrhizas) have resulted in plant varieties which depend on the maintenance of high soil nutrient status for adequate productivity. There is an increasing awareness that high applications of P fertilizers should not continue unchecked, for several reasons. The finite reserves of rock phosphate need to be used efficiently and high P applications result in wasteful losses through the soil profile. This movement can lead to excessive growth of (sometimes-toxic) algae in rivers and reservoirs and to reduced water quality. High P application also often leads to low colonization of roots by VA mycorrhizal fungi, with consequent reductions in their contribution to nutrient absorption in field situations as well as in plant breeding programs (Smith, Robson and Abbott, 1992).

Agricultural management practices making the most of mechanisms leading to effective absorption and utilization of P by plants, including the contribution of mycorrhizal associations, need to be developed. This will require an understanding of the control of mycorrhizal colonization by the plant, including interactions with environmental variables. In addition, we need to know more about the cellular

* Permanent address: Department of Plant Science, University of Western Australia, Nedlands, WA 6009, Australia.

G. Gissel-Nielsen and A. Jensen (eds.), Plant Nutrition – Molecular Biology and Genetics, 303–310.

mechanisms involved in transfer of nutrients between the symbionts, as these play significant roles in determining the efficiency of the symbiosis in terms of photosynthate expended in the acquisition of P and hence the responsiveness of plants to colonization.

2. Analysis of Colonization and Responsiveness in Host Plants

Quantitative differences between cultivars and genotypes of VA mycorrhizal host plants certainly exist in the extent of both colonization and responsiveness of the plant in terms of growth or nutrient uptake. These variations are of particular interest in programs to enhance nutrient efficiency, but only a few studies have been directed towards analysis of the genetic bases of these quantitative traits. In one example (Baon *et al.*, 1993) responsiveness of barley to P supply was shown to be correlated with responsiveness to mycorrhizal colonization. More work along these lines needs to be done, with awareness of the interesting, but potentially confounding, effects of differences in root length, colonization, fertilizer application and interactions between them. Increased attention to the selection of genotypes for future work to determine the mechanisms that influence nutrient efficiency will also be important. These might include land races and wild relatives which probably still retain high mycorrhizal dependence. Experimental approaches need to compare mycorrhizal effects on genetically similar genotypes which exhibit extremes of tolerance or intolerance of nutrient deficiencies or toxicities, particularly if mechanisms underlying them are known.

2.1. PLANT CONTROL OF COLONIZATION

In nature, the absolute (qualitative) ability of potential host plants to form mycorrhizas with any VA mycorrhizal fungus is generally unspecific and is expressed at the taxonomic level of family or, in a few cases, genus. Colonization involves a number of morphologically well-defined steps which, together with functional modifications, are presumed to be under the control of suites of genes in both symbionts. The ancient origin of the symbioses (probably Devonian, based on occurrence of fossil mycorrhizas and molecular phylogeny of VA mycorrhizal (Glomalean) fungi, see Smith and Read, 1997), probably means that common genes permitting and controlling colonization occur in most land plants. The taxonomic position (and therefore the genetic make-up) of the host plant does have an important influence on the details of mycorrhizal structures formed by these fungi, with two major types (*Arum* and *Paris*) currently recognised (Gallaud, 1905; Smith and Smith, 1996, 1997). The main distinguishing feature is the way in which hyphae grow within the roots of the different plants. In *Arum*-type mycorrhizas intracellular arbuscules arise as lateral branches from intercellular hyphae, whereas in *Paris*-type mycorrhizas they originate from intracellular hyphal coils. Development of arbuscules is always variable and influenced by environmental conditions and, particularly in *Paris*-type plants, may be totally absent. The functional significance of the differences between the two types is currently unknown, but has been the subject of some speculation particularly in relation to the route of carbon transfer from plant to fungus and consequently to possible variations in nutritional efficiency of the symbioses. Most research has centred on *Arum*-type mycorrhizas, as these predominate in crop plants and have come to be viewed as 'typical' VA mycorrhizas, despite the fact that *Paris*-type mycorrhizas probably occur in more higher plant families. Developmental and physiological comparisons are urgently required in order to understand the extent of variation. The economically important Gramineae and Solanaceae are reported to include genera forming both types (Smith and Smith, 1997) and may offer opportunities for detailed investigation in genetically well characterised plants. As more information is gathered, *Arum* and *Paris* types may turn out to be extremes in a continuum of variations in structure and development which have hitherto not been recognised as separate types of mycorrhiza formed by Glomalean fungi.

We have initiated research into the development of the two major mycorrhizal types, using time-course experiments and confocal microscopy to determine the changes in relative contribution of intercellular hyphae, intracellular coils and arbuscules to the surface areas of interface between fungus and plant. This will provide base-line data for studies of nutrient exchange between the symbiotic partners (Dickson *et al.*, 1998; Dickson and Kolesik, in preparation). So far data from arbuscules in *Allium* (*Arum*-type) and intracellular coils in *Lilium* (*Paris*-type) indicate that despite the very extensive branching of the *Arum*-type arbuscules, there may be little difference in surface area between these two structures. This is an interesting and intuitively surprising fact which requires confirmation for different developmental stages and in different species.

2.2. MUTANTS WITH ALTERED MYCORRHIZAL PHENOTYPES

The isolation of plant mutants now offers the promise of determining some of the mechanisms controlling colonisation as well as mapping and cloning the genes involved. In all the mutants so far isolated mycorrhizal colonization is blocked at the epidermis or is severely restricted within the root cortex (e.g. Duc *et al.*, 1989; Gianinazzi-Pearson, 1996; see Peterson and Bradbury, 1995; Smith and Read, 1997). In our research we have screened a mutagenised (fast neutron bombardment) population of *Lycopersicon esculentum* Mill. and isolated a mutant with highly reduced VA mycorrhizal colonization by *Glomus mosseae, G. intraradices* and *Gigaspora margarita* (Barker *et al.*, 1998). The mutation is recessive and has been designated *rmc*. Using confocal microscopy, we have shown that even when subjected to very high inoculum potential in nurse pots (Rosewarne *et al.*, 1997), the fungi remain restricted to the surface of the roots of the mutant, with small differences between *G. intraradices* and *Gi. margarita* in terms of development of external mycelium and complex branching 'appressoria'. These alterations in fungal morphogenesis indicate that recognition between the symbionts does occur, but that subsequent penetration of the root cells is blocked by a mechanism that has not yet been determined. This mutant is potentially important, because it is the first in a non-legume and was detected without a preliminary screen for defective nodulation which could well have restricted the number and type of mutations to those that that are common to mycorrhizal and rhizobial symbioses. We can be fairly sure that the *rmc* mutation in tomato is in a biochemical pathway common to all mycorrhizal plants. We aim to identify mutations that modify other steps in the colonization process, providing material for cellular, molecular and genetic dissection of the mechanisms of control. The amenability of tomato for molecular-genetic characterization (only matched by *Medicago truncatula* and *Lotus japonicus* among the legumes from which mycorrhizal mutants have been isolated) should enable us to map and clone the mutated gene(s), and, together with phenotypic analyses, identify the biochemical bases of mechanisms of normal mycorrhizal symbiosis.

3. Molecular investigation of mycorrhizal interfaces

3.1. ORGANIZATION OF THE INTERFACES.

Increased uptake of P, and other nutrients such as Zn, by mycorrhizal plants is well established at the whole plant level but the mechanisms that operate at the molecular and cellular levels are only now beginning to be pieced together to give a picture of the way fungal and plant processes are integrated at the cellular and molecular levels to provide a soil-to-fungus-to-plant pathway of nutrient uptake. This pathway involves the fungal interface with the soil, where nutrients are absorbed by external hyphae and then translocated to the intraradical fungal structures (intercellular hyphae, coils and arbuscules) that form interfaces with the plant. These symbiotic interfaces vary in details of structure and probably function (see Smith and Smith, 1990; Smith 1995; Bonfante and Perotto, 1995 and references therein), but have several essential features in common. An apoplastic interfacial compartment is always bounded by the plasma membranes of fungus and plant and their activities control the bidirectional transfer of nutrients between the symbionts. Transfer of P or other nutrients from fungus to plant involves two steps; efflux from the fungal partner into the interfacial compartment, followed by uptake by the plant and this needs to be taken into account in formulating hypotheses about the transfer of nutrients in the interfaces. Furthermore, it has been proposed that transfer of P and sugars may occur on structurally different interfaces (see Gianinazzi-Pearson *et al.*, 1991; Smith and Smith, 1996), but more evidence is required to substantiate this idea. The interfacial apoplastic compartments contain varying amounts of fungal and plant wall material, as well as molecules that appear to be specifically deposited there, but whose functions are as yet unclear (see Bonfante and Perotto, 1995). Changes in composition may influence the physico-chemical conditions in the apoplast and hence the activities of the key membrane transport proteins discussed below.

3.2. H^+-ATPases

Activity of P-type H^+-ATPases has been used as a marker to indicate energization of the plasma membrane in different interfaces between mycorrhizal fungi and plant cells. Cytochemical methods (e.g. Gianinazzi-Pearson *et al.*, 1991) indicate that activity is pronounced in hyphae growing on the surface of the roots and in intercellular hyphae, but is slight or absent on the arbuscular branches. The membranes of the root epidermal cells show some activity, but this is much more pronounced on the plant membrane surrounding the arbuscules (periarbuscular membrane or PAM) in cortical cells. In contrast, the cortical cells adjacent to intercellular hyphae showed no activity. These data are consistent with a role for external hyphae and epidermal cells in active uptake of nutrients from soil and for the periarbuscular membrane in absorbing nutrients from the interfacial apoplast in cells where P is delivered via hyphae and arbuscules. Additional evidence supporting a key role for plant H^+-ATPases in mycorrhizal processes comes from the finding that a gene in barley encoding an H^+-ATPase is upregulated during mycorrhizal colonization, but is unaffected by P nutrition or by colonization of roots by other fungi (Murphy *et al.*, 1997). Work to determine the relative expression of encoding different members of the large family of P-type H^+-ATPases during mycorrhizal colonization in tomato is in progress (Rosewarne *et al.*, unpublished). At the same time Ferrol *et al.* (1998) have cloned 5 putative P-type H^+-ATPases from a mycorrhizal fungus, *Glomus mosseae*, so that we should soon have the tools to investigate the expression of genes in both plant and fungus.

3.3. PHOSPHATE TRANSPORTERS

High affinity phosphate transporters have been cloned from a number of different plant species, including tomato (e.g C. Liu *et al.*, 1998; H. Liu, 1998; Rosewarne *et al.*, in preparation and see Schachtman *et al.*, 1998) providing tools for the investigation of changes in expression during symbiotic development. Using tomato roots synchronously colonised by *Glomus intraradices* (see Rosewarne *et al.*, 1997 for methods) we have shown that transcripts of the high affinity P transporter LePT1 remained at a relatively constant level between 4 and 12 days in mycorrhizal tomato roots, but increased markedly in non-inoculated roots (see preliminary report in Smith *et al.*, 1998 in press). Work by others has shown similar mycorrhizal effects in *Medicago*. Furthermore, expression is influenced by P supply in both *Medicago* and tomato (C. Liu *et al.*, 1998; H. Liu *et al.*, 1998), so that upregulation of LePT1 in non-mycorrhizal plants may be an effect of phosphate starvation, with the gene being particularly highly expressed in the epidermal cells of the root (H. Liu *et al.*, 1998). Importantly, in-situ hybridization in mycorrhizal tomato roots has shown localization of LePT1 transcripts in cortical cells containing arbuscules (Rosewarne *et al.*, in preparation), which is consistent with transfer of P to the plant across the arbuscular interfaces. However, the homologous gene in *Medicago* probably does not show this pattern of expression (Maria Harrison, personal communication), suggesting that there may be differences between plant species in the way they respond to mycorrhizal colonization. A fungal P transporter GvPT has also been cloned from *Glomus versiforme*. The gene is highly expressed in the external mycelium where the fungus is known to absorb P and other nutrients actively, but not in the fungal structures within the root (Harrison and van Buuren, 1995)

Taken in combination, the information is beginning to indicate that H^+-ATPases and P transporters are organised in a way that would promote one-way transfer of P, polarised in the direction of the plant (Smith *et al.*, in press). Active uptake from soil by the fungus would be via the transporter encoded by GvPT, with the necessary H^+ gradient provided by one of the fungal P-type H^+-ATPases. After delivery to the arbuscules P must efflux to the interfacial apoplast, by a mechanism that is currently not known. Plant P transporters (e.g. LePT1) could then transfer P to the cortical cells, with the proton motive force (PMF) generated by plant H^+-ATPases on the PAM. The lack of expression of GvPT in the intraradical fungus would prevent the fungus from reabsorbing P from the interfacial apoplast and hence also promote one-way transfer.

It will be worth looking for spatial and temporal differences in the expression of different members of families of transport proteins in both symbionts, depending on the availability of nutrients to different tissues. For P especially, the poor mobility in soil and consequent depletion close to absorbing cells means that to be effective, expression of P transporters must be localised in tissues (plant or fungal) where P is present in the adjacent soil solution. It is not surprising therefore, that genes encoding some plant P transporters are highly expressed in cells of the root epidermis, close to root tips (C. Liu *et al.*, 1998). It is these cells that are positioned, by apical growth of roots, in regions of soil from which P has not as yet been depleted. Mycorrhizal hyphae may well obtain P from sites spatially distant from the root, and/or from small soil pores where roots cannot reach. These hyphae deliver the P to root cortical cells, and it will be interesting to determine how far the distribution of mycorrhizal colonization and in particular the distribution of arbuscules affects the expression of P transporters behind the root tips.

3.4. PHYSICO-CHEMICAL CONDITIONS IN THE INTERFACIAL COMPARTMENTS

The activities of the membrane proteins (H^+-ATPases and transporters) will both influence and respond to the physico-chemical conditions in the cellular and apoplastic compartments of the symbiotic interfaces. Ayling and co-workers are using both microelectrodes and pH sensitive fluorescent dyes to probe these environments in the surface layers of cells in mycorrhizal and non-mycorrhizal roots (Ayling *et al.*, 1997,1998). So far evidence indicates that there are no major changes in pH of the compartments or potential differences (PDs) between them induced by colonization, with one exception. Hyphae associated with roots have a transmembrane PD (*Glomus mosseae*) of around -130 mV and a cytoplasmic pH (*Scutellospora calospora*) of about 7.5, which is unaffected by changes in external pH and therefore under metabolic control. The interfacial apoplast between intercellular hyphae and root cells is acid. Fungal wall potential does, however, appear to change. In external hyphae it is about -12 mV, but appressoria and hyphae closely associated with the roots have wall potentials of about -32 mV, which is close to the value for plant cell walls. The functional significance of this difference is not clear, but it may indicate the development of a common apoplastic compartment between the symbionts.

4. Conclusions

A combination of different experimental approaches is being used to obtain the information needed to understand variations in VA mycorrhizal colonization and responsiveness of the plant to it. By linking structural, developmental, molecular and microphysiological studies, we aim to determine how transport of nutrients from fungus to plant occurs and which are the key regulatory steps in the process. When this information can be related to the expression of particular genes we will be in a position to apply our knowledge in targeted plant breeding programs to increase mycorrhizal contributions to nutrient efficiency, promoting low P inputs. In the meantime, some progress towards this aim will certainly be made by careful study of the mycorrhizal responsiveness of carefully selected plant genotypes. A complementary approach will be to investigate the effects of different fungal species and genotypes on P acquisition and interactions with host plant genotype (see Jakobsen, this volume).

5. Acknowledgments.

Our research is funded by the Australian Research Council and forms part of a program in the Centre for Plant Root Symbioses, in The University of Adelaide.

6. References

Ayling, S.M., Smith, S.E., Smith, F.A. and Kolesik, P. (1997) Transport processes at the plant-fungus interface in mycorrhizal associations: physiological studies, in T. Ando, K. Fujita, T., Mae, H. Matsumoto, S. Mori and J. Sekiya. (eds.), *Plant nutrition for sustainable food production and environment*. Kluwer Academic Publishers, Dordrecht, Boston, London, pp 33-35.

Ayling, S.M., Smith, S.E., Reid, R.J. and Smith, F.A. (1998) Changes in the wall potential of *Scutellospora calospora* associated with colonisation of *Allium porrum* roots are not accompanied by equivalent changes in the host. *Can. J. Bot.* **76**, in press.

Baon, J.B., Smith, S.E. and Alston, A.M. (1993) Mycorrhizal responses of barley cultivars differing in P efficiency. *Pl. Soil* **157**, 97-105.

Barker, S.J., Stummer, B., Gao, L., Dispain, I., O'Connor, P.J. and Smith, S.E. (1998) A mutant in *Lycopersicon esculentum* Mill. with highly reduced mycorrhizal colonisation: isolation and preliminary characterization. *Plant J.***15**, 791-767.

Bonfante, P and Perotto, S. (1995) Strategies of arbuscular mycorrhizal fungi when infecting host plants. *New Phytol.* **130**, 3-21.

Dickson, S. and Kolesik, P. Quantification of the volume and surface area of mycorrhizal fungi using laser scanning confocal microscopy. *New Phytol.* submitted

Dickson, S., Kolesik, P., Smith, F.A. and Smith, S.E. (1998) Quantification of the surface area of *Arum*- and *Paris*-type VA mycorrhizal structures using laser scanning confocal microscopy. Abstracts of Second International Conference on Mycorrhiza. Uppsala, Sweden. p 52.

Duc, G., Trouvelot, A., Gianinazzi-Pearson, V. and Gianinazzi, S. (1989) First report of non-mycorrhizal plant mutants (myc-) obtained in pea (*Pisum sativum* L.) and faba bean (*Vicia faba* L.). *Plant Sci.* **60**, 215-222.

Ferrol, N., Barea, J-M. and Azcón Aguilar, C. (1998) Cloning and expression analysis of P-type H^+-ATPase genes in the arbuscular mycorrhizal fungus *Glomus mosseae*. Abstracts of Second International Conference on Mycorrhiza. Uppsala, Sweden. p 62.

Gallaud, I. (1905) Etudes sur les mycorrhizes endotrophs. *Rev Gen Bot* **17**, 5-48, 66-83, 123-135, 223-239, 313-325, 425-433, 479-500.

Gianinazzi-Pearson, V. (1996) Plant cell responses to arbuscular mycorrhizal fungi: getting to the roots of the symbiosis. *Plant Cell* **8**, 1871-1883.

Gianinazzi-Pearson, V., Smith, S.E., Gianinazzi, S. and Smith, F.A. (1991) Enzymatic studies on the metabolism of vesicular-arbuscular mycorrhizas V. Is H^+-ATPase a component of ATP-hydrolysing enzyme activities in plant/fungus interfaces? *New Phytol.* **117**, 61-74.

Harrison, M.J. and Van Buuren, M.L. (1995) A phosphate transporter from the mycorrhizal fungus *Glomus versiforme*. *Nature* **378**, 626-629.

Liu, C., Muchhal, U.S., Uthappa, M., Kononowicz, A.K. and Raghothama, K.G. (1998) Tomato phosphate transporter genes are differentially regulated in plant tissues by phosphorus. *Plant Physiol.* **116**, 91-99.

Liu, H., Trieu, A.T., Blaylock, L.A. and Harrison, M.J. (1998) Cloning and characterisation of two phosphate transporters from *Medicago truncatula* root: regulation in response to phosphate and to colonisation by arbuscular mycorrhizal (AM) fungi. *MPMI* **11**, 14-22.

Murphy, P.J., Langridge, P. and Smith, S.E. (1997) Cloning plant genes differentially expressed during colonisation of roots of *Hordeum vulgare* L. by the vesicular-arbuscular mycorrhizal fungus *Glomus intraradices* Schenck and Smith. *New Phytol.* **135**, 291-301.

Peterson, R.L. and Bradbury, S.M. (1995) Use of plant mutants, intraspecific variants and non-hosts in studying mycorrhiza formation and function. in A. Varma and B. Hock (eds), *Mycorrhiza, Structure, Function, Molecular Biology and Biotechnology*. Springer Verlag, Berlin and Heidelberg, pp. 157-180.

Rosewarne, G. Barker, S.J. and Smith, S.E. (1997) Production of near synchronous colonisation in tomato for developmental and molecular analysis of mycorrhiza. *Mycol. Res.* **101**, 966-970.

Rosewarne, G., Schachtman, D.P., Smith, F.A., Barker, S.J. and Smith, S.E. (in preparation). A *Lycopersicon esculentum* phosphate transporter (LePT1) implicated in P nutrition of VA mycorrhizal plants.

Schachtman, D.P., Reid, R.J. and Ayling, S.M. (1998) Update on phosphorus uptake. Phosphate uptake by plants: from soil to cell. *Plant Physiol.* **116**, 447-453.

Smith, F.A. and Smith, S.E. (1996) Mutualism and parasitism: biodiversity in function and structure in the "arbuscular" (VA) mycorrhizal symbiosis. *Adv. Bot. Res.* **22**, 1-43.

Smith, F.A. and Smith, S.E. (1997) Structural diversity in (vesicular)-arbuscular mycorrhizal symbioses. *New Phytol.* **137**, 373-388.

Smith, S.E. 1995 Discoveries discussions and directions in mycorrhizal research. in A. Varma and B. Hock (eds), *Mycorrhiza: Structure, Function, Molecular Biology and Biotechnology*. Springer Verlag, Berlin and Heidelberg, pp 3-24.

Smith, S.E. and Smith, F.A. (1990) Structure and function of the interfaces in biotrophic symbioses as they relate to nutrient transport. *New Phytol.* **114**, 1-38.

Smith, S.E. and Read, D.J. (1997) *Mycorrhizal Symbiosis* Edition 2. Academic Press, London.

Smith, S.E., Robson, A.D. and Abbott, L.K. (1992) The involvement of mycorrhizas in assessment of genetically dependent efficiency of nutrient uptake and use. *Pl. Soil* **146**, 169-179.

Smith, S.E., Rosewarne, G., Ayling, S.M., Dickson, S., Schachtman, D.P., Barker, S.J., Reid, R.J. and Smith, F.A.. (1998) Phosphate transfer between vesicular-arbuscular (VA) mycorrhizal symbionts: insights from confocal microscopy, microphysiology and molecular studies, in J.P. Lynch and J. Deikman (eds), *Phosphorus in Plant Biology: Regulatory Roles in Molecular, Cellular, Organismic and Ecosystem Processes*. American Society of Plant Physiologists. In press.

PLANT RESPONSES TO INADEQUATE AND TOXIC MICRONUTRIENT AVAILABILITY: GENERAL AND NUTRIENT-SPECIFIC MECHANISMS

H. SCHAT
Department of Ecology and Ecotoxicology of Plants
Faculty of Biology, Vrije Universiteit, De Boelelaan 1087,
1081 HV Amsterdam, The Netherlands

1. Introduction: general concepts and definitions

By definition, mineral nutrients have specific and essential functions in plant metabolism. For each given nutrient, the minimal adequate supply rate varies strongly, both between and within species, usually in a nutrient-specific way. The maximally tolerable supply rates are equally variable and often positively correlated with the minimally required ones. Depending on the quantity in which it is required, any nutrient can be classified as either a macronutrient or a micronutrient. To date, iron (Fe), manganese (Mn), copper (Cu), zinc (Zn), nickel (Ni), molybdenum (Mo), boron (B), and clorine (Cl) have been idendified as universal plant micronutrients (Marschner, 1995). Both the sufficiency and toxicity thresholds for micronutrient supply are lower than those for macronutrient supply, usually by several orders of magnitude, except for Cl, which is not very toxic.

For each particular nutrient, there seems to be a certain range of supply rates in which the growth performance is maintained at a constant level, even when all the other mineral and non-mineral resources are provided at luxurious, but non-toxic rates. The tendency to maintain a constant performance over a range of different environmental conditions is often referred to as 'homeostasis'. Nutrient homeostasis can be defined as the tendency to keep internal nutrient concentrations between the limits of sufficiency and toxicity. Nutrient homeostasis is primarily based on regulated uptake. However, plant-internal nutrient concentrations usually tend to increase with external availability, even within the supply range where a constant growth performance is maintained. This apparent ability to tolerate excessive nutrient burdens relies on the operation of plant-internal sequestration and compartmentation mechanisms, by means of which interferences with sensitive metabolic processes are prevented. These mechanisms are usually considered to represent an essential component of nutrient homeostasis ('intracellular homeostasis').

The frequencies of the terms 'stress' and 'stress tolerance' are ever increasing in recent physiological and ecological literature. In most cases, the term stress seems to denote conditions under which plant performance is negatively affected, either temporarily or permanently. In this way, both inadequate and toxic nutrient supply rates may be taken to represent 'nutrient stress'. In general, conditions of stress appear to induce large-scale alterations of the normal gene transcription patterns, resulting in

G. Gissel-Nielsen and A. Jensen (eds.), Plant Nutrition – Molecular Biology and Genetics, 311–326.

accumulation of heat shock proteins ('stress proteins'), increased capacities of 'stress enzymes', accumulation of 'stress metabolites', altered phytohormone levels, etcetera (see below). The combined occurrence of these effects upon disruption of the homeostatic condition is often referred to as the 'stress response'. The stress response seems to show only a limited degree of specificity with respect to the nature of the 'stressor', and is believed to be adaptive, in that it would counteract or even reverse the deleterious effects of the stressor on plant performance. This idea has been expressed as the 'general adaptation syndrome (GAS)' hypothesis (Leshem and Kuiper, 1996).

The term 'stress tolerance' is often used to indicate an inducible state of increased resistance to stressors, mediated by the stress response (see above). It is also used in the sense of an inherent attribute common to plant species which are native to low-resource environments, such as infertile soils or arctic climates (Grime, 1979). These connotations are fundamentally different, however. Wild plant populations are genetically adapted to their native environments, regardless of whether these are productive or unproductive. There are no reasons to suppose that plants from permanently infertile or toxic soils would 'perceive' their own environment as more stressful than plants from productive soils would do. Likewise, there are neither reasons to suppose that 'acquired stress tolerance', mediated by the general stress response, would play a prominent role in evolutionary adaptation to unproductive soils.

2. Micronutrient homeostasis in yeast: copper and zinc

Regulated uptake seems to be the major component of nutrient homeostasis. The mechanisms underlying controlled micronutrient uptake in plants are far from understood to date. At any rate, a minimal regulatory system would be expected to include: (1) membrane transporters, (2) sensor proteins, (3) sensor-responsive transcriptional regulation of transporter expression. Evidently, independent homeostasis of different nutrients would require nutrient-specific transporters, nutrient-specific sensors, as well as nutrient-specific transcriptional regulation of transporter activities. The well-studied regulatory mechanisms underlying Cu and Zn uptake in *Sacharomyces cerevisiae* provide clear-cut examples.

Cu uptake in yeast is mediated by at least two highly Cu-specific plasmamembrane transporters, encoded by *CTR1* and *CTR3*. Expression of these genes requires transcriptional activation by Mac1p, which has a Cu binding domain acting as a Cu sensor. As long as its Cu sensing domain remains unoccupied, Mac1p binds to the promotors of *CTR1* and *CTR3*, and activates transcription. Occupation of the Cu sensing domain abolishes its binding the these promotors, and triggers its proteolytic degradation. It is noteworthy that the transcriptional activity of Mac1p is specifically regulated by Cu, and not by any other metal (Labbe *et al.*, 1997; Zhu *et al.*, 1998). Two more genes have been shown to be controlled by Mac1p, viz. *FRE1* and *FRE2*, which encode the two plasmamembrane surface Fe(III)/Cu(II) reductases, Fre1p and Fre2p. Mac1p activates transcription of *FRE1* under Cu starvation, and is essential for the downregulation of *FRE2* expression under excessive Cu exposure. Both genes are also under Fe-responsive transcriptional control. The major role of both reductases is to mobilize Fe, but they have been shown to be involved in Cu uptake as well (Georgatsou *et al.*, 1997). Zn uptake in yeast is equally strongly regulated. It is largely mediated by two specific plasmamembrane transporters, viz. Zrt1p and Zrt2p. The *ZRT1* and *ZRT2* genes are both controlled by the specifically Zn-responsive transcriptional activator Zap1p. This protein has a histidine and cysteine-rich domain, which could act as a Zn sensor. *ZAP1* expression has been shown to be under positive autoregulatory control, which probably accelerates the transcriptional response to decreasing Zn availability (Zhao and Eide, 1997).

Transcriptional activation or repression of membrane transporter genes is certainly not the only way to control nutrient uptake rates. Micronutrient transporter activities may also be under additional post-transcriptional control. For example, Cu, when excessively available, specifically triggers the proteolytic degradation of its own high-affinity transporter in yeast, Ctr1p (Ooi *et al.*, 1996). Furthermore, yeast mutants lacking functional proteasome subunits or ubiquitine ligases are often hypersensitive to Cu, Zn, or Cd, which also points at a very prominent role for proteolysis in micronutrient homeostasis.

As shown by the above examples and other studies, yeast cells are well capable of regulating micronutrient uptake in a nutrient-specific way, particularly at the transcriptional level. However, the nutrient-specificity of membrane transporters is high, but not absolute. Low-affinity transporters for a given nutrient may act as high-affinity transporters for another one. Thus, under conditions of excessive availibility of a given nutrient, it doesn't seem to be a valid option to repress all the transporters that are capable of mediating its uptake, because this would probably cause deficiencies in other ones. This might be the main reason why internal nutrient concentrations are eventually bound to increase with increasing external availability.

Excessive nutrient accumulation can be buffered by internal sequestration mechanisms. Again, the mechanisms of micronutrient sequestration in *Saccharomyces cerevisiae* are better known than those in plants, particularly for Cu. In yeast Cu is sequestered by high-affinity binding to the metallothioneins Cup1p and Crs5p. *CUP1* and *CRS5* are both controlled by the Cu-responsive transcriptional activator Ace1p. Like Mac1p, Ace1p has a Cu sensing domain. Occupation of this domain allows for the binding of Ace1p to the *CUP1* and *CRS5* promotors and subsequent transcription of these genes. Ace1p-mediated transcription of *CUP1* already occurs at picomolar Cu concentrations. Other metals have been shown to bind to the sensing domain, but this does not lead to transcriptional activation (Dameron *et al.*, 1993; Dobi *et al.*, 1995). The corresponding Cu-sensing transcriptional activator in *Candida glabrata*, called Amt1p, has been shown to be under positive autoregulatory control, which probably accelarates the transcriptional response to increasing Cu availability (Zhou and Thiele, 1993). The third gene which is known to be controlled by Ace1p is *CRS4*, which encodes the cytosolic CuZnSOD. Crs4p seems to play an essential, but ill-understood role in Cu homeostasis, which is unrelated to its superoxide scavenging activity (Cizewski Culotta *et al.*, 1995). The precise mechanisms of Zn sequestration in yeast have not been elucidated yet. Mutants lacking vacuoles or functional V-ATPase are hypersensitive to zinc (Ramsay and Gadd, 1997). Thus, vacuolar compartmentation mediated by proton gradient-dependent transporters seems to be a major component of intracellular Zn homeostasis. There is no evidence of any Zn-dependent transcriptional regulation of vacuolar transporters, however.

3. Micronutrient homeostasis in plants: iron, copper and zinc

Yeast mutant complementation studies have identified a number of micronutrient transporter genes in *Arabidopsis thaliana*. *ZIP1* and *ZIP3* probably encode plasmamembrane transporters mediating uptake of, preferentially, Zn (Grotz *et al.*, 1998). *IRT1* encodes a plasmamembrane transporter mediating Fe uptake (Eide *et al.*, 1995), and *COPT1* encodes a putative plasmamembrane transporter mediating Cu uptake (Kampfenkel *et al.*, 1995). All of these genes exhibit significant sequence identity with one or more of their functional analogs in yeast. In *Arabidopsis ZIP1*, *ZIP3*, and *IRT1* are exclusively expressed in roots, whereas another member of the family, *ZIP4*, is also expressed in shoots. *ZIP* gene expression is specifically induced under conditions of Zn deficiency (Grotz *et al.*, 1998). *IRT1* is specifically expressed in roots under conditions of Fe deficiency (Eide *et al.*, 1995). The expression patterns of *COPT1* have not been established thus far. These examples clearly show that plants, just like yeast, possess nutrient-specific transporters, which are sensitively regulated at the transcriptional level, in a highly nutrient-specific way. The underlying mechanisms of nutrient sensing and nutrient-specific transcriptional activation of transporter genes in plants have not been elucidated yet. Transcription factors combining the functions of sensing and transcriptional activation, such as in yeast (see above), have not been identified thus far.

The mechanisms of micronutrient sequestration and compartmentation in plants are incompletely known. Plant genomes contain a number of genes encoding cystein-rich metallothionein-like proteins (plant MTs). In *Arabidopsis* MT1 is constitutively expressed in roots, but inducible in shoots. MT2 shows the opposite pattern. Expression is induced by Cu, and, though less effectively, Cd and Zn (Zhou and Goldsbrough, 1995a,b; Murphy and Taiz, 1995; Murphy *et al.*, 1997). MT expression has been shown to be required for normal 'wild type' tolerance to Cu, but not for normal tolerance to Cd or Zn (P.B. Goldsbrough, pers. comm.). Fe sequestration seems to depend on ferritin expression, which is under Fe-responsive transcriptional control (Gaymard *et al.*, 1996). Zn sequestration probably relies on vacuolar compartmentation, just like in yeast (see below). These examples indicate that the micronutrient sequestration mechanisms operating in plants are, again, basically similar to those in yeast. Anyway, plants possess additional metal sequestering peptides, viz. (gamma-glutamylcysteinyl)$_n$-glycines (n = 2-11), also known as cadystins, class III metallothioneins, or phytochelatins (PCs). These non-protein peptides have also been found in *Schizosaccharomyces pombe* and a number of other fungi, but not in *Saccharomyces cerevisiae*. Their synthesis from glutathione is catalysed by the enzyme phytochelatin synthase (PC-synthase), which is a constitutively expressed metal-activated enzyme. A broad range of heavy metals, both essential and non-essential ones, are capable to activate the enzyme *in vivo*, thereby triggering the onset of PC synthesis (Grill *et al.*, 1985, 1987, 1989). PC-based metal sequestration has been considered to play a role in metal micronutrient homeostasis. However, PC synthase-deficient *Arabidopsis* mutants are hypersensitive to Cd and Hg, but not or hardly to any essential heavy metal micronutrient (Howden *et al.*, 1992). In conformity with this, *in vivo* inhibition of PC synthesis in *Silene vulgaris*, by means of the gamma-glutamylcysteine synthetase inhibitor buthionine sulfoximine, decreases Cd tolerance, but fails to affect Zn tolerance and Cu tolerance (De Knecht *et al.*, 1992; Schat and Vooijs, 1997a). As yet, the only unambiguously established functions of PCs are Cd and Hg detoxification.

Yeast mutant complementation studies have shown to be a powerful tool in analysing micronutrient homeostasis in plants. However, yeast-based models are subject to important limitations. Plants are multicellular organisms showing tissue and organ differentiation. The nutrient requirements and sensitivities will doubtlessly vary from tissue to tissue and organ to organ. This evidently provides the whole plant with additional possibilities for homeostasis. There is ample evidence of regulated nutrient transport from root to shoot and visa versa. Also, controlled organ turnover via regulated senescence, or regulated nutrient withdrawal from senescing organs are likely to represent essential components of plant nutrient homeostasis. The precise regulatory mechanisms involved are poorly known to date. Plant hormones, such as ABA, jasmonate, and ethylene are doubtlessly involved in the coordination of the underlying organ-specific and tissue-specific transcriptional and post-transcriptional responses.

4. Micronutrient stress

Micronutrient stress can be defined as the condition of a growth-limiting - either inadequate or toxic - level of availability of a given micronutrient. Micronutrient stress, particularly toxicity stress, has been shown to induce a number of responses with a low degree of specificity with regard to the nature of the stressor (the so-called 'general stress response', or 'general adaptation syndrome'). Typical components of the stress response are (1) accumulation of 'stress metabolites', such as free amino acids (proline), quaternary ammonium compounds (glycine betaine), sugar alcohols (sorbitol, pinetol), various sugars, and/or polyamines (putrescine), (2) upregulation of the capacities of 'stress responsive enzymes', such as SOD (superoxide dismutase), APX (ascorbate peroxidase), DHAR (dehydro-ascorbate reductase), GPX (glutathione peroxidase), GSSGR (glutathione reductase), gamma-GCS (gamma-glutamylcysteine synthetase), POD (aspecific peroxidase), TRX (thioredoxine), ME (malic enzyme), ICDH (isocitrate dehydrogenase), G-6-PDH (glucose-6-phosphate dehydrogenase), GDH (glutamate dehydrogenase), and many others, and (3) strongly increased expression of 'heat shock proteins', or 'stress proteins'.

Proline, glycine betaine, sorbitol, and pinetol are often called 'osmolytes'. This reflects the common assumption that their accumulation under stress would primarily serve to decrease the cellular osmotic potential and, therefore, to prevent or counteract plant water imbalance. They are also known as 'compatible solutes', which refers to *in vitro* demonstrations of uninhibited enzyme activities in the presence of extremely high concentrations of these solutes. In addition, proline and glycine betaine have been shown to protect enzymes from inactivation by heat, chilling, salinity, or dilution *in vitro*, through physical mechanisms unrelated to the osmotic potential as such (Aspinall and Paleg, 1981; Paleg *et al.*, 1981; Selinioti et al., 1987; Krall *et al.*, 1989). Finally, several stress metabolites, including proline and putrescine, have been shown to act as oxygen free radical scavengers *in vitro* (Smirnoff and Cumbes, 1989). The *in vivo* significance of all these potentially protective mechanisms, however, is far from clear. Even the presumed importance of the osmoregulatory function is not undisputable (Hare *et al.*, 1998).

Many stress responsive enzymes are directly or indirectly involved in the cellular antioxidant defense systems. Some of them primarily serve to scavenge superoxide anions (SOD), or hydrogen peroxide (APX, GPX). Others, such as APX, DHAR, GSSGR, and gamma-GCS, are involved in the regeneration or synthesis of reduced ascorbate or glutathione, which play pivotal roles in the first, second and third 'lines of defence' against reactive oxygen species and lipid radicals (De Vos and Schat, 1992). Many other stress enzymes exhibit dehydrogenase activity. The functional significance of their upregulation under stress might lie in NAD(P)H regeneration and/or proton balance maintenance.

Stress proteins, also called heat shock proteins (HSPs), are members of a heterogeneous family of non-catalytic proteins. Most of them seem to act as chaperones of proteins and RNAs. They have been shown to function in the targeting, transport, folding, polymerization, and breakdown of proteins, ribosomal complex stabilization, RNA processing, mRNA transport, mRNA stabilization under stress (heat shock granule formation), and several other processes (Nover and Höhfeld, 1996).

Heat shock protein (HSP) expression is controlled by so-called heat shock factors (HSFs), a family of transcriptional activators that bind to the 'heat shock response element' (HSE), which is present in the promotor sequences of HSP-encoding genes, and several others. HSF proteins may be constitutively present, or (partly) stress-inducible. They are activated by dephosphorylation, cytoplasm to nucleus transport, and, particularly, trimerization. HSF trimerization seems to be mainly controlled by HSP70. This protein binds to monomeric HSF, thereby preventing HSF trimerization, but also to denaturated proteins. Stress-induced accumulation of denaturated proteins is believed to cause a decrease in the concentration of free HSP70, which would trigger HSF trimerization, and subsequent transcriptional activation of *HSP* genes (Baler *et al.*, 1996; Schöffl *et al.*, 1998). The precise signalling mechanisms for HSP expression under micronutrient toxicity are far from clear. Heavy metals, particularly Cu, Ni and Zn, and many non-essential ones, might be sufficiently reactive to produce protein denaturation by themselves. However, HSP accumulation upon toxic metal exposure occurs much faster in leaves than in roots (P. Stolt, pers. comm.), even though the initial rates of metal accumulation and growth inhibition are much higher in the roots. This clearly points to more indirect signalling mechanisms (see below).

The transcriptional activation mechanisms underlying the various stress enzyme responses in plants are largely obscure. Many of the enzymes involved in the cellular antioxidant machinery in yeast appear to be controlled by members of the Yap family of transcriptional activators, particularly Yap1p. These dimeric bZIP proteins are constitutively expressed and their transcriptional activity seems to be regulated by phosphorylation and transport between nucleus and cytoplasm. This transport depends on a cysteine-rich domain, which is believed to act as a redox sensor (Kuge *et al.*, 1997). Eight members of this family have been identified thus far, all of them with specific, but overlapping functions. Yap1p is activated by hydrogen peroxide and other oxidizing chemicals (Fernandes *et al.*, 1997). Heavy metals, both micronutrients and non-essential ones, are potent inducers of stress enzymes, regardless of whether they are capable (Cu, Fe, Mn, As, Cr), or incapable to undergo redox-cycling *in vivo* . Also metals with a stable valency, such as Cd, Pb, or Ni have been shown to induce increased levels of superoxide, hydrogen peroxide, or lipid peroxidation, possibly through interfering with electron transport chains, or glutathione-based antioxidant activity, or by increasing the Fenton-availability of catalytic metals, such as Cu or Fe (De Vos and Schat, 1992; Moran *et al.*, 1994).

The transcriptional mechanisms involved in stress metabolite accumulation in plants are completely unknown. Proline accumulation under heat, chilling, drought, or salinity is generally believed to be triggered by the development of cellular water deficits (Aspinall and Paleg, 1981). Its occurrence under conditions of heavy metal toxicity stress (Cu, Zn, Cd, Pb) has been claimed to result from metal influx into the leaves as such (Kastori *et al.*, 1992). However, toxic heavy metal exposure may cause severe plant water imbalance (Barcelo and Poschenrieder, 1990). In addition, metal-stressed plants do not accumulate proline in the leaves under conditions of high air humidity (> 95% RH), not even at extremely high leaf metal influx rates (Schat *et al.*, 1997). This clearly confirmes that water loss may be the universal signal for stress-induced proline accumulation.

As implied by the above, different elements of the stress response in plants appear to be triggered by different signals, viz. protein denaturation or misfolding, dehydration, and, probably, activated oxygen species. Also, the transcriptional responses to these signals are mediated by different transcription factors, viz. HSFs, redox-sensing factors, and, probably, water potential-responsive factors. Thus, the frequently reported coincident occurrence of osmolyte accumulation, stress enzyme expression and HSP expression, regardless of the nature of the stressor, seems to be explained by a coincident induction of the different transcriptional signals involved. This is not surprising. Almost any stressor would be expected to cause water deficit, protein denaturation, and oxygen activation, albeit in different ways. Moreover, water deficits have been shown to induce oxygen activation and lipid peroxidation, possibly through decompartmentation of catalytic metals (Moran *et al.*, 1994), and would be expected to affect protein hydration and, therefore, tertiary and quaternary protein structure. Reactive oxygen species may directly damage proteins, and induce peroxidative degradation of membrane lipids, leading to osmolyte (e.g. potassium) leakage from the cells, and, consequently, water imbalance and loss of turgor (Girotti, 1985; De Vos *et al.*, 1989; De Vos and Schat, 1991).

Acquired stress resistance, mediated by the stress response, is non-stressor-specific, at least to some extent. Induction of the stress response by heat shock, for example, has been shown to increase the resistance to other stressors, such as salinity, or Cd (Neumann *et al.*, 1994; Schöffl *et al.*, 1998). In case of heat shock-induced heavy metal resistance, it may be relevant that all the metallothionein encoding genes investigated thus far appear to contain heat shock response elements in their promotor regions. Heat shock-mediated MT expression has been demonstrated in many organisms, including *Arabidopsis* (Murphy and Taiz, 1995). Thus, heat shock-induced resistance to heavy metals might as well result from expression of MTs, which are usually not classified as HSPs. This may particularly apply to Cd, because this metal combines the properties of a high affinity to plant MT2 and several yeast MTs with a low tendency to induce MT expression, at least in plants and baker's yeast (Zhou and Goldsbrough, 1994; Sewell *et al.*, 1995).

5. Genetic adaptation

Persistent exposure of natural populations to inadequate or toxic micronutrient availability would be expected to provoke evolutionary adaptation, provided that the appropriate genetic variation is available in the populations in question. The so-called 'metallophytes', i.e. plant species occurring on strongly metal-enriched soil (e.g. ore outcrops, mine waste deposits, smelter waste deposits), provide striking examples of micro-evolutionary adaptation to toxic heavy metal availability. Most of these species are 'facultative' metallophytes: they occur on both normal and metalliferous soil types. Well-known examples in western and central Europe are *Festuca ovina, F. rubra, Agrostis capillaris, A. gigantea, A. stolonifera, A. canina, Deschampsia cespitosa, D. flexuosa, Minuartia verna, Thlaspi caerulescens*, and *Silene vulgaris*. All these species have been shown to exhibit a very pronounced inter-population variation in the degree of heavy metal tolerance. Plants from metalliferous sites are often five- to fifty-fold more tolerant to particular heavy metals than plants from non-metalliferous sites (Schat and Ten Bookum, 1992a). These high-level tolerances have been shown to be heritable in almost all cases studied thus far.

The transmission genetics of heavy metal tolerance, in so far as it segregates in crosses between plants from normal soil ('non-tolerants') and plants from metal-enriched soil ('tolerants'), have been extensively studied in *Silene vulgaris*. The segregation of Cu tolerance, Zn tolerance, and Cd tolerance in these crosses appeared to be largely governed by either one major gene, or two addditive genes, dependent on the level of tolerance of the tolerant parent (Schat and Ten Bookum, 1992b; Schat *et al.*, 1993; Schat *et al.*, 1996). In general, the inheritance of adaptive high-level metal tolerance appears to be governed by single major genes in other metallophyte species as well (Macnair, 1983; Macnair *et al.*, 1992). The levels of tolerance, however, may be under additional control of a low number of 'modifiers', i.e. genes that are hypostatic to the major genes that control the very occurrence of tolerance, relative to the non-metallicolous parent plants (Macnair, 1983; Schat and Ten Bookum, 1992b). The amount of intergenic variation with respect to the major tolerance genes among different metallicolous populations appears to be low. In general, F2 crosses between equally tolerant plants from different geographically isolated mines do not segregate (Schat *et al.*, 1993; 1996). No more than two loci for Cu tolerance, two for Zn tolerance, and one or two for Cd tolerance have been found among plants from a total of four Cu-tolerant, five Zn-tolerant, and three Cd-tolerant isolated *S. vulgaris* mine populations. Even populations belonging to different subspecies, i.e. ssp. *vulgaris* and ssp. *maritima*, usually don't show intergenic variation for either of these tolerances (Schat *et al.*, 1996), which demonstrates that this apparent low degree of intergenic variation must have resulted from independent parallel evolution at different localities, rather than from descent from a common tolerant ancestor. Apparently, the *S. vulgaris* genome seems to contain only a few specific loci with the appropriate mutability to produce adaptive high-level tolerance.

Adaptive high-level tolerances are highly metal-specific. Co-segregation analyses in *S. vulgaris* revealed that the tolerances to Cu and Zn, Cu and Cd, and, at least largely, Zn and Cd are under non-pleiotropic genetic control. Some of these tolerances, however, produce 'co-tolerance' to one or more chemically related metals. For example, Cu-tolerant plants exhibit a degree of tolerance to Ag and Hg, which co-segregates consistently with Cu-tolerance in crosses to non-tolerants. Likewise, Zn-tolerant plants show a degree of tolerance to Co and Ni, which co-segregates with one particular Zn tolerance gene in crosses to non-tolerants. Anyway, these apparent, probably non-functional 'co-tolerances' are only up to three times higher than the constitutive levels in non-metallicolous plants, which is much less than the corresponding functional tolerances in metallicolous plants (Schat and Vooijs, 1997b).

The genes and gene products that produce the high-level tolerances in metallicolous plants have not been identified or functionally characterised thus far. In general, high-level tolerance could either rely on reduced uptake, or increased plant-internal sequestration. High-level arsenate tolerance in grasses represents the only clear-cut example of metal tolerance through reduced uptake. Arsenate uptake in plants is known to be mediated by one or more phosphate transporters, and high-level arsenate tolerance in mine populations of *Holcus lanatus* has been shown to be genetically correlated with reduced high-affinity phosphate uptake (Meharg and Macnair, 1992). High-level tolerances to Cu, Zn, and Cd, on the other hand, are usually not due to reduced uptake of the metal in question. Cu uptake in copper mine plants may be slightly lower than in non-mine plants, such as in *Silene vulgaris*, but the maximally tolerable uptake rates are several-fold (up to five-fold) higher (De Vos *et al.*, 1991; 1992; Schat and Kalff, 1992). The lower uptake in the Cu-tolerant plants might in fact represent a consequence, rather than a cause of high-level tolerance (Strange and Macnair, 1991). Zn and Cd uptake rates in zinc mine plants are not considerably different from those in non-metallicolous plants (Harmens *et al.*, 1993; De Knecht, 1994). It may be surprising that high-level heavy metal tolerance is generally not effected by reduced uptake, particularly in case of essential nutrients, such as Cu and Zn, for which specific plasmamembrane transporters have been shown to exist (Kampfenkel *et al.*, 1995; Grotz *et al.*, 1998). However, the expression of these specific transporters might be completely suppressed already at relatively low levels of exposure, and uptake under higher levels of exposure might be largely mediated by incompletely specific transporters of other nutrients (see above).

As follows from the above, adaptive high-level heavy metal tolerances must be due to specifically increased capacities for plant-internal sequestration, in general. Increased retention of metals in the root system has been considered to represent a primary mechanism of tolerance in non-hyperaccumulating metallicolous plants. Metallicolous plants often exhibit increased retention in the roots, indeed, particularly of Zn and Cd (De Knecht *et al.*, 1991; Harmens *et al.*, 1993a). However, split-root experiments with *S. vulgaris* have demonstrated that the dose-response relationships for metal-imposed root growth inhibition are largely independent of the metal burden and the growth performance of the shoot, both in tolerant and non-tolerant plants (Harmens *et al.*, 1993a). Root growth inhibition apparently results from direct effects of the metals on the roots. Thus, the lower root growth inhibition in the tolerant plants must be due to the a more effective metal sequestration in the roots, rather than to the associated reduction of metal accumulation in the shoot as such. In addition, increased heavy metal tolerance in metallicolous plants has been amply demonstrated at the levels of isolated leaf discs, leaf mesophyll protoplasts, and pollen (Repp, 1963; Baumeister *et al.*, 1967; Searcy and Mulcahy, 1985). In other words, high-level tolerance is probably expressed in every tissue or cell, and increased retention of metals in the roots seems to be a consequence of the tolerance mechanism operating in the root cells, rather than a primary mechanism of tolerance as such.

Increased expression of metallothioneins obviously represents a potential mechanism for high-level heavy metal tolerance at the cellular level. A number of yeast mutants displaying hypertolerance to Cu or Cd have been shown to exhibit inducible and/or constitutive overexpression of *CUP1* (Sewell *et al.*, 1995). Comparative studies into MT expression in metallicolous and non-metallicolous higher plants, however, have not been published thus far. Interecotypic variation in Cu tolerance in the non-metallophyte *Arabidopsis thaliana*, however, is strictly correlated with MT2 expression, which could be taken as a positive indication of some role for MTs in metallophytes as well. However, Cu tolerance in *Arabidopsis* is an inducible trait. It is manifested as a (partial) recovery of growth, several hours upon Cu exposure, of which the magnitude depends on the degree of induced MT2 transcription (Murphy and Taiz, 1995). High-level Cu-tolerance in *Silene vulgaris*, on the other hand, does not require induction by Cu or any other agent. It is evident within minutes upon exposure, for example from the absence of K efflux from the roots, which occurs almost instantaneously in Cu-sensitive plants (De Vos *et al.*, 1991). This evidently precludes any involvement of Cu-induced gene transcription. Constitutive MT overexpression remains a possibility, though.

Overexpression of components of the phytochelatin-based sequestration system would be another conceivable basis for high-level heavy metal tolerance. Studies on Cd-hypersensitive fission yeast mutants have revealed that efficient PC-based Cd sequestration does not only depend on PC synthase activity as such, but also on the vacuolar compartmentation of Cd-PC complexes, which is mediated by an ABC-type transporter (Ortiz *et al.*,1992), and on the stabilization of these complexes by incorporation of acid-labile sulfide (Speiser *et al.*, 1992). Also, the chain length of the PCs could be important, because the stability of metal-PC complexes tends to increase with chain length (Matsumoto *et al.*, 1990). However, the maximally inducible PC synthase activities in crude root extracts are the same in metallicolous and non-metalicolous *S. vulgaris* (De Knecht *et al.*, 1995). The *in vivo* PC chain length distributions and the amounts of acid-labile sulfide incorporated in the metal-PC complexes are identical, too (De Knecht *et al.*, 1994). The *in vivo* rates of PC accumulation in roots under metal-exposure, however, are much lower in the metallicolous plants (De Vos *et al.*, 1992; De Knecht *et al.*, 1994; Harmens *et al.* 1993b). Furthermore, *in vivo* inhibition of PC synthesis by buthionine sulfoximine (BSO) does not appreciably affect the levels of metal tolerance in the metallicolous plants (De Knecht *et al.*, 1992; Schat and Vooijs, 1997a). The latter holds true even for Cd tolerance (De Knecht *et al.*, 1992), which has been shown to depend on PC synthesis in the non-metallicolous plants (see above). It may be argued that the rates of vacuolar compartmentation of metal-PC complexes could be decisive nevertheless, because the BSO-imposed suppression of PC synthesis in the roots is not complete, due to shoot to root glutathion transport (De Knecht *et al.*, 1995). However, this hypothesis does not account for the consistently lower rates of PC accumulation in the tolerant plants. These cannot be explained by a higher PC turnover, because PC degradation, which presumably takes place in the vacuole, proceeds at the same rate in both plant types (De Knecht *et al.*, 1995). This implies that the lower PC accumulation in the tolerant plants must be explained by a lower availability of cellular metals for PC synthase activation. Obviously, this can only be taken as circumstantial evidence of increased activities of alternative non-PC-based sequestering mechanisms. Thus, it is highly unlikely that high-level metal tolerance in *Silene vulgaris* would depend on PC-dependent sequestration.

High-level heavy metal tolerance in metallophytes, particularly to Cd and Zn, has often been suggested to depend on accelerated vacuolar sequestration (Mathys, 1977; Harmens *et al.*, 1994). Comparisons of vacuolar metal contents in metallicolous and non-metallicolous plants, however, have not been published thus far. Anyway, Cd and, when excessively available, Zn have been shown to accumulate primarily in vacuoles, both in metallicolous and non-metallicolous plants (Mullins *et al.*, 1985; Vögeli-Lange and Wagner, 1989; Harmens, 1993; Vázquez *et al.*, 1994).

The major Cd species transported across the tonoplast is probably Cd-PC (Vögeli-Lange and Wagner, 1989; Salt and Rauser, 1995), at least in non-metallicolous plants (see above). Cd-PC tonoplast transport is probably mediated by an ABC-type transporter, like in fission yeast (Ortiz *et al.*, 1992). *In vitro* experiments with purified oat root tonoplast vesicles have revealed Cd^{2+}/H^+ antiport activity (Salt and Wagner, 1993). The Cd affinity of this antiporter, however, may to be too low to account for substantial transport *in vivo*. Direct measurements of Cd transport in metallophyte tonoplast vesicles are not available to date. At any rate, it is unlikely that either increased Cd-PC transport, or increased Cd/proton antiport activity would be responsible for high-level Cd tolerance in metallicolous plants.

The mechanisms of vacuolar Zn transport in plants have hardly been studied to date. Recent studies with *S. vulgaris* tonoplast vesicles suggest that there are at least two different Zn-transporting systems. One of them is a high-affinity/low-capacity system which probably operates both in non-tolerant and Zn-tolerant plants. The other one is a low-affinity/high-capacity system, which is only apparent in the Zn-tolerant plants (Verkleij *et al.*, 1998). Both systems depend on Mg-ATP and seem to transport Zn in the free ionic form. The high-affinity system dissipates the transmembrane electric potential, whereas the low-affinity system does not (A.N. Chardonnens and P.L.M. Koevoets, pers. comm.). Both systems are constitutively present, irrespective of the plant Zn exposure levels. These results strongly suggest that high-level Zn tolerance in *S. vulgaris* is due to (increased) expression of a constitutive low-affinity tonoplast transport system. However, the genetic correlation between tonoplast transporter activity and Zn tolerance has not been definitely established yet.

In conclusion, high-level heavy metal tolerances seem to be based on constitutive, rather than inducible increases of metal-specific cellular sequestration capacities. This could explain the frequently reported increased requirements for particular essential metals in highly metal-tolerant metallophytes (Ernst, 1975; Mathys, 1975; De Vos *et al.*, 1993). At any rate, genetic adaptation to metalliferous soil is not associated with constitutive or inducible overexpression of elements of the general stress response. For example, when grown at adequate Cu supply, Cu-tolerant *S. vulgaris* does not show any constitutive or inducible overexpression of typical stress-responsive enzymes, such as SOD, catalase, peroxidases, and glutathion reductase. On the contrary, increased expression of these enzymes in tolerant plants requires Cu exposure levels that are highly toxic or even deadly to non-tolerant plants (De Vos, 1991). On the other hand, within the low supply range, tolerant plants might already exhibit stress responses at supply rates that are fully adequate for non-tolerant plants (De Vos *et al.*, 1993). Natural selection for specific metal tolerances apparently leads to a shift of the whole dose-response curve for growth with a concomittant shift of the threshold exposure levels for stress response induction.

References

Aspinall, D. and Paleg, L.G. (1981) Proline accumulation: physiological aspects. In: *The Physiology and Biochemistry of Drought Resistance in Plants* (Paleg, L.G. and Aspinall, D., eds). Academic Press, Sidney, pp. 205-241.

Baler, R., Zou, J. and Voellmy, R. (1996) Evidence for a role of Hsp 70 in the regulation of the heat shock response in mammalian cells. *Cell Stress Chap.* **1**, 33-39.

Barcelo, J. and Poschenrieder, C. (1990) Plant water relations as affected by heavy metal stress: a review. *J. Plant Nutr.* **13**, 1-37.

Baumeister, W., Ernst, W.H.O. and Rüther, F. (1967) Zur Soziologie und Ökologie europaeischer Schwermetallpflanzen *Forschungsber. Land. Nordrhein-Westphalen* **1803**, 1-46.

Cizewski Culotta, V., Joh, H.D., Lin, S.J., Hudak Slekar, K., Strain, J. (1995) A physiological role of *Saccharomyces cerevisiae* copper/zinc superoxide dismutase in copper buffering. *J. Biol. Chem.* **270**, 29991-29997.

Dameron, C.T., George, G.N., Arnold, P., Santhanagopalan, V. and Winge, D.R. (1993) Distinct metal binding configurations in ACE1. *Biochemistry* **32**, 7294-7301.

De Knecht, J.A. (1994) *Cadmium tolerance and phytochelatin production in* Silene vulgaris. PhD Thesis, Vrije Universiteit, Amsterdam.

De Knecht, J.A., Koevoets, P.L.M., Verkleij, J.A.C., and Ernst, W.H.O. (1992) Evidence against a role for phytochelatins in naturally selected increased cadmium tolerance in *Silene vulgaris*. *New Phytol.* **122**, 681-688.

De Knecht J.A., Van Baren, N., Ten Bookum, W.M., Wong Fong Sang, H.W., Koevoets, P.L.M., Schat, H. and Verkleij, J.A.C. (1995) Synthesis and degradation of phytochelatins in cadmium-sensitive and cadmium-tolerant *Silene vulgaris*. *Plant Science* **106**, 9-18.

De Knecht, J.A., Van Dillen, M., Koevoets, P.L.M., Schat, H., Verkleij, J.A.C. and Ernst, W.H.O. (1994) Phytochelatins in cadmium-sensitive and cadmium-tolerant *Silene vulgaris*: chain length distribution and sulfide incorporation. *Plant Physiol.* **104**: 255-261.

De Vos, C.H.R. (1991) *Copper-induced oxidative stress and free radical damage in roots of copper tolerant and sensitive* Silene cucubalus. PhD Thesis, Vrije Universiteit, Amsterdam.

De Vos, C.H.R. and Schat, H. (1991) Free radicals and heavy metal tolerance. In: *Ecological Responses to Environmental Stresses* (Rozema, J. and Verkleij, J.A.C., eds). Kluwer Academic Publishers, Dordrecht, pp. 22-30.

De Vos, C.H.R., Schat, H., De Waal, M.A.M., Vooijs, R. and Ernst, W.H.O. (1991) Increased resistance to copper-induced damage of the root cell plasmamembrane in copper tolerant *Silene cucubalus*. *Physiol. Plant.* **82**, 523-528.

De Vos, C.H.R., Schat, H., Vooijs, R. and Ernst, W.H.O. (1989) Copper-induced damage to the permeability barrier in roots of *Silene cucubalus*. *J. Plant Physiol.* **135**, 164-169.

De Vos, C.H.R., Ten Bookum, W.M., Vooijs, R., Schat, H. and De Kok, L.J. (1993) Effect of copper on fatty acid composition and peroxidation of lipids in the roots of copper tolerant and sensitive *Silene cucubalus*. *Plant Physiol. Biochem.* **31**, 151-158.

De Vos, C.H.R., Vonk, M.J., Vooijs, R. and Schat, H. (1992) Glutathione depletion due to copper-induced phytochelatin synthesis causes oxidative stress in *Silene cucubalus*. *Plant Physiol.* **98**, 853-858.

Dobi, A., Dameron, C.T., Hu, S., Hamer, D. and Winge, D.R. (1995) Distinct regions of Cu(I) cntdot ACE1 contact two spatially resolved DNA major groove sites. *J. Biol. Chem.* **270**, 10171-10178.

Eide, D., Broderius, M., Fett, J. and Guerinot M.-L. (1995) A novel iron-regulated metal transporter from plants identified by functional expression in yeast. *Proc. Natl. Acad. Sci. USA* **93**, 5624-5628.

Ernst, W.H.O. (1975) Physiology of heavy metal resistance in plants. In: *Proc. Int. Conf. on Heavy Metals in the Environment, Toronto*, Vol. 2. CEP Consultants, Edinburgh, pp. 121-136.

Fernandes, L., Rodrigues Pousada, C. and Struhl, K. (1997) Yap, a novel family of eight bZIP proteins in *Saccharomyces cerevisiae* with distinct biological functions. *Mol. Cell. Biol.* **17**, 6982-6993.

Gaymard, F., Boucherez, J. and Briat, J.F. (1996) Characterization of a ferritin mRNA from *Arabidopsis thaliana* accumulated in response to iron through an oxidative pathway independent of abscisic acid. *Biochem. J.* **318**, 67-73.

Girotti, A.W. (1985) Mechanisms of lipid peroxidation. *J. Free Rad. Biol. Med.* **1**, 87-95.

Georgatsou, E., Mavragiannis L.A., Fragiadakis, G.S. and Alexandraki, D. (1997). The yeast Fre1p/Fre2p cupric reductases facilitate copper uptake and are regulated by the copper-modulated Mac1p activator. *J. Biol. Chem.* **272**, 13786-13792.

Grill, E. , Winnacker, E.-L. and Zenk, M.H. (1985) Phytochelatins: the principle heavy-metal complexing peptides of higher plants. *Science* **230**, 674-676.

Grill, E., Winnacker, E.-L. and Zenk, M.H. (1987) Phytochelatins, a class of heavy-metal-binding peptides from plants, are functionally analogous to metallothioneins. *Proc. Natl. Acad. Sci. USA* **84**, 439-443.

Grill, E., Löffler, S., Winnacker, E.-L. and Zenk, M.H. (1989) Phytochelatins, the heavy-metal-binding of plants, are synthesized from glutathione by a specific gamma-glutamylcysteine dipeptidyl transpeptidase (phytochelatin synthase). *Proc. Natl. Acad. Sci. USA* **86**, 6838-6842.

Grime, J.P. (1979) *Plant Strategies and Vegetation Processes*, Wiley, Chichester.

Grotz, N., Fox, T., Connolly, E., Park, W., Guerinot, M.L. and Eide, D. (1998) Identification of a family of zinc transporter genes from Arabidopsis that respond to zinc deficiency. *Pro. Natl. Acad. Sci. USA* **95**, 7220-7224.

Hare, P.D., Cress, W.A. and Van Staden, J. (1998) Dissecting the roles of osmolyte accumulation during stress. *Plant Cell Env.* **21**, 535-553.

Harmens, H. (1994) *The physiology of zinc tolerance in* Silene vulgaris. PhD Thesis, Vrije Universiteit, Amsterdam.

Harmens, H., Cornelisse, E., Den Hartog, P.R., Ten Bookum, W.M., and Verkleij, J.A.C. (1993b) Phytochelatins do not play a key role in naturally selected zinc tolerance in *Silene vulgaris* (Moench) Garcke. *Plant Physiol.* **103**, 1305-1309.

Harmens, H., Gusmão, N.G.C.P.B., Den Hartog, P.R., Verkleij, J.A.C. and Ernst, W.H.O. (1993a) Uptake and transport of zinc in zinc-sensitive and zinc-tolerant *Silene vulgaris*. *J. Plant Physiol.* **141**, 309-315.

Harmens, H., Koevoets, P.L.M., Verkleij, J.A.C. and Ernst, W.H.O. (1994) The role of low molecular weight organic acids in the mechanism of increased zinc tolerance in *Silene vulgaris* (Moench) Garcke. *New Phytol.* **126**, 615-621.

Howden, R., Goldsbrough, P.B., Andersen, C.R. and Cobbett, C.S. (1995) Cadmium-sensitive *cad1* mutants of *Arabidopsis thaliana* are phytochelatin deficient. *Plant Physiol.* **107**, 1059-1066.

Kampfenkel, K., Kushnir, S., Babiychuk, E., Inzé, D. and Van Montagu, M. (1995) Molecular characterization of a putative *Arabidopsis thaliana* copper transporter and its yeast homologue. *J. Biol. Chem.* **270**, 28479-28486.

Kastori, R., Petrovic, M. and Petrovic, N. (1992) Effect of excess lead, cadmium, copper and zinc on water relations in sun flower. *J. Plant Nutr.* **15**, 2427-2439.

Krall, J.P., Edwards, G.E. and Andreo, C.S. (1989) Protection of pyruvate, Pi dikinase from maize against cold lability by compatible solutes. *Plant Physiol.* **89**, 280-285.

Kuge, S., Jones, N. and Nomota, A. (1977) Regulation of yAP-1 nuclear localization in response to oxidative stress. *EMBO Journal* **16**, 1710-1720.

Labbe S., Zhu, Z., and Thiele D.J. (1997) Copper-specific transcriptional repression of yeast genes encoding critical components in the copper transport pathway. *J. Biol. Chem.* **272**, 15951-15958.

Leshem, Y.Y. and Kuiper, P.J.C. (1996) Is there a GAS (general adaptation syndrome) response to various types of environmental stress? *Biol. Plant.* (Prague) **38**, 1-18.

Macnair, M.R. (1983) The genetic control of copper tolerance in the yellow monkey flower *Mimulus guttatus*. *Heredity* **50**, 283-293.

Macnair, M.R., Cumbes, Q.J. and Meharg, A.A. (1992) The genetics of arsenate tolerance in Yorkshire fog *Holcus lanatus*. *Heredity* **69**, 325-335.

Marschner, H. (1995) *Mineral Nutrition of Higher Plants*, Academic Press, London.

Matsumoto, Y., Okada, Y., Min, K.-S., Onosaka, S. and Tanada, K. (1990) Amino acids and peptides XXVII. Synthesis of phytochelatin-related peptides and examination of their heavy metal-binding properties. *Chem. Pharm. Bull.* **38**, 2364-2368.

Mathys, W. (1975) Enzymes of heavy-metal-resistent and non-resistent populations of *Silene cucubalus* and their interaction with some heavy metals *in vitro* and *in vivo*. *Physiol. Plant.* **33**, 161-165.

Mathys, W. (1977) The role of malate, oxalate, and mustard oil glucosides in the evolution of zinc-resistance in herbage plants. *Physiol. Plant.* **40**, 130-136.

Meharg, A.A. and Macnair, M.R. (1992) Genetic correlation between arsenate tolerance and the rate of influx of arsenate and phosphate in *Holcus lanatus*. *Heredity* **69**, 336-341.

Moran, J.F., Becana, M., Iturbe-Ormaetxe, I., Frechilla, S., Klucas, R.V. and Aparicio-Tejo, P. (1994) Drought induces oxidative stress in pea plants. *Planta* **194**, 346-352.

Mullins, M., Hardwick, K. and Thurman D.A. (1985) Heavy metal location by analytical electron microscopy in conventionally fixed and freeze-substituted roots of metal-tolerant and non-tolerant ecotypes. In: *Proc. Int. Conf. Heavy Metals in the Environment, Athens*, Vol 1. CEP Consultants, Edinburgh, pp. 43-46.

Murphy, A. and Taiz, L. (1995) Comparison of metallothionein expression and nonprotein thiols in ten *Arabidopsis* ecotypes. *Plant Physiol.* **109**, 945-954.

Murphy, A., Zhou, J., Goldsbrough, P.B. and Taiz, L. (1997) Purification and immunological dentification of metallothioneins 1 and 2 from *Arabidopsis thaliana*. *Plant Physiol.* **113**, 1293-1301.

Neumann, D., Lichtenberger, O., Günther, D., Tsiersch, K. and Nover, L. (1994) Heat-shock proteins induced heavy-metal tolerance in higher plants. *Planta* **194**, 360-367.

Nover, L. and Höhfeld, I. (1996) Hohe Temperaturen. In: *Stress bei Pflanzen* (Brunold, Ch., Rüegsegger, A., Brändle, R., eds). Verlag Paul Haupt, Bern, pp. 49-70.

Ooi, C.E., Rabinovich, E., Dancis, A., Bonifacino, J.S. and Klausner, R.D. (1996). Copper-dependent degradation of the *Saccharomyces cerevisiae* plasmamembrane copper transporter ctr1p in the apparent absence of endocytosis. *EMBO Journal* **15**, 3515-3523.

Ortiz D.F., Kreppel, L., Speiser, D.M., Scheel, G., McDonald, G. and Ow, D.W. (1992) Heavy metal tolerance in the fission yeast requires an ATP-binding cassette-type vacuolar membrane transporter. *EMBO Journal* **11**, 3491-3499.

Paleg, L.G., Douglas, T.J., Van Paal, A. and Keech, D.B. (1981) Proline, betain and other organic solutes protect enzymes against heat inactivation. *Aust. J. Plant Physiol.* **8**, 107-114.

Repp, G. (1963) Die Kupferresistenz des Protoplasmas höherer Pflanzen auf Kupferertzböden. *Protoplasma* **57**, 643-659.

Salt, D.E. and Rauser, W.E. (1995) MgATP-dependent transport of phytochelatins across the tonoplast of oat roots. *Plant Physiol.* **107**, 1293-1301.

Salt, D.E. and Wagner, G.J. (1993) Cadmium transport across tonoplast vesicles from oat roots. Evidence for Cd^{2+}/H^+ antiport activity. *J. Biol. Chem.* **268**, 12297-12302.

Schat, H. and Kalff, M.M.A. (1992) Are phytochelatins involved in differential metal tolerance or do they merely reflect metal-imposed strain? *Plant Physiol.* **99**, 1475-1480.

Schat, H., Kuiper, E., Ten Bookum, W.M. and Vooijs, R. (1993) A general model for the genetic control of copper tolerance in *Silene vulgaris*: evidence from crosses between plants from different tolerant populations. *Heredity* 70, 142-147.

Schat, H., Sharma, S.S. and Vooijs, R. (1997) Heavy metal-induced accumulation of free proline in a metal-tolerant and a nontolerant ecotype of *Silene vulgaris*. *Physiol. Plant.* **101**, 477-482.

Schat, H. and Ten Bookum, W.M. (1992a) Metal-specificity of metal tolerance syndromes in higher plants. In: *The Vegetation of Ultramaffic (Serpentine) Soils* (Baker, A.J.M., Reeves, R.D. and Proctor, J., eds). Intercept Ltd, Andover, pp. 337-352.

Schat, H. and Ten Bookum, W.M. (1992b) The genetic control of copper tolerance in *Silene vulgaris*. *Heredity* **68**, 219-229.

Schat, H. and Vooijs, R. (1997a) Effects of decreased cellular glutathione levels on growth, membrane integrity and lipid peroxidation in roots of copper-stressed *Silene vulgaris*. In: *Contaminated Soils* (Prost, R., ed.), INRA, Paris, CD-Rom, 057.PDF.

Schat, H. and Vooijs, R. (1997b) Multiple tolerance and co-tolerance to heavy metals in Silene vulgaris: a co-segregation analysis. *New Phytol.* **136**, 489-496.

Schat, H., Vooijs, R. and Kuiper, E. (1996) Identical major gene loci for heavy metal tolerances that have independently evolved in different local populations and subspecies of *Silene vulgaris*. *Evolution* 50, 1888-1895.

Schöffl, F., Prändle, R. and Reindl, A. (1998) Regulation of the heat-shock response. *Plant Physiol.* **117**, 1135-1141.

Searcy, K.B. and Mulcahy, D.C. (1985) The parallel expression of metal tolerance in pollen and sporophytes of *Silene dioica* (L.) Clairv., *S. alba* (Mill.) Krause and *Mimulus guttatus* DC. *Theor. Appl. Genetics* 69, 597-602.

Selinioti, E., Nikolopulos, D. and Manetas, Y. (1987) Organic cosolutes as stabilisers of phosphoenol pyruvate carboxylase in storage: an interpretation of their action. *Aust. J. Plant Physiol.* **14**, 203-210.

Sewell, A.K., Yokoya, F., Yu, W., Miyagawa, T., Murayama, T. and Winge, D.R. (1995) Mutated yeast heat shock transcription factor exhibits elevated basal transcriptional activation and confers metal resistance. *J. Biol. Chem.* **270**, 25079-25086.

Smirnoff, N. and Cumbes Q.J. (1989) Hydroxyl radical scavenging activity of compatible solutes. *Phytochemistry* **28**, 1057-1060.

Speiser, D.M., Ortiz, D.F., Kreppel, L., Scheel, G., McDonald, G. and Ow, D.W. (1992) Purine biosynthetic genes are required for cadmium tolerance in schizosaccharomyces pombe. *Mol. Cell. Biol.* **12**, 5301-5310.

Strange, J. and Macnair, M.R. (1991) Evidence for a role for the cell membrane in copper tolerance of *Mimulus guttatus* Fisher ex DC. *New Phytol.* **119**, 383-388.

Vázquez, M.D., Poschenrieder, C., Barcelo, J., Baker, A.J.M. and Cope, G.H. (1994) Compartmentation of zinc in roots and leaves of the zinc hyperaccumulator *Thlaspi caerulescens* J & L Presl. *Botanica Acta* **107**, 243-250.

Verkleij, J.A.C., Koevoets, P.L.M., Blake-Kalff, M.M.A. and Chardonnens, A.N. (1998) Evidence for an important role of the tonoplast in the mechanism of naturally selected zinc tolerance in *Silene vulgaris*. *J. Plant Physiol.* **153**, 188-191.

Vögeli-Lange, R. and Wagner, G.J. (1989) Subcellular localization of cadmium and cadmium-binding peptides in tabacco leaves. *Plant Physiol.* **92**, 1086-1093.

Zhao, H. and Eide, D.J. (1997). Zap1p, a metalloregulatory protein involved in zinc-responsive transcriptional regulation in *Saccharomyces cerevisiae*. *Mol. Cell. Biol.* **17**, 5044-5052.

Zhou J. and Goldsbrough, P.B. (1995a) Functional homologes of fungal metallothionein genes from *Arabidopsis*. *Plant Cell* **6**, 875-884.

Zhou, J. and Goldsbrough, P.B. (1995b) Structure, organization and expression of the metallothionein gene family in *Arabidopsis*. *Mol. Gen. Genet.* **248**, 318-328.

Zhou, P. and Thiele, D.J. (1993) Rapid transcriptional autoregulation of a yeast metalloregulatory transcription factor is essential for high-level copper detoxification. *Genes & Development* **7**, 1824-1831.

Zhu, Z., Labbe, S., Pena, M.M.O. and Thiele, D.J. (1998). Copper differentially regulates the activity and degradation of yeast Mac1 transcription factor. *J. Biol. Chem.* **273**, 1277-1280.

MAGNESIUM TRANSPORT AND ALUMINIUM TOLERANCE

KATHY M. FOWLER, COLIN J. W. MACDIARMID, KEITH D. RICHARDS AND RICHARD C. GARDNER

School of Biological Sciences
University of Auckland
Private Bag 92019
Auckland
New Zealand.

1. Introduction

Aluminium toxicity is a major factor limiting crop productivity internationally. Under acidic soil conditions, Al is converted from insoluble forms into soluble Al. The ion is toxic to plants and causes an immediate cessation of root growth. The mechanism of this inhibition has not been clearly established, despite a plethora of suggestions (reviewed by Kochian, 1995).

The goal of our research is to identify genes that confer Al tolerance in crop plants, with the long term aim of engineering Al tolerance in a range of susceptible species. To date, only one gene conferring Al tolerance has been identified. Recently, de la Fuente et al (1997) showed that expression of a citrate synthase gene from *Pseudomonas aeruginosa* conferred Al tolerance on tobacco plants and cultured papaya cells. This was the first report of successful engineering of Al tolerance in plants.

We are using brewers yeast (*Saccharomyces cerevisiae*) as a model system for Al toxicity studies and have shown that over expression of two magnesium transport genes confers Al tolerance on yeast (reviewed below). Here we present evidence that two bacterial Mg transport genes also confer Al tolerance on yeast and describe preliminary results of expressing these Mg transport genes in *Arabidopsis thaliana*.

2. Yeast as a model system for studying Al tolerance

Brewers yeast provides a useful model system for the study of Al tolerance. Yeast grows well at acidic pH, and is the easiest and best-characterised eukaryote in which to undertake genetic studies. We have shown that Al is toxic to yeast in appropriate defined media - in addition to lowering the pH and phosphate concentration, reducing the Mg concentration to low levels proved to be key to obtaining sensitivity to Al (MacDiarmid and Gardner, 1996). The ability of high levels of Mg to ameliorate Al toxicity in yeast was reminiscent of the situation in plants, where a range of cations can block the effect of Al on root elongation. However, in yeast Ca and K had no such effect, and amelioration seemed to be Mg-specific.

Several other results pointed to the importance of Mg transport in Al toxicity. The Mg uptake system in yeast has previously been shown to transport a range of divalent

G. Gissel-Nielsen and A. Jensen (eds.), Plant Nutrition – Molecular Biology and Genetics, 327–332.

cations, including cobalt. We showed that Al blocked uptake of Co very effectively. We also showed that a mutant with reduced activity of the Mg transport system was more sensitive to Al than wild-type yeast. These results led to the hypothesis that Al blocks Mg uptake in yeast (Macdiarmid and Gardner, 1996). It has been possible to isolate a range of Al-sensitive mutants of yeast (Schott and Gardner, 1977). These mutations proved to be very common and to occur in a wide range of different genes, some of which influenced a range of stress responses. It appears that, in yeast, there is a basal level of tolerance to Al which has a complex genetic basis.

Toxic levels of Al have been shown to induce at least two stress genes in yeast, *SED1* and *HSP150* (Ezaki et al. 1998). The *HSP150* gene appears to play a role in protecting against Al stress, since disrupted *hsp150* strains were sensitive to Al. This is the first demonstration that genes induced by Al stress play a protective role against that stress. Both *hsp150* and *sed1* mutants were also sensitive to oxidative stress, suggesting that there is a link between these two stresses in yeast. Similar links have been proposed in plants (see Richards et al. 1998).

3. Over expression of Mg transport genes confers Al tolerance

We recently identified two yeast genes that give Al tolerance when over expressed (Macdiarmid and Gardner, 1998). The two genes, called *ALR1* and *ALR2*, are closely related to each other, and are homologous to a bacterial gene family known as CorA, whose function is to transport Mg into bacterial cells (see Smith and Maguire,1998). The CorA family of transporters is unusual, with only three transmembrane-spanning domains and a large hydrophilic extracellular domain. The mechanism by which these genes transport Mg is unclear (Smith and Maguire, 1998).

Phenotypic and physiological analysis of mutant yeast strains suggested that the two *ALR* genes transport Mg in yeast (Macdiarmid and Gardner, 1998). Disrupted strains required extremely high levels of Mg in the media for growth. Over-expressing strains were tolerant to both Al and Ga, but conferred increased sensitivity to a range of cations (Co, Zn, Mn, Ni, Cu, Ca and La). This result suggested that the Mg transport system also transported these ions, in agreement with previous competition studies. Co uptake measurements of mutant strains confirmed this interpretation, with a higher rate of Co uptake in over-expressing strains and a significantly reduced uptake in the disrupted strain. The over-expressing strains showed some residual Co transport in the presence of a level of Al sufficient to inhibit the wild-type strain. The simplest interpretation of these results is that Al toxicity in yeast occurs because Al inhibits the activity of the divalent cation uptake system, encoded by the *ALR* genes. The major function of this uptake system under normal growth conditions is to transport Mg into the cell. Hence under normal growth conditions it appears that Al toxicity equates with Mg deficiency in yeast. At present we cannot determine whether the inhibition of Mg transport by Al is a direct effect on the transport proteins, or whether Al acts via an indirect effect such as by altering the membrane potential.

Expression of bacterial Mg transport genes in yeast

As a confirmation of the hypothesis above, we set out to express two Mg transport genes from bacteria in yeast and test whether they confer Al tolerance. Maguire and coworkers have identified two Mg transport genes in bacteria that are unrelated to the CorA family (reviewed in Smith and Maguire, 1998). The MgtA/MgtB genes in *Salmonella typhimurium* and *E. coli* are P-type ATPases with 10 transmembrane domains. They comprise the inducible, high affinity transport system for Mg. Another set of genes, called MgtE, are a novel family of transporters with 5 transmembrane domains that are found in some bacterial species. Both classes of Mg transport genes transport a range of divalent cations in addition to Mg.

4.1 CONSTRUCTION OF YEAST PLASMIDS EXPRESSING MgtA AND MgtE
The MgtA gene was amplified from *Salmonella typhimurium* (obtained from D. Love, University of Auckland), and the MgtE gene from *Providencia stuartii* (obtained from the New Zealand Reference Collection at the Institute of Environmental Science). The primers (added restriction sites are italicised) and PCR conditions were as follows.
MgtA1: *GAATTC*ACTATGCTAAAAATCATTACCCGCCA
MgtA2: *TCTAGA*TTACTGCCAGCCATAACGTCTG
MgtE1: *GAATTC*AATATGAGCCAATCTTATTTGCCTG
MgtE2: *TCTAGA*TTAGGCGAACTCCGTCATTACC.
Amplification used the Expand High Fidelity Kit (Boehringer) in a Gene Amp PCR System 2400 (Perkin Elmer). For *MgtA*, the run parameters were: 10 cycles of 15 s at 94 degrees C, 30 s at 55 degrees C and 1 min at 72 degrees C. After 10 cycles, the 72 degrees C elongation time was extended by a further 15 s at each cycle. In the final cycle, the elongation temperature was held for 7 min. The same parameters were used to amplify *MgtE* but the annealing temperature was reduced to 50 degrees C. The *MgtA* and *MgtE* PCR products were cloned and sequenced to confirm their identity and inserted behind the *GAL1* promoter in the yeast vector pYES3. Both plasmids were introduced into the yeast strain FY833 and an isogenic *ALR1* disruption mutant CM45 (see MacDiarmid and Gardner, 1998). CM45 is derived from S288C, and shows no expression of the *ALR2* gene; disruption of the *ALR1* gene produced the same phenotype as disruption of both *ALR* genes (C. MacDiarmid, unpublished results).

4.2 MgtA AND MgtE ARE EXPRESSED IN YEAST
Both the MgtA and MgtE genes complemented the Mg growth defect of the yeast *alr1* strain. CM45 strains containing the MgtE or MgtA genes were able to grow on 10 and 20 microM Mg, respectively. In contrast, the strain containing the vector required 250 mM for growth (standard yeast plates contain 4 mM Mg). The two bacterial genes are considerably less efficient at complementation than the *ALR1* gene, which allows growth at 200 microM Mg even in a single copy. However, the complementation by the bacterial genes occurred on glucose plates and was not significantly improved on galactose medium. This result suggests that transcriptional activation of the gal1 promoter did not influence the amount of protein product produced. Effective function of the bacterial genes in yeast may be limited by translation or protein localisation in the yeast membrane.

To confirm that expression of the bacterial genes was occurring, we tested the metal tolerance phenotype of strains expressing MgtA and MgtE. Both MgtA and MgtE transport a range of divalent cations in addition to Mg (Smith and Maguire, 1998). These experiments needed to be performed in a strain with wild-type *ALR1* function, since low levels of Mg are required for metal sensitivity. Expression of MgtA and MgtE both altered sensitivity to a range of metals (see Table 1). MgtE increased the sensitivity of the strain to Zn, Ni, Co and Mn, while MgtA conferred slightly increased tolerance to at least some of these cations. These phenotypic changes in metal tolerance are consistent with the bacterial genes being expressed in yeast. However, the results differed from expectation in two ways. First, the pattern of changes seen is not consistent with the relative substrate preferences for these enzymes, as deduced from competitive inhibition studies (see Smith and Maguire, 1998). The differences may reflect the different assay systems used, or the different hosts used for expression. Second, expression of *MgtA* conferred tolerance to cations, when the simple expectation from increased metal uptake would be increased sensitivity. However, this simple relationship does not always hold for over expression of transport proteins; differences in transport kinetics and internal pool sizes, along with mis-localisation of the over-expressed protein in internal membranes, could result in unexpected changes in cation tolerance.

Table 1. Expression of *MgtA* and *MgtE* alter the divalent cation sensitivity of yeast

Strain/Plasmid	Zn	Ni	Mn	Co
FY833/pYES3-*ALR2*	---	---	---	---
FY833/pYES3-*MgtE*	---	--	-	--
Fy833/pYES3-*MgtA*	+	(+)	+	wt

Spots containing five-fold serial dilutions of an exponentially growing yeast culture were plated on SC plates containing a range of metal concentrations (see MacDiarmid and Gardner, 1998). Growth was compared with FY833 containing pYES3 in all cases. --- = strain showed much greater sensitivity, -- = strain showed moderate sensitivity, - = strain showed a small degree of sensitivity, + = strain showed a small degree of tolerance, (+) = strain showed a very slight degree of tolerance, wt = same as wild-type strain.

4.3 MgtA AND MgtE CONFER AL TOLERANCE ON YEAST

These phenotypic changes demonstrated that the bacterial genes were expressed to some degree in yeast. We next addressed the question of whether the yeast lines over expressing bacterial Mg transporters conferred tolerance to Al.

For wild type strains expressing either MgtA and MgtE, a small increase in Al tolerance was observed. Under stringent conditions of Al selection (LPM plates with pH 3.5, 100 microM Mg; see MacDiarmid and Gardner 1996) both bacterial genes allowed growth of FY833 cells on plates containing 30 microM Al, with the MgtA lines growing slightly better that the MgtE lines. Cells containing the vector only did not grow at this Al concentration.

The Al tolerance of strains expressing MgtA and MgtE was highly reproducible, occurring with over 20 individual transformed colonies from each plasmid, chosen from six independent transformations of the disrupted strain. However, the level of Al tolerance conferred is very low, since under the same conditions, FY833 cells containing pYES-*ALR2* plasmids are not inhibited until the concentration reaches 150 microM Al. This small degree of tolerance was not unexpected, since both bacterial genes also provide relatively poor complementation of the Mg transport deficiency, and are probably poorly expressed in yeast. We have observed that conferring Al tolerance appears to require a higher level of expression than complementing Mg deficiency (see MacDiarmid and Gardner, 1998).

These results suggest that we have expressed two bacterial Mg transport genes in yeast, albeit at relatively low levels. Expression of either Mg transport gene in yeast conferred increased tolerance to Al, supporting the hypothesis that Al toxicity equates with Mg deficiency.

5. Mg transport genes and Al tolerance in plants.

In yeast, the results to date are entirely consistent with the idea that Al toxicity is caused by a block in Mg uptake. In plants, there is some evidence supporting the hypothesis that Al blocks Mg uptake in plants. However, Al has been shown to block root elongation in medium completely lacking Mg, so that simple Mg deficiency seems an unlikely explanation for Al toxicity (see discussion in MacDiarmid and Gardner,1996).

We are currently attempting to determine whether over expression of Mg transport genes in plants confers Al tolerance. Four magnesium transport genes have been introduced to Arabidopsis: the yeast *ALR* genes, the Salmonella Mgt*A* gene and the *Pseudomonas* MgtE gene. All the transgenic plants contained the gene driven by the cauliflower mosaic virus 35S promoter. For two of the genes, MgtE and *ALR2*, the levels of transcript obtained in all the transgenic lines were very low, barely detectable by Northern hybridisation. For the other two genes, good levels of transcript were detected in the best lines (up to 0.5% of total mRNA). However, for both genes, shortened transcripts were detected. For *ALR1*, a short 2 kb transcript formed the great majority of the hybridising RNA, with only a small amount of the expected 3.3 kb mRNA found. For MgtA, a reasonable amount of apparently full-length transcript was present. These MgtA containing plants are currently under investigation.

The difficulties with expression of prokaryote and yeast transport genes in plants has led us to initiate the isolation of magnesium transport genes from plants. The approach being taken is to use an *Arabidopsis* cDNA library to complement the Mg deficiency of an *alr1alr2* yeast strain. These experiments will allow us to determine whether over expression of Mg transport genes in plants can confer Al tolerance.

References

de la Fuente JM, Ramirez-Rodriguez V, Cabrera-Ponce JL, Herrera-Estrella L (1997). Aluminum tolerance in transgenic plants by alteration of citrate synthesis. Science 276:1566-1568

Ezaki B, Gardner RC, Ezaki Y, Kondo H, Matsumoto H (1998). Protective roles of two aluminum (Al)-induced genes, HSP150 and SED1 of Saccharomyces cerevisiae, in Al and oxidative stresses. FEMS Microbiol Lett 159:99-105

Kochian, LV (1995). Cellular mechanisms of aluminum toxicity and resistance in plants. Annu Rev Plant Physiol Plant Mol Biol 46: 237-260

MacDiarmid CW, Gardner RC (1996). Al toxicity in yeast. A role for Mg? Plant Physiol112:1101-1109

MacDiarmid CW, Gardner RC (1998). Overexpression of the Saccharomyces cerevisiae magnesium transport system confers resistance to aluminum ion. J Biol Chem 273:1727-1732

Richards KD, Schott EJ, Sharma YK, Davis KR, Gardner RC (1998). Aluminium induces oxidative stress genes in *Arabidopsis thaliana*. Plant Physiol. 116: 409 - 418

Schott EJ, Gardner RC (1997). Aluminum-sensitive mutants of Saccharomyces cerevisiae. Mol Gen Genet 254:63-72

Smith RL, Maguire ME (1998). Microbial magnesium transport: unusual transporters searching for identity. Mol Microbiol 28:217-226.

GENETIC VARIABILITY IN CESIUM RESISTANCE IN RICE (*ORYZA SATIVA* L.)

HIROSHI HASEGAWA[1] and YUTAKA OKUMOTO[2]

[1] School of Environmental Science, The University of Shiga Prefecture, Hikone, Shiga 522-8533, Japan

[2] Graduate School of Agriculture, Kyoto University, Sakyo-ku, Kyoto 606-8502, Japan

1. Introduction

It is of practical importance to investigate genetics of ion uptake by plant roots, because breeding of crops with efficient uptake and use of nutrient ions is one of the goals for low-input crop production. In particular, many attentions have been paid on the uptake of nitrogen (Wray 1988, Crawford 1995). Recent advances have revealed that both nitrate and ammonium are absorbed into plant cells mediated by ion-specific transporter(s) in the plasma membrane of root epidermis cells. The genes encoding nitrate and ammonium transporter have been cloned and sequenced in several plant species (Tsay et al. 1993, Ninnemann et al. 1994, Trueman et al. 1996, Quesada et al. 1997, Tanaka et al. 1998).

For genetic improvement of nutrient uptake, it is also necessary to manipulate genes involving regulation of an ion transporter. Electrophysiological studies have shown that H^+-ATPase, a proton pump, plays an important role in the transport of both anions and cations across the plasma membrane of root epidermal cells. Nitrate uptake in plants is drived by electrogenic proton cotransport (Crawford 1995). However, there might be difficulty in genetic improvement of energetics of nitrate uptake because of multiplicity of plasma membrane H^+-ATPases (Sheahan et al. 1993).

Several cesium resistant mutants were isolated in *Arabidopsis thaliana* (Sheahan et al. 1993), and rice mutants defective in nitrate uptake were also shown to be cesium resistant (Hasegawa 1996). Electrophysiological study of a cesium resistant *Arabidopsis* mutant demonstrated differential response to K^+-induced membrane depolarization between the mutant and the wild type (Maathuis and Sanders 1996). From these results, we assume that cesium resistance might be a useful parameter for detecting genetic variability in the H^+-ATPase activity as well as ion uptake ability. The aims of this study are to identify rice genes involving regulating factors of nitrate uptake using

G. Gissel-Nielsen and A. Jensen (eds.), Plant Nutrition – Molecular Biology and Genetics, 333–339.

cesium resistance as a selective marker. For the first step of this purpose, screening for cesium resistant genotypes and selection for cesium resistant mutants were carried out.

2. Materials and Methods

2.1. SCREENING FOR CESIUM RESISTANT GENOTYPES

Fifty-four rice (*Oryza sativa* L.) genotypes including 6 Japanese modern lowland cultivars, 5 Japanese upland cultivars, 21 Japanese old and native cultivars and 22 foreign (outside of Japan) cultivars were used. Two mutants defective in nitrate uptake reported by Hasegawa (1996) were also used.

Seedlings were cultivated with 0 and 20 μ M cesium chloride (CsCl) for 14 days from sowing in a biotron controlling at 25±1℃ with fluorescent lamps (16h lighting). CsCl concentration for the screening was determined based on the preliminary test, in which four rice cultivars were cultivated 0 to 50 μ M CsCl for 14 days as same condition as described above. First leaf length on the 14th day after sowing was measured and relative first leaf length (RFLL), defined as percentage of the length of cesium-treated seedlings to that of control, was calculated as a parameter of cesium sensitivity.

2.2 ISOLATION AND CHARACTERIZATION OF CESIUM RESISTANT MUTANTS

M2 seeds mutagenized by gamma-ray (200 Gy) on dry seeds of a *japonica* cultivar, Nipponbare, were allowed to germinate and were cultivated with 20 μ M CsCl in a growth chamber controlled at 25±1℃ under natural daylight supplemented with fluorescent lamps (16h lighting). Four weeks after sowing, surviving seedlings were selected as cesium resistant variants and were transferred to hydroponic culture without cesium. In M3, seedlings of selfed progeny of the variants were cultivated with 0 and 20 μ M CsCl for 14 days for reexamining cesium resistance. Further genetic and physiological studies have been carried out using the progeny of CSR9, one of the selected mutants, in M4 and the following generations. For the genetic analysis, CSR9 and Nipponbare were crossed reciprocally. F2 seedlings were cultivated with 20 μ M CsCl for 14 days from sowing and the first leaf length was measured.

2.3. ION UPTAKE MEASUREMENT

For ion uptake measurement, seedlings were cultivated with deionized water in a biotron at 25±1℃. Cesium uptake from 20 μ M CsCl supplemented with calcium 400 μ M $CaSO_4$ was measured at 28±1℃ on the 14th day after sowing. Uptake of ions from 250 μ M solution of $CsNO_3$ or KNO_3 (supplemented with 400 μ M $CaSO_4$) was measured at 28±1℃ on the 15th day from sowing. In the case of ion uptake from nitrate solution, seedlings were placed in 100 μ M KNO_3 + 400 μ M $CaSO_4$ solution for 24h at 25±1℃ before the experiment for the induction of nitrate transporter. Uptake of each ion was determined by the depletion from the solution for 3h. Cesium concentration were determined with an atomic absorption spectrometer (Shimadzu AA-6400G, Japan) and concentration of potassium and nitrate was determined with an ion chromatography (Dionex, DX-300, USA).

2.4. MOLECULAR STUDIES

Roots of CSR 9 and Nipponbare were excised from field-grown plants and then the plants were transferred to the culture with deionized water in a greenhouse for a week. Newly developed roots were collected for the molecular analysis.

Simplified differential display method performed to investigate whether cDNA polymorphism was detected between CSR9 and the wild type or not. Extraction of mRNA and synthesis of first-strand cDNA were carried out by the methods as described by Yoshida et al. (1994). The second-strand cDNA was synthesized with PCR using twenty different 10-base RAPD primers (A01 to A20, Operon Tec.) in each reaction. PCR products were resolved by electrophoresis in 1.0% agarose gel and were stained with ethidium bromide.

3. Results

3.1. CESIUM SENSITIVITY AND SCREENING FOR CESIUM RESISTANT GENOTYPES

Prior to the screening for cesium resistant genotypes, seedlings of two *japonica* cultivars, Nipponbare and Koshihikari, and two *indica* cultivars, IR28 and Leuang Tawng, were cultivated with 0 to 50 μ M CsCl from sowing. First leaf length was decreased with increase in cesium concentration. Seedling growth was almost completely inhibited when seedlings were cultivated with 50 μ M CsCl. At the concetration of 20 μ M CsCl, clear difference in cesium response was observed between *japonica* cultivars and *indica* cultivars.

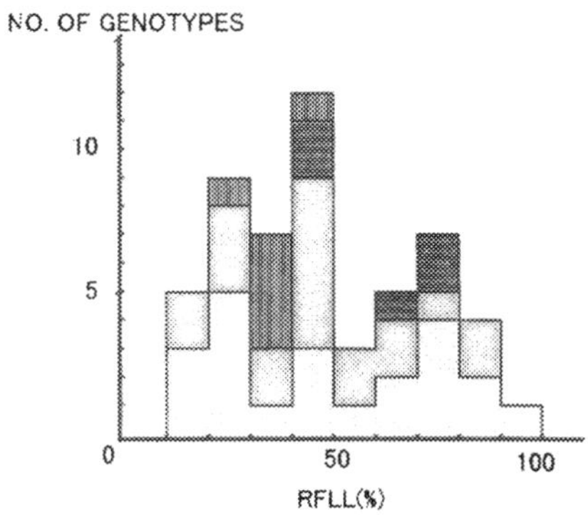

Fig.1 Relative first seedling length (RFLL) in 54 genotypes evaluated on 14th day after sowing. Seedlings were cultured with 0 and 20 μ M CsCl.
Japanese modern lowland cultivars; Japanese old and native cultivars; Japanese upland cultivars; Foreign cultivars (outside of Japan)

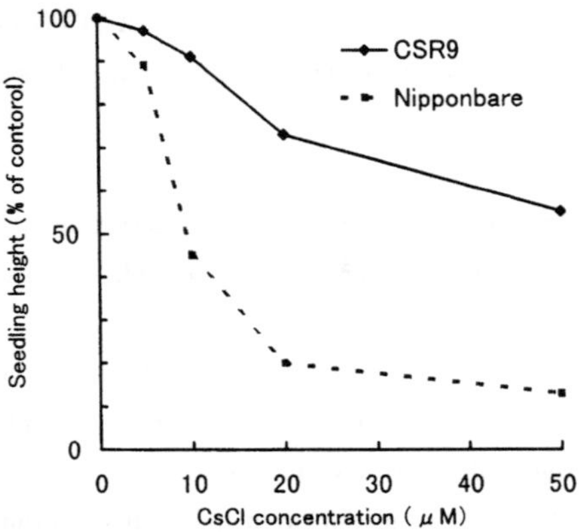

Fig. 2 Difference in Cs sensitivity between CSR9 and the wild type cultivar, Nipponbare. Seedlings were cultivated with CsCl and the seedling height was measured on 14th day after sowing.

As shown in Fig. 1, there was great variability in cesium sensitivity expressed as RFLL. In the genotypes tested, RFLLs in 14-day-old seedlings ranged from 10.6 to 91.1%. Cesium resistant group, in which RFLL was more than 80%, included 5 genotypes, f145 (India), Bengara (Japanese native), Nanking-koutou (China), Eigo (Japanese native) and Kasalath (India). On the other hand, 4 genotypes, Muskakdanti (India), Lemout (U.S.A.), Bonnet 73 (U.S.A.) and Kitadani 2 (Japanese native) were very sensitive to cesium and their RFLLs were less than 20%. From Japanese modern lowland cultivars and U.S.A. cultivars, no genotypes classified as a higher RFLL group (RFLL was 50% or higher) were found. RFLLs of two mutants defective in nitrate uptake (NUE13-1 and NUE36-4) were higher than that of the wild type cultivar (data not shown), Nipponbare.

3.2. ISOLATION AND CHARACTERIZATION OF CESIUM RESISTANT MUTANTS

Of 10,800 M2 seedlings screened, 23 seedlings were still growing with 20 μ M CsCl after 4 weeks from sowing. In M3, the selfed progenies of only 6 selected seedlings in M2 recovered cesium resistance and were identified as a cesium resistant mutant. Other selected seedlings in M2 were died after transplanting to hydroponic culture without cesium or did not recover the resistance in M3.

Clear difference in cesium sensitivity was observed between CSR9 and Nipponbare (Fig.2). In particular, when the seedlings were cultured 50 μ M CsCl, first leaf length in CSR9 was above 50% of that grown without cesium. First leaf length of F2 population from the reciprocal crosses bewteen CSR9 and Nipponbare was measured, but no clear segregation of cesium resistants was observed (data not shown).

Difference in ion uptake from the roots was also found between CSR9 and Nipponbare. In CSR9, uptake of ions analyzed in this experiment were lower than that in Nipponbare except for cesium uptake from 250 μ M $CsNO_3$ (Table 1). Cesium uptake from 250 μ M CsCl in CSR9 was 20%

higher than that of Nipponbare. In CSR9, nitrate uptake from different cation salt resulted in same reduction (65 to 75% of that of Nipponbare), although nitrate uptake from cesium salt was lower than that from potassium salt. Uptake of cesium from 20 μ M CsCl and potassium in CSR9 was also reduced to around 70% of that in Nipponbare.

Table 1 Ion uptake ability of CSR9 and the wild type cultivar, Nipponbare. Each ion measurement was replicated 5 times (m±s.d).

	Uptake (μ mol·g^{-1}fw·h^{-1})	
Ion	CSR9	Nipponbare
Cs^+ 1)	0.269±0.027 (65.1*)	0.413±0.039
Cs^+ 2)	2.161±0.426 (122.3)	1.767±0.446
K^+ 3)	1.827±0.157 (74.3)	2.460±0.365
NO_3^- 2)	0.823±0.139 (65.4)	1.258±0.090
NO_3^- 3)	1.405±0.161 (70.6)	1.991±0.095

* Percent of that of Nipponbare
1) uptake from 20 μ M CsCl + 400 μ M $CaSO_4$
2) uptake from 250 μ M $CsNO_3$ + 400 μ M $CaSO_4$
3) uptake from 250 μ M KNO_3 + 400 μ M $CaSO_4$

3.3. MOLECULAR STUDY

In each reaction, one to eight PCR products were obtained. Of the 20 RAPD primers used in this experiment, eleven primers gave difference in amplification between CSR9 and Nipponbare. In each reaction with a primer giving amplification polymorphism, one band specific to Nipponbare with strong signal intensity was found. PCR products specific to CSR9 were also found, but those had weak signal intensity. As for other nine primers, one gave no amplification and eight showed no difference in PCR products between CSR9 and Nipponbare.

4. Discussions

Cesium is a toxic analog ion for potassium in green algae (Tester 1988) and higher plants (Sheahan et al. 1993). Investigations on the cesium effects on root epidermal cells have been shown to be important to understand the ion uptake kinetics (Sheahan et al. 1993, Bellando et al. 1995, Maathuis and Sanders 1996). In the present study, genetic variability in cesium resistance in rice was revealed from two experiments, screening for cesium resistant genotypes and selection of cesium resistant mutants.

Great variability in cesium response was found among 54 genotypes surveyed in this study. Cesium resistant cultivars were found among Japanese native and indica cultivars. On the other hand, cesium sensitive cultivars were also found among Japanese native, and indica cultivars.

Japanese native cultivars included two extremes and were shown to be a promising genetic resource for cesium-related characteristics. Great variability in cesium response was also found among the genotypes (most of them were *indica*) collected from outside of Japan. As for cesium resistance in rice, there seems to be neither phylogenic nor ecotype relationship, as reported in chlorate resistance, a well-known marker for nitrate assimilation ability (Morishima and Oka 1981, Hasegawa et al. 1995). However, the results that no higher RFLL genotypes were found from Japanese modern lowland cultivars and U.S.A cultivars might be related to physiological change of root for adapting the modern paddy field management in both country.

In higher plants cesium resistant mutants have been reported in two species. Sheahan et al. (1993) reported that in cesium resistant *Arabidopsis* mutants uptake of rubidium, another potassium analog, reduced to 30 to 60% of that of the wild type and that the resistance was controlled by a single recessive nuclear gene. Electrophysiological studies on one of *Arabidopsis* mutants demonstrated that the mutant affect the potassium uptake (Maathuis and Sanders 1996). Hasegawa (1996) reported that three rice mutants with decreased level of nitrate uptake had also cesium tolerance, suggesting that these mutants are due to the change of energistics of nitrate uptake. Present experiments demonstrated that uptake of potassium and nitrate in CSR9 reduced to 65 to 75% of that of the wild type except for cesium uptake from 250 μ M $CsNO_3$ and that cesium reduced nitrate uptake. Furthermore, RFLLs of rice mutants defective in nitrate uptake (Hasegawa 1996) were higher than that of the wild type. These results support the view that cesium affects both cation and anion uptake (Sheahan et al. 1993). Differential response of cesium uptake between from 20 μ M CsCl and from 250 μ M $CsNO_3$ suggested the change of kinetics of high affinity cesium uptake, although effect of different cesium salt was not negligible. Simplified differential display method revealed the PCR products specific to the wild type cultivar, Nipponbare. It is reasonable that cesium resistance in CSR9 is due to the deletion of the PCR products specific to the wild type. Further molecular and electrophyiological investigations are needed for characterization of cesium resistant mutants.

Finally, radiocesium pollution by collapse of an atomic energy station is one of the today's serious environmental problems. Plants showing high capacity of cesium uptake are needed for the removement of radiocesium from the contaminated soil. Interspecific difference in the uptake of radiocesium was reported (Buysse et al. 1996) and great intragenetic variability in cesium resistance in rice was demonstrated in the present study. It seems to be hopeful to obtain genotypes with higher capacity of cesium uptake in rice and other plant species, although cesium uptake was not measured in this study.

5. References

Bellando, M., M.T. Marre, S. Sacco, A.Talarico, A.Venegoni and E. Marre (1995) Transmembrane potential-mediated coupling between H^+ pump operation and K^+ fluxes in *Elodea densa* leaves hyperpolarized by fusicoccin, light or acid load. Plant Cell Environ 18: 963-976.

Buysse, J., K. Van den Brande and R. Merckx (1996) Genotypic differences in the uptake and distribution of radiocaesium in plants. Plant Soil 178,265-271.

Crawford, N.M. (1995) Nitrate: nutrient and signal for plant growth. Plant Cell 7: 859-868.

Hasegawa, H. (1996) Selection for mutants with low nitrate uptake ability in rice (*Oryza sativa*). Physiol. Plant. 96,. 199-204.

Hasegawa, H., O. Yatou and M. Ichii (1995) A chlorate-hypersensitive, high chlorate uptake mutant in rice (*Oryza sativa* L.). Breed. Sci. 45, 229-232.

Maathuis, F.J.M. and D. Sanders (1996) Characterization of *csi52*, a Cs^+ resistant mutant of *Arabiddopsis thaliana* altered in K^+ transport. Plant J. 10, 579-589.

Morishima, H. and H.I. Oka (1981) Phylogenic differentiation of cultivated rice. XXII. Numerical evaluation of the Indica-Japonica differentiation. Jpn. J. Breed. 31, 402-413.

Ninnemann, O., J.-C. Jaouniaux and W.B. Frommer (1994) Identification of a high affinity NH_4^+ transporter from plants. EMBO J. 13, 3464-3471.

Sheahan,J.J., L. Ribeiro-Neto and M.R. Sussman (1993) Cesium insensitive mutants of *Arabidopsis thaliana*. Plant J. 3, 647-656.

Quesada, A., A.Krapp, A. Daniel-Vedele, L.J. Trueman, E. Fernandez, B. Forde and M. Caboche (1997) PCR identification of a *Nicotiana plumbaginifolia* cDNA homologous to the high affinity nitrate transporters of the *crnA* family. Plant Mol. Biol. 34, 265-279.

Tanaka,T., H. Hasegawa and S. Ida (1998) Isolation of a rice homologue of high-affinity nitrate transporter cDNA. Plant Cell Physiol. 39, Suppl. (in press).

Tester, M. (1988) Potassium channels in the plasmalemma of *Chara corallina* are multi-ion pores: Voltage-dependent blockade by Cs^+ and anomous permeabilities. J. Membrane Biol. 105: 87-94.

Wray, J.L. (1988) Molecular approaches to the analysis of nitrate assimilation. Plant Cell Environ. 11: 369-382.

Yoshida, K.T., S. Naito and G. Takeda (1994) cDNA cloning of regeneration-specific genes in rice by differential screening of randomly amplified cDNAs using RAPD markers. Plant Cell Physiol. 35, 1003-1009.

INTERACTION OF NUTRIENT AND WATER DEFICIENCY ON THE DEVELOPMENT OF RYE (*SECALE CEREALE* L) GENOTYPES

BORBÁLA HOFFMANN, *GÁBOR GALIBA

Department of Plant Production, Georgikon Faculty,

Pannon University of Agricultural Sci., 8360 Keszthely, Hungary

*Agricultural Research Inst. of the Hungarian Academy of Sci.,

2462 Martonvásár, Brunszvik str. 2., Hungary

Abstract

Seedlings of 15 rye (*Secale cereale* L.) cultivars and lines were evaluated regarding the dry matter accumulation and the content of essential nutrient elements, grown in modified Hoagland nutrient solution in the phytotron of the Agricultural Research Institute in Martonvásár. Osmotic or/and nutrient stresses were imposed by applying PEG or/and 8^{-1} strength nutrient solution.

The dry matter accumulation of cultivars can be evaluated according to absolute and relative quantities: the most dry matter was accumulated by the variety Lovászpatonai, but it was reduced by 47% compared to its control. On the other hand S838587 accumulated the same quantity (82 mg) of shoot dry matter in control and stressed treatments as well.

Regarding the nutrient elements examined, the osmotic stress had generally little effect. The nutrient deficiency and the combined stress decreased the N content of shoots equally. The decrease was much smaller in the roots. In opposition, the P_2O_5 content of the shoots was always higher than that of the roots. It was the K_2O content of plants that decreased the most in all genotypes becausse of stresses and the extent of change in roots overpassed the shoots. The differences found in dry matter accumulation and nutrient content of rye seedlings verified the existence of variance in the reaction to stresses among the studied genotypes.

G. Gissel-Nielsen and A. Jensen (eds.), Plant Nutrition – Molecular Biology and Genetics, 341–347.

1. Introduction

Abiotic factors are considered to be the main sources of yield reductions (Boyer, 1982). Limited soil moisture influences nutrient availability for plants, on the other hand water use efficiency is reported to be improved by adequate fertilization. Thus the effects of nutrient supply on plant growth and yield production under conditions of drought stress are very complex.

In past decades the primary approach for alleviation of environmental stress focused on modification of the environment (soil amelioration, fertilizer use, irrigation, etc.). Economic and ecological constraints make the approach of genetic improvement of stress resistance a viable alternative. The most powerful breeding approach to assess the value of certain genotypes regarding the environmental stress tolerance was the multi-year, multi-location field trials. But the large genotype-environment interaction makes this empirical approach inefficient and expensive (Acevedo and Fereres 1993).

In wheat experiments Schinkel (1991) found significant genotype-nitrogen interaction variances. Direct selection for high performance under low N supply was generally superior to that of indirect selection under high N supply. Studying the interactive effects of N-, P- and K-nutrition on the development of maize seedlings grown under conditions of limited water supply Studer (1993) assumed, that P-fertilization reduced the adverse effects of water stress on plant development, whereas N-fertilized plants experienced large reduction in shoot growth when water stress was imposed.

A research programme has been set up in France to breed for winter wheat cultivars which could use more efficiently nitrogen fertilizers. Le Gouis and Pluchard (1995) selected some cultivars which had almost the same yield at low and high nitrogen levels.

Rye is known as the most tolerant cereal species to unfavourable environments. It is revealed in the old Hungarian practice: rye is produced on areas where wheat production is unprofitable because of soil characteristics (coarse-textured, poor in nutrients and low water-retention capacity) and climate conditions. Furthermore, rye's adoption in sustainable agriculture depends upon its suitability for growing under low inputs (Oettler, 1996).

That is why we decided to study the adaptation of rye genotypes with different origin to osmotic or/and nutrient stresses.

2. Materials and methods

Seedlings of 15 rye (*Secale cereale L.*) varieties and lines (Lovászpatonai, Kisvárdai 1, Kisvárdai alacsony, Danko, Motto, Carokurz, Halo, L311, L201, L312-301, L306-311, Syn8389=Syn1, Syn838587=Syn2, Syn8392=Syn3, Syn8890=Syn4) were grown in half-strength modified Hoagland nutrient solution for two weeks prior to the stress period in plant growth chambers in the phytotron of the Agricultural Research Institute in Martonvásár.

The nutrient solution contained the following chemicals in mM: $KNO_3$5; $Ca(NO_3)_2$*$4H_2O$ 5; $MgSO_4$*7 H_2O 2; KH_2PO_4 1; NaFe-EDTA 0.1 (Sigma). Micronutrient concentration (in μM) were H_3BO_3 11.5; $MnCl_2$*4 H_2O 4.6;

$ZnSO_4*7H_2O$ 0.2; $Na_2MO_4*2H_2O$ 0.12 and $CuSO_4*5H_2O$ 0.08. The pH of the nutrient solution originally was 5.8 and did not change till the renewal. The solution was renewed weekly. Day/night (16/8 hours) cycles of temperature and relative humidity of the ambient air were 15/10 ^{0}C and 60/75 % respectively, with 300 $\mu molm^{-2}s^{-1}$ photosynthetic photon flux during the day.

Osmotic or/and nutrient stress was imposed at the beginning of the third week by applying PEG 4000 (Sigma) at 15% concentration or/and 8^{-1} strength modified Hoagland nutrient solution. Samples were taken after one week of stress treatment, one sample consisted of 6 plants. The fresh and dry weight and the content of essential nutrient elements of shoots and roots were measured.

For determination of mineral content the tissues were wetdigested using H_2SO_4 and H_2O_2 (1:2 by volume). K_2O- and P_2O_5-concentration was determined colorimetrically and N by Kjedal method.

Statistical analyses were carried out using the SPSS 7.5 statistical analyses system.

3. Results

3.1. DRY MATTER ACCUMULATION

The dry matter accumulation can be evaluated according to absolute and relative quantities. Regarding the absolute quantities (*Figure 1.*) the cultivar Lovászpatonai accumulated the most dry matter in all treatments (I.-IV and V.-VIII. are shoot and root control, osmotic-, nutrient- and combined sterss) 247, 202, 131, 120 mg, respectively.

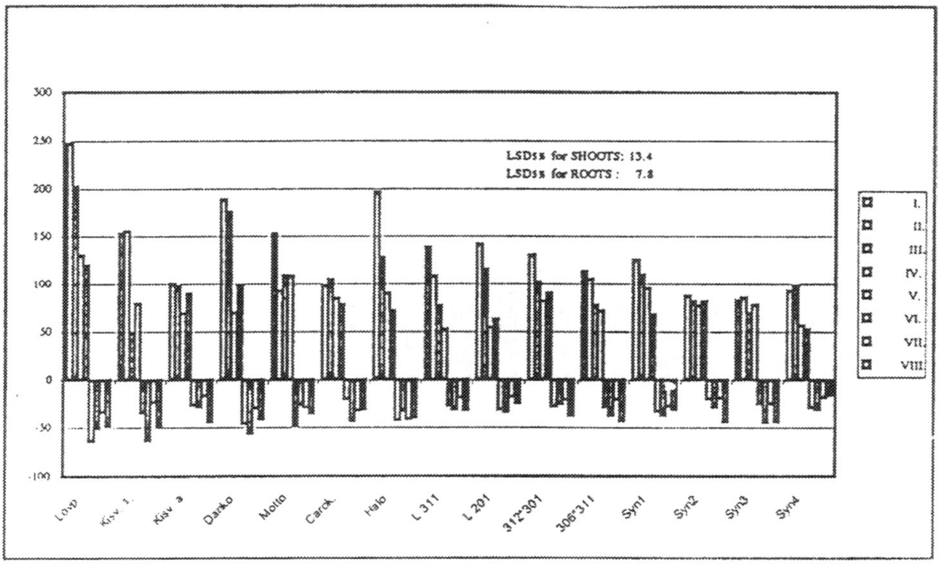

Figure 1. Shoot and root dry matter (mg).
I.-IV and V.-VIII. are shoot and root control, osmotic-, nutrient- and combined sterss.

Danko, Halo and Kisvardai1 reached also high values in control and in osmotic sterss. If we consider the relative quantities: the dry matter accumulation of stressed plants in percentage of the non-stressed plants, we find a different order. The shoot dry matter accumulation was less decreased because of stresses in case of Carokurz, Syn838587 and Syn8392. The greatest reductions were found by Halo, L201, 311 and Lovászpatonai. In the mean of cultivars the osmotic stress caused the smallest decrease, while the nutrient deficiency and the combined stress had similar effect, these reached 62% and 63%.

The shoot and root dry matter accumulation were closely related: the most shoot dry matter was accumulated by the cultivars having the greatest root dry matter. But the stresses had different effects on shoot and root growth: the osmotic and combined stresses increased the root dry matter accumulation especially by the cultivars having the smallest shoot dry matter reduction (Carokurz, Syn838587 and Syn8392). This trend can be seen in the change of the shoot/root ratio (not shown) as well.

3.2. N-CONTENT OF SHOOTS AND ROOTS

The N-content of shoots (*Figure 2.*) in the control treatment was fairly equalized (about 40 g/kg). The osmotic stress had little effect on it but we found signifficant decrease by L201, 312*301 and Syn2. The nutrient deficiency and the combined stress decreased it equally by 57% in the mean of the genotypes. The least decrease was measured in case of L311, Halo, Carekurz and Kisvárdai, reaching 52, 49, 47 and 45 % of their control, respectively.

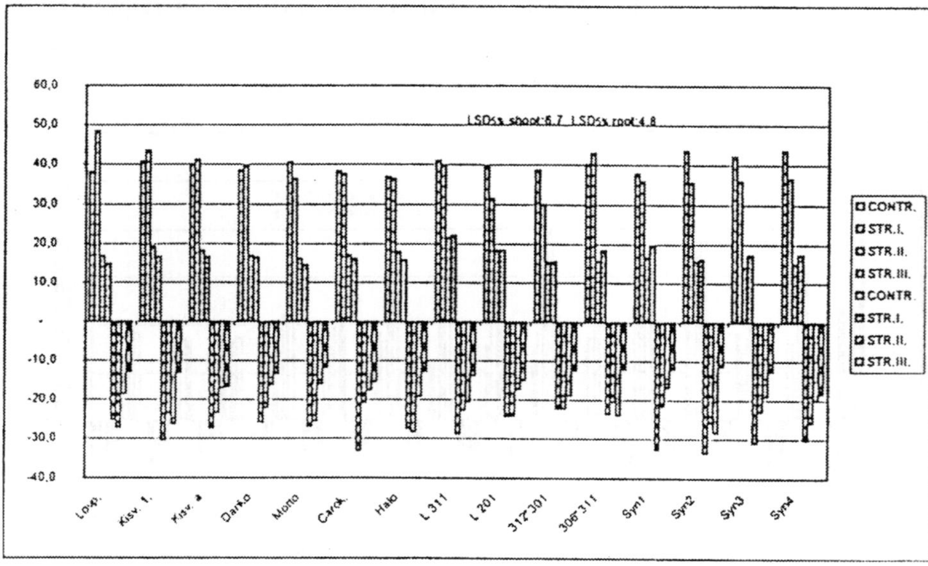

Figure 2. N-content of shoots and roots (g/kg).
STR.I.: osmotic-, STR.II.: nutrient-, STR.III.: combined stress treatment.

The N-content of roots was decreased by 14% because of PEG treatment, so it was more reduced than the N-content of shoots, but there were some exception e.g. L201 and 312*301. But the nutrient deficiency caused in all genotypes a remarkably smaller N-content decrease in the roots than in the shoots (the mean of the genotypes for roots 71% and for shoots 43% of their control) and - inconsistent with shoots - the combined stress further reduced the N-content of roots.

3.3. P_2O_5-CONTENT OF SHOOTS AND ROOTS

The P_2O_5-content of shoots was generally not changed because of the osmotic stress but Syn1, Lovászpatonai and Danko showed significantly higher values after stress. The nutrient deficiency reduced the amount of P_2O_5 by half in the shoots, only L311 and 306*311 reached 78 and 62% of their control, while Syn2 and Halo had only 41 and 43%. The parallel application of osmotic and nutrient stresses increased the P_2O_5-content of shoots to the level of the control or even above: Syn2 and Kisv.1/a reached only 86 and 87% (*Figure 3.*).

Roots had higher P_2O_5 values than shoots in control plants (the mean of the genotypes are 17 and 14.7 g/kg, respectively). The osmotic stress caused a considerable increase in the P_2O_5 concentration in most of the genotypes: L201, L311 and 312*301 had 306, 275 and 239% of their control, but Kisv.a and Syn4 reached only 85 and 93%, respectively. The consequence of the nutrition deficiency is the reduction of the P_2O_5-content to the 25% of the control in the mean of the genotypes – there were not great differences among the genotypes in this respect. The combined stress caused an increase – like in the case of shoots – but only to the half of the control values.

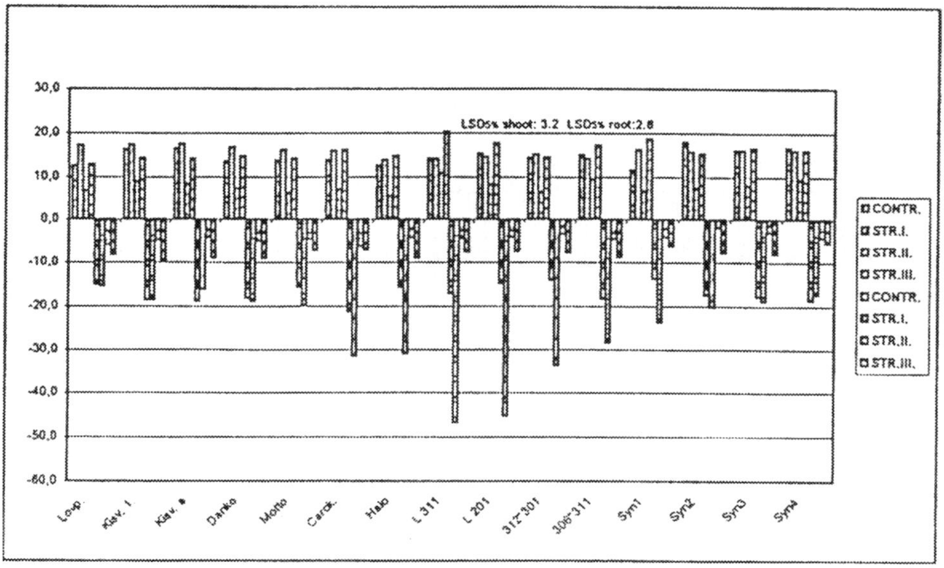

Figure 3. P_2O_5-content of shoots and roots (g/kg).
STR.I.: osmotic-, STR.II.: nutrient-, STR.III.: combined stress treatment.

3.4. K_2O-CONTENT OF SHOOTS AND ROOTS

Similarly to the N- and P_2O_5, the K_2O content of shoots did hardly change because of osmotic stress, though significant decrease was found in L201, 306*311, L311 and Syn4, while Danko and Syn1 had higher values compared to their control. It was the K_2O content of plants that decreased the most in all cultivars and lines as a cosequence of nutrient deficiency (the mean of genotypes was 31% of the control). The combined stress slightly increased this value (35%). There were no great diferences among the cultivars and lines in this respect (*Figure 4.*).

The K_2O content of roots changed similarly to the shoots, but the extent of decrease was even greater especially in case of the combined stressed plants, which caused a further reduction in K_2O content of roots compared to the nutrient stressed ones. From the point of K_2O content of roots the osmmotic stress was best tolerated by Syn4 and Danko (104% and 97%), while Halo, 312*301, 306*311 and L201 reached only 58- 61- 61- and 62%, respectively. The nutrient deficiency caused the less decrease in cultivars Carokurz and Lovp. (45% and 41%), while the greatest decrease was measured in Syn2, 312*301 and 306*311 having 16- 21- and 22% of their contol. The combined stress gave similar result: Danko and Kisv1 proved to be the most tolerant, while Syn4, Syn1, Syn2 and Carocurz showed the lowest values: 14- 15- 16- and 16% of their control, respectively.

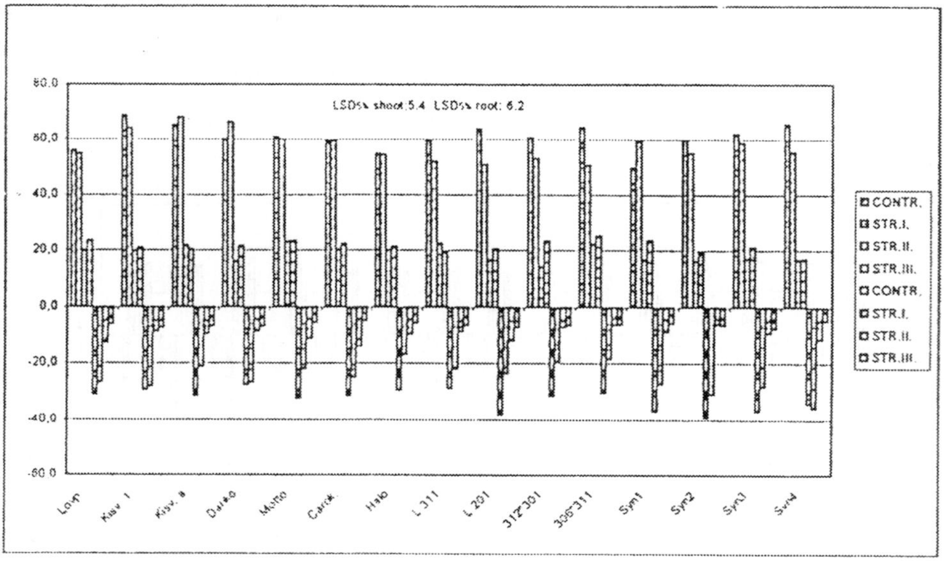

Figure 4. K_2O-content of shoots and roots (g/kg).
STR.I.: osmotic-, STR.II.: nutrient-, STR.III.: combined stress treatment.

Accordind the analyses of variance results were significant at $p=0.001$ level.
This study was supported by the grants: FKFP-0036/1997 and OTKA T02118.

5. Discussion

According to the data shown, we have to discriminate between absolute and relative quantities of accumulated dry matter. In contrast with Silva and Gabelman's results (1993), our data indicate, it is not possible to conclude from the absolute quantity of accumulated dry matter to the sensitivity for stresses, but the „relative quantity" (%), that is the change in the dry matter accumulation, is of signalling value.

The shoot and root dry matter accumulation are closely related: the most shoot dry matter was accumulated by the cultivars having the greatest root dry matter. But the stresses had different effects on shoot and root growth: the shoot dry matter accumulation of stessed plants never exceded that of the control's, while the root dry matter accumulation of some - but not all - genotypes was significantly greater compared to its control. These findings are supported by Clark (1983), who measured similar shoot and plant dry weight in a low-P stress tolerance study. Haberle at al (1996) also founded root proportion decrease after fertilization in al examined cultivars.

Regarding the nutrient elements examined, the osmotic stress had generally little effect. The nutrient deficiency and the combined stress decreased the N content of shoots equally. The nutrient deficiency caused a remarkably smaller N-content decrease in the roots but it was further reduced by the combined stress.

In opposition, the P_2O_5 content of the shoots was always higher than that of the roots. The nutrient deficiency decreased, but the contemporary application of osmotic and nutrient stresses increased the P_2O_5-content of shoots to the level of the control or even above. The osmotic stress caused a considerable increase in the P_2O_5 concentration of roots in most of the genotypes. The consequence of the nutrition deficiency was the reduction of the P_2O_5-content, while the combined stress caused some increment. Among the studied minerals the amount of K_2O has decreased the most in all genotypes becausse of stresses and the extent of change in roots was the grater. Studer (1993) had similar findings.

The differences found in dry matter accumulation and nutrient content verified the existence of variance in the reaction to stresses among the studied rye genotypes.

6. References

Acevedo, E. and Fereres, E. (1993): Resistance to abiotic stresses. In: M. D. Hayward, N. O. Bosemark and I. Romagosa (eds), *Plant Breeding, Principles and Prospects*, Chapman and Hall, London, pp. 406-421.

Boyer, J. S. (1982): Plant productivity and environment. *Science* **218**, 443-448.

Clark, R. B. (1983): Plant genotype differences in the uptake, translocation and use of mineral elements required for plant growth. *Plant and Soil* **72**, 175-196.

Haberle, J., Svoboda, P., Matula, J. (1996): A nondestructive study of the effect of soil and fertilization on three winter wheats. *Cereal Research Com.* **24**, 477-483.

Le Gouis J., Pluchard P. (1995) Genetic variability for nitrogen use efficiency in winter wheat (*Triticum aestivum L.*) Adaptation in Plant Breeding. *XIV Eucarpia Congress.* University of Jyväskylä, Finland. Proceeding, 87

Oettler, G., (1996): Variation and covariation of agronomic traits and quality in triticale at low nitrogen input. *Plant Breeding*, **115**, 445-450.

Schinkel, B., (1991): Untersuchungen zur nutzbaren genetischen Variation für Stickstoff-Effizienz bei Winterweizen. *Diss. Univ.* Hochenheim, Stuttgart.

Silva, A., Gabelman, W. (1993): Screening maize inbred lines for tolerance to low-P stress condition. In: P. Randall et al. (eds), *Genetic aspects of plant mineral. nut.* Kluwer Ac. Pub., The Netherlands, pp. 233-239.

Studer, C. N. (1993): Interactive effects of N-, P-, K-nutrition and water stress on the development of young maise plants (*Zea mays* L.). Dissertation, ETH, Zürich.

POTENTIAL FOR BREEDING *MEDICAGO* SPP. FOR ZINC EFFICIENCY TO INCREASE TOLERANCE TO *RHIZOCTONIA SOLANI*

T.C. STREETER[1], R.D. GRAHAM[1], A.W.H. LAKE[2] and Z. RENGEL[3]
[1] *University of Adelaide, PMB 1, Glen Osmond SA 5064, Australia*
[2] *South Australian Research Development Institute, GPO Box 397, Adelaide SA 5001, Australia*
[3] *University of Western Australia, Nedlands WA 6907, Australia*

1. Introduction

Breeding for micronutrient-efficient crops, including Zn efficiency, will assist in alleviating micronutrient deficiencies not only in the human population but also in livestock. To date, applying modern breeding techniques to adapt annual medic plants to soils of low nutritional status has not been attempted. In breeding pasture plants for improved Zn efficiency for example, not only will the amount of pasture be improved so will the quality of the pasture, alleviating any potential Zn deficiency disorders in foraging animals.

In order to breed for Zn-efficient medics a number of factors need to be considered. According to Graham (1984), to breed for nutritional characteristics, in this case Zn, it is necessary to demonstrate that:

- the soil is capable of supplying Zn to an improved genotype;
- genetic diversity in Zn efficiency exists in the germplasm; and
- there are sound agronomic and economic reasons for pursuing a breeding solution to Zn deficiency.

Firstly, in considering soil availability of Zn, it has already been demonstrated that Zn deficiency is largely due to low availability of Zn and not the low absolute amount of Zn present in the soil (Graham, 1978), with the total amount of soil-Zn adequate for thousands of crops. To get an insight into genetic diversity of Zn efficiency in medics, 19 genotypes of eight annual medic species were screened in a controlled environment in the experiment presented below. Agronomic and economic reasons for pursuing a breeding solution to Zn deficiency are also discussed.

2. Genetic Diversity for the Zn Efficiency Trait in Medics

2.1 METHODS

A factorial design combining 19 medic genotypes (*M. truncatula* genotypes Caliph, SA12591, SA18289 and Z942; *M. tornata* genotypes Rivoli, Tornafield, SA20278, SA20248 and SA20521; *M. littoralis* genotypes Harbinger and Herald; *M. noeana* genotype SA15485; *M. polymorpha* genotypes Santiago and SA4188; *M. scutellata* genotype Sava; *M. rugosa* genotype Paraponto, and progeny from a cross between *M.*

G. Gissel-Nielsen and A. Jensen (eds.), Plant Nutrition – Molecular Biology and Genetics, 349–352.

littoralis x *M. tornata* genotypes Z1064, Z1065 and Z1068) were screened for Zn efficiency at two levels of Zn (0 and 0.9 mg Zn/kg) in three replicates.

Plants were grown in a Zn-deficient sand (Mt Compass sand) with a basal solution added (in mg/kg): 350 NH_4NO_3; 57 KH_2PO_4; 90 K_2SO_4; 90 $MgSO_4.7H_2O$; 5.5 $MnSO_4.4H_2O$; 0.5 $H_2MoO_4.H_2O$; 1 $CoSO_4.7H_2O$; 0.15 $NiSO_4.7H_2O$; 5 $CuSO_4.5H_2O$; 0.23 H_3BO_3; 3.5 $FeSO_4.7H_2O$; and 0.9 $ZnSO_4.7H_2O$. $CaCO_3$ (0.3%) was mixed thoroughly with the dry soil before basal nutrients were added. Four germinated seeds were sown per pot of 1 kg soil and watered to 12% w/w. Seven days after sowing, plants were thinned (two plants/pot) and inoculated with *Rhizobium meliloti*. Plants were watered every second day for the duration of the experiment and were harvested 42 days after sowing. Plants were ranked as Zn-efficient based on the definition of Graham *et al.* (1992):

$$\text{Zn efficiency} = \frac{\text{Shoot dry weight in Zn - deficient soil}}{\text{Shoot dry weight in Zn supplemented soil}} * 100\%$$

2.2 RESULTS AND DISCUSSION

Genotypes considered the most Zn-efficient (based on minimal yield difference between low and high Zn levels in soil) include Rivoli, SA20278, SA20248 and SA20521 (Fig. 1). These genotypes were selected from trials on light, sandy soils in South Australia that are Zn-deficient. It is likely that the Zn efficiency trait (or traits) were retained in the selection process offering an obvious advantage in adaptation to these soils.

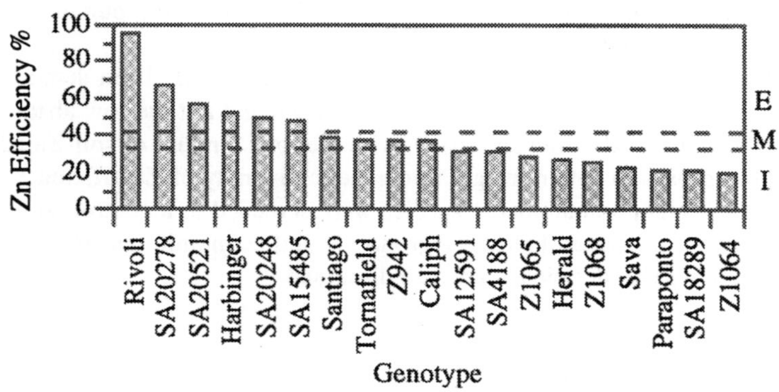

FIGURE 1. Zinc efficiency ranking (E=efficient, M=medium, I=inefficient) of 19 medic genotypes grown for 42 days. The ranking was based on the ratio of shoot dry matter production at Zn deficiency (0 Zn added) and Zn sufficiency (0.9 mg Zn/kg). The mid-point of the medium efficiency interval corresponds to the median Zn efficiency value; the upper and lower boundaries of the medium-efficiency interval were constructed by adding to, or subtracting from, the median point the value of the standard error of the genotype effect.

Genotypes of *M. tornata* appear to be Zn-efficient by virtue of their larger and more extensive root system (data not shown). These genotypes were able to extract a high amount of Zn per plant from the soil and yield well on Zn-deficient soils (Streeter, 1998). However, they did not reach yield potential under Zn-deficient conditions. Genotype Rivoli on the other hand was able to reach maximum yield potential under Zn-deficient conditions, but its potential was low (Streeter, 1998).

In medics, Zn deficiency depressed root dry weight to a greater extent than shoot dry weight as the shot/root ratios of most medic genotypes were higher in the 0 Zn treatment compared with 0.9 mg Zn/kg soil (data not shown). The shoot/root ratio was also greatest in Zn-inefficient genotypes (Z1064, Z1065 and Z1068; data not shown). This situation may occur if Zn deficiency in medics has a greater impact on root cells compared to cells in the shoot, or if Zn deficiency inhibits the export of assimilates from the shoot to the roots. The decrease in root dry matter accumulation correlates with work on canola where Grewal *et al.* (1997) observed impaired root growth in Zn-inefficient genotypes in soils of low Zn status.

These results imply at least two mechanisms for Zn efficiency can occur in *Medicago* species and were observed in only one screening for Zn efficiency. With a broader screening for Zn efficiency a number of other mechanisms may be observed. Crossing the two most Zn-efficient genotypes in the present screening (e.g. SA20278 and Rivoli), with the aim of isolating progeny having the three factors underlying the Zn-efficiency trait: high uptake of Zn, high yield potential, and the capacity to reach maximum yield potential under Zn-deficient conditions, is a reasonable expectation.

3. Agronomic and Economic Reasons for Pursuing a Breeding Solution to Zn Deficiency in Medics

Currently, many medic pastures are deficient in Zn, with recommended levels of Zn fertiliser inadequate for maintaining optimum growth in medics (Streeter, 1998). Zinc deficiency in medics is an ongoing problem, and can have quite a severe impact on profitability, both through reductions in pasture and subsequent crop yield and through additional fertiliser costs needed to correct deficiencies. Hence, breeding for Zn efficiency in medics will reduce Zn fertiliser costs. Breeding for plants which are able to explore a greater area of soil will also increase the capacity to scavenge other diffusion-limited nutrients and therefore reduce the cost of micronutrient fertilisers even further.

The fungus *Rhizoctonia solani* is responsible for "bare patch" disease in Australia, and is a major problem for farmers adopting minimum tillage in areas with sandy soils. In the past, the only effective way to control this disease was by two or more tillage operations before sowing (De Boer *et al.*, 1991).

The use of Zn fertilisation has been shown to increase medic tolerance to infection by *R. solani* (Streeter, 1998). Preliminary work on Zn-efficient medic genotypes has demonstrated a stable nutrient concentration in the shoots under Zn-deficient conditions when infected with *R. solani*. Zinc-inefficient genotypes were more sensitive to infection by *R. solani* when deficient in Zn, with a parallel reduction in concentration of most nutrients in the shoot. This implies root pruning by *R. solani* is affecting nutrient uptake to a point where plant growth is inhibited more severely in Zn-inefficient than Zn-efficient genotypes.

With Zn-efficient genotypes, a higher amount of Zn is also expected to be loaded into the seed, improving seed quality and therefore increasing seedling vigour and tolerance to disease further in the subsequent generations. These genotypes will be better able to persist and yield well despite infection by *R. solani*, thus leading to greater farming profitability.

Medics are the main source of essential mineral elements required by grazing livestock, and it therefore follows that improving the Zn content of the forage and seed will assist in overcoming Zn deficiency in animals. Marginal Zn deficiency is typically observed in sheep and cattle on pastures containing less than 20 mg Zn/kg dry weight and is characterised by suboptimum growth, reduced fertility and mild skin disorders (White, 1993). As White (1993) points out, it should be of evolutionary significance that the concentration of Zn required for optimum pasture growth is similar to the requirements for sheep growth. It may therefore follow that a Zn-deficient medic pasture results in Zn-deficient sheep and/or cattle. Breeding for Zn-efficient medics will prevent this from occurring.

4. References

De Boer, R. F., Kollmorgen, J.F., Macauley, B.J. and Franz, P.R. (1991) Effects of cultivation on Rhizoctonia root rot, cereal cyst nematode, common root rot and yield of wheat in the Victorian Mallee. Aust. J. Exp. Agric. 31: 367-372.

Graham, R.D. (1978) Nutrient efficiency objectives in cereal breeding. Proc. 8th International Colloquium of Plant Analysis and Fertilizer Problems, Auckland. pp. 165-170.

Graham, R.D. (1984) Breeding for nutritional characteristics in cereals. Adv. Plant Nutr. 1: 57-102.

Graham, R.D., Ascher, J.S. and Hynes, S.C. (1992) Selecting zinc-efficient cereal genotypes for soils of low zinc status. Plant and Soil 146: 241-250.

Grewal, H.S., Graham, R. D. and Rengel, Z. (1997) Genotypic variation in zinc efficiency and resistance to crown rot disease (*Fusarium graminearum* Schw Group 1) in wheat. Plant and Soil 186: 219-226.

Streeter, T.C. (1998) Role of Zn Nutritional Status on Infection of *Medicago* Species by *Rhizoctonia solani*, PhD thesis submitted at The University of Adelaide.

White, C.L. (1993) The zinc requirements of grazing ruminants. In: Zinc in Soils and Plants. (ed) A.D. Robson, Kluwer Academic Publishers, pp. 197-206.

CYTOSOLIC Ca^{2+} ACTIVITIES IN INTACT WHEAT ROOT APICAL CELLS SUBJECTED TO ALUMINIUM TOXICITY

W. H. ZHANG and Z. RENGEL
Department of Soil Science & Plant Nutrition, The University of Western Australia, Nedlands WA 6907, Australia

1. Abstract

Aluminium toxicity in plants appears to be related to interactions between Al and Ca, including inhibition of Ca uptake by blocking the plasma membrane Ca^{2+} channels; displacement of apoplasmic Ca^{2+} and disturbance of intracellular Ca^{2+} homeostasis. Several lines of evidence have indicated that inhibition of Ca^{2+} uptake and desorption of apoplasmic Ca^{2+} may not be primary causes of Al toxicity in plants. The involvement of disturbance of intracellular Ca^{2+} homeostasis of intact root apical cells in Al toxicity remains to be demonstrated experimentally. With the Ca^{2+}-sensitive fluorescent probe Fluo-3 and the confocal laser scanning microscopy technique, a sustained increase in Fluo-3 fluorescence was observed in intact root apical cells of wheat in the presence of $AlCl_3$, suggesting Al elicited an increase in cytosolic Ca^{2+} activity. The increase in cytosolic Ca^{2+} activities was greater in Al-sensitive than in Al-tolerant isogenic wheat lines. These findings provide unambiguous evidence in support of disturbance of intracellular Ca^{2+} homeostasis being involved in Al toxicity in plants. This increase in the cytosolic Ca^{2+} activities may serve as a trigger to alter a number of biochemical processes by disrupting Ca^{2+}-mediated signal transduction system and/or Ca^{2+}-dependent cytoskeletal dynamics.

2. Introduction

Aluminium, as the most abundant metal in the earth's crust, becomes hydrolysed to the toxic trivalent form (Al^{3+}) under acidic conditions. The initial and most dramatic symptom of Al stress in plants is inhibition of root growth, which can be observed within 1-2 h after exposure to Al (Ryan et al., 1993; Jones and Kochian, 1995; Horst, 1995). Aluminium-induced inhibition of root growth often precedes, or coincides with, a decline in cell division (Rengel, 1992a; Horst, 1995). Therefore, inhibition of cell elongation rather than cell division is responsible for the initial inhibition of root growth, while prolonged exposure to Al probably results in inhibition of both cell elongation and division. Inhibition of root growth requires the root apex to be directly exposed to Al (Ryan et al., 1993; Sivaguru and Horst, 1998), indicating that the root apex is a critical target of Al toxicity. Despite decades of extensive research, the primary cause of Al toxicity still remains to be identified (Delhaize and Ryan, 1995; Horst, 1995; Kochian, 1995).

G. Gissel-Nielsen and A. Jensen (eds.), Plant Nutrition – Molecular Biology and Genetics, 353–358.

3. Inhibition of Ca^{2+} influx across the plasma membrane

Interactions between Al and Ca have long been implicated in Al toxicity in plants (Rengel, 1992a; Kochian, 1995). Inhibition of Ca^{2+} influx into plant cells, as demonstrated in intact plant cells (Huang et al., 1992; Ryan and Kochian, 1993; Reid et al., 1995; Jones et al., 1995), protoplasts (Rengel and Elliott, 1992) and membrane vesicles (Huang et al., 1996; White, 1998) is one of the most distinct and earliest effects of Al on plants. Like other trivalent cations, Al can block the plasma membrane Ca^{2+} channels (Ding et al., 1993; Ding and Pickard, 1993; Piñeros and Tester, 1995, 1997; Rengel et al., 1995). The blockade of Ca^{2+} channels may provide a molecular basis for the Al-induced inhibition of Ca^{2+} influx into plant cells. The inhibition of Ca^{2+} influx into the root apex preceded the inhibition of root growth and any visible symptoms of Al toxicity (Huang et al., 1992; Ryan and Kochian, 1993; Jones et al., 1995; Reid et al., 1995). Importantly, Al-inhibited Ca^{2+} influx into the root apex of Al-sensitive more than Al-tolerant genotypes (Huang et al., 1992; Ryan and Kochian, 1993). These findings have led to the hypothesis that inhibition of Ca^{2+} influx by Al is a cause of Al toxicity in plants (Huang et al., 1992; Rengel, 1992a, b). However, further studies have revealed that inhibition of root growth in the presence of low concentration of Al can occur without inhibiting Ca^{2+} influx, and addition of other cations (Mg^{2+} and Na^{+}) improved root growth but the Ca^{2+} influx remained inhibited (Ryan et al., 1993; Jones et al., 1995; Reid et al., 1995). Thus, inhibition of Ca^{2+} influx seems unlikely to be a critical event in triggering Al toxicity in plants.

4. Displacement of apoplasmic Ca^{2+} by Al

In addition to inhibition of Ca^{2+} influx, Al interacts with the cell wall (Horst, 1995; Blamey, 1998), resulting in displacement of Ca^{2+} by Al (Kinraide et al., 1994; Ryan et al., 1997). Aluminium binds more strongly to pectin than does Ca (Blamey, 1998). As Ca^{2+} is involved in cross-linking the pectic molecules in the cell wall (Carpita and Gibeaut, 1993), the displacement of pectin bound Ca^{2+} would disrupt physical properties of cell wall, such as extensibility, rigidity and permeability (Horst, 1995; Blamey, 1998). Consistent with the Ca-displacement hypothesis is that Al toxicity symptoms can be effectively ameliorated by elevated external Ca^{2+} levels (Kinraide et al., 1994). However, an addition of other cations, which also displace the apoplasmic Ca^{2+}, significantly improves the Al-affected root growth (Kinraide et al., 1994; Ryan et al., 1997), suggesting that displacement of apoplasmic Ca by Al is unlikely to be a mechanism of Al toxicity in plants. The ameliorative effect of Ca^{2+} and other cations on plants suffering from Al stress has been interpreted as the results of a decrease in activities of Al at the plasma membrane surface due to shielding or neutralising the membrane surface charges (Kinraide et al., 1994; Horst, 1995; Ryan et al., 1997).

5. Disturbance of cytosolic Ca^{2+} homeostasis by Al

Cytosolic free Ca^{2+} ions, acting as second messengers, play a pivotal role in transduction of hormonal and environmental signals to the responsive elements of cellular metabolism in plant cells (Gilroy et al., 1993; Poovaiah and Reddy, 1993; Bush, 1995; Webb et al., 1996). Many environmental stresses, such as salinity (Bush, 1996;

Cramer and Jones, 1996), cold (Knight et al., 1996), hypoxia (Bush, 1996) and oxidative stress (Price et al., 1994) have been shown to induce changes (usually increases) in cytosolic Ca^{2+} levels that precede the physiological responses. Similarly, disturbance of cytosolic Ca^{2+} homeostasis has been proposed as a primary cause of Al toxicity (Rengel, 1992a, b; Rengel et al., 1995). Some evidence in support of this hypothesis is emerging (Nichol and Oliveira, 1995; Lindberg and Strid, 1997; Jones et al., 1998). In animal cells, Al interactions with the phosphoinositide signal transduction pathway and subsequent disturbance of cytoplasmic Ca^{2+} homeostasis appear to be a primary mechanism of Al toxicity (Shi and Haug, 1992; Haug et al., 1994). In a recent study, Jones and Kochian (1995) showed that Al inhibited H_2O_2-generated increase in inositol 1,4,5-trisphosphate (IP_3) as a result of inhibition of the plasma membrane lipid phosphatidylinositol 4,5-bisphosphate by specifically targeting the phospholipase C (PLC). It is worthwhile noting that IP_3 is involved in mediation of intracellular Ca^{2+} release through opening the tonoplast Ca^{2+} channels in plants (Alexandre and Lassales, 1992). The putative inhibition of intracellular Ca^{2+} release and blockade of the plasma membrane Ca^{2+} channels by Al would make plant cells unable to generate the Ca^{2+} spikes and waves that are required for cell division and elongation (Rengel, 1992a; Hepler, 1994; Jones and Kochian, 1995). Moreover, the disruption of the Ca^{2+} signalling system may also interfere with growth-related cytoskeletal dynamics (Jones and Kochian, 1995).

Direct measurement of cytosolic Ca^{2+} activity ($[Ca^{2+}]_c$) and monitoring its response to short-term exposure to Al is an essential step in testing the hypothesis that disruption of cytosolic Ca^{2+} homeostasis is a primary cause of Al toxicity in plants (Rengel, 1992b). Recent development of Ca^{2+}-sensitive fluorescent probes and Ca^{2+} imaging techniques have allowed direct measuring of the cytosolic Ca^{2+} activity (Read et al., 1992; Webb et al., 1996). Indeed, studies using Ca^{2+}-sensitive fluorescent dyes and Ca^{2+} imaging techniques have indicated Al-induced changes in cytosolic Ca^{2+} levels in several plant systems. For instance, Al-induced increases in $[Ca^{2+}]_c$ in excised barley roots (Nichol and Oliveira, 1995) and protoplasts isolated from wheat roots (Lindberg and Strid, 1997). On the other hand, Jones et al. (1998) showed that Al induced a rapid decrease in $[Ca^{2+}]_c$ in BY-2 tobacco cell cultures. However, Nichol and Oliveira (1995) only investigated long-term (2 days) effects of Al on the $[Ca^{2+}]_c$ using the excised barley roots, while an Al-elicited oscillating increase in $[Ca^{2+}]_c$ in the wall-free protoplasts (Linderberg and Strid, 1997) may not be physiologically pertinent to the intact root apical cells. Moreover, the Al-induced decrease in $[Ca^{2+}]_c$ of the tobacco cell culture (Jones et al., 1998) may have limited implications in Al toxicity in plants as physiology and morphology of the cell culture may differ significantly from those of intact root cells. Therefore disturbance of cytosolic Ca^{2+} homeostasis as a trigger of Al toxicity syndrome remains to be substantiated experimentally in intact plant roots.

Using the Ca^{2+}-sensitive fluorescent dye Fluo-3 and confocal laser scanning microscopy, we investigated the short-term (<1 h) effect of Al on $[Ca^{2+}]_c$ of intact root apical cells, a critical site of Al stress, employing near-isogenic wheat lines differing in Al tolerance at a single locus. We have overcome problems associated with loading of Fluo-3 acetoxymethyl (AM) ester into intact plant cells (Cork, 1986), which is a prerequisite for directly imaging the intracellular Ca^{2+} activities, by developing a non-invasive 'cold-loading' technique (Zhang et al., 1998).

6. Aluminium elicited an increase in intracellular Fluo-3 fluorescence

Intact wheat roots cold-loaded with Ca^{2+}-sensitive Fluo-3 showed strong fluorescence from the root apical cells when imaged under a confocal laser scanning microscope (Zhang et al., 1998). There was a significant increase in the intracellular fluorescence intensity in the presence of Ca^{2+}-ionophore A23187 and hydrogen peroxide (Zhang et al., 1998). Hydrogen peroxide can induce a rapid increase in $[Ca^{2+}]_c$ in both plant (Price et al., 1994) and animal cells (Klyubin et al., 1996). Taken together, these results suggest that the intracellular fluorescence of the root apical cells is directly related to the cytosolic Ca^{2+} activity.

Exposure of the roots to 50 µM $AlCl_3$ (pH 4.2) led to a sustained and reversible increase in the fluorescence in both Al-sensitive (ES8) and Al-tolerant (ET8) wheat lines. The intracellular fluorescence increased 48 % in ES8 and 27 % in ET8 roots after 1-h treatment with 50 µM $AlCl_3$ (Zhang and Rengel, 1998). The fluorescence of ES8 root cells increased by 18 % when exposed to 2.6 µM $AlCl_3$ for 1 h, while the same treatment caused a slight decrease in the fluorescence of ET8 root cells (Zhang and Rengel, 1998). The differential increase in the root cell fluorescence (i.e. $[Ca^{2+}]_c$) in ES8 and ET8 wheat roots may lie in the ET8 roots being capable of complexing toxic Al species in the rhizosphere and or apoplasm by releasing malate (Delhaize et al., 1993). Alternatively, Al-induced greater increase in $[Ca^{2+}]_c$ may result from a higher symplasmic Al activity in ES8 than ET8 roots.

Since Al incompletely inhibits Ca^{2+} influx due to only partial blocking of the plasma membrane Ca^{2+} channels (see above), Ca^{2+} influx from the apoplasm as well as the release of Ca^{2+} from the intracellular stores (e.g., vacuoles and ER) might be responsible for the increased $[Ca^{2+}]_c$. Regardless of the source of Ca^{2+}, the increase in the root cell $[Ca^{2+}]_c$ (Zhang and Rengel, 1998) may provide a mechanistic explanation for induction of callose synthesis in wheat root tips in the presence of Al (Zhang et al., 1994), because an increase in $[Ca^{2+}]_c$ is a prerequisite for initiating synthesis of callose in plant cells (Kauss, 1985).

7. Response of root cell fluorescence to hydrogen peroxide

Jones and Kochian (1995) showed that Al inhibited a hydrogen-peroxide-elicited increase of IP_3 in wheat root tip cells, indicating that disruption of IP_3-mediated signal transduction system could be a mechanism involved in Al toxicity syndrome in plants. A rapid and marked increase in the fluorescence of root cells was evoked when H_2O_2 (10 mM) was added to the incubation solution (Zhang and Rengel, 1998). Pre-treatment of roots with 100 µM $AlCl_3$ for 2 h reduced the H_2O_2-induced increase in fluorescence by about 10 %, while the 2-h pre-treatment of roots with 1 mM $LaCl_3$ inhibited the H_2O_2-induced increase in fluorescence by about 40 % (Zhang and Rengel, 1998). Similar La concentration has been widely used to search for causes of the increase in $[Ca^{2+}]_c$ in response to environmental stimuli in plants (Price et al., 1994; Knight et al., 1996). The less inhibition of H_2O_2-elicited increase in $[Ca^{2+}]_c$ by Al than La may suggest that the plasma membrane Ca^{2+} channels have been inhibited less by Al than La (Zhang and Rengel, 1998).

8. Concluding remarks

Aluminium induced an increase in the $[Ca^{2+}]_c$ in intact root apical cells of wheat. This increase in the $[Ca^{2+}]_c$ is strongly correlated with inhibition of root growth in both Al-sensitive and Al-tolerant near-isogenic wheat lines. These findings provide unambiguous evidence to support the involvement of cytosolic Ca^{2+} homeostasis in Al toxicity in plants. The Al-related sustained increase in $[Ca^{2+}]_c$ could disrupt Ca^{2+}-dependent metabolic processes, which are directly or indirectly involved in regulation of cell division and elongation (Rengel, 1992b; Hepler, 1994), leading to the observed inhibition of root growth. The role of disruption of the cytosolic Ca^{2+} homeostasis by Al in the overall Al toxicity syndrome might be tested by experimentally manipulating cytosolic Ca^{2+} activities using caged Ca^{2+} and Ca^{2+} buffers while monitoring the response of roots to Al.

9. Acknowledgement

This study was financially supported by the Australian Research Council.

10. References

Alexandre J and Lassales J P 1992 Intracellular Ca^{2+} release by $InsP_3$ in plants and effect of buffers on Ca^{2+} diffusion. Phil. Trans. Roy. Soc. London B, 338, 53-61.

Blamey F C P 1998. The role of the root cell wall in aluminium toxicity. In New Concepts of Plant Nutrient Acquisition (N. Ae, J Arihara, K Okada and Srinivasa eds). Cambridge University Press (in press).

Bush D S 1995 Calcium regulation in plant cells and its role in signalling. Annu. Rev. Plant Physiol. Mol. Biol. 46, 95-122.

Bush D S 1996 Effects of gibberellic acid and environmental factors on cytosolic calcium in wheat aleurone cells. Planta 199, 89-99.

Carpita N C and Gibeaut D M 1993 Structural models of primary cell walls in flowering plants: consistency of molecular structure with physical properties of the walls during growth. Plant J. 3, 1-30.

Cork R J 1986 Problems with the application of quin-2 AM to measuring cytoplasmic free calcium in plant cells. Plant Cell Environ. 9, 157-160.

Cramer G R and Jones R L 1996 Osmotic stress and abscisic acid reduce cytosolic calcium activities in roots of *Arabidopsis thaliana.* Plant Cell Environ. 19, 1291-1298.

Delhaize E and Ryan P R 1995 Aluminum toxicity and tolerance in plants. Plant Physiol. 107, 315-321.

Delhaize E, Ryan P R and Randall P J 1993 Aluminum tolerance in wheat (*Triticum aestivum* L.). II. Aluminum-stimulated excretion of malic acid from root apices. Plant Physiol. 103, 695-702.

Ding J P and Pickard B G 1993 Mechanosensory calcium-selective cation channels in epidermal cells. Plant J. 3, 83-110.

Ding J P, Badot P-M and Pickard B G 1993 Aluminium and hydrogen ions inhibit a mechanosensory calcium-sensitive cation channels. Aust. J. Plant Physiol. 20, 771-778.

Gilroy S, Bethke P C and Jones R L 1993 Calcium homeostasis in plants. J. Cell Sci. 106, 453-462.

Hepler P K 1994 The role of calcium in cell division. Cell Calcium 16, 322-330.

Haug A, Shi B and Vitorello V 1994 Aluminium interaction with phosphoinositide-associated signal transduction. Arch. Toxicol. 68, 1-7.

Horst W J 1995 The role of the apoplast in aluminium toxicity and tolerance of higher plants: A review. Z. Pflanzenernähr. Bodenkd. 158, 419-428.

Huang J W, Pellet D M, Papernik L A and Kochian L V 1996 Aluminium interactions with voltage-dependent calcium transport in plasma membrane vesicles isolated from roots of aluminium-sensitive and -resistant wheat cultivars. Plant Physiol. 110, 561-569.

Huang W J, Shaff J E, Grunes D L and Kochian LV 1992 Aluminium effects on calcium fluxes at the root apex of aluminium-tolerant and -sensitive wheat cultivars. Plant Physiol. 98, 230-237.

Jones D L and Kochian L V 1995 Aluminium inhibition of the inositol 1,4,5-trisphosphate signal transduction pathway in wheat roots: a role in aluminium toxicity. Plant Cell 7, 1913-1922.

Jones D L, Shaff J E and Kochian LV 1995 Role of calcium and other ions in directing root hair tip growth in *Limnobium stoloniferum*. Planta 197, 672-680.

Jones D L, Kochian L V and Gilroy S 1998 Aluminium induces a decrease in cytosolic calcium concentration in BY-2 tobacco cell culture. Plant Physiol. 116, 81-89.

Kauss H 1985 Callose biosynthesis as a Ca^{2+} regulated process and possible relations to the induction of other metabolic changes. J. Cell Sci. Suppl. 2, 89-103.

Kinraide T B, Ryan P R and Kochian L V 1994 Al^{3+}-Ca^{2+} interactions in aluminium rhizotoxicity. II. Evaluating the Ca^{2+}-displacement hypothesis. Planta 192, 104-109.

Klyubin I V, Kirpichnikova K M and Gamaley I A 1996 Hydrogen peroxide-induced chemotaxis of mouse peritoneal neutrophils. Eur. J. Cell Biol. 70, 347-351.

Knight H, Trewavas A J and Knight M R 1996 Cold calcium signalling in Arabidopsis involves two cellular pools and a change in calcium signature after acclimation. Plant Cell 8, 489-503.

Kochian L V 1995 Cellular mechanisms of aluminium toxicity and tolerance in plants. Annu. Rev. Plant Physiol. Plant Mol. Biol. 46, 237-260.

Lindberg S and Strid H 1997 Aluminium induces rapid changes in cytosolic pH and free calcium and potassium concentrations in root protoplasts of wheat (*Triticum aestivum*). Physiol. Plant. 99, 405-414.

Nichol B E and Oliveira L A 1995 Effects of aluminium on the growth and distribution of calcium in roots of an aluminium-sensitive cultivar of barley (*Hordeum vulgare*). Can. J. Bot. 73, 1849-1858.

Piñeros M and Tester M 1995 Characterisation of a voltage-dependent Ca^{2+}-sensitive channel from wheat roots. Planta 195, 478-488.

Piñeros M and Tester M 1997 Calcium channels in higher plant cells: selectivity, regulation and pharmacology. J. Exp. Bot. 48, 551-577.

Poovaiah B W and Reddy A S N 1993 Calcium and signal transduction in plants. Crit. Rev. Plant Sci. 12, 185-211.

Price A H, Tailor A, Ripley S J, Griffiths A, Trewavas A J and Knight M R 1994 Oxidative signals in tobacco increase cytosolic calcium. Plant Cell 6, 1301-1310.

Read N D, Allan W T G, Knight H, Knight M R, Malho R, Russell A, Shacklock P S and Trewavas A J 1992 Imaging and measurement of cytosolic free calcium in plant and fungal cells. J. Microsc. 166, 57-86.

Reid R J, Tester M and Smith F A 1995 Calcium/aluminium interactions in the cell wall and plasma membrane of *Chara*. Planta 195, 362-368.

Rengel Z 1992a Role of calcium in aluminium toxicity. New Phytol. 121, 499-513.

Rengel Z 1992b Disturbance of cell Ca^{2+} homeostasis as a primary trigger of Al toxicity syndrome. Plant Cell Environ. 15, 931-938.

Rengel Z and Elliott D C 1992 Mechanism of aluminium inhibition of net $^{45}Ca^{2+}$ uptake by *Amaranthus* protoplasts. Plant Physiol. 98, 632-638.

Rengel Z, Piñeros M and Tester M 1995 Transmembrane calcium fluxes during Al stress. Plant Soil 171, 125-130.

Ryan P R, DiTomaso J M and Kochian L V 1993 Aluminium toxicity in roots: an investigation of spatial sensitivity and the role of the root cap. J. Exp. Bot. 44, 437-446.

Ryan P R and Kochian L V 1993 Interaction between aluminium toxicity and calcium uptake at the root apex in near-isogenic lines of wheat differing in aluminium tolerance. Plant Physiol. 102, 975-982.

Ryan P R, Reid R J and Smith F A 1997 Direct evaluation of the Ca^{2+}-displacement hypothesis for Al toxicity. Plant Physiol. 113, 1351-1357.

Shi B and Haug A 1992 Aluminium interferes with signal transduction in neuroblastoma cells. Pharmacol. Toxicol. 71, 308-313.

Sivaguru M and Horst W J 1998 The distal part of the transition zone is the most aluminum-sensitive apical root zone of maize. Plant Physiol. 116, 155-163.

Webb A A R, McAinsh M R, Taylor J E and Hetherington A M 1996 Calcium ions as intracellular second messengers in higher plants. Adv. Bot. Res. 22, 45-96.

White P J 1998 Voltage-dependent Ca^{2+} uptake by right-side-out plasma membrane vesicles derived from maize shoots. J. Plant Physiol. 152, 17-24.

Zhang G, Hoddinott and Taylor G J 1994 Characterisation of 1,3-(ß-D-glucan) (callose) synthesis in roots of *Triticum aestivum* in response to aluminium toxicity. J. Plant Physiol. 144, 229-234.

Zhang W H, Rengel Z and Kuo J 1998 Determination of intracellular Ca^{2+} in cells of intact wheat roots: Loading of acetoxymethyl ester of Fluo-3 under low temperature. Plant J. 15, 147-152.

Zhang W H and Rengel Z 1998 Aluminum induces an increase in cytoplasmic calcium in intact wheat root apical cells. Plant Physiol. (submitted).

MOLECULAR BREEDING OF BARLEY FOR DROUGHTED LOW INPUT AGRICULTURAL CONDITIONS

B.P. FORSTER[1], R.P. ELLIS[1], A.C. NEWTON[1], R. TUBEROSA[2], D. THIS[3], A.S. EL-GAMAL[4], M.H. BAHRI[5], and M.BEN SALEM[6]
[1]Cell and Molecular Genetics Department, Scottish Crop Research Institute, Invergowrie, Dundee, DD2 5DA, Scotland, UK
[2]Dipartimento di Agronomia, Università delgi Studi, Bolgna, Via Filippo Re 6, 40126 Bolgna, Italy
[3]INRA-Montpellier, Genetique et Amelioration des Plantes, 2 place Viala, 34060 Montpellier cedex, France
[4]Barley Research Department, Field Crops Research Institute, 8 El-Gamaa Street, Giza 12619, Egypt.
[5]Department d'Agronomie et d'Amelioration des Plantes, Ecole Nationale d'Agriculture de Meknes, PBS40 Meknes, Morocco.
6Labo. De Physiologie Vegetale, INRAT 2049, rue Hedi Karray, Ariana, Tunis, Tunisia

1. Abstract

Molecular genetic markers and other enabling technologies are now sufficiently developed for exploitation in breeding programmes. They allow the possibility of accelerating, and improving the efficiency, of breeding for abiotic and biotic stress tolerances. Barley is a good model species in which to demonstrate this. Barley is a diploid species in which genetic analysis is relatively easy, its short life cycle and inbred nature has also provided for good physiological research. The results of genetic and physiology studies can now be exploited more efficiently in breeding using the tools of contemporary biotechnologies. A programme of work has been initiated aimed at developing more stable yields in barley (*Hordeum vulgare*) for droughted, low input agricultural conditions of Mediterranean rim countries. Various molecular breeding approaches are to be compared which vary in gene donors, recipients and methods. Two of the approaches exploit wild barley (*H. spontaneum*) as a source of genetic variation for abiotic stress tolerance, the third uses an adapted landrace. Results from controlled environment experimentation will be compared with field performance in naturally stressed environments of N. Africa. The work will provide a testing ground for the application of ideas and technologies developed from research.

2. Background

Drought is a serious problem limiting yields in Mediterranean rim countries. In recent years traditional cereal growing areas of Europe e.g. the Iberian Peninsula, south east

G. Gissel-Nielsen and A. Jensen (eds.), Plant Nutrition – Molecular Biology and Genetics, 359–363.

England and northern France, have become affected by drought due to changes in rainfall patterns. Irrigation is a possible solution, but the limited water availability in such areas is more likely to be directed towards cash crops such as vegetables and fruits than to cereals. Enlarging the cereal production area is another possibility, but this would mean bringing into production less favourable lands which would demand even more stress tolerant lines, and more irrigation. Breeding cereals with greater yield stability is a more feasible option.

Barley is one of the oldest crop species being recorded in archaeological sites as far back as 8,000BC. The divergence of the crop species, *Hordeum vulgare* L., from its wild progenitor, *H. spontaneum* Koch, has lead to differences in adaptation to biotic and abiotic stresses. Modern breeding practices typically aim for uniformity, a practice that has restricted genetic variation in the modern cultivars. In recent years there has been a change in emphasis towards yield stability, particularly with respect to unpredictable environments and changing climate. This objective requires the development of new genotypes carry genes currently outside the primary gene pool of modern cultivars. There is a rich source of genetic variation for drought tolerance available in the wild relatives and primitive landraces of barley. This variation can be accessed and exploited with great efficiency using contemporary tools of genetics and other biotechnologies of molecular breeding.

3. Objectives

The general objective is to compare three molecular breeding approaches for their efficiencies in developing better adapted barley for droughted, low input agriculture of Mediterranean rim countries. Each approach involves genetic mapping of molecular genetic markers and quantitative trait loci (QTL) analyses for traits associated with drought tolerance and drought relief induced mildew susceptibility (DRIMS).

Molecular genetic marker protocols offer a powerful means for germplasm characterisation, identification and screening of agronomically valuable genes. The limited genetic diversity within cultivated species has lead to an urgent and careful re-evaluation of landrace and wild species with the objective of capturing and exploiting useful variation. Barley is an ideal species for such studies because; 1) it is an important crop; 2) genetic and physiological studies in barley have revealed genetic loci and physiological traits associated with stress responses; 3) genotypes and useful sampling sites have been identified; and 4) enabling biotechnologies have been developed in barley for molecular breeding and are being used.

Landrace and barley cultivars are classed as belonging to the species *Hordeum vulgare* L, and are intercrossable with the wild progenitor species, *H. spontaneum* Koch, all three forms share a common genome (HH, 2n=14). Landrace material and wild barley in particular exhibit huge variation for a number of traits, and are of interest in improving cultivars via gene introgression. Past efforts have concentrated on introgression of major, single genes e.g. for mildew resistance. These have met with limited success because of linkage drag, i.e. the introduction of a beneficial gene is linked to agronomically unacceptable traits. Also the genetic background may have to be adapted to allow sufficient gene expression, these problems can take many years of breeding effort to overcome. As yet no-one has attempted introgressing a QTL from wild into cultivated barley as this has additional problems. QTL effects are difficult to quantify and large errors can be associated in estimating them. The new era of molecular breeding aims to overcome these problems. Genetic markers can be exploited that aid the rapid detection of rare recombinants where linkage with undesired genes is broken in desirable genetic backgounds.

Wild barley has its main centre of diversity in the Middle East Fertile Crescent where it occupies a wide range of habitats ranging from swamps to deserts, mountains to sub-sea levels, varying daylengths, and from temperatures below freezing to in excess of 30°C. Wild barley is interfertile with cultivated barley and normal meiotic recombination takes place. Wild barley is therefore seen as a valuable source of genes for crop adaptation to stressed environments. Recent work on wild barley has detected associations between molecular markers and physiological responses to both experimentally imposed stress and site-of-origin ecogeography. Collection sites and lines have been identified for adaptation to various environments. Markers that account for large portions of the variation for salt and drought tolerance have been mapped (other abiotic stressors are currentlybeing assessed). Markers include isozymes, RLPs, AFLPs and SSRs. The effectiveness of these markers is to be studied in two molecular breeding programmes using field performance as a parameter of success. The main difference between the approaches is the choice of recipient line, in one contemporary N. Africa lines are used, in another a European cultivar is used. The aim here is to study the effects of wild barley genes in two contrasting genetic backgrounds.

Some barley landraces developed in Mediterranean countries show a substantial tolerance to drought. For example, the line known as Tadmor, which has been selected out of a Syrian landrace, is well adapted to the dry conditions of the Southern and Eastern rim of the Mediterranean where it exhibits high yield stability. A segregating population of F8 recombinant inbred lines (RILs) has been developed by CIMMYT and ICARDA from the cross Tadmor X Er/Apm. Er/Apm is an ICARDA line adapted to moderate water deficit conditions, but is considerably more drought susceptible in terms of yield. The Tadmor x ER/APM cross has been shown to represent contrasting drought tolerance characteristics such as osmotic adjustment, plant architecture, growth habit and chlorophyll content. The current genetic map constructed from the RILs comprises over 100 genetic markers including RFLPs, RAPDs and some morphological traits. Data from controlled environment studies will be compared with performance of RILs in naturally stress field conditions in three N. African countries.

Drought relief induced mildew susceptibility (DRIMS) has been built into the programme. Commercial barley germplasm expressing the highly effective, and only durable mildew resistance gene *mlo*, sometimes show temporary loss of resistance following relief of water stress. Mildew is the most serious foliar disease of barley in Europe, and the problem of DRIMS has been reported in Europe, there are also concerns elsewhere e.g. N. Africa. Clearly it would be desirable to avoid the development of drought tolerant barleys which suffer from mildew susceptibility on relief of water stress. The genetic control of this trait is to be studied.

2.1 KEY QUESTIONS TO BE ADDRESSED

1) Which genetic markers and physiological traits are retained and which are lost in the various approaches?
2) Do the markers and physiological traits associated with drought and DRIMS correspond to those inherited from cultivated, landrace or wild species germplasm?
3) What are the similarities/contrasts in lines selected in the three N. African country test sites.
4) Are genes/physiological traits associated with drought tolerance and DRIMS also associated with yield penalties?
5) Can undesirable linkages be broken?
6) How important are background effects? Are genes from wild barley expressed equally in European and N. African genetic backgrounds?
7) What is the potential for widening the genepool for barley to include landrace and wild species variation?
8) How can more efficient molecular breeding strategies be developed?

3. Technologies and Methodologies to be deployed

The work does not involve genetic transformation. As yet transformation in barley is not routine, and candidate genes for drought and DRIMS are not known. Since the genetic control of drought tolerance and DRIMS is under polygenic control a transgenic approach is further ruled out at the moment. Instead the work capitalises on: 1) the identification of germplasm carrying desired traits which can be crossed onto cultivated barley; 2) the identification of genetic and physiological markers associated with traits; 3) well developed genetic maps; 4) well developed marker technologies; and 5) well developed doubled haploid production protocols. These factors combine to offer effective molecular breeding approaches; these are briefly described below.

The approaches using wild barley as sources of genes for drought and DRIMS involve traditional crossing of wild barley lines onto cultivated barley lines. In order to speed up the programme, the F1 seed is excised three weeks after pollination and sprouted *in vitro*. Once plantlets have developed roots they are grown on for doubled haploid production. Uninucleate microspores are collected from the developing ears and cultured using established protocols that promote microspore embryogenesis. During this process the initial haploid nucleus of the microspore spontaneously doubles to produce doubled haploid embryos which develop into plants. The spontaneous doubling provides a rapid means of generating homozygous plants that are true breeding and can be tested repeatedly. Embryo rescue and doubled haploidy can be performed at all stages in a breeding programme, in addition to DHs produced from F1 the work also involves DH production from F1BC1 material in which the F1 plants have been backcrossed to the recipient cultivar. This is done to increase the proportion of the recipient, cultivar DNA. The doubled haploids produced will be screened for the presence of genetic markers associated with drought tolerance genes donated by the wild parent and screened for the background genetic constitution of the recipient parent.

Undesirable weedy characteristics inherited from the wild species e.g. spontaneous shattering of the ripened ear will also be selected against. The material will be simultaneously tested in naturally drought field conditions of the three N. African partner countries (Egypt, Morocco and Tunisia), and the genetic data compared to field performance. Genetic maps and QTL analyses will be further developed from the results.

Recombinant inbred lines developed from a barley landrace are used in the third workpackage. This workpackage concentrates on physiological traits rather than on molecular marker selection. One of the advantages of this programme is that it uses adapted material as a source of genetic variation, selection against non-agronomic characteristics of the wild species is therefore not a problem. In addition the material has a life cycle suited to the target area (vernalisation and photoperiodic sensitivities are not potential problems as they are with wild barley introgression). Physiological responses to drought under controlled environmental conditions will be compared to responses of field grown plants. Measurements will be taken on carbon metabolism, osmotic adjustment, plant architecture and chlorophyll content. High density genetic mapping will be performed around key areas of the genome which are known to carry genes for drought responses, this approach has the potential to reveal candidate genes.

4. Acknowledgements

The work described above is the subject of a proposal submitted to the EU INCO programme (International Cupertino with third countries), the contract for the work is currently being negotiated. The Scottish Crop Research Institute receives grant-in-aid from the Scottish Office Agriculture, Fisheries and Environment Department.

MAIZE BREEDING FOR TOLERANCE TO LOW NITROGEN AND DROUGHT STRESS IN THAILAND

Y. CHANTACHUME[1], P. KRUDLOYMA AND T. MANUPEERAPAN
[1] Corresponding author
Maize breeder, Nakhon Sawan Field Crops Research Center (NSW FCRC), Takfa, Nakhon Sawan, Thailand 60190

Abstract

The project of maize breeding for tolerance to low N was established at NSW FCRC in 1995 with the objectives of developing inbred lines, hybrids and populations tolerant to low N. A block of land of approximately 0.7 ha was used as a testing block for low N tolerance. Before using the block, two successive crops of maize were planted with no fertilizer in 1995 early rainy season (ER) and 1995 late rainy season (LR) to create the low N condition. The evaluation for tolerance to low N was started in 1996ER. The breeding scheme used was screening of inbred lines and early generation lines from NSW FCRC hybrid maize development project and the selected lines were recombined into a population NSLN1C_0 which will be used as a population source for tolerance to low N. These selected lines were also crosses with two elite inbred lines Nei 9202 and Nei9008 and the top-cross hybrids were evaluated under low N condition. In summary, twenty three top-cross hybrids were selected based on tolerance to low N and will be tested in various environments in 1998 ER.

Breeding for drought stress tolerance was initiated in 1977. The objectives of this project are to develop inbred lines, hybrids and populations with tolerance to drought stress. Two populations namely KK-DR(S)C_5 and NS-DR(S_2)C_1 were developed using S_1 and S_2 recurrent selection, respectively. The S_1 and S_2 progenies were evaluated under artificial stress and normal water conditions during cycles of the population improvement and the tolerant progenies were selected based on drought index and some other important agronomic traits under stress condition such as anthesis and silking interval, number of ears per plant, wilt score and shelling percentage, etc. Three maize varieties were developed from each of the two populations. Commercial hybrids and inbred lines evaluations for drought stress tolerance were also conducted. There were approximately 150 top-cross hybrids developed from the selected tolerant lines crossing with 2 elite inbred lines Nei 9202 and Nei 9008 and these top-cross hybrids were evaluated for tolerance to drought in 1998 dry season.

G. Gissel-Nielsen and A. Jensen (eds.), Plant Nutrition – Molecular Biology and Genetics, 365–370.

There were at least 2-3 sets of drought and low N trials from CIMMYT each year for evaluation of the stress tolerance of inbred lines and hybrid maize.

Introduction

Drought and low N are two of the major limiting factors of maize production in Asia. The report from the 6th Asian Regional Maize Workshop in India in 1995 indicated that about half of the total maize production area in Asia has been affected by both stresses. The estimated losses of yield due to drought and low N stress were in the range of 10-75% and 10-50%, respectively (Logrono and Lothrop, 1996). In Thailand, during 1975 to 1994, the estimated affected planting area and yield loss due to the two stress conditions were approximately 0.1-0.2 million hectares and 0.2-0.3 million metric tons per year, respectively, which equal to the value of 30-50 million US$ per year (Office of Agricultural Economics, 1985-1995).

Genetic variation of maize population (Lafitte and Edmeades, 1994) and inbred lines (Balko and Russell, 1980) in response to N supply was observed, and it seem, therefore, possible to breed varieties or hybrids with tolerance to low N. Gain in grain yield under drought from four recurrent selection studies in low land tropical maize have averaged about 100 kg/ha/year or about 5% per year (Edmeades et al., 1995). Development of maize germplasm tolerance to both stresses should give a long term benefit. CIMMYT has established networks of collaborating scientists interested in the area of drought and low N tolerant maize. There are 20 and 15 countries participated in the drought and low N net work, respectively (Edmeades et al., 1996).

For Thailand, drought and low N projects have been started in 1977 and 1995, respectively (Chantachume et al., 1996). The objectives of this paper are to present the results and progresses of the projects.

Materials and methods

1. Low Nitrogen

In the early rainy season of 1996, exotic and local maize germplasms were screened and selected for tolerance to low N at Nakhon Sawan Field Crops Research Center (NSW FCRC), Phra Phutthabat Field Crops Experiment Station (PPB FCES), Ban Mai Samrong Field Crops Experiment Station (BMS FCES) and Phetchabun Field Crops Experiment Station (PBN FCES). The experiments were conducted under both high N (50 kgN/ha) and low N (25 kgN/ha and 0 kgN/ha). The data collected from the two conditions were combined and an index (low N index) (Chantachume et al., 1996) of grain yield was calculated.

Low N index (LI) = (X1/X2) / (Y1/Y2)

X1 = Mean yield of genotype "X" under stress (low N),

X2 = Mean yield of genotype "X" under nonstress (high N),
Y1 = Mean yield of all genotypes under stress,
Y2 = Mean yield of all genotypes under nonstress,
LI => 1 = tolerance to low N, LI < 1 = low N susceptibility.
Germplasms used for screening came from;

1.1 309 S1 lines of a good adapted open pollinated variety "NS 1" (Chantachume et al., 1989)
1.2 1,002 selfing lines (S3-S7) of NSW FCRC hybrid program
1.3 CIMMYT maize populations and topcrosses
1.4 Hybrid maize from private and government sectors and the Asian Regional Maize Projects (ARMP)

2. Drought

The objectives of this experiment are to select drought-tolerant inbred lines for the production of hybrids tolerant to drought. However, selection for drought tolerance based only on grain yield under drought may have little success because of low genetic variation for the trait, and could reduce yield potential. Use of secondary traits of adaptive value, whose observed genetic variation is increased under drought, should increase selection efficiency (Bolanos and Edmeades, 1996). Two hundred and fifty six inbred lines were tested for drought tolerance under artificial drought stress condition during three weeks before to three weeks after anthesis at NSW FCRC in 1996 dry season. The experimental design used was 16x16 simple lattice. Seeds of each inbred line was sown in a 2.5 m-row with one plant per hill in a spacing of 75x25 cm. Grain yield and some important agronomic characters related to tolerance to drought stress for example, 50% anthesis and 50% silking date (50% of the plants within a plot have their silk or tassel emerged), wilt scores, plant recovery scores, number of ear per plot, root and stalk lodging and plant and ear height were measured. The wilt scores ranged from 1 to 5 (1 = the best and 5 = the worst)and were carried out one and two weeks after anthesis while the plant recovery was scored one week after irrigation which was started at three weeks after anthesis.

Results and discussion

1. Low Nitrogen

Only the results from the screening of S1 lines from "NS 1" will be described in this paper. 309 S_1 lines from NS 1 populations were evaluated under stress and non stress conditions. Four sets of experiment were conducted using the Simple Lattice Design 10x10, 10x10, 8x8 and 7x7. Seeds of each line were sown in two rows of 5 m length with the spacing of 25 cm between hills. Correlation between some important agronomic traits including grain yield, anthesis and silking interval (ASI), number of ears per plant, nitrate reductase activity (NRA), leaf senescence (LS) and low N index (LI) were calculated. The minimum, maximum and mean grain yield under both conditions of some of the tested and selected S_1 lines are shown in Table 1. The lines were selected based on LI, mean grain yield under stress, number of ears per plant and ASI. The mean grain yield under stress were in the range of 23-60% of that observed under non-stress conditions whereas the mean yield of the 36 selected lines under stress were 31-81% of that found under nonstress conditions. The mean yield of the control variety "NS1" under stress was 28-69% of that under nonstress. The LI of the 36 selected lines were 1.34-1.42 whereas those of the starting material were 1.02-1.05. The studies of the relationship between these traits demonstrated that the number of ears per plant under stress was highly significant correlated to the other traits including LI, grain yield under both conditions and ASI. Mean grain yield under both conditions were highly significant correlated, however, there was no correlation between grain yield under nonstress and LI (Table 2). Therefore, based on this experiment, LI, grain yield, number of ears per plant and ASI under stress could be used as a selection criteria for tolerance to low N.

Table 1. Minimum, maximum and mean of grain yield (kg/ha) under stress condition averaged from 99 NS 1-S1 lines of the first experiment, tested at NSW FCRC 1996ER.

Traits	Min	Max	Exp. mean	NS 1	9 selected	C.V. (%)
LI	0.23	2.72	1.03**	1.29	1.36	38
Y1	131	2,331	1,150**	2,806	1,631	32
Y2	2,075	7,744	5,019**	9,987	5,331	13
Ear/plant	0.19	1.26	0.69**	0.86	0.80	22
ASI[1]	0.79	8.09	2.40**	1.34	1.59	59
LS[2]	1.94	4.08	2.87**	3.00	2.44	15
NRA[1]	0.48	1.46	0.79	0.53	0.77	25

[1] Transformed data (square root), Y1 = yield (-N), Y2 = yield (+N), [2] Leaf senescence

** significant different at $P < 0.01$

Table 2. Correlation coefficient (r) between grain yield and some important traits of the tested 99 NS 1-S1 lines, tested in the first experiment under low N condition at NSW FCRC 1996ER.

Traits	Y1	Y2	ASI	Ear/plant	NRA	E. aspect[1]
LI	0.78**	- 0.09	- 0.33**	0.57**	- 0.10	- 0.53**
Y1		0.52**	- 0.46**	0.76**	- 0.09	- 0.74**
Y2			- 0.27	0.46**	- 0.01	- 0.40**
ASI				- 0.52**	0.09	0.33**
Ear/plant					- 0.03	- 0.58**
NRA						0.05

[1] Ear aspect, Y1 = yield (-N), Y2 = yield (+N)

2. Drought

Only the results from the screening of inbred lines are presented. The grain yield means of the 256 tested lines were in the range of 0-1,869 kg/ha while the mean of the 50% anthesis and silking dates were 72 and 74 days, respectively (Table 3). The maximum, minimum and mean of the wilt and plant recovery scores and some important traits of these lines were identified (Table 3). In contrast to the results of Bolanos and Edmeades (1996), there was no relationship between anthesis and silking interval (ASI) and grain yield, however, there was a trail of negative correlation between these traits (data not shown). The occurrence of this non-significant correlation may be because of the genetic back ground of the tested inbred lines which was rather narrow and some of them were related. Highly significant correlation between grain yield and both of the wilt and plant recovery scores (Table 4) were observed. Thirty one inbred lines were selected based on the wilt and plant recovery scores and grain yield. Based on this experiment, the wilt and plant recovery score should be used as selection criteria for tolerance to drought.

Table 3. Mean, minimum and maximum of grain yield (Kg/ha) and agronomic traits averaged from 256 inbred lines, tested for drought tolerance at NSW FCRC in dry season 1996.

	Days to 50%		Height (cm)		No. of	Wilt score		Plant[1]	Grain
	Anthesis	Silk	Plant	Ear	ear[2]	I	II	recov.	yield
Min.	63	66	35	17	0	0.1	1.0	0.8	0
Max.	88	91	111	52	12	5.2	5.2	5.3	1,869
Mean	72	74	75	34	4.5	3.6	3.6	3.6	375
C.V.(%)	7.8	7.0	22.5	23.3	54.6	22.1	24.9	24.7	87.5
LSD(.05)	11.0	10.2	33.4	15.7	4.8	1.6	1.7	1.8	586.9

[1] Plant recovery, [2] Number of ear per plot

Table 4. Correlation coefficients (r) between grain yield and wilt score and plant recovery of 256 inbred lines, tested for drought tolerance at NSW FCRC in dry season 1996.

Traits	Grain yield
Wilt score I	-0.40**
Wilt score II	-0.59**
Plant recovery	-0.54**

n = 512

Acknowledgments

We would like to thank Dr. S.K. Vasal, CIMMYT-ARMP, Thailand, for his consultation and supports to our maize breeding program. We grateful to the CIMMYT Maize Program for providing the genetic materials evaluated in these projects.

References

Balko, L. G. and Russell, W. A. (1980) Effects of rate of nitrogen fertilizer on maize inbred lines and hybrid progeny. I. Prediction of yield response. Maydica, 25: 65-79.

Bolanos, J. and Edmeades, G. O. (1996) Mid-season drought as a selection environment for tropical maize improvement. *In* The Symposium of Developing Drought and Low-Nitrogen Tolerant Maize. (Book of abstracts). CIMMYT, El Batan, Mexico.

Chantachume, Y., Kitbomrong, J., Krudloyma., P., and Chuva, S. (1989) Corn variety " Nakhon Sawan 1". pp. 74-81. In Thailand National Corn and Sorghum Program Annual report 1989. Bangkok, Thailand.

Chantachume, Y., Manupeerapan, T., Grudloyma, P., Tongchuay, S., Noradachanon, S. and De Leon, C. (1996) Selection for low N tolerance in the Thai maize breeding program. pp. 511-513. *In* G.O. Edmeades, M. Banziger, H.R. Mickelson, and C.B. Pena-Valdivia (Eds). The Symposium of Developing Drought and Low-Nitrogen Tolerant Maize. 25-29 Mar. 1996. CIMMYT, El Batan, Mexico.

Edmeades, G. O., Banziger, M., Chapman, S. C., Ribaut, J. M. and Bolanos, J. (1995) Recent advances in breeding for drought tolerance in maize. Paper presented at Sichuan Academy of Agricultural Sciences, Chengdu, P.R. China in July 1995.

Edmeades, G. O., Banziger, Beck, D. L., Chapman, S. C. and Cortes, M. (1996) Drought and low N testing networks: past, present and future. pp. 405-414. *In* G.O. Edmeades, M. Banziger, H.R. Mickelson, and C.B. Pena-Valdivia (Eds). The Symposium of Developing Drought and Low-Nitrogen Tolerant Maize. 25-29 Mar. 1996. CIMMYT, Mexico.

Lafitte, H. R. and Edmeades, G. O. (1994) Improvement for tolerance to low soil nitrogen in tropical maize. I. Selection criteria. Field Crops Res., 39: 1-14.

Logrono, M. L. and Lothrop, J. E. (1996) Impact of drought and low N on maize production in South Asia. pp. 39-43. *In* G.O. Edmeades, M. Banziger, H.R. Mickelson, and C.B. Pena-Valdivia (Eds). The Symposium of Developing Drought and Low-Nitrogen Tolerant Maize. 25-29 Mar. 1996. CIMMYT, El Batan, Mexico.

Office of Agricultural Economics, (1975-1995) Agricultural statistics of Thailand. Crop year 1984/85-1994/1995. Ministry of Agriculture and Co-operative. Bangkok, Thailand.

THE PHYSIOLOGY OF PHOSPHORUS NUTRITION IN PLANTS - CAN WE ASSEMBLE THE COMPONENTS INTO A PRACTICAL MODEL

JAMES DUNLOP
AgResearch Grasslands,
Private Bag 11008, Palmerston North, New Zealand.

Abstract

There is a large body of data on the physiology of components related to phosphate response and phosphorus efficiency in plants. However plant physiologists are not yet in a position to give unequivocal advice to breeders and molecular biologists on the development of phosphorus efficient plants. Difficulties arise from gaps in the understanding of some physiological processes and the difficulty in integrating the large number of interacting components that are involved. Further physiological studies are required and can be justified but thought needs to be given the topics to be studied and the methods to be used. A multi-disciplined approach is required. Use of standardised methods would add value to the studies. Ways to model the complex body of information are required. Some suggestions are made on the last two points.

Introduction

Phosphorus (P) is a major plant nutrient of particular significance to many regions of the world. However fertiliser phosphate is also a potential pollutant. It is an additional cost and, for some farmers, it is prohibitively expensive. The need for its prudent use has stimulated considerable research into the physiology of phosphorus nutrition. A wide range of physiological properties and processes which are involved in or have a direct impact on the acquisition and utilisation of phosphate (PO_4) have been investigated by many groups. A number of researchers have envisaged programmes, guided by physiological information, to breed PO_4 efficient plant cultivars but very few such plants have been developed. The growing power of molecular genetics suggests that genetic engineering has the potential to overcome difficult breeding problems but the term engineering implies an ability to generate accurate designs which will function reliably. Is plant physiology able to meet this challenge?

G. Gissel-Nielsen and A. Jensen (eds.), Plant Nutrition – Molecular Biology and Genetics, 371–377.

The Challenge

The physiology of P nutrition is the subject of many papers and has been frequently reviewed (the reader is directed to Bieleski and Ferguson, 1983; Clarkson and Grignon, 1991, and many chapters in Waisel, Eshel and Kafkafi, 1996). A list of research themes and topics which focus on acquisition of PO_4 by roots, including interactions of roots with soil, is given in Table 1. The list is not exhaustive and, more importantly, it does not consider the large subject areas of PO_4 biochemistry, growth and the functions primarily associated with the shoot. Its value is limited to indicating the range of factors which are relevant. For the majority of these topics our knowledge is incomplete with significant gaps in understanding. Furthermore the difficulty these gaps create when we attempt to form a holistic model of P nutrition are matched by the complexity arising from the myriad interactions that exist between the various components. Figure 1, which indicates some of the known interactions, conveys the impression of a tangled skein of partially understood processes that defies our ability to make qualitative, let alone quantitative, predictions. This may explain why plant physiologists have not been successful in giving unequivocal instructions to plant breeders or genetic engineers on how they might produce PO_4 efficient cultivars despite the large body of literature on processes that have an impact on PO_4 response and PO_4 efficiency. We lack understanding of some vital components and we need a method to handle the complexity of the processes involved.

Table 1 Major themes and topics in studies of the physiology of roots related to phosphate and associated interactions with soil.

Soil exploration	**Influx into roots**	**Translocation**
root morphology	kinetics and dual (multiple) mechanisms	movement to xylem vessels
root hairs	regulation by P status	xylem anatomy & P distribution
mycorrhizal associations	H^+- symports (and Na^+ - symports?)	redistribution via phloem
proteoid roots	PO_4 transporter genes	
root proliferation in P rich sites microsites	influence of the cell wall	
	influences of the shoot	
Soil modification	**Efflux from roots**	**Partitioning of plant PO_4**
rhizosphere organisms	efflux rates	constant $[PO_4]_{i,\ cytoplasm}$
PO_4 solubilising exudates	Cmin	vacuole as a reservoir
phosphatases	ion channels for PO_4	biochemical forms of PO_4
Root anatomical differences		
variation of influx along length		
variation in H^+ efflux		
root hair growth		
periderm development		

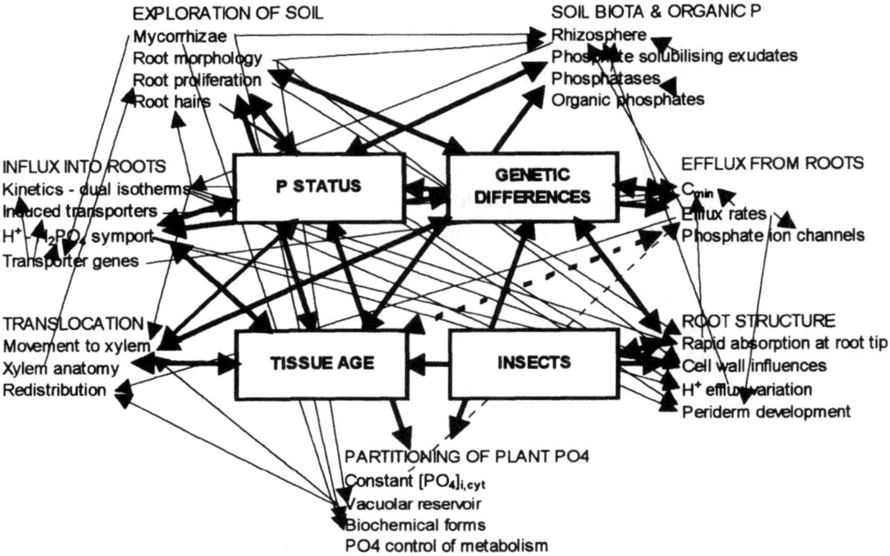

Fig. 1. Processes contributing to or influencing acquisition of phosphate by roots including interactions of roots with soil but excluding phosphate biochemistry, growth and the functions primarily associated with the shoot. Arrows are drawn to indicate known interactions. The four labelled boxes are examples of topic areas which also interact with phosphate acquisition.

GAPS IN OUR KNOWLEDGE

An exhaustive list of aspects of P nutrition things on which more information is needed would be very long and shorter lists reveal the bias of the author. I consider that the following items are particularly important -

1. The mechanisms by which P status of the plant controls PO_4 influx. The success of attempts to increase the maximum rate of PO_4 influx depends on information on whether plants will permit the potential increase in rate to be expressed. Progress on this topic will soon come from studies on the expression and function of genes coding for high affinity PO_4 transporters (Clarkson and Hawkesford, 1993).
2. The control of PO_4 transport between the root and shoot. Benefits from additional PO_4 absorption will depend on its appropriate distribution within the plant. This may not occur without intervention. Research suggests that transport of PO_4 to the shoot may be important in the control of absorption by the root (Clarkson and Grignon, 1991).
3. Mapping of influx and efflux along the length of root systems. Many studies of influx and efflux tacitly assume that they are uniform along the length of roots. This is known to be untrue (e.g. Hay, Dunlop and Hopcroft, 1986). This could lead to gross errors when extrapolating data to roots in soil as PO_4 availability will often vary greatly along the length of the root. Furthermore if the zones of efflux are distant from the zones of influx the balance between influx and efflux may be

substantially different than the situation in well stirred solutions. In particular, data for Cmin would have little value.

4. How should data obtained from hydroponic experiments be extrapolated to plants growing in soil? Soil presents a very different environment from hydroponics, the medium used in many physiological experiments (Da Silva, and Gabelman, 1993). Differences of consequence between the two may include the diffusion of PO_4 to the root and the development of depletion zones; the diffusion of other ions, particularly protons from the root; competition from soil micro flora and fauna; effects of soil on root morphology; the proliferation of roots in zones of high P availability and mycorrhizal associations.
5. What data obtained using model species can be safely applied to other species? Experiments on the physiology of P nutrition have focused on a relatively limited number of species. For some phenomena we have little information on how well they apply to other species.

These are illustrative examples. There are others which may be more important.

THE COMPLEXITY

Because the processes which contribute to the P nutrition of plants are connected by a web of interactions, it is not possible to change one without inducing a change in a number of others. Consequently it is far from simple to predict results of altering one process on the overall P nutrition of the plant. The difficulty is compounded by the range of the processes involved extending beyond the principal area of expertise of an individual. Our viewpoint needs to embrace the fact that plant functions are integrated.

A Role for Plant Physiology

The examples of P efficient genotypes currently available to us have been identified principally by using measures of shoot dry matter and P content rather than basic physiological characters (e.g. see papers in Maranville, 1993). Because plant physiology is not yet ready to guide breeders, it seems that breeding should proceed using dry matter and P content. However this approach is not always practical. Areas of uniformly P deficient soil sufficiently large to screen the numbers of plants required are not always available. For some crops, screening in pots, as a method of ensuring uniform P availability, results in selection of material that grows well in pots but not other environments. Furthermore it can be difficult to determine whether a genotype that performs well under low P does so because of some aspect of its P nutrition or another property not directly related to P and its productivity may still be limited by P. Practical ways to determine whether a genotype's P nutrition had robust advantages would be valuable. Plant breeders would welcome development of criteria that quickly and simply identify plant characters which have a good correlation with superior P nutrition. This would be valuable for screening the large numbers of plants that are necessary in the breeding of many species. Therefore there is still potential for plant physiology to contribute to the development of P efficient species.

The "Integrated screening programme" (ISP, Grime et al., 1997) is an ecological study, which provides encouragement for physiologists aspiring to contribute to the breeding of nutrient efficient plants. A wide variety of traits in 43 species were analysed to unravel causal relationships with performance in different habitats, a complex task which has parallels to understanding variation in P nutrition. Statistical analysis showed that the 67 traits measured in the different species clustered into three groups (axes) which related to the ecological strategy of the species. Axis 1, "showed remarkably consistent trends", with marked correlations with foliar concentrations of nutrients. This aggregation of traits suggests that components contributing to P nutrition may not be distributed randomly but will be clustered to some degree thus decreasing the complexity. Furthermore ISP used laboratory measurements on plants grown under controlled conditions. Its success indicates that laboratory studies can be relevant to the field. However it is important to note that the measurements were very carefully chosen and standardised. It would be to valid to use the approach of ISP to evaluate traits thought to be associated with P efficiency. However cultivars from one species which differ in their P efficiency may not be available in sufficient numbers to allow valid statistical analysis. This ecological study, along with others, suggests a fundamental trade-off between plant traits allowing fast growth and traits allowing resource conservation and storage. Genetic manipulation of agricultural plants to improve P efficiency is aimed at tinkering with this evolutionary trade-off. Plant physiology is required to assist with information on modifications that will provide what evolution has not.

INCREASING THE RELEVANCE OF PHYSIOLOGICAL RESEARCH

There are substantial economic, environmental and humanitarian imperatives to warrant renewed commitment to breeding cultivars with improved P nutrition (Schaffert, 1993; Brown, 1997). However it is important that the way to achieve this is carefully considered and the programmes designed accordingly. The following features will be important to achieving success -

1. The creation of teams with a breadth of expertise ranging from molecular genetics to agronomy and plant breeding (Schaffert, 1993; Imsande, 1998). Phosphorus nutrition is multi-faceted with many aspects which do not fall within the expertise of a single discipline. Progress is most likely to be made by teams which encompass the range of relevant skills and can interact openly and willingly. Team members will need to be willing to extend themselves to understand the others in the team and they may need to suspend their prejudices and preoccupation with their individual specialities. Equitable funding is likely to be important.
2. The use of experimental approaches which provide incisive data relevant to plants growing in agricultural situations (Graham, 1984). Plants growing in soil can be

difficult experimental subjects and often there is no practical way to use them to determine the physiological information that is required. However it is important that experiments always bear the end purpose in mind and use methods that will produce data that is as relevant as the practicalities allow.

3. As a corollary of the preceding two points, the assumptions implicit in the choice of traits studied or the measurement techniques employed should be evaluated and discussed explicitly. These assumptions could be considered as being analogous to recognition sites on cell membranes - they can either facilitate or inhibit interaction. The use of standardised measurements would be valuable.
4. The ability to integrate information on a wide range of traits involved in P nutrition and the interactions that occur between them needs to be developed. The ISP may provide a useful model to do this. The applicability of other knowledge engineering methods should also be investigated. For example the developing discipline of "fuzzy cognitive mapping" (or "conceptual mapping") (Taber, 1991) may provide a rigorous way to obtain answers to "what if" questions.
5. The identification of screening parameters that are quick and accurate in identifying valuable material. Molecular markers e.g. quantitative trait loci (QTL's) promise to provide tests with the specificity and simplicity that are required. However development of appropriate markers would require physiologists to determine the characters that would be useful for nutrient efficient plants.

Conclusion

Thirty five years ago, Epstein (1963), Gerloff (1963) and Vose (1963) pointed out the prospects for breeding crop plants with improved mineral nutrition. Since then there have been some notable successes but for P nutrition, overall progress has been limited (Graham, Ascher and Hynes, 1993). Molecular biology is greatly expanding our understanding of some aspects of the P nutrition of agricultural plants and providing potentially valuable genetic components but it may be some time before it is able to provide plants with improved mineral nutrition to agricultural users (Clarkson and Hawkesford, 1993). Plant physiologists working in this area would do well to consider the opportunities and difficulties involved and review their strategies where necessary. The many phenomena which impinge on phosphate nutrition indicate the need for multi-disciplined teams (Schaffert, 1993). There is need for additional information for some phenomena and this will have greater value if it is obtained using methods that allow direct application with data from other researchers. In addition to data providing subsets of P nutrition understanding, we need to acquire the ability to integrate the subsets to give a model that is useful in a practical sense. For any one species, these tasks constitute a very considerable amount of work. However the humanitarian, environmental and economic rewards from developing crop plants that reduce the need for phosphatic fertilisers or increase the effectiveness of their use justify the investment that will be required.

References

Bieleski, R.L. and Ferguson, I.B. (1983) Physiology and metabolism of phosphate and its compounds. in A. Lauchli and R.L. Bieleski (eds.) *Inorganic Plant Nutrition, Encyclopedia of Plant Physiology, New Series, Vol. 15A,* Spinger-Verlag, Berlin, pp 422-449.

Brown, L. (1997) Facing the challenge of food scarcity: Can we raise the grain yields fast enough? in T. Ando, K. Fujita, T. Mae, H. Matsumoto, S. Mori and J. Sekira (eds.) *Plant Nutrition for Sustainable Food Production and Environment,* Kluwer Academic Publishers, Dordrecht, pp. 15-24.

Clarkson, D.T. and Grignon, C. (1991) The phosphate transport system and its regulation in roots, in C. Johansen, K.K. Lee, and K.L. Sahrawat, *Phosphorus nutrition of grain legumes in the semi-arid tropics,* ICRISAT, Pancheru, pp.49-62.

Clarkson, D.T. and Hawkesford, M.J. (1993) Molecular biology approaches to plant nutrition, *Plant and Soil* **155/156**, 21-31.

Da Silva, A.E., and Gabelman, W.H. (1993) Screening maize inbred lines for tolerance to low-P stress condition, in P.J. Randall, E. Delhaize, R.A. Richards, and R. Munns (eds.) *Genetic Aspects of Plant Mineral Nutrition,* Kluwer Academic Publishers, Dordrecht, pp.233-239.

Hay, M.J.M., Dunlop, J. and Hopcroft, D.H. (1986) Phosphate uptake and anatomy of unthickened and secondarily thickened adventitious root of field-grown white clover (*Trifolium repens)* L.), *New Phytologist* **103**. 659-668.

Epstein, E. (1963) Selective ion transport in plants and its genetic control, in *Desalination Research Conference. National Academy of Sciences - National Research Council Publication 942* pp.284-298.

Gerloff, G.C. (1963) Comparative mineral nutrition of plants, *Annual Reviews of Plant Physiology* **14**, 107-124.

Graham, R.D. (1984) Breeding for nutritional characteristics in cereals, *Advances in Plant Nutrition* **1**, 57-102.

Graham, R.D., Ascher, J.S. and Hynes, S.C. (1993) Selecting zinc-efficient cereal genotypes for soils of low zinc status, in P.J. Randall, E. Delhaize, R.A. Richards, and R. Munns (eds.) *Genetic Aspects of Plant Mineral Nutrition,* Kluwer Academic Publishers, Dordrecht, pp. 349-358.

Grime, J.P. et al. (1997) Integrated screening validates primary axes of specialisation in plants, *Oikos* **79**, 259-281.

Imsande, J. (1998) Iron, sulfur, and chlorophyll deficiencies: A need for an integrative approach in plant physiology. *Physiologia Plantarum* **103**, 139-144.

Maranville, J.W., Baligar, V.C., Duncan, R.R. and Yohe, J.M. (1993), *Adaptation of Plants to Soil Stresses,* INTSORMIL Publication No. 94-2, University of Nebraska, Lincoln.

Schaffert, R.E. (1993) Discipline interactions in the quest to adapt plants to soil stresses through genetic improvement, in J.W. Maranville, V.C. Baligar, R.R. Duncan and J.M. Yohe (eds.), *Adaption of Plants to Soil Stresses,* INTSORMIL Publication No. 94-2, University of Nebraska, Lincoln, pp. 1-13.

Taber, R. (1991) Knowledge processing with fuzzy cognitive maps, *Expert Systems with Applications* **2**, 83-87.

Vose, P.B. (1963) Varietal differences in plant nutrition, *Herbage Abstracts* **33**, 1-13.

Waisel, Y., Eshel, A. and Kafkafi, U. (1996) *Plant Roots. The Hidden Half. (2nd ed.)* Marcel Dekker., New York.

THE EFFECTIVENESS OF MARKER-ASSISTED INTROGRESSION IN BACKCROSS BREEDING FOR NUTRIENT TRAITS OF SELF-FERTILIZING CROPS

Takuro Ishii
Dep. of Crop Breeding, Chugoku National Agri. Exp. Sta., Fukuyama, Hiroshima, 721-8514, Japan

1. Introduction

Marker-assisted introgression in backcross breeding could advantageously be used, because in this method the genome of the recurrent parent as well as genes to be transferred from the donor parent can be correctly selected.

It has been pointed both experimentally (Young and Tanksley 1989) and theoretically (Hospital *et. al.* 1992) that, when a major gene is to be transferred using this method, the number of backcrossings can be decreased to about two.

Recently, many quantitative trait loci (QTLs) of nutrient traits such as tolerance to phosphorus deficiency, alkali soils and so on, have been detected, and backcross breeding with marker-assisted introgression (MABC method) for quantitative nutrient traits is now going to be practised in actual breeding projects. Visscher *et. al.* (1996) and Hospital and Charcosset (1997) have proposed theoretically optimum procedures for transferring QTL alleles from donor parents, but their works were not based on the actual breeding schemes.

In the present paper, the effectiveness of MABC for quantitative traits is investigated under the actual plant breeding schemes by means of Monte Carlo computer simulation. The effectiveness of MABC was compared with that of traditional backcross breeding method (BC method) based on phenotype selection.

G. Gissel-Nielsen and A. Jensen (eds.), Plant Nutrition – Molecular Biology and Genetics, 379–385.

2. Methods

Backcross breeding method simulated was composed of several backcrossings and subsequent selfings (Figure 1). The number of backcrossings was 1 to 4, and the number of selfings was 1 to 5. Using this method, 3 to 10 QTL alleles were tried to transfer. This scheme is considered best, because to get the best single genotype, not to improve the mean genotypic value of a population is important in breeding self-fertilizing crops.

In selection in MABC, individuals were assessed in the first step based on the marker score summed over the objective flanking or single marker(s), and then based on the recovery of 72 markers which were located randomly in the genome of the recurrent parent. Flanking markers mean the markers which exist on both sides of the QTL. While single markers mean the marker on the one side of the QTL.

On the other hand, selection in traditional BC method was conducted based

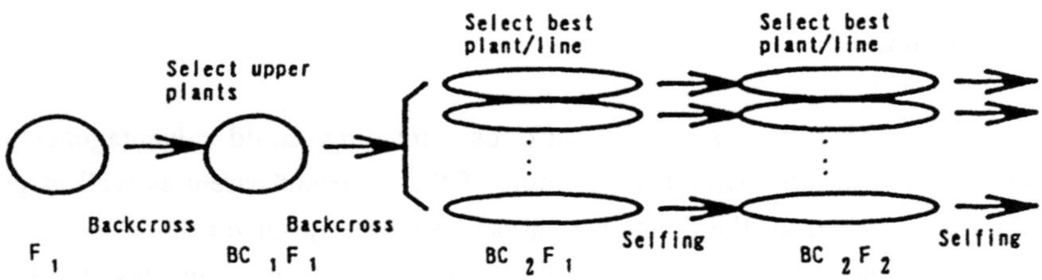

Figure 1. System of backcross breeding simulated

TABLE 1. Parameters used in simulations

	MABC	BC
Genotype and	(++,+−,−−)*	(++,+−,−−)
genotypic value	=(1, 0.5, 0)	=(1, 0.5, 0)
Linkage between	0.01,0.1	−
markers and QTLs		
Linkage between QTLs	0.1,0.3,0.5	0.1,0.3,0.5
Heritability	−	0.5,0.8
Selection intensity	0.05,0.1,0.2,0.5	0.1
Population size	50,100,200,500,1000	100

* + : desirable gene, − : undesirable gene

on only the phenotypic value of the quantitative trait to be transferred.

Parameters used in simulations are listed in Table 1. The genetic effect was additive and all QTLs had the same effect. Most simulations were performed under linkage between QTLs of 0.5, population size of 100 and selection intensity of 0.1.

The effectiveness of selection was measured by the genetic value per QTL (Max) of the individuals which showed the highest marker score (in MABC) or the best phenotypic value (in BC) and by the recovery of recurrent parent genome (Rec). Rec of individuals obtained in BC was also calculated using the same 72 markers as in MABC.

One hundred replications were run for each case of the simulation. The computer program for simulations was written in Fortran77 and run on a HP-9000 computer.

3. Results and Discussion

The effects of the number of backcrossings and selfings in MABC are shown in Figure 2. As the number of backcrossings increased, Max decreased and Rec increased. On the other hand, as the number of selfings increased, both Max and Rec increased, but the responses did not increase markedly when the number of selfings was more than three. With Max and Rec being taken into consideration simultaneously, advance by selection appeared to be optimum

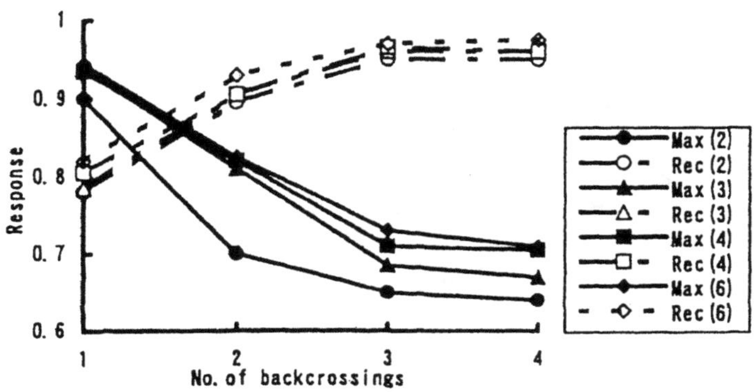

Figure 2. Effect of number of backcrossings and selfings on the selection response of MABC. (the value in the parentheses shows the no. of selfings, QTL=5, r1=0.1)

in BC_2F_4 generation. In most simulation cases, while the selection response increased or decreased compared with the case shown in Figure 2, this tendency, that is, advance by selection was optimum in BC_2F_4 generation, was observed. Then, the effectiveness of MABC was evaluated on the result of BC_2F_4 generation.

Table 2 shows the selection responses of MABC and BC. As the number of QTLs increased, Max decreased markedly. When flanking markers were used and linkage between markers and QTLs (r1) was 0.1, 100% of the aimed QTL alleles were transferred in case of QTLs=3, whereas only 60% in case of QTLs=10. When r1 was 0.01, the result was more or less improved but not prominently. But when linkage between QTLs (r2) was 0.1, 90% of the aimed genes were transferred even in case of QTLs=10 (Figure 3). So it is rather

TABLE 2. Selection responses of MABC and BC

QTLs	Selection effect	MABC				BC	
		Flanking markers		Single marker		h^2=0.5	h^2=0.8
		r1=0.1	r1=0.01	r1=0.1	r1=0.01		
3	Max	1.00	1.00	0.95	1.00	0.81	0.95
	Rec	0.93	0.95	0.93	0.95	0.88	0.88
4	Max	0.92	0.99	0.86	0.97	0.68	0.88
	Rec	0.91	0.92	0.93	0.93	0.88	0.88
5	Max	0.82	0.90	0.77	0.87	0.62	0.79
	Rec	0.91	0.92	0.91	0.92	0.88	0.88
10	Max	0.60	0.67	0.57	0.65	0.45	0.59
	Rec	0.90	0.90	0.90	0.91	0.88	0.88

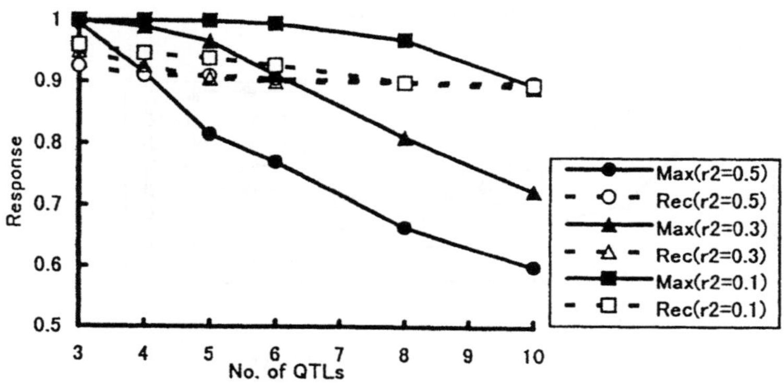

Figure 3. Effect of number of QTLs on the selection response (r1=0.1)

difficult to transfer more than a few QTL alleles when r2 is loose, but it is possible to transfer when r2 is tight.

When single markers were used, Max decreased about 5% at r1=0.1 and about 2% at r1=0.01 (Table 2). These decreases are not so large as Edwards and Page (1994) pointed for recurrent selection in allogamous crops. As r1 became smaller, advantage of flanking over single markers was diminished. When r1 is 0.01, single markers are sufficiently reliable.

Rec was more than 90% in MABC, a little higher than 88% which was retained in BC in the absence of selection for Rec (Table 2). Rec in MABC was not very high, particularly in the case when many QTL alleles should be transferred. It was because selection was first done based on the markers linked with QTLs and then based on the genomic background which was similar to the recurrent parent. The superiority of MABC to BC was not cancelled even for traits with a heritability (h^2) as high as 0.8, where the effectiveness of MABC was 1 to 5% higher than BC, depending on the number of QTL alleles to be transferred (Table 2). As heritabilities of most quantitative traits are considered to be much lower, the effectiveness of MABC is expected to be much higher than BC.

In MABC with a population size of 100, a selection rate of 0.1 was the most effective (Figure 4).

Effectiveness of selection was not markedly improved with population sizes larger than 200 (Figure 5).

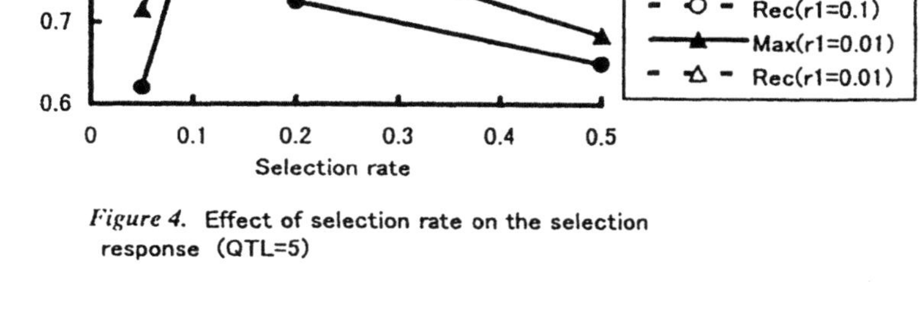

Figure 4. Effect of selection rate on the selection response (QTL=5)

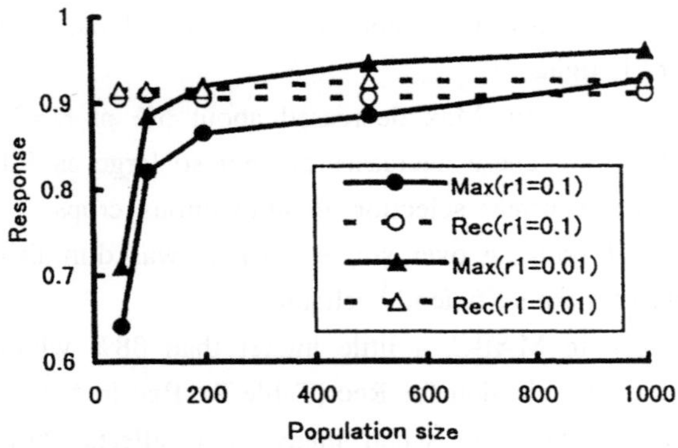

Figure 5. Effect of population size on the selection response (QTL=5)

Since heritabilities of most quantitative traits are low, selection efficiency is not high when traditional BC method is used. In this study, it has been shown that MABC for quantitative traits is also effective as in the case when a major gene is to be transferred. MABC is considered to be a new breeding method for quantitative traits. Recently many QTLs have been mapped such as phosphorus deficiency tolerance, and as more and more QTLs will be detected in near future, MABC method is expected to be applied in the actual plant breeding projects for quantitative traits.

The difference between MABC for quantitative traits and for traits controlled by major genes is that the former is more difficult to select both QTLs and genomic background simultaneously. In this paper, BC_2F_4 generation is thought to be optimum (Figure 2). But when genomic background is more important, more backcrossings will be needed, though the genetic gain for genes to be transferred will decrease.

Since MABC makes use of linkage disequilibrium, high response is expected at the early generations, but the response is thought to be decreased at the later generations. The cost problem also being taken into consideration, the strategies which combine MABC and BC may be required.

In these simulations, the positions of QTLs are assumed to be evident, but in many cases the positions are not known with certainty and only estimated. The studies on the cases that the positions of QTLs are treated as stochastic will be necessary.

4. Reference

Edwards,M.D. and Page,N.J.(1994) Evaluation of marker-assisted selection through computer simulation,*Theor. Appl.Genet*.88,376-382.

Hospital,F., Chevalet,C. and Mulsant,P.(1992) Using Markers in Gene Introgression Breeding Programs, *Genetics* 132,1199-1210.

Hospital,F. and Charcosset,A.(1997) Marker-Assisted Introgression of quantitative Trait Loci,*Genetics* 147,1469-1485.

Visscher,P.M.,Haley,C.S. and Thompsom,R.(1996) Marker-Assisted Introgression in Backcross Breeding Programs,*Genetics* 144,1923-1932.

Young,N.D. and Tanksley,S.D.(1989) RFLP analysis of the size of chromosomal segments retained around the *Tm-2* locus of tomato during backcross breeding, *Theor. Appl. Genet*.77,353-359.

INFLUENCES OF FERTILISER LEVEL ON BREAD-MAKING QUALITY IN SWEDISH GROWN WHEATS

E. JOHANSSON[1], P. OSCARSON[1], P. STOLT[1] AND G. SVENSSON[2]
[1]Department of Plant Breeding Research, The Swedish University of Agricultural Sciences, S-268 31 Svalöv, Sweden and [2]Svalöv Weibull AB, S-268 81 Svalöv, Sweden

Abstract

The grain proteins in hexaploid wheat are of primary importance for the bread-making quality of the flour. Both the protein concentration and the protein composition are of importance for the volume and texture of the baked loaves. In the present study it has been shown that the protein concentration, the gluten strength and the bread volume vary depending on the applied fertiliser level. Higher fertiliser level led to higher protein concentration, lower gluten strength and higher bread volumes. As well the fertiliser level as the cultivar and cultivation year was shown to significantly influence the protein concentration, gluten strength and bread volume. No variation in amount of different proteins and protein subunits could be detected in cultivars grown hydroponically with different N regimes. The possibilities of breeding cultivars with good and stable bread-making qualities when cultivated in low-input agriculture is also discussed in the present paper.

Introduction

Bread-making is one field of application for wheat (Triticum aestivum L.), and thereby the bread-making quality is an important character (Pomeranz 1988). Bread dough properties result from a balance between different components as starch, gluten proteins, lipids, water and so forth, as well as from interactions between these components (Kasarda 1989). Of these components, the proteins have been shown to have the most fundamental role for the processing of the leavened bread (e.g. Kasarda 1989). Both the concentration of the proteins as well as the composition of particular gliadins and subunits of glutenin have been found to be of importance for the bread-making quality (Finney and Barmore 1948, Payne et al. 1987, Johansson et al. 1993, Johansson and Svensson 1995, Johansson 1996). As well the protein concentration as the composition of the grain proteins reflects processes relating to N metabolism in wheat (Jenner et al. 1991). An important factor for a high protein concentration in the grains is availability of soil N for the plant during grain-filling (Simpson et al. 1983). Also, environmental factors, and genotype - environment interactions have been found to affect the gluten strength (Johansson et al 1998). Gluten strength are dependent on the protein composition and strongly correlated with the bread-making quality

G. Gissel-Nielsen and A. Jensen (eds.), Plant Nutrition – Molecular Biology and Genetics, 387–390.

(Johansson and Svensson 1998). Differences in the availability of soil N for the plant during grain-filling might thus affect the protein concentration as well as relative amounts of different proteins and protein subunits in the grains.

The aim of the present study was to investigate the variation in protein concentration, gluten strength and bread-making quality in relation to fertiliser level. Another purpose was to investigate if the variation in these characters could be explained by variation in protein composition.

Materials and methods

The plant material studied comprised field grown wheat cultivars, with different fertiliser levels applied, from a range of years. The cultivars were grown and tested for quality according to Johansson and Svensson (1995) at Svalöf Weibull AB, Landskrona, Sweden. The different fertiliser levels applied were 0, 70, 70+70 and 140 kg N/ha Two of the cultivars were grown hydroponically according to the method described by Oscarson et al (1995). Nitrate was added once daily in exponentially increasing doses (Ingestad and Lund 1979) according to:

$$N_t = N_0 \times e^{RA x t}$$

The total N added were for the different N regimes of the hydroponically grown plants; 20, 29, 38, 47, 56 and 65 mg N per plant.

The hydroponically grown plants were analysed by sodium dodecyl sulphate-polyacryl amide gel electrophoresis (SDS-PAGE; Johansson et al. 1993) reversed phase-high performance liquid chromatography (RP-HPLC; Johansson and Svensson 1996) and enzyme linked immuno sorbent assay (ELISA; Howes et al. 1991) for composition of wheat storage proteins.

Analysis of variance (ANOVA; SAS 1985) was carried out in order to determine the importance of different years, cultivars and fertilizer levels for the variation in protein concentration, gluten strength, and bread volume. Spearman rank correlations and regression analyses (SAS 1985) were used to calculate the correlations among protein concentration, gluten strength and bread volume. Standard errors were calculated in order to estimate the variation within wheats containing subunits 2+12 and 5+10. For evaluation of the data from the RP-HPLC and ELISA analyses, the GLM procedure (SAS 1985) was used to calculate the significance of different parameters to the results.

Results

Large variations in protein concentration as well as in gluten strength and bread volume were found when different nitrogen fertiliser level were applied. Higher fertiliser levels led to higher protein concentration, lower gluten strength and higher bread volume. A significant negative correlation ($P<0.005$) were found between the gluten strength and the protein concentration. However, during specific years, as for example 1994 and 1995, the protein concentration were increased by higher fertiliser levels, although the gluten strength was not decreased to the same extent. When analysis of variance were carried out in different wheat materials in order to determine the importance of different years, cultivars and fertilizer levels for the variation in protein concentration, gluten strength and bread volume, highly significant influences were found in almost all cases.

Variation in protein composition could be detected between the two wheat cultivars grown hydroponically, when analysed with SDS-PAGE and RP-HPLC. However, no significant differences in the composition of proteins between cultivars of the same type, grown with different N regimes could be detected in the cultivars grown hydroponically when analysed with SDS-PAGE, RP-HPLC and ELISA.

Discussion

One desire from the Swedish baking industry is to have the possibility to buy Swedish flour of the same bread-making quality from year to year. This study shows the importance of the genotype as well as of the environment and their interactions in determining bread-making quality. Large variations in bread-making quality due to weather conditions and fertiliser levels have been shown. By altering the fertiliser level, the bread-making quality can be changed to fit the specific cultivation year. Late fertiliser applications have been shown to have an strong impact on the protein concentration, gluten strength and bread-making quality. Earlier investigations on field grown wheat have shown that differences in N fertilisation are changing protein quantities and proportions (Wieser and Seilmeier 1998). By using hydroponically grown wheats with different N regimes, we were not able to show any changes in the amount of different proteins and protein subunits in the present study. Also in an investigation of wheats grown during different years and with different baking quality, it has been difficult to correlate amounts of specific proteins and protein subunits to the differences in gluten strength (Johansson et al 1998b). Instead extractability of large protein polymers have been shown to be an important character. The later character has so far not been investigated for the hydroponically grown wheats in this study.

When breeding wheat cultivars for a low-input agriculture, it seems to be of importance to use cultivars with a relatively low gluten strength. This is due to the Swedish bread-making method with a fixed and relatively gentle dough mixing (Johansson and Svensson 1995). Cultivar stability of the bread-making quality parameters are also of importance for cultivars aimed to low-input agriculture, as well as a high competitive strength in N uptake and utilisation. Cultivars containing the high molecular weight (HMW) subunits 2+12 have been shown to have lower gluten sterngth and to be more stable in gluten strength compared to those containing HMW gluten subunits 5+10 (Johansson et al 1998a). The 2+12 containing cultivars might thereby be more suitable for low-input agriculture compared to the ones containing 5+10. Differences in N uptake capacity between cultivars have not been shown so far (Oscarson et al 1995) although differences in remobilisation of N from the vegetative tissues to the grain have been indicated. This character might also be of importance for wheats grown in low-input agriculture.

Acknowledgements: This work was supported by the Cerealia Foundation R&D, The Knut and Alice Wallenberg Foundation and Svalöf Weibull AB.

References

Finney, K.F. and Barmore, M.A. (1948) Loaf volume and protein content of hard winter and spring wheats, *Cereal Chem.* **25,** 291-312.
Howes, N.K., Ng, P.K.W., Bushuk, W. and Zawistowski, J. (1991) in W Bushuk and R Tkachuk (eds.) *Gluten Proteins* American Association of Cereal Chemists, St Paul, pp 459-466.
Ingestad, T. and Lund, A.B. (1979) Nitrogen stress in birch seedlings. 1. Growth technique and growth, *Physiol. Plant.* **45,**137-148.
Jenner, C.F., Ugalde, T.D. and Aspinall, D. (1991) The physiology of starch and protein deposition in the endosperm of wheat, *Aust. J. Plant Physiol.* **48,** 211-226.
Johansson, E. (1996) Quality evaluation of D zone omega gliadins, *Plant Breed.* **115,** 57-62.
Johansson, E. and Svensson, G. (1995) Contribution of the high molecular weight subunit 21* to breadmaking quality of Swedish wheats, *Cereal Chem.* **72,** 287-290.
Johansson, E. and Svensson, G. (1996) Environmental influences on storage proteins in wheat, in C.W. Wrigley (ed.) *Proc. 6th Int. Gluten Workshop* , Cereal Chemistry Division, Royal Australia Chemical Institute, North Melbourne, Australia. pp 457-460.
Johansson, E. and Svensson, G. (1998) Variation in bread-making quality: effects of weather parameters on protein concentration and quality in some Swedish wheat cultivars grown during the period 1975-1996, *J. Sci. Food Agric.* **78,** 109-118.
Johansson, E., Henriksson, P., Svensson, G. and Heneen, W.K. (1993) Detection, chromosomal location and evaluation of the functional value of a novel high M_r glutenin subunit found in Swedish wheats, *J. Cereal Sci.* **17,** 237-245.
Johansson, E., Svensson, G. and Tsegaye, S. (1998a) Genotype and environment effects on bread-making quality of Swedish grown cultivars containing HMW glutenin subunits 2+12 and 5+10 (Submitted).
Johansson, E., Nilsson, H,., Mazhar, H., Skerritt, J., MacRitchie, F. and Svensson, G. (1998b) Influences on storage proteins on bread-making quality by the cultivation year in Sweden, (manuscript).
Kasarda, D.D. (1989) Glutenin structure in relation to wheat quality, in Y Pomeranz (ed.), *Wheat is unique* American Association of Cereal Chemists, St Paul, pp 277-302.
Oscarson, P., Lundborg, T., Larsson, M. and Larsson, C.-M. (1995) Genotypic differences in nitrate uptake and nitrogen utilization for spring wheats grown hydroponically, *Crop Sci.* **35,** 1056-1062.
Payne, P.I., Nightingale, M.A., Krattiger, A.F. and Holt, L.M. (1987) The relationship between HMW glutenin subunit composition and bread-making quality in British-grown varieties, *J. Sci. Food Agric.* **40,** 51-65.
Pomeranz, Y. (1988) Composition and functionality of wheat flour components, in Y. Pomeranz (ed.), *Wheat: chemistry and technology, vol II*, St Paul, American Association of Cereal Chemists, pp 219-370.
SAS User's Guide (1985) Statistics 1985 SAS Institute Inc. Cary, NC.
Simpson, R.J., Lambers, H. and Dalling, M.J. (1983) Nitrogen redistribution during grain growth in wheat (*Triticum aestivum* L.) Development of quantitative model of the translocation of nitrogen to the grain, *Plant Physiol.***71,** 7-14.
Wieser, H. and Seilmeier, W. The influences of nitrogen fertilisation on quantities and proportions of different protein types in wheat flour, *J. Sci. Food Agric.* **76,** 49-55.

GENETIC VARIATION OF NITROGEN-EFFICIENCY IN FIELD EXPERIMENTS WITH OILSEED RAPE (*BRASSICA NAPUS L.*)

Bettina KESSEL, Heiko C. BECKER
Institut of Agronomy and Plant Breeding,
Georg-August-University Göttingen, D-37075 Göttingen

1. Introduction

Oilseed rape (*Brassica napus L.*) is an important crop to diversify cereal dominated crop rotations. However, oilseed rape has a relatively poor nitrogen (N) efficiency (AUFHAMMER et al. 1994, SCHJOERRING et al. 1995). In order to obtain maximum seed yield, high rates of nitrogen fertilizer have to be applied. Moreover, the relatively small recovery of applied nitrogen in the seed involves risks for leaching of nitrogen to the groundwater (LICKFETT 1993). Breeding for improved efficiency for uptake and utilization of nitrogen has never been done in this crop. The aim of this study is to investigate the genetic variation in N-efficiency in a genetically broad material of winter oilseed rape.

2. Materials and Methods

Seventy genotypes from four different groups (lines, hybrids, resyntheses (the genomes of *Brassica campestris* and *Brassica oleracera* have been brought together artificially into one organism) and crosses between Falcon msl (male sterility Lembke) and resyntheses) were investigated under field conditions 1996/97 at the locations Göttingen and Einbeck in Northern Germany. The experiment was performed with two N treatments, two replications and plot sizes of 7,9 m^2 in Einbeck and 4,5 m^2 in Göttingen. N fertilizer was given as KAS in a 'high' (220 kgN/ha) and a 'reduced' (0 kgN/ha in Göttingen, 100 kgN/ha in Einbeck) treatment.

Several characters which possibly are related to N-efficiency were evaluated at the beginning and end of flowering and at maturity: date of flowering, chlorophyll content, plant height, fresh weight, dry weight, seed yield, N amount in the seeds, straw yield and its N amount, harvest index, N harvest index, N content in dropped leaves and oil content in the seed.

G. Gissel-Nielsen and A. Jensen (eds.), Plant Nutrition – Molecular Biology and Genetics, 391–395.

To get the N content in dropped leaves, the leaves freshly dropped during two days were collected at each N-supply and dryed at 60°C. N and oil contents were measured with NIRS (near-infrared-reflexion-spectroscopy) (REINHARDT 1992).

3. Results

All characters showed significant interactions between genotype and nitrogen application except chlorophyll content.

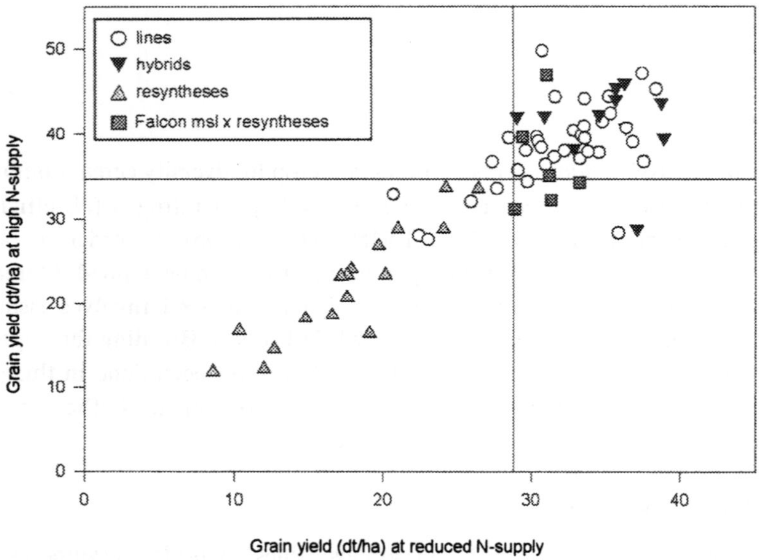

Fig.1: Grain yield at the two N levels reduced and high (two locations 1997)

The hybrids are the relatively highest yielding genotypes even with reduced nitrogen supply.

Genotype groups are clearly distinguishable in their yield level (Fig. 1). The yield of the resyntheses is always low and highly correlated between the two N levels. For the other genotypes the correlation between the two N levels is low. This distinction of the genotype groups can also be seen for N yield of the seeds (Fig. 2 and 3).

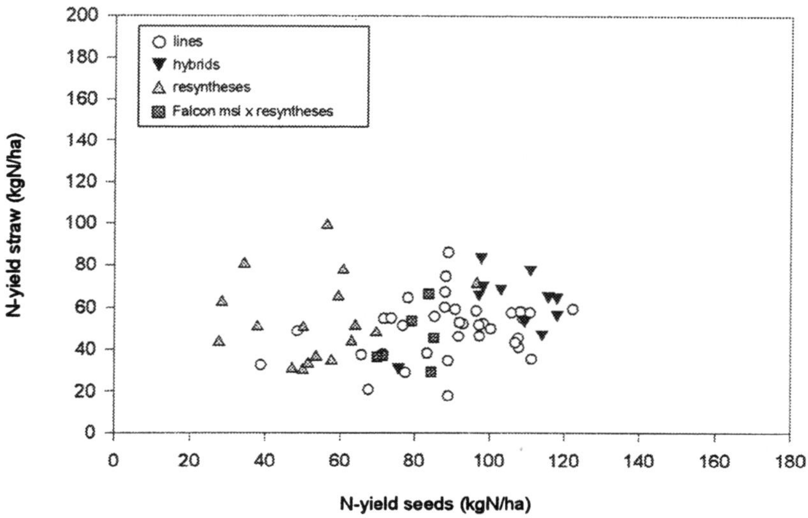

Fig.2: N yield of seeds and straw at reduced N-supply (100 kgN/ha) Einbeck 1997

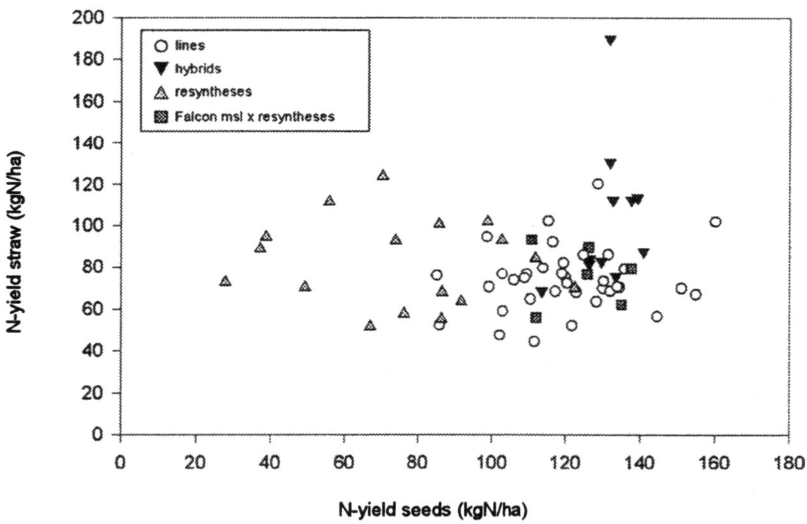

Fig.3: N yield of seeds and straw at high N-supply (240 kgN/ha) Einbeck 1997

The N yield of straw showed a relatively greater variation at reduced N-supply than at high N-supply. The genotypic variation was significant for both N yield of seed and straw at both N levels. The N yields of seed and straw were not correlated.

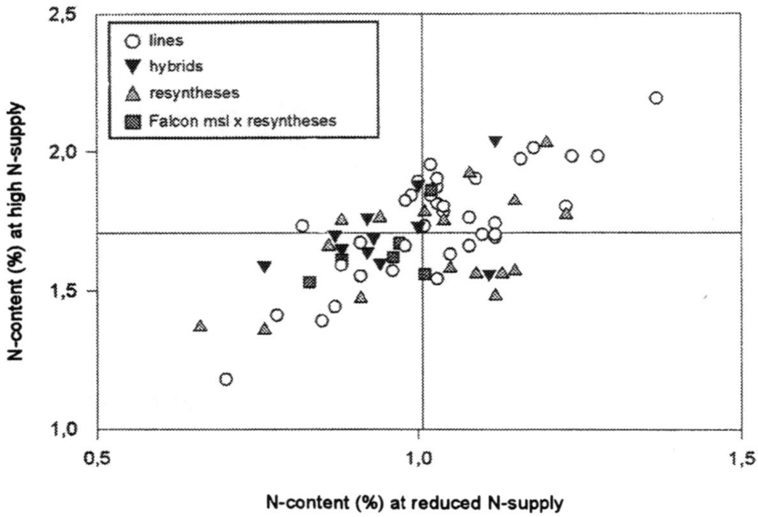

Fig.4: N content in dropped leaves (two locations 1997)

N content of dropped leaves showed a large variation at both N levels and is specific for each genotype (Fig. 4). No general difference can be observed among genotype groups.

4. Outlook

A breeding strategy for improved N-efficiency can be based on selection for low N content in dropped leaves and for high N harvest index by reducing N yield of straw. Large genetic variation was observed for both characters. The experiments will be continued including old landvarieties at several locations under different climatic conditions.

5. References

AUFHAMMER, W., KÜBLER, E. and BURY, M. 1994: Stickstoffaufnahme und Stickstoffrückstände von Hauptfrucht- und Ausfallrapsbeständen. J. Agronomy & Crop Science **172**, 255-264

LICKFETT, T. 1993: Auswirkungen verminderter Produktionsintensität in zwei Rapsfruchtfolgen auf Elemente des N-Haushaltes im System Boden-Pflanze. Diss., Göttingen

REINHARDT, T.-C. 1992: Entwicklung und Anwendung von Nah-Infrarot-spektroskopischen Methoden für die Bestimmung von Öl-, Protein-, Glucosinolat-, Feuchte- und Fettsäure-Gehalten in intakter Rapssaat. Diss., Göttingen

SCHJOERRING, J.K., BOCK, J.G.H., GAMMELVIND, L., JENSEN, C.R. and MOGENSEN, V.O. 1995: Nitrogen incorporation and remobilization in different shoot components of field-grown winter oilseed rape (*Brassica napus L.*) as affected by rate of nitrogen application and irrigation. Plant and Soil **177**, 255-264

NEW RICE VARIETIES TO CONTROL RICE ALLERGEN

Hideo MAEDA, Takuro ISHII, Shuichi IIDA and Hiroshi NEMOTO
Department of Crop Breeding, Chugoku National Agricultural Exp. Stn.
Fukuyama 721-8514, Japan

1. Introduction

Allergy due to intake foods becomes an important problem in Japan. It occurs atopic dermatitis mainly in children and many foods resulting from allergy has reported such as eggs, milk, soybeans and so on. Recently, patients of rice allergy are increasing and the symptoms are very serious compared with other food allergies. Rice is a staple food for Japanese people, development of new rice variety to control rice allergen is required.

The main allergenic protein in rice grain has been identified to be the 16-kDa albumin. In 1990, rice mutant lines were screened for lacking or low content of 16-kDa polypeptide and 4 mutant lines were selected from gamma-rays irradiated or chemical mutagen (EMS) treated populations (Nishio and Iida 1993). These mutant lines were divided into two groups. Two mutant lines lacked 16-kDa polypeptide, and these plants were sterile. The other 2 mutant lines contained half amount of 16-kDa protein compared with that of the original cultivars and these plants of this type were fertile. These mutant lines were investigated agronomic characers, one of them showed the same level of yield as the original cultivar. This mutant line was designated as 'LA-1', however, it has floury character in the endosperm and lodging resistance is very weak.

To improve these problems, we started crossing 'LA-1' and japonica breeding lines.

2. Methods

2.1. SCREENING OF ALLERGENIC PROTEIN

The allergenic protein components of breeding lines were investigated according to the method of Nishio and Iida (1993). Matured dry seeds of selected breeding lines

G. Gissel-Nielsen and A. Jensen (eds.), Plant Nutrition – Molecular Biology and Genetics, 397–400.

were crushed with pillers and immersed in extraction solution containing 20% grycelin, 5% mercaptoethanol, 8M urea, 4% sodium lauril sulfate (SDS) and 0.125M TRIS-HCl buffer (pH 6.8). After incubation at room temperature for 16 h, the suspension was centrifuged at 10,000 rpm for 5 min., and 5μl of supernatant was subjected to SDS polyacrylamide gel electrophphoresis (SDS-PAGE). The electropnoresis was carried out on a 15% polyacrylamide gel, and the gel was stained by Coomassie Brilliant Blue R-250.

2.2. BREEDING PROCESS

To improve lodging resistance and floury character in the endosperm, we started crossing between 'LA-1' and eight Japanese breeding lines in 1991. These F_1 plants were grown in the green house in winter, and F_2 populations were planted to paddy field next year. From F_2 to F_4 generations were maintained as bulked populations. After that, F_5 populations were independently planted to the paddy field, and we picked up plants that have good plant form. The contents of 16-kDa protein of selected F_5 plants were investigated by SDS-PAGE. Out of them, low allergen F_6 lines were grown as independent lines and investigated agronomic characters. From F_6 to F_7, selected lines were investigated more important agronomic characters such as yielding, disease resistance, cooking quality and so on. At last, three breeding lines were selected and designated as 'Chugoku 161', 'Chugoku 162' and 'Chugoku 163'.

TABLE 1. Agronomic characters of low allergen breeding lines and major japonica rice cultivars

Variety name	Heading date	Maturing date	Culm length (cm)	Panicle length (cm)	Yielding (kg/ha)
Chugoku 161	7.Aug	19.Sep	70	19.3	5910
Chugoku 162	14.Aug	24.Sep	80	20.5	6290
Chugoku 163	24.Aug	11.Oct	78	21.9	6750
Akitakomachi	2.Aug	16.Sep	79	19.9	5440
LA-1	6.Aug	24.Sep	83	19.7	5660

These data were investigated in 1996.

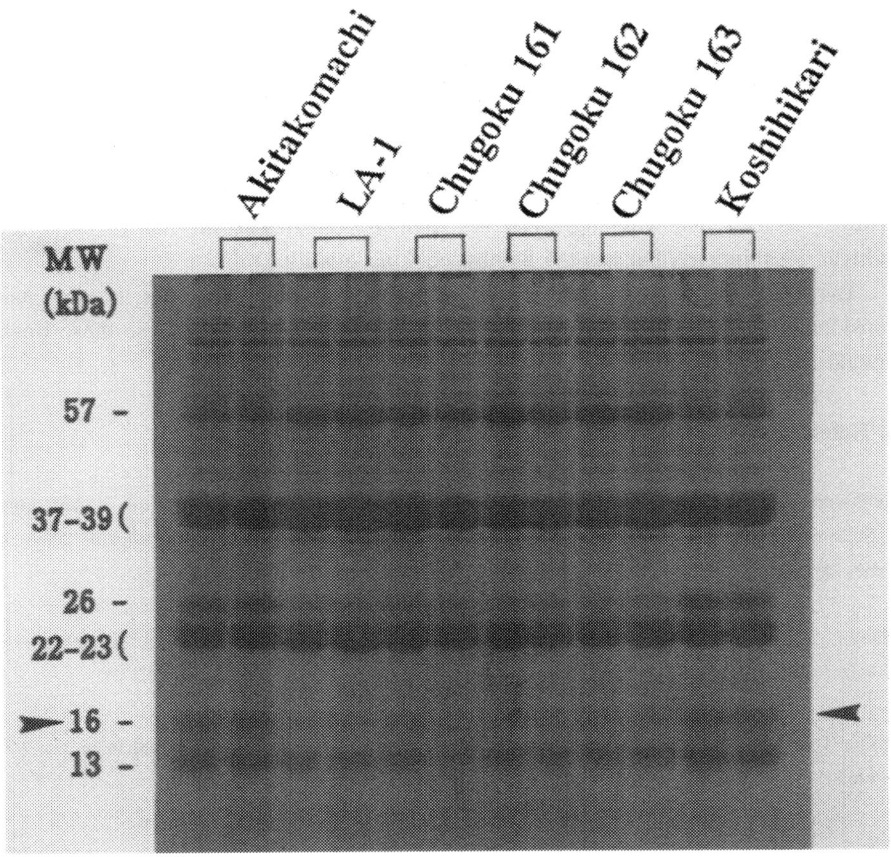

Figure 1. The allergenic protein components of three breeding lines and parental varieties. Total seed proteins stained by Coomassie Brilliant Blue. Arrow head indicates the 16-kDa allergenic polypeptide.

3. Results and discussion

'Chugoku 161', derived from cross between 'LA-1' and Japanese breeding line 'Chugoku 125', is early maturing variety. 'Chugoku 162', derived from 'LA-1' and Japanese cultivar 'Aoinokaze', is medium maturing. 'Chugoku 163', derived from high yielding variety 'Chugoku 116' and 'LA-1', is late maturing. The 16-kDa protein components of these breeding lines were shown in Figure 1. They showed half amount of the 16-kDa and 26-kDa polypeptides and contained a high level of the 57-kDa polypeptide.

Agronomic characters of these lines were indicated in Table 1. 'Chugoku 162' and 'Chugoku 163' were high yielding varieties compared with 'LA-1' about 15%, 25% each. The floury character in the endosperm was slightly improved and lodging resistance was well improved in three breeding lines compared with 'LA-1'. Their taste of cooked rice was slightly better than 'LA-1', it was inferior to the original cultivar 'Koshihikari' that has the highest cooking quality.

The effectiveness for clinical use of these low allergen lines is now under consideration. When the test shows clear result, we will be able to use these lines as practical varieties.

4. References

Kumamaru T, Satoh H, Iwata N, Omura T, Ogawa M, Tanaka K (1988) Mutants for rice storage proteins. 1. Screening of mutants of rice storage proteins of protein bodies in the starchy endosperm. Theor Appl Genet 76:11-16

Matsuda T, Nomura R, Sugiyama M, Nakamura R (1991) Immunochemical studies on rice allergenic proteins. Agric Biol Chem 55:509-513

Nishio T and Iida S (1993) Mutants having a low content of 16-kDa allergenic protein in rice (*Oryza sativa* L.). Theor Appl Genet 86:317-321

COMPARISON OF OLD AND MODERN SPRING BARLEY GERMPLASM UNDER LOW AND HIGH NITROGEN FERTILISATION

J.A. OKENO and S. SCHITTENHELM
Institute of Crop Science
Federal Agricultural Research Centre (FAL), Bundesallee 50, D-38116 Braunschweig, Germany

Abstract

The aims of this study were to (1) compare old and modern spring barley (*Hordeum vulgare* L.) germplasm for grain yield, grain N content, straw N content, grain N yield, straw N yield, and biological N yield, (2) estimate the general combining ability (GCA) and specific combining ability (SCA) effects in the inheritance of these characters, and (3) establish the relationships between grain yield and the N related characters at low and high N fertilisation. Eight two-rowed old and modern spring barley cultivars together with their intra-group half diallel crosses were evaluated in the field in 1995 and 1996 on a siltic sandy soil in Braunschweig, Germany at low and high N fertilisation (0 and 80 kg N ha^{-1}). The modern germplasm was superior in grain yield, grain N yield and biological N yield but inferior in grain N content and straw N yield. The genotypic variation noted for most of the characters studied indicated that adequate genetic diversity existed in modern and old germplasm. Except for grain yield, the GCA and SCA effects were not significant. Positive correlations with grain yield suggested that grain N yield and biological N yield could be used to indirectly select for high yield.

1. Introduction

Comparisons of old and modern spring barley (*Hordeum vulgare* L.) cultivars have shown that there were genetic changes in agronomic characters following selection for high yield (Ekman, 1981; Riggs *et al.*, 1981; Wych and Rasmusson, 1983; Schittenhelm *et al.*, 1996). Nitrogen (N) fertiliser is important to a cereal crop in the synthesis of N compounds and dry matter (DM). Nitrogen in the plant can be expressed as a concentration e.g., grain N content, straw N content, or as a yield e.g., grain N yield, straw N yield, and biological N yield. The N and DM partitionings both depend on energy and are therefore interrelated physiological processes. The interrelatedness of N and DM partitionings suggests that selection for high yield in spring barley might have caused changes in the above-mentioned N related characters. The direction and magnitude of change would depend on their relationships with grain yield. Reports comparing old and modern spring barley cultivars relative to N related characters are lacking. However, experiments with wheat (*Triticum*

G. Gissel-Nielsen and A. Jensen (eds.), Plant Nutrition – Molecular Biology and Genetics, 401–407.

aestivum L.) have shown that the modern cultivars are superior in N use efficiency, N utilisation efficiency, N harvest index, grain N yield, and biological N yield but inferior in grain N content (Austin *et al.* 1980; Slafer *et al.* 1990).

Selection of parents to combine to give a superior progeny in cultivar development is based on the assessment of their GCA and SCA effects. Reports on GCA and SCA effects of N related characters in spring barley are lacking. But experiments with maize (*Zea mays* L.) and faba beans (*Vicia faba* L.) have shown that N uptake efficiency, N utilisation efficiency, and N harvest index are under control of additive and non-additive genetic effects (Pollmer *et al.* 1979; Stelling *et al.* 1996). Selection of progenies in a plant breeding program is usually done at non-stressed N environments. Any excess N not taken up by the plant leaches into the ground and pollutes the water reservoir (Löffler and Busch, 1982). Therefore, assessment of N related characters in spring barley at low and high N fertilisation will be important in breeding due to economic as well as ecological reasons. The aims of this study were to (1) compare old and modern spring barley germplasm for grain yield, grain N content, straw N content, grain N yield, straw N yield, and biological N yield, (2) estimate the GCA and SCA effects in the inheritance of these characters, and (3) establish the relationships between grain yield and the N related characters at low and high N fertilisation.

2. Materials and methods

The genetic materials studied consisted of eight old and eight modern two-rowed spring barley cultivars and their 8 × 8 half-diallel intra-group crosses. The old cultivars were 'Heines Hanna', 'Pflugs Intensiv', 'Pflugs Extensiv', 'Ackermanns Bavaria', 'Heils Franken', 'Bethge II', 'Gull', and 'Archer', released between 1895 and 1921. The modern cultivars were represented by 'Alexis', 'Apex', 'Cheri', 'Defra', 'Golf', 'Baronesse', 'Prinzesse', and 'City', registered in Germany between 1982 and 1991.

The field experiment was conducted in the 1995 and 1996 growing seasons on a siltic sandy soil at Braunschweig, Germany. Parents and F_3 crosses were grown together in an experiment laid out in a randomised complete-block design with split-plot arrangement and consisting of three replications and two N fertilisation levels (0 and 80 kg N ha^{-1}). The N levels were assigned to the main plot, and the genotypes to the sub-plots. The planting density was maintained at 400 seeds m^{-2}. In March of each year, 70 kg P_2O_5 ha^{-1}, 140 kg K_2O ha^{-1}, and 21 kg Mg ha^{-1} were applied to the experimental sites. Seeds dressed with 'Abavit UF' (Hoechst Schering AgrEvo GmbH, Frankfurt, Germany) were sown on 5 April 1995 and 15 April 1996. The 4 x 1.35 m plots, consisting of six rows per plot were drilled with 21 cm spacing between rows using a plot seed drill (Wintersteiger, Ried, Austria). In both seasons, the experiments followed oats (*Avena sativa* L.). Weeds, foliar diseases and insects were controlled with recommended chemicals. Calcium ammonium nitrate (27% N) was added in two split applications to the high N plots during the growing period using a Hege 33 plot fertiliser-spreader (H.-U. Hege Saatzuchtmaschinen, Waldenburg, Germany).

Grain yield was determined from whole plots at harvest. Grain and straw samples were dried at 65 °C for 72 h and milled through a 0.5 mm screen. N content in grain and straw was assessed by near infrared reflectance spectroscopy (NIRS) using a monochromator instrument of the type NIRSystems 6500 (now FOSS NIRSystems, Silver Springs, U.S.A.) and calibrated by standard Kjeldahl technique. Grain N yield and straw N yield were determined by multiplying grain N content with grain yield and straw N content with straw yield, respectively. Biological N yield was got from the sum of grain N yield and straw N yield. All the measurements were adjusted on a dry matter (DM) basis.

The analysis of variance (ANOVA) across years for each N level was performed using the PLABSTAT computer program (Utz, 1991) following the combined ANOVAs proposed by Carmer *et. al.* (1989). The combining ability analyses were performed exclusively on the crosses using DIALLEL-SAS program developed by Zhang and Kang (1997). Year and replication effects were assumed to be random whereas genotype effect was assumed to be fixed according to Method 4, Model I (Griffing, 1956). Simple phenotypic correlations between characters were calculated by means of SPSS (SPSS Inc., 1993).

3. Results and Discussion

Cultivar 'Baronesse' (modern) produced the highest grain yield with a mean of 55.3 dt h^{-1} across the N levels. Its mean grain N yield and biological N yield were also high because these are equation characters of grain yield. But the grain N content of 'Baronesse' was comparatively low, pointing to the inverse relationship between grain yield and grain N content (Table 1). Between-group comparisons showed that the modern germplasm was superior in grain yield, grain N yield, and biological N yield but was inferior in grain N content and straw N yield. Straw N content was the same for both germplasm types. The modern germplasm lodged less, tillered more, and produced more spikes m^{-2}, which improved its grain yield (Austin *et al.* 1980; Riggs *et al.* 1981; Wych and Rasmusson, 1983). The superiority of the modern germplasm was more pronounced at high N level. Due to high harvest index, high N level increased the grain yield of the modern germplasm with a greater margin than that of the old germplasm. Grain N yield and biological N yield behaved similarly when N was increased. The lower straw N yield of modern germplasm resulted from its comparative low straw yield. The low straw yield of modern germplasm meant that a smaller amount of N from a limited straw N reservoir was subjected to a larger grain biomass and thereby 'diluted' more to give low grain N content. It was shown in wheat (*Triticum aestivum* L.) too, that modern cultivars have higher grain N yield and biological N yield and lower grain N content than the old cultivars (Austin *et al.* 1980). Slafer *et al.* (1990) later confirmed some of the findings of Austin *et al.* (1980) but differed from theirs relative to biological N yield in that they did not observe differences.

The significant, intra-group, genotypic variation observed in the various characters in both N levels proposed that adequate genetic diversity existed in spring barley (data not shown), which should ensure further selection progress. The individual GCA estimate for the modern cultivars 'Baronesse' and 'City' suggested that they can be used as a source of variability to improve grain yield both at low and high N fertilisation (Table 2). To improve grain N yield and biological N yield, 'City' could be used in low N while 'Baronesse' in high N fertilisation.

TABLE 1. Means for grain yield, grain N content, straw N content, grain N yield, straw N yield, and biological N yield of eight modern and eight old spring barley cultivars and their crosses at low and high N grown in 1995 and 1996.

Cultivar	Grain yield		Grain N content		Straw N content		Grain N yield		Straw N yield		Biol. N yield	
	Low N	High N	Low N	High N	Low N	High N	Low N	High N	Low N	High N	Low N	High N
	dt ha^{-1}		%				kg N ha^{-1}					
Modern												
Alexis	37.8	60.3	1.40	1.53	0.42	0.43	52.7	92.6	10.1	21.2	62.7	113.7
Apex	36.5	56.8	1.46	1.60	0.34	0.40	53.3	90.9	9.1	19.0	62.4	109.9
Cheri	35.7	59.3	1.48	1.52	0.42	0.40	53.5	90.1	11.0	18.2	64.5	108.3
Defra	32.9	55.7	1.54	1.57	0.43	0.42	50.9	86.9	11.2	19.6	62.1	106.4
Golf	40.0	63.0	1.39	1.56	0.37	0.37	55.7	98.0	11.0	20.0	66.7	118.0
Baronesse	46.8	63.8	1.29	1.58	0.30	0.35	60.8	101.4	9.9	17.7	70.7	119.1
Prinzesse	36.0	61.7	1.39	1.55	0.36	0.39	50.3	95.4	9.9	21.1	60.2	116.4
City	37.6	64.1	1.35	1.52	0.41	0.39	51.0	97.4	12.0	20.5	62.9	117.9
Old												
Heines Hanna	30.1	42.5	1.76	1.91	0.36	0.44	53.4	81.1	11.2	25.2	64.6	106.3
Pflugs Intensiv	32.1	46.4	1.72	1.89	0.39	0.43	55.2	87.7	11.8	21.3	67.0	109.0
Pflugs Extensiv	31.3	47.4	1.63	1.88	0.36	0.48	50.9	89.3	9.0	25.2	59.9	114.5
Ackerm. Bavaria	29.7	45.6	1.72	1.95	0.36	0.44	51.3	88.8	12.1	23.8	63.3	112.6
Heils Franken	33.1	44.6	1.50	1.85	0.36	0.44	49.8	83.0	12.1	24.5	61.9	107.2
Bethge II	32.0	46.5	1.71	1.92	0.37	0.45	54.9	89.5	11.0	24.0	65.9	113.5
Gull	32.4	45.5	1.62	1.84	0.38	0.46	52.2	83.6	10.4	23.0	62.6	106.5
Archer	27.6	40.8	1.54	1.78	0.39	0.50	42.5	72.6	13.0	33.3	55.5	105.9
LSD (0.05)	5.7	6.7	0.10	0.10	0.04	0.05	8.5	11.5	2.7	7.6	9.6	11.7
Group means												
Modern parents	37.9	60.6	1.41	1.55	0.38	0.39	53.5	94.1	10.5	19.6	64.0	113.7
Old parents	31.0	44.9	1.65	1.88	0.37	0.45	51.3	84.5	11.3	25.0	62.6	109.5
Modern crosses	40.8	64.0	1.40	1.57	0.36	0.39	57.1	100.5	10.6	20.7	67.6	121.2
Old crosses	31.7	44.9	1.71	1.81	0.38	0.47	54.6	81.3	11.9	27.5	66.6	108.6
LSD (0.05)	3.0	5.0	0.16	0.30	0.04	0.02	7.7	15.3	3.3	5.6	10.3	10.4

TABLE 2. General combining ability (GCA) estimates for grain yield, grain N content, straw N content, grain N yield, straw N yield, and biological N yield of eight modern and eight old spring barley cultivars at low and high N grown in 1995 and 1996.

	Grain yield		Grain N content		Straw N content		Grain N yield		Straw N yield		Biol. N yield	
Cultivar	Low N	High N	Low N	High N	Low N	High N	Low N	High N	Low N	High N	Low N	High N
					Modern							
Alexis	-0.8	-0.7	1.3	-1.7	5.4**	4.4**	-0.3	-1.5	2.4*	3.4**	0.3	-0.2
Apex	-2.4*	-2.3*	2.5*	0.6	-0.7	-1.7	-1.4	-1.8	-2.0*	-0.5	-1.6	-1.6
Cheri	-2.6*	-3.6**	2.3*	0.4	6.2**	4.7**	-1.3	-3.0**	2.9**	-0.1	-0.4	-2.5*
Defra	-3.5**	-4.0**	2.8**	1.2	2.1*	3.6**	-2.0*	-3.1**	1.3	0.6	-1.4	-2.4*
Golf	0.6	1.0	-1.8	-2.3*	-3.6**	-4.4**	0.0	-0.4	-1.8	-3.5**	-0.4	-1.4
Baronesse	4.5**	5.0**	-8.0**	1.3	-8.4**	-5.2**	1.0	5.1**	-4.2**	-1.7	-0.2	3.7**
Prinzesse	0.7	2.7**	3.3**	1.4	-2.4*	-2.5*	1.8	3.3**	0.3	0.4	1.6	2.8**
City	3.6**	2.0*	-2.2*	-0.8	1.5	1.1	2.3*	1.2	1.2	1.3	2.2*	1.4
					Old							
Heines Hanna	-1.5	-3.6**	1.3	-1.7	5.4**	4.4**	-0.9	1.8	0.3	1.4	-0.6	2.0*
Pflugs Intensiv	1.2	0.8	2.5*	0.6	-0.7	-1.7	0.0	2.7**	0.0	-3.5**	0.0	0.5
Pflugs Extensiv	-1.6	1.1	2.3*	0.4	6.2**	4.7**	-2.7**	1.0	-1.5	-1.1	-2.6*	0.3
Ackermanns Bavaria	-2.5*	1.2	2.8**	1.1	2.1*	3.6**	-0.2	2.2*	-1.3	-0.8	-0.6	1.3
Heils Franken	1.3	0.6	-1.8	-2.3*	-3.6**	-4.4**	-0.3	-2.4*	-0.9	-2.2*	-0.5	-2.8**
Bethge II	1.8	1.6	-8.0**	1.3	-8.4**	-5.2**	2.3*	-0.6	0.1	-1.5	1.8	-1.2
Gull	1.2	3.9**	3.3**	1.4	-2.4*	-2.5*	2.2*	0.9	0.4	-1.5	1.8	0.0
Archer	0.1	-5.6**	-2.2*	-0.8	1.4	1.1	-0.3	-5.4**	2.9**	9.2**	0.6	-0.2

*, ** Significant at 0.05 and 0.01 probability levels, respectively.

TABLE 3. Phenotypic correlations of grain yield with grain N content, straw N content, grain N yield, straw N yield, and biological N yield of modern and old spring barley germplasm at low and high N grown in 1995 and 1996.

Character	Germplasm group	1995		1996	
		Low N	High N	Low N	High N
Grain N content	Modern parents	-0.70**	-0.62**	-0.24	0.01
	Modern crosses	-0.39**	-0.30**	0.16	0.07
	Old parents	-0.26	0.15	0.42*	-0.32
	Old crosses	-0.34**	0.02	-0.01	-0.18
Straw N content	Modern parents	-0.59**	-0.52**	-0.24	-0.61**
	Modern crosses	-0.55**	-0.26*	0.14	-0.29**
	Old parents	-0.58**	-0.34	0.66**	-0.34
	Old crosses	-0.37**	-0.35**	0.39**	-0.15
Grain N yield	Modern parents	0.89**	0.88**	0.92**	0.86**
	Modern crosses	0.94**	0.87**	0.93**	0.87**
	Old parents	0.86**	0.84**	0.92**	0.79**
	Old crosses	0.72**	0.70**	0.87**	0.51**
Straw N yield	Modern parents	0.02	-0.02	0.42*	-0.23
	Modern crosses	0.04	0.03	0.52**	0.30**
	Old parents	-0.11	-0.16	0.57**	-0.40
	Old crosses	0.19	-0.07	0.59**	-0.12
Biological N yield	Modern parents	0.82**	0.79**	0.88**	0.77**
	Modern crosses	0.89**	0.79**	0.89**	0.82**
	Old parents	0.83**	0.83**	0.89**	0.40
	Old crosses	0.71**	0.59**	0.84**	0.35**

*, ** Significant at 0.05 and 0.01 probability levels, respectively.

The correlations between grain yield and grain N content were significantly negative in the modern germplasm in low and high N and in the old crosses in low N in 1995 (Table 3). In 1996, the significant positive correlations between the two characters were detected only in the old cultivars. These observations suggested that the association between grain yield and grain N content was determined by the genetic composition of the evaluated material and the environmental fluctuations. Negative correlation between grain yield and grain N content was reported in wheat (Slafer *et al.* 1990). In most cases in our study, grain yield was negatively correlated with straw N content. The gene(s) which condition the accumulation and withholding of N in the straw could have hindered the translocation of N and hence reduced DM synthesis in the grain. High positive correlations were detected between grain yield and grain N yield and between grain yield and biological N yield. N yield and biological N yield are equation-characters of grain yield and therefore these observations were expected. High positive correlations between grain yield and grain N yield and between grain yield and biological N yield were reported in wheat (Slafer *et al.* 1990; May *et al.* 1991). These previous results and ours indicated that grain N yield and biological N yield could be used to develop high yielding breeding populations in wheat and barley. Straw N yield was, in most cases, not related to grain yield. But in 1996, significant positive correlations were observed between the two characters in low N in all the germplasm groups and in high N in the modern crosses. These observations indicate that when N was limiting and the season more cool and wet, grain yield of spring barley could also be predicted from straw N yield.

4. Acknowledgements

J.A. Okeno is grateful to the German Academic Exchange Service (DAAD) for the award of a Ph.D. Scholarship and to the Prof. Werner Schulze-Stiftung for the funds to attend the symposium.

5. References

Austin, R.B., Bingham, R.D., Evans, L.T., Ford, M.A., Morgan, C.L., and Taylor, M. (1980) Genetic improvements in winter wheat yields since 1900 and associated physiological changes, *J. Agric. Sci., Camb.* **94,** 675–689.

Carmer, S.G., Nyquist, W.E., and Walker, W.M. (1989) Least significance differences for combined analyses of experiments with two- or three-factor treatment designs, *Agron. J.* **81,** 665–672.

Ekman, R. (1981) Biomass component studies in barley, their correlation to some yield characters and estimation of durable effect from 50 years of barley breeding, in M.J.C. Asher (ed.), *Proc. 4th Intl. Barley Genetics Symp.*, Edinburgh Press, Edinburgh, pp. 104–111.

Griffing, B. (1956) Concept of general and specific combining ability in relation to diallel crossing systems, *Aust. J. Biol. Sci.* **9,** 463–493.

Löffler, C.M., and Busch, R.H. (1982) Selection for grain protein, grain yield and nitrogen partitioning efficiency in hard red spring wheat, *Crop Sci.* **22,** 591–596.

May, L., Van Sanford, D.A., MacKown, C.T., and Cornelius, P.L. (1991) Genetic variation for nitrogen use in soft red x hard red winter wheat populations, *Crop Sci.* **31,** 626–630.

Pollmer, W.G., Eberhard, D., Klein, D., and Dhillon, B.S. (1979) Genetic control of nitrogen uptake and translocation in maize, *Crop Sci.* **19,** 82–87.

Riggs, T. J., Hanson, P. R., Start, N. D., Miles, D. M., Morgan, C. L., and Ford, M. A. (1981) Comparison of spring barley varieties grown in England and Wales between 1880 and 1980, *J. Agric. Sci., Camb.* **97,** 599–610.

Schittenhelm, S., Okeno, J.A., and Friedt, W. (1996) Prospects of agronomic improvement in spring barley based on a comparison of old and new germplasm, *J. Agronomy and Crop Sci.* **176,** 295–303.

Slafer, G.A., Andrade, F.H., and Feingold, S.E. (1990) Genetic improvement of bread wheat (*Triticum aestivum* L.) in Argentina: relationships between nitrogen and dry matter, *Euphytica* **50,** 63–71.

SPSS Base System Syntax Reference Guide Release 6.0 (1993), SPSS Inc. Chicago, IL.

Stelling, D., Wang, S.-H., and Römer, W. (1996) Efficiency in the use of phosphorus, nitrogen and potassium in topless faba beans (*Vicia faba* L.) - variability and inheritance, *Plant Breeding* **115,** 361–366.

Utz, H.F. (1991) PLABSTAT - A computer program for statistical analysis of plant breeding experiments, Version 2F. Institute of Plant Breeding, Seed Science and Population Genetics, University of Hohenheim, Stuttgart, Germany.

Wych, R.D., and Rasmusson, D.C. (1983) Genetic improvement in malting barley cultivars since 1920, *Crop Sci.* **23,** 1037–1040.

Zhang, Y., and Kang, M.S. (1997) DIALLEL-SAS: A SAS Program for Griffing's diallel analyses, *Agron. J.* **89,** 176–182.

THE APPLICATION OF AMPLIFIED FRAGMENT LENGTH POLYMORPHISM (AFLP) FOR BREEDING MN EFFICIENCY IN DURUM WHEAT (*TRITICUM TURGIDUM* L. VAR. *DURUM*)

H. KHABAZ-SABERI, S.J. BARKER[1], R.D. GRAHAM and A.J. RATHJEN

Department of Plant Science, The University of Adelaide, Glen Osmond, South Australia 5064

[1]Department of Plant Science, Faculty of Agriculture, The University of Western Australia, Nedlands WA 6907, Australia

Abstract

Manganese deficiency is a major constraint on alkaline soil, particularly for durum wheat, which is more intolerant than either bread wheat or barley. Genetic variation exists in current germplasm of durum wheat. In addition the development of a consistent selection criterion and the recent information on the most probable model for genetic control of Mn efficiency (two genes with additive effect) has made breeding for Mn efficiency feasible. The development of a precise and accurate screening technique will accelerate the breeding and also the genetic study of Mn efficiency. This technique preferably should be accurate, fast, less tedious and supplement the current pot bioassay. Marker assisted selection (MAS) is probably the most commonly claimed application of molecular markers in plant breeding. To identify markers linked to a locus conferring Mn efficiency in durum wheat Amplified Fragment Length Polymorphism (AFLP) analysis was combined with a Bulked Segregant Analysis (BSA) in order to increase the efficiency of identifying of markers linked to loci of interest. A total of 64 primer combinations was used to compare the bulk samples with the two parents in an F_2 population of a cross of Mn-efficient (*Stjocri 2*) x Mn-inefficient (*Hazar*) durum wheat. Two Primers (Pst1+ACA/Mse1+CAA) and (Pst1+ACC/MSE1+CAG) each revealed one AFLP, where the band was present in the efficient bulk and efficient parent but absent in the inefficient bulk and inefficient parent. The AFLPs identified will be tested for cosegregation of the trait by progeny

G. Gissel-Nielsen and A. Jensen (eds.), Plant Nutrition – Molecular Biology and Genetics, 409–416.

testing the entire F_2 population. If linkage to one or two Mn efficiency loci is confirmed, these markers will provide a valuable selection tool for efficient progeny in further segregating populations bred from the same efficient progenitor.

Introduction

Manganese deficiency is widespread throughout the world and is possibly the most severe micronutrient disorder in South Australia (SA). It is a major disorder limiting the growth and development of durum wheat as a new crop in SA. The poor agronomic solution to Mn deficiency, arising mainly from the low availability of both native and applied Mn in alkaline soil, from the low residual value of the applied fertiliser and also from the possibility of yield loss due to unrecognised subclinical deficiency (Robson and Snowball, 1986), has made the genetic solution of breeding for Mn efficiency attractive. Genotypic differences to Mn deficiency have been reported in diverse crop species (Graham, 1988). The presence of considerable genotypic variation in durum wheat for Mn efficiency (Mn-efficient genotypes, *Stojocri 2* with a relative yield of 58% compared to 15% for *Yallaroi* durum wheat check cultivar) and development of a selection criterion (Mn content of 35-day-old seedlings grown under controlled environment using Mn-deficient soil) has made the breeding for Mn efficiency feasible (Khabaz Saberi et al., 1997). Mn efficiency in barley appeared to be simply inherited and controlled by a single, dominant major gene, in a cross of *Weeah* (Mn-efficient) and *Galleon* (Mn-inefficient) (Longnecker et al., 1988). For soybean, a study of F_2 and F_2-F_3 derived families carried out under field conditions suggested a digenic mode of inheritance, although there were low estimates of narrow-sense heritability (Graham et al., 1995). The study of F_1 hybrid, F_2 and F_2-derived families from a cross between *Stojocri 2* and *Hazar* under controlled-environmental condition revealed that the trait is controlled most probably by two loci with additive effect (Khabaz Saberi et al., 1998). Although adequate, the current pot bioassay for Mn efficiency is time and resource consuming, and development of an alternative accurate, fast procedure would accelerate breeding for this trait. The potential uses of molecular marker in plant breeding have been discussed by Langridge (1994). To identify markers linked to a locus conferring Mn efficiency in durum wheat, AFLP analysis, a novel fingerprinting technique (Vos et al., 1995) was combined with Bulked Segregant Analysis (BSA) (Michelmore et al., 1991) and applied to bulked samples of efficient and inefficient F_2 progenies derived from *Stojocri 2/Hazar* and compared with parents using 64 primer combinations in this study.

Materials and Methods

PLANT MATERIAL

DNA samples were obtained from the F_2 population of a cross between Stojocri 2 by Hazar, and also from the parents. The leaf samples were taken from the regrowth of F_2 seedlings 20 days after harvesting the shoot (for Inductively Coupled Plasma Atomic Emission Spectrometry (ICPAES) analysis of Mn and other elemental status) and transplanting into normal potting mix.

DNA PREPARATION

The mini prep DNA preparation procedure was followed as described by (Langridge et al., 1996).

AFLP AND BULK SEGREGANT ANALYSIS (BSA)

Bulk Segregant Analysis (Michelmore *et al.*, 1991) was used in identification of potential AFLP markers. The DNA (10 µg) samples of five extreme Mn efficient and five extreme Mn-inefficient F_2 plants of the Stojocri 2/Hazar cross were collected and pooled separately. One microgram of the pooled DNA samples were used as bulked Mn inefficient (BI) and bulked Mn efficient (BE) in addition to DNA from Hazar and Stojocri 2, respectively, as inefficient parent (IP) and efficient parent (EP) in AFLP analysis. The primer combinations (8x8=64) leading to potential markers were applied later to each individual DNA samples to find the best primer combination.

AFLP Analysis

The AFLP method developed by with some modification was followed (Vos *et al.*, 1995) and comprised in three steps: (1) Preparation of template DNA (2) Selective amplification of template DNA (3) Gel analysis of amplified fragments and autoradiography.

(1) Preparation of template. The template DNA was produced in four steps including: Restriction digest, Annealing of adaptors, Ligation of adaptors, Pre-amplification of DNA

(1.1) Restriction digest. 1 μg genomic DNA of EB, IB, EP and IP samples were digested using two restriction enzymes (MseI and PstI). Five units of each enzyme were used in a total reaction volume of 50 μl. The reactions were incubated at 37 °C for 3 hrs.

(1.2) Annealing of adaptors. stock solutions containing both MseI adaptors (MseI and MseII) and PstI adaptors (PstI and PstII) were prepared at 50 μM and 5 μM respectively. Adaptors were heated at 90°C for 3 min and then left to anneal (double stranded) at room temperature for 30 min prior to use.

(1.3) Ligation of adaptors. One microliter of each adaptor (MseI and PstI) was used in a final reaction volume of 10 μl containing 1μl reaction buffer (10xRL buffer), 1.2 μl 10 mM ATP and 1 μl T4 DNA ligase (1 μ/μl). Ten microliters of solution was added to each digest for ligation of adaptors then incubated at 37°C for 3 hrs and left at 4°C overnight. DNA was precipitated by adding 129 μl ethanol and 6 μl 4.8 M sodium acetate (pH 3.2) and placing in liquid nitrogen for 5 min, followed by centrifuging for 5 min. The precipitated DNA pellet was washed with 1 ml 70% ethanol and then dried in a speed-vac for 10 min followed by re-suspension in 60 μl 0.1 M TE.

(1.4) Pre-amplification of DNA. Pre-amplification was performed using primers specific to the PstI and MseI adaptors including one selective nucleotide. One microliter (75 ng/μl) of each of MseI+I primer and PstI+I primer were used in a total reaction volume of 21 μl of buffer mix containing 4 μl dNTPs (1.25 mM), 2.5 μl 10x Taq buffer, 1.5 μl $MgCl_2$ (25 mM), 0.2 μl Taq Polymerase (5 μ/μl). Four microliters of template DNA (R-L DNA) were added to the buffer, mixed and subjected to PCR. The pre-amplification PCR conditions consisted of 20 cycles of 94°C for 30 seconds, 56°C for 1 minutes and 72°C for 1 minute. Following PCR, the DNA sample (template DNA) was diluted 1: 5 in sterilised water before being used in selective amplification.

(2) Selective amplification of template DNA. In selective amplification the similar primers with three selective bases were used. The PstI primers were end labelled followed by selective PCR.

(2.1) End labelling of primer I. A total of 8 PstI and 8 MseI primers were used giving 64 possible primer combinations. One microliter of PstI primers (50 ng/μl) was used in a total reaction volume of 10 μl including 1.5 μl $^{32}P\gamma ATP$ (10μCi/ml), 1μl 10x PNK buffer and 0.2 μl T4 PNK (10μ/ml) followed by incubation at 37°C for 30-60 min.

(2.2) Selective PCR. One microlitere of end-labelled PstI-1 was used in a total reaction volume of 18 μl including 2 μl 10x Taq buffer, 1.2 μl $MgCl_2$ (25 mM), 3.3 μl dNTPs (1.25 mM), 0.5 μl PstI-1 (50 ng/μl), 0.5 μl MseI-I (50 ng/μl), 0.2μl Taq Polymerase (5μ/μl). Two microliters of template DNA was added to the 18 μl of the buffer mix before selective PCR. The PCR reaction conditions for selective amplification consisted

of one cycle at 94°C for 30 seconds, 65°C for 30 seconds and 72°C for 1 minute followed by 9 cycles over which the annealing temperature was decreased by 1°C per cycle with a final step of 25 cycles of 94°C for 30 seconds, 56°C for 30 seconds and 72°C for 1 minute.

(3) Gel analysis of amplified fragments. After selective PCR, samples were denatured by adding 20 µl ficoll and heating for 5 minutes at 90 °C. Two microliters of each sample were loaded on 6% denaturing polyacrylamide gels for separation of amplified fragments. The gels were transferred to 3 MM paper for drying and auto-radiography was carried out with Fuji RX medical X-ray film at room temperature for 24-48 hours.

Results and Discussion

GENETICS OF MANGANESE EFFICIENCY IN A CROSS OF EFFICIENT BY INEFFICIENT DURUM WHEAT

The number genes and their mode of action was previously determined (Khabaz Saberi *et al.*, 1998) and seemed to be controlled by additive action of 2 genes in a cross of Stojocri 2 by Hazar. The analysis was carried out by comparison of observed variance to expected variance of F_2 and $F_{2:3}$ populations based on one and two gene models using shoot Mn content as selection criterion and the pot bioassay developed by Huang et al. (1994).

AFLP ANALYSIS

A total of 64 primer combinations was used to compare the bulk samples with the two parents. Approximately 60-70 distinguishable bands were observed for each of primer combination and on average 50-60 of these bands were polymorphic between the parents. Primers (Pst1+ACA/Mse1+CAA) and (Pst1+ACC/MSE1+CAG) each revealed one AFLP, where a band was present in the efficient bulk and efficient parent but absent in inefficient bulk and inefficient parent (Fig. 1).

Application of AFLP technique is probably the most powerful DNA fingerprinting technique to detect polymorphism due to the presence or absence of restriction enzyme sites. The most important feature of the AFLP technique is the number of markers which can be screened in each experiment (10 to 20 primer combinations). In our research, the use of two restriction enzymes (PstI and MseI) and 64 primer combinations provided a high number of selectively amplified DNA fragments. The reproducibility of AFLP is another feature of this technique as compared to other PCR based finger printing techniques such as RAPD analysis which is based on use of

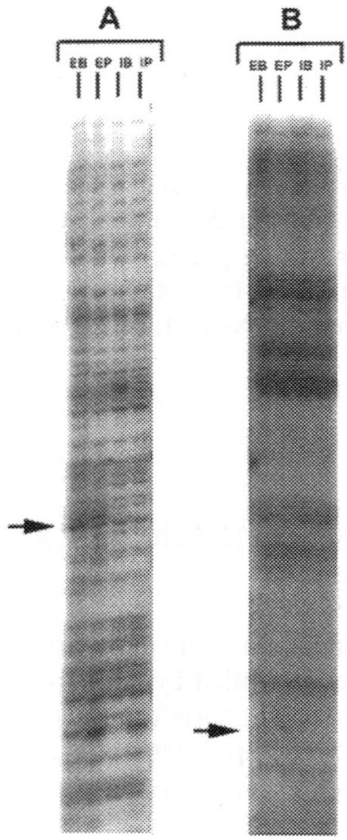

Fig. 1 AFLP markers linked to Mn efficiency locus. Bulked Segregant Analysis (BSA) is presented as a set of four lanes: Mn efficient bulk (E_B), Mn efficient Parent (E_P), Mn inefficient bulk (I_B) and Inefficient parent (I_P). The AFLP marker (Pst1+ACA/Mse1+CAA) (**A**), and (Pst1+ACC/MSE1+CAG) (**B**), identified by BSA and indicated by an arrow, are present in E_B and EP but absent in I_B and I_P.

random primers. Two important features are combined in primers used to obtain AFLP markers. (1) complementary characteristic of primer to adaptors which creates a higher specific primer annealing and (2) their selectivity; changing the 3' nucleotides allows amplification of a different set of DNA fragments from pre-amplified DNA samples. Combining the power of BSA which is a useful tool allowing screening for markers tightly linked to the character of interest with AFLP analysis will provide an efficient approach to identify the markers close to desired locus.

The AFLP markers identified in Mn EP and EB will be used to screen the segregating population which will lead to a better understanding of the genetics of Mn efficiency in the cross of Stjocri 2/Hazar. If the markers are linked to one or two Mn efficiency

genes, these markers have the potential of being used to select for efficient individuals in further segregating populations bred from the same efficient progenitor. Combined AFLP and BSA was also applied in analysis of disease resistance in *Populus* spp (Cervera et al., 1996). Three AFLP marker closely linked to *Melampsora larici-populina* Locus (Mer) were identified using the above technique to screen segregating family. The identified marker is currently used in *Populus* spp breeding program (Cervera et al., 1996).
To facilitate screening of progeny derived from the cross of a Mn efficient progenitor by a simple PCR based assay, cloning and sequencing of an AFLP marker could be the next approach to be undertaken. Cloning and sequencing of the AFLP markers has the potential of converting them into a Sequence Characterised Amplified Region (SCAR) for further PCR screening of progeny derived from same efficient progenitor or those sharing the same Mn efficient gene. The use of AFLP markers in other genotypes with different levels of Mn efficiency will probably lead to the identification of additional loci conferring Mn efficiency in durum wheat and will help better understand the genetic basis of Mn efficiency in durum wheat.

Acknowledgments

I wish to thank to Ass. Prf. P. Langridge; University of Adelaide, and Dr S.J. Barker for providing the facilities for AFLP analysis. Special thanks to J.M. Kretscher and M. Pallotta for their advice and assistance with laboratory work. Financial support of joint IRAN/ICARDA and CRC-MPB for this research is highly appreciated.

References

Cervera, M. T., Gusmao, J., Steenackers, M., Peleman, J., Storme, V., Vanden Broeck, A., Van Montagu, M., and Boerjan, W. (1996) Identification of AFLP molecular markers for resistance against *Melampsora larici-populina* in Populus. *Theor Appl gent* **93**, 733-737.

Graham, M. J., Nickell, C. D., and Hoeft, R. G. (1995) Inheritance of tolerance to manganese deficiency in soybean. *Crop Sci* . **35**, 1007-1010.

Graham, R. D. (1988) Genotypic differences in tolerance to manganese deficiency. In *Manganese in soils and plants* (Edited by R. D. Graham, R. J. Hannam, & N. c. Uren), pp. 261-276. Kluwer Academic Publishers, Dordrecht, The Netherlands.

Huang, C., Webb, M. J., and Graham, R. D. (1994) Manganese efficiency is expressed in barley growing in soil system but not in solution culture. *J Plant Nutr* **17**, 83-95.

Khabaz Saberi, H., Graham, R. D., and Rathjen, A. J. (1997) Genotypic variation for manganese efficiency in durum wheat (*Triticum turgidum* L. var. *durum*). In (T. Ando, K. Fujita, T. Mae, H. Matsumoto, S. Mori, & J. Sekiya), Kluwer Academic Publishers. Tokyo, Japan. 33-35.

Khabaz Saberi, H., Graham, R. D., and Rathjen, A. J. (1998) Inheritance of Mn efficiency in durum wheat (*Triticum turgidum* L. var. *durum*). *J Plant Nutr* **in press**,

Langridge, P. (1994) Molecular markers in wheat breeding and research. In *7th Assembly Wheat Breeding Society of Australia* (Edited by J. G. Paull, I. S. Dundas, & K. W. Shepherd), The University of Adelaide, Adelaide.

Langridge, p., Karakousis, A., and Nield, J. (1996) Course Manual of Practical Workshop in Basic Recombinant DNA Techniques Application to Plant Breeding. Centre for Cereal Biotechnology, Waite Agriculture Research Institute, The University of Adelaide, Adelaide, SA.

Michelmore, R. W., Paran, I., and Kesseli, R. V. (1991) Identification of markers linked to disease-resistance genes by bulked segregant analysis: a rapid method to detect markers in specific genomic regions by using segregating populations. *Proc Natl Acad Sci USA* **88**, 9828-9832.

Robson, A. D., and Snowball, K. (1986) Nutrient deficiency and toxicity symptoms. In *Plant analysis-An interpretation manual* (Edited by D. J. Reuter, & J. B. Robinson), pp. 13-19. Inkata press, Melborne.

Vos, P., Hogers, R., Bleeker, M., Reijans, M., Lee, T. V. D., Hornes, M., Frijters, A., Pot, J., Peleman, J., Kuiper, M., and Zabeau, M. (1995) AFLP: A new technique for DNA fingerprinting. *Nucleic Acids Res* **23**, 4407-4414.

LOCALISATION OF GENES FOR ZINC AND MANGANESE EFFICIENCY IN WHEAT AND RYE

R. SCHLEGEL[1], A. OZDEMIR[2], I. TOLAY[2], I. CAKMAK[2], H. SABERI[3] and M. ATANASOVA[1]

[1]*Institute of Wheat & Sunflower Research, BG-9520 General Toshevo/Varna, Bulgaria*

[2]*Cukurova University, Dep. of Soil Sciences, TR-01330 Adana, Turkey*

[3]*Waite Agricultural Res. Institute, The University of Adelaide, 05064 Glen Osmond, Australia*

Key words: wheat, rye, wheat-rye introgression, nulli-tetrasomics, manganese efficiency, zinc efficiency, gene mapping

Summary

Among the most highly adapted and advanced hexaploid and tetraploid wheat cultivars in the world genetic diversity is limited, particularly for characters like micronutrient efficiency. A screening for zinc and manganese efficiency was conducted in greenhouse and controlled environment cabinets using different varieties of hexaploid wheat, diploid rye, triticales, wheat-*Agropyron*, wheat-*Aegilops* and several wheat-alien chromosome addition series. There were considerable differences in zinc efficiency between wheat and its relatives. The Bulgarian wheat variety 'Trakia' was identified as the most zinc- and manganese-efficient genotype (84 % and 96 %), although there is no significant correlation between zinc and manganese efficiency.

By using a series of nulli-tetrasomic lines of the hexaploid wheat cv. 'Chinese Spring', it could be demonstrated that in wheat zinc efficiency is under genetic control. The lines N1B/T1D and N5D/T5A showed a significant increase for zinc efficiency. It was concluded that chromosomes 1D and 5A carry major genetic information for this character. Moreover, chromosomes of homoeologous groups 1 and 5 as well as those of the A genome seem essential for a good plant performance under zinc deficiency.

In addition, individual chromosomes of *Secale* carry major genes for this character. The alien genetic information was clearly expressed in the wheat genetic background. It was revealed by previous experiments that genes from the rye chromosome region 5RL2.3 may contribute to a high copper and iron efficiency. Other genes on the chromosomes/arms 1RS and 7RS of rye heavily increased zinc efficiency while genes present on chromosome 6R resulted in increase of manganese efficiency of wheat, respectively.

Introduction

Zinc (Zn) and manganese (Mn) deficiency are widespread micronutrient problems in wheat production, particular in calcareous soils of arid and semiarid regions (Reuter et al. 1988; Graham and Welch 1996). In wheat, Zn deficiency occurs on large areas of different countries, resulting in severe decreases in grain yield as found in India (Takkar et al. 1989), Australia (Graham et al. 1992) and Turkey (Cakmak et al. 1996). Zn deficiency reduces not only grain and/or dry matter production but also nutritional

G. Gissel-Nielsen and A. Jensen (eds.), Plant Nutrition – Molecular Biology and Genetics, 417–424.

quality of grains. Moreover, Zn and Mn deficiency increases sensitivity of cereals to different root diseases (Sparrow and Graham 1988; Willhelm et al. 1988).
There is a genetic variation in both Zn and Mn efficiency in cereals. Graham et al. (1983) in Australia and Bansal et al. (1991) in India showed existence of a considerable diversity among and between wheat and barley cultivars in Mn-deficient calcareous soils. Similarly to Mn deficiency, cereal species and cultivars greatly differ in Zn efficiency, when grown in Zn-deficient calcareous soils (Graham et al. 1992; Cakmak et al. 1997a, 1998). Only few results are available on the genetics of Zn efficiency. In studies with different wheat amphiploids, addition, substitution and translocation lines Schlegel et al. (1997, 1998) showed that the rye genome, and within the rye complement, the chromosomes 1R and 7R carry major genetic information for this character. This paper presents additional evidence for substantial variation among rye, wheat and their wild relatives in both Zn and Mn efficiency, and provides information on the genetic control of micronutrient efficiency. In the paper, the term efficiency is used in order to reflect the ability of a genotype to grow and yield better under Zn and Mn deficiency in comparison to other genotypes.

Material and methods

The wheat, rye, wild species, amphidiploid and aneuploid accessions derived from the Genetic Stocks Centre of the Institute of Wheat Sunflower Research, General Toshevo (Bulgaria). All the aneuploid and amphidiploid material was microscopically checked for the correct chromosome number before growing and testing. *Zinc efficiency* - Pot experiments were carried out under greenhouse conditions at Adana (Turkey) with 2.2 kg soil/plastic pot with (10 mg Zn/kg soil as $ZnSO_4$) and without Zn application. All experiments were carried in three replications. Zinc deficient soil was used from Eskisehir (Turkey) where severe Zn deficiency occurs in wheat (Cakmak et al., 1996). All pots were randomised every 4-5 days and watered daily to about field capacity using deionised water. The Zn efficiency of plants was calculated as the ratio of dry matter yield at Zn fertilisation to the yield produced without Zn fertilisation (in %). Plants were harvested after 40 days of growth. The shoots were dried at 70 °C to determine the dry weight. *Manganese efficiency* - To avoid the confounding effect of seed Mn content, genotypes with the same Mn content were selected in this study. A calcareous soil from a chronically Mn deficient site on Eyre Peninsula at Wangary, South Australia, with approximately 80% $CaCO_3$ and pH 8.5 was used. The application of 15 and 100 mg Mn /kg dry soil representing deficient and sufficient level of Mn supply respectively was employed for this study. The experiments were conducted in a growth cabinet set at 15 °C day/10 °C night. Plants were harvested 35 days after sowing, shoots were dried at 85 °C, and digested in 70% nitric acid prior to analysis for mineral elements by ICP spectrometry (see Khabaz Saberi et al., 1997).

Results and Discussion

The plant reaction to zinc deficiency may be detected already in a early developmental stage. Most pronounced effects are reduced shoot growth and leaf size. It is accompanied by degeneration of chlorophyll resulting in chlorotic and finally in whitish-brown necrotic patches on leaf blades. The morphological modifications observed are directly correlated with the shoot dry matter production (r=+0.74**). Therefore, the shoot dry matter production (SDW) was mainly considered for the further investigations. The Zn efficiency of rye generally excels hexaploid wheat (Tab. 1). However, a genetic variation is visible between the accessions. Among the hexaploid wheats there is also quite a big differentiation, best demonstrated by the varieties 'Bolal' and 'Trakia'. The high Zn efficiency of 'Trakia' depends not on any alien introgression as cytological studies revealed. When the manganese efficiency was studied, again the same Bulgarian variety 'Trakia' showed an exceptionally high value (Fig. 1), although there is no general correlation between Zn and Mn efficiency. It might be concluded that 'Trakia' represents a particular wheat genotype which strongly confers the ability to take up Zn as well as Mn under deficient soil conditions. Higher Zn and Mn efficiency might be related to enhanced uptake rate of Zn and Mn under deficient conditions. Several organic compounds are known to be released from roots and involved in solubilisation and uptake of sparingly soluble Mn compounds in soils, such as organic acids and phytosiderophores (Marschner 1988). Enhanced exudation of such compounds from roots of 'Trakia' under Zn and Mn deficiency might be the basis of its high Zn and Mn efficiency.

Table 1 Comparison of different wheat and rye genotypes for zinc and manganese efficiency (-DW/+DW; high values = high efficiency)

Wheat	**Efficiency**	**(%)**	**Rye**	**Efficiency**	**(%)**
variety	**Mn**	**Zn**	**variety**	**Mn**	**Zn**
Avalon		49	Danae	64	84
Bezostaya 1		50	Graser		67
Bolal	68	30	Imperial	90	63
Borenos		64	Inbred	98**	61
Carola		42	King II		89**
Cheyenne		55	Pico		65
Chinese Spring	72	55	Pluto		90**
Dagdas	74	48	Petka	78	81
Fakon		53			
Gerek 79		45			
Giza		74			
Holdfast		48			
Seri		36			
Trakia	96**	84**			
Trakia-allopl.		64			
Viking		60			
Vilmorin		73			
Mean	**77**	**54**		**83**	**77**

** - significant at P = 5%

However, there were no correlation between the Mn efficiency and the Mn concentration and Mn content in the shoot tissue neither in wheat nor in rye (Fig. 1). The same was true for several octoploid triticales and other intergeneric amphidiploids (Fig. 2) which also showed a high Mn efficiency. Also the Zn concentrations in shoots or leaves were not related to Zn efficiency (Cakmak et al. 1997a,b). It seems likely that internal utilisation of Zn and Mn differ between genotypes.
Only in some defined wheat-alien chromosome substitution and in addition lines there was a slight but not significant correlation between Mn efficiency and Mn shoot concentration (r=+0.65) but a clearly significant correlation between the Mn efficiency and the Mn shoot content (r=+0.97**) (Fig. 2). The situation that in common varieties where these sorts of correlations are absent but present in defined genetic tester lines may be explained by a certain degree of genetic variability within breeding material which obscures clear-cut genotypic effects.
In addition, Fig. 2 provides evidence that chromosome 3D of hexaploid wheat 'Chinese Spring' can strongly improve the Mn efficiency of durum wheat 'Langdon' (*Triticum durum-Triticum aestivum* substitution 3D(3A)). The rye chromosome 6R showed the same effect in hexaploid wheat 'Holdfast', while the added chromosome 2R did not change the Mn efficiency. A study of Graham (1987), using wheat-rye additions, revealed that chromosome 2R and with lesser effects 6R contribute to Mn efficiency, while triticale lacking chromosome 2R show lower efficiency. Thus at least rye chromosome 6R can be confirmed by the present data carrying a major gene or genes for Mn efficiency.

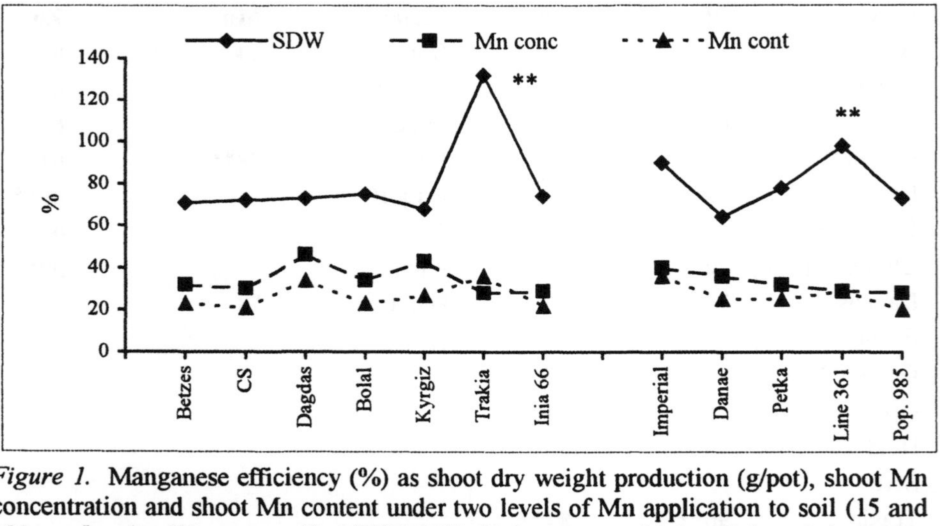

Figure 1. Manganese efficiency (%) as shoot dry weight production (g/pot), shoot Mn concentration and shoot Mn content under two levels of Mn application to soil (15 and 100 mg/kg dry Wangary soil) (-DW/+DW; high values = high efficiency) in barley (Betzes), wheat (Chinese Spring, Dagdas, Bolal, Kyrgiz, Trakia, Inia 66) and rye (Imperial, Danae, Petka, Inbred line 361, population 985) grown in a phytotron

In order to reveal individual chromosomes of wheat contributing to Zn efficiency the series of nulli-tetrasomics of the variety 'Chinese Spring' was considered. Data for zinc efficiency are given in Fig. 3. Nearly all of the nulli-tetrasomics showed a higher zinc efficiency as compared to the control 'Chinese Spring'. However, statistically significant effects were only found in N1B/T1D and N5D/T5A. When all the tetrasomic chromosomes were compiled, it turned out that the group 5 chromosomes contributed most to high zinc efficiency, while, on the other hand, the chromosomes of the A genome gave the best averages (Tab. 2). The latter is confirmed by the nullisomy of A-genome chromosomes. The lowest values of zinc efficiency were observed when they were nullisomic. The zinc efficiency of 'Chinese Spring' wheat (34 %), which was used as a control, corresponds to the data from previous experiments (Schlegel et al. 1998), showing that 'Chinese Spring' represents a zinc-inefficient wheat genotype.

Since the nulli-tetrasomic lines N1B/T1D and N5D/T5A showed the highest zinc efficiency, it is concluded that chromosomes 1D and 5A confer major genetic information for zinc efficiency. It matches the results on wheat-alien chromosome additions, substitutions and translocations. Chromosomes of homoeologous group 1 and 5 (but also group 7) of *Secale cereale, Agropyron intermedium, Hordeum vulgare* and *Haynaldia villosa* have been shown to contribute most to zinc efficiency (Schlegel and Cakmak 1997a,b, Cakmak et al. 1998).

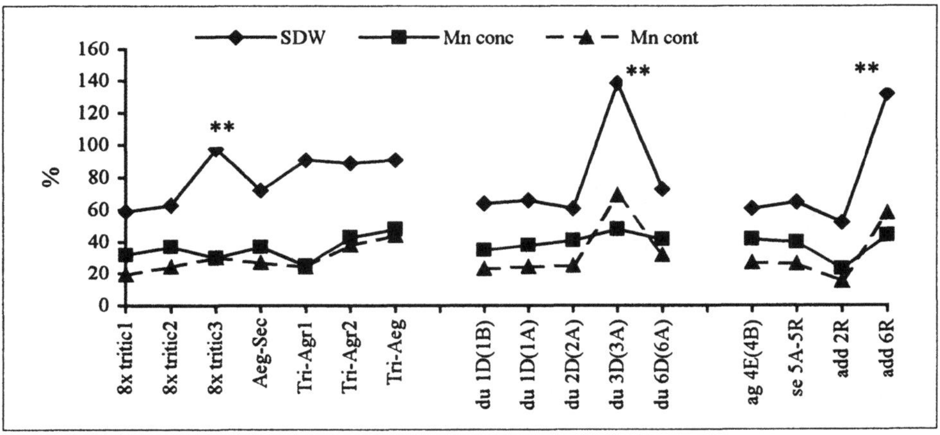

Figure 2. Manganese efficiency (%) as shoot dry weight production (SDW - g/pot), shoot Mn concentration and shoot Mn content under two levels of Mn application to soil (15 and 100 mg/kg dry Wangary soil) (-DW/+DW; high values = high efficiency) in octoploid *Triticum-Secale* amphidiploids (tritic1-3), a tetraploid *Aegilops-Secale* amphidiploid (Aeg-Sec), two octoploid *Triticum-Agropyron* amphidiploids (Tri-Agr1-2), a Triticum-*Aegilops* amphidiploid (Tri-Aeg) and in *Triticum durum-Triticum aestivum* substitutions (du 1D(1B), du 1D(1A), du 2D(2A), du 3D(3A), 6D(6A)), a wheat-*Agropyron elongatum* substitution line 4E(4B), a wheat-rye translocation line 5AS.5RL (se 5A-5R) and two wheat-rye addition lines (add 2R, add 6R) grown in a phytotron

Tetrasomy for chromosome 1D resulted in increased zinc efficiency only when chromosome 1A was disomically present; nullisomy for chromosome 1A and tetrasomy for chromosome 1D did not change the efficiency (Fig. 4). The same was true when a 1D(1A) disomic substitution in tetraploid wheat was investigated. In this instance the chromosome 1D did not increase zinc efficiency in the absence of 1A

(Schlegel & Cakmak 1998). Also in group 5 the highest zinc efficiency was observed for disomy 1A and tetrasomy 5A. In both cases therefore, a complementary gene interaction is supposed.

Table 2. Mean zinc efficiency (-DW/+DWx100) of tetrasomic chromosomes averaged over the homoeologous groups and genomes (individual values of chromosomes can be taken from Fig. 3)

	Homoeologous Group							
Genome	1	2	3	4	5	6	7	Mean
A								60
B								58
D								48
Mean	53	50	50	53	68	53	49	

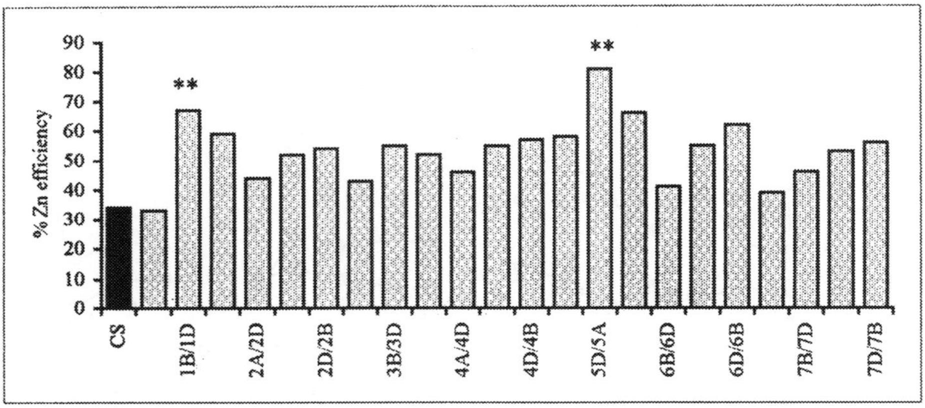

Figure 3. Zinc efficiency (-DW/+DWx100) of hexaploid wheat (Chinese Spring, CS) as control and its nulli-tetrasomic aneuploids (** - significant at P = 5%)

In general, in hexaploid wheat the presence of the A genome seems to be a minimum prerequisite for reasonable zinc efficiency. In good agreement with this, we recently revealed that by experimental addition of A genomes to *Tiriticum durum,* the synthetic amphiploids showed higher Zn efficiency, and among many wheat accessions, e.g. *Triticum polonicum* (AABB), *T. dicoccoides* (AABB), *T. monococcum* (AA), *T. compactum* (AABBDD) or *T. aestivum* (AABBDD), the einkorn wheats usually succeeded the others (Cakmak et al., unpub.). Chromosomes of the A genome (disomic or tetrasomic) contributed more to higher zinc efficiency than those of either B or D genomes (Tab. 2). This is supported by data from disomic *Triticum durum-Triticum aestivum* substitution lines. None of the D-genome chromosomes of hexaploid wheat 'Chinese Spring' significantly improved the zinc efficiency of tetraploid wheat 'Langdon' (Schlegel & Cakmak 1998). The study of various sets of wheat-alien addition lines confirms a major effect of the rye genome (Fig. 4). The rye chromosomes 1R and 7R seem to carry genes which may promote the uptake of Zn under conditions of limited Zn availability. In addition, Fig. 4 shows that rye genotypes may differ in their response. Chromosome 7R of the rye variety 'King', which has been the donor for wheat-rye addition set 'Holdfast-King II', led to the highest value among all addition lines.

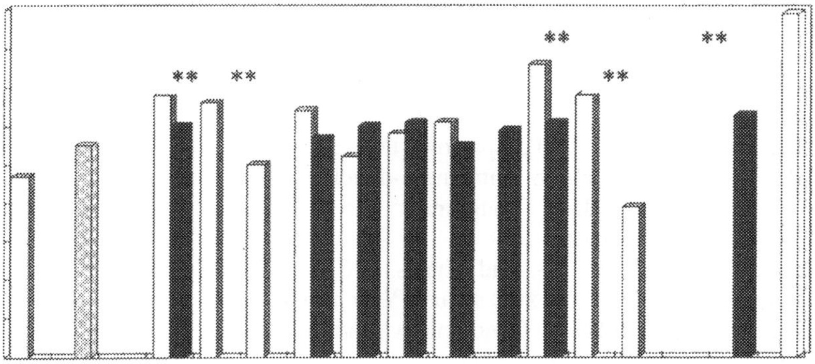

Figure 4. Zinc efficiency (-DW/+DW) of hexaploid wheat (Holdfast; Ho; Chinese Spring, CS) and ryes (Imperial, Im - dimed culmn; King II, Ki - unfilled column) and several disomic wheat-rye chromosome addition lines (Holdfast-King II and Chinese Spring-Imperial) under Zn-deficient growth

By including in the experiment wheat-rye telosomic additions, it was possible to associate the genetic effect not only with an individual chromosome of rye but even with a chromosome arm. The telocentric rye chromosome additions of the critical chromosomes 1R and 7R demonstrate that only the short arms increase the Zn efficiency of wheat when they were compared with entire chromosome (Fig. 4). It is concluded that relevant genetic information for improvement of Zn efficiency is associated with rye chromosome arms 1RS and 7RS. However, an increase of the gene dosage of 1RS by using double and triple wheat-rye translocation lines, i.e. comparing 2, 4 or 6 dosages of 1RS, did not improve the Zn efficiency (Schlegel and Cakmak 1997b). No additive gene action can be supposed from the recent study.

Acknowledgements
This research was supported by the NATO-Science for Stability Project and The Scientific and Technical Council of Turkey (TUBITAK).

References

Bansal, R.L., Nayyar, V.K. and Takkar, P.N. (1991) Field screening of wheat cultivars for manganese efficiency, *Field Crops Res.* 29, 107-112.

Cakmak, I., Yilmaz, A., Kalayci, M., Ekiz, H., Torun, B., Erenoglu, B. and Braun, H.-J. (1996) Zinc deficiency as a critical nutritional problem in wheat production in Central Anatolia *Plant and Soil* 180, 165-172.

Cakmak, I., Ekiz, H., Yilmaz, A., Torun, B. Köleli, N., Gültekin, I., Alkan, A. and Eker, S. (1997a) Differential response of rye, triticale, bread wheat and durum wheats to zinc deficiency in calcareous soils, *Plant and Soil* 188, 1-10.

Cakmak, I., Oztürk, L. Eker, S., Torun, SB., Klafa, H. and Yilmaz, A. (1997b) Concentration of zinc and activity of copper/zinc-superoxide dismutase in leaves of rye and wheat cultivars differing in sensitivity to zinc deficiency, *J. Plant Physiol.* 151, 91-95.

Cakmak, I. , B. Torun, B. Erenoglu, L. Ozturk, H. Marschner, M. Kalayci and A. Yilmaz, 1998, Morphological and physiological differences in the response of cereals to zinc deficiency, *Euphytica* 100, 349-357.

Graham, R. D. (1987) Triticale - a cereal for micronutrient deficient soils, *Triticale Topics*, ed. by Triticale Grain Assoc. Australia, 6-7.

Graham, R.D., Davies, W.J., Sparrow, D.H.B. and Ascher, J.S. (1983) Tolerance of barley and other cereals to manganese-deficient calcareous soils of South Australia, in M.R. Saric and B.C. Loughman (eds.), *Genetic Aspcets of Plant Nutrition*, M.Nighoff, Netherlands.

Graham, R.D., Ascher, J.S. and Hynes, S.C. (1992) Selelcting zinc-efficient cereal genotypes for soils of low zinc status, *Plant and Soil* 146, 241-250.

Khabaz Saberi, H. (1997) Genotypic variation for Mn efficiency in durum wheat (*Triticum turgidum* L. *var. durum*), in T. Ando, *Plant Nutrition - for sustainable food production and environment*, Kluwer Acad. Publ. (Tokyo), 289-290.

Marschner, H. (1988) Mechanisms of manganese acquisition by roots from soil, in R. D. Graham *et al Manganese in soils and plants*, Kluwer Academic Publ., Dordrecht, Netherlands, 191-204.

Schlegel, R. and Cakmak, I. (1997a) Physical mapping of rye genes determining micronutrient efficiency in wheat, in T. Ando, *Plant Nutrition - for sustainable food production and environment*, Kluwer Acad. Publ. (Tokyo), 287-288.

Schlegel, R. and Cakmak, I. (1997b) Micronutritional efficiency in crop plants - a new challenge for cytogenetic research, in T. Lelley (ed.), *Current Topics in Plant Cytogenetics Related to Plant Improvement*, WUF Univ. Verlag, Wien (Austria), 91-102.

Schlegel, R. and Cakmak, I. (1998) Zinc efficiency in nulli-tetrasomics of hexaploid wheat 'Chinese Spring', *EWAC Newsletter* 12, 22-25.

Schlegel, R., Cakmak , I., Torun, B., Eker, S. and Koleli, N. (1997) The effect of rye genetic information on zinc, copper, manganese and iron concentration of wheat shoots in zinc deficient soil, *Cer. Res. Comm.* 25, 177-184.

Schlegel, R., Cakmak, I., Torun, B., Eker, S., Tolay, I., Ekiz, H., Kalayci, M. and Braun, H.-H. (1998) Screening for zinc efficiency among wheat relatives and their utilisation for alien gene transfer, Euphytica 100, 281-286.

Sparrow, D.H., and Graham, R.D. (1988) Susceptibility of zinc-deficient wheat plants to colonization by *Fusarium graminearum* Schw. Group I., *Plant and Soil* 112, 261-266.

Takkar, P.N., Chibba, I.M., and Mehta, S.K. (1989) Twenty years of Coordinated Research of Micronutrients in Soil and Plants (1967-1987), *Indian Inst. Soil Sci., Bhopal, IISS, Bull.*

Willhelm, N.S., Graham, R.D. and Rovira, A.D. (1988) Application of different sources of manganese sulfate decreases take-all (*Gaeumannomyces graminis var. tritici*) of wheat grwon in a manganese-deficient soil, *Austr. J. Agric. Res.* 39, 1-10.

STRATEGIES FOR THE INVESTIGATION OF N-EFFICIENCY IN OILSEED RAPE

B. SEIFFERT, W. ECKE, G. LOHAUS[1], M. WALLBRAUN, Z. ZHOU[1] and C. MÖLLERS

Institut für Pflanzenbau und Pflanzenzüchtung, Von-Siebold-Str. 8, 37075 Göttingen, Germany

[1] Albrecht-von-Haller-Institut für Pflanzenwissenschaften, Untere Karspüle 2, 37073 Göttingen, Germany

1. Introduction

Breeding for increased nitrogen efficiency in winter rapeseed is gaining increased attention for several reasons. Ecologically, the possible leaching of nitrate from crop residues in autumn and winter into the groundwater represents a problem. Economically, N-fertilizer is applied in rapeseed in amounts up to 200 kg N/ha and breeding of cultivars that give high yields with a reduced amount of N fertilizer could help maintaining profit margins. Politically, EC regulations concerning the subsidy of the rapeseed crop and environmental aspects of rapeseed cultivation may become more restrictive in the future.

Because of its high oil and protein content in the seeds rapeseed is a particulary interesting crop for studying factors of N-efficiency. Different factors of N-efficiency can also be studied on the molecular level since rapeseed can be transformed relatively easy by *Agrobacterium tumefaciens* and several genes

G. Gissel-Nielsen and A. Jensen (eds.), Plant Nutrition – Molecular Biology and Genetics, 425–432.

involved in the uptake and assimilation of nitrate and ammonium have already been cloned.

2. Factors influencing N-efficiency

There are at least two ways to define N-efficiency: a. the ability of a genotype to give an outstanding yield under limited N supply, and b. the ability of a genoytpe to convert applied fertilizer N more efficiently into yield (Sattelmacher et al. 1994). Fig. 1 shows an example of two rapeseed cultivars tested in field experiments at 6 different locations in north-western Germany, that respond differently to the applied N fertilizer. The cv. 'Bristol' gives an above average yield under limited N supply, whereas the cv. 'Lizard' efficiently converts applied fertilizer N to yield.

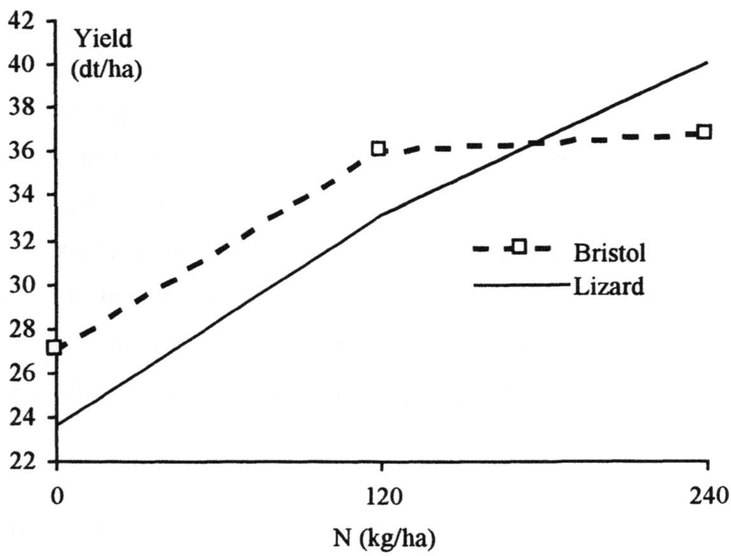

Fig. 1: Yield of the two winter rapeseed cultivars 'Bristol' and 'Lizard' grown under different N regimes (mean of 6 locations)

Genetic variation for N-efficiency is multi-factorial. Some of the factors possibly involved in N-efficiency of rapeseed are shown in Fig. 2. The relative importance of these factors may vary with the developmental stage of the plant. Genotypic differences in N-efficiency are caused by genetic variation of these factors and their complex interaction (the black box in Fig. 2). To date, there have been only a few investigations on genetic variation for N-efficiency in rapeseed (Yau and Thurling 1987a,b). However, these and our own preliminary results indicate that genotypic variation exists in *Brassica napus* (see also Kessel and Becker, this issue) for most, if not for all of the factors shown in Fig. 2.

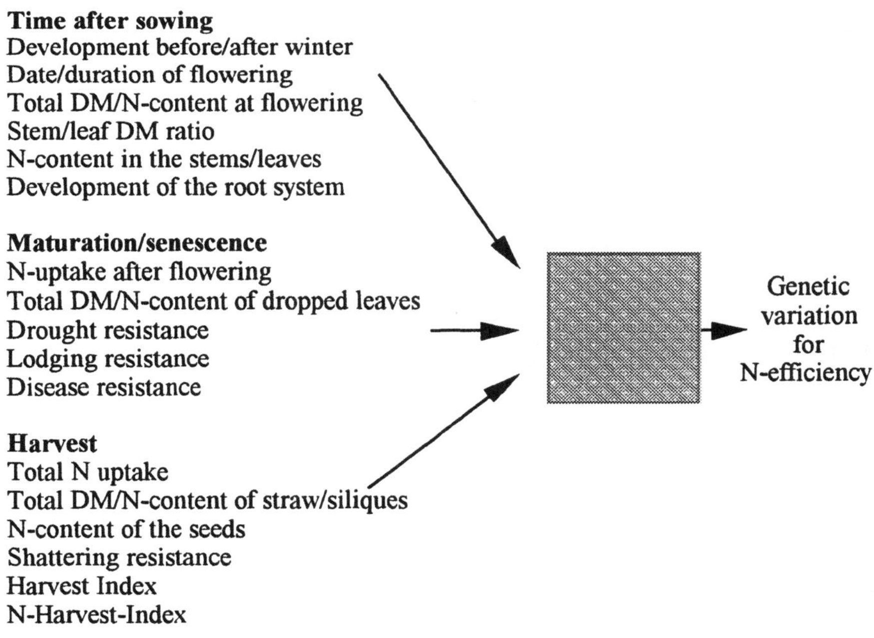

Fig. 2: Factors possibly involved in N-efficiency of rapeseed

3. Breeding for improved N-efficiency

By screening *Brassica napus* germplasm it is possible to identify existing variability for 'yield under limiting N supply' and for 'efficient conversion of N into yield'. However, it is much more labour intensive to reliably evaluate the variabilty for some of the factors shown in Fig. 2. Whereas the determination of some traits, e.g. the N-content of the dropped leaves, can be performed with a large number of genotypes, this is impossible to do for other traits, e.g. root system characteristics. Furthermore, the relative contribution of most of the factors shown in Fig. 2 to 'N-efficiency' is probably relatively small. The major challenge will be to identify factors that contribute substantially to the trait 'N-efficiency'. The advantage of field experiments is that after 2-3 years of multi-locational field trials reliable data for genotypic variability concerning 'yield under limiting N-supply' and for the different above mentioned factors will be available. Through recombination and selection, genotypes with a more distinct expression of single factors can be produced for further genetic studies.

4. Molecular strategies to investigate N-efficiency

There are two molecular strategies to investigate N-efficiency: the generation of transgenic plants with modified activities of putative relevant enzymes and - the mapping of QTL's in rapeseed populations segregating for one or a few factors of N-efficiency.

4.1. GENERATION OF TRANSGENIC PLANTS

Due to the limited knowledge about the many factors involved in N-efficiency, their physiological background and complex interaction, it is impossible to follow a straight forward molecular strategy to improve N-efficiency. However, most of the principal genes involved in nitrate and ammonium uptake and assimilation have been cloned from either *Arabidopsis*, *Nicotiana* or *Brassica napus* (Table 1) and can be used under control of the 35S promotor in transgenic plants to up- or downregulate the respective enzyme activities and probably also the level of certain metabolites like nitrate, glutamate/glutamine, etc. in the rapeseed plant.

Table 1: Examples of cloned genes of the nitrate and ammonium uptake and assimilation pathway

Gene	Origin	Name	Reference
NH_4^+ transporter	*A. thaliana*	AMT1, AMT3	Ninnemann et al. (1994)
NO_3^- transporter	*A. thaliana*	ATNRT21	Zhang and Forde (1998)
Fd-GOGAT	*A. thaliana*	ATFDGSYN	Suzuki et al. (1997)
Nitrate reductase	*B. napus*	pBnNR1405 pBnNR1412	Fukuoka et al. (1996)
Cytosolic glutamine synthetase	*B. napus*	BnGSR1 BnGSR2	Schock et al. (1994)
Plastidic glutamine synthetase	*B. napus*	BnGS18	Ochs et al. (1993)

As an alternative to using homologous gene constructs, new genes may be introduced into rapeseed to create novel metabolic pathways. We have started transformation experiments using an asparagine synthetase gene (AsnA) from *E. coli* (Patent file WO 91/11524). In contrast to the plant asparagine synthetase, AsnA directly accepts ammonium as a substrate. This gene also has the advantage that it should not be feed-back regulated by the plant. In rapeseed, glutamine is the predominant amino acid in the phloem (Lohaus 1995, Möllers et al. 1996). However, asparagine is the more efficient N-transporter with a N/C ratio of 0.5 compared to 0.4 of glutamine. The expression of the AsnA gene in rapeseed could deregulate nitrate assimilation and thus reveal some of the regulatory bottlenecks.

From genetically engineered and modified enzyme activities and metabolite levels, however, it can not be expected to improve the complex trait N-efficiency. At best, a modification of one of the several factors involved in N-efficiency can be achieved. It is rather likely that transgenes are either tolerated by the plants without larger changes in the N-metabolism or that a complete deregulation with pleiotropic effects occur. In the first case, the conclusions from a specifically engineered enzyme activity could be that the expression level of the respective gene and the metabolite concentration is not critical for N-efficiency. In the case of a major deregulation of the N-metabolism with gross changes it will be difficult to distinguish the cause from the effect.

4.2 MAPPING QTL FOR N-EFFICIENCY

Using molecular markers and populations segregating for traits related to N-efficiency (Fig. 2) like 'N-content in the seeds', 'total N uptake' or the 'yield ratio between different levels of N supply' it should be possible to genetically map the QTL controlling the observed phenotypic variation. In an independent study, not conducted to investigate N-efficiency, a number of QTL for traits possibly involved in N-efficiency (Table 2) have already been mapped in a segregating doubled haploid population (Uzunova 1995, Weißleder 1996). This population is currently used in new field trials to specifically analyse N-efficiency related traits. In addition, the genes listed in Table 1 have to be regarded as candidate genes for QTL's for N-efficiency since an allelic variation at the respective gene loci may lead to genetic differences in this trait. By genetically mapping these genes in the same populations used for QTL mapping, it may be possible to associate individual gene loci with specific QTL, thereby identifying the genes underlying the mapped QTL. An analysis of the allelic variation at these loci and the phenotypic variation caused by this would lead to a better understanding of the genetics of N-efficiency and could help to identify more specific approaches to improve N-efficiency by genetic engineering.

Table 2: Number and linkage group of mapped QTL's for some factors possibly related to N-efficiency in rapeseed in a segregating DH population (Weißleder 1996)

Factor	Mapped QTL's	
	number	linkage group
Date of flowering	5	2, 4, 8, 9, 14
Duration of flowering	4	2, 9, 15, 15
Plant length at flowering	4	2, 8, 16, 18
Diameter of the stem	2	5, 18
Harvest-Index	1	5
N-content in the seeds	1	15

Acknowledgements

The financial support of the Deutsche Forschungsgemeinschaft, the Deutsche Bundesstifung Umwelt, the European Commission (AIR 3 CT94 1941) and of the company AgrEvo is gratefully acknowledged.

References

Fukuoka, H., Ogawa, T., Minami, H., Yano, H., Ohkawa, Y. (1996) Developmental stage-specific and nitrate-independent regulation of nitrate reductase gene expression in rapeseed, Plant Physiol. **111**, 39 - 47.

Lohaus, G. (1995) Vom Source zum Sink: Phloemtransport verschiedener Kohlenstoff- und Stickstoffverbindungen. Dissertation Universität Göttingen. ISBN 3-89588-216-X.

Möllers, C., Böhne, T., Wallbraun, M., and Lohaus, G. (1996) Protein content and amino acid composition of the seeds and and the phloem sap of one *Brassica carinata* and different *B. napus* L. genotypes. Proceedings of the Symposium on Breeding of Oil and Protein Crops. 5 - 8. August 1996, Zaporozhye, Ukrainia, 148 - 154.

Ninnemann, O. W., Jauniauex, J. C., Frommer, W. B. (1994) Identification of a high affinity ammonium transporter from plants, EMBO J. **13**, 3464 - 3471.

Ochs, G., Schock, G., Wild, A. (1993) Chloroplastic glutamine synthetase from *Brassica napus*, Plant Physiol. **103**, 303 - 304.

Sattelmacher, B., Horst, W.J., Becker, H.C. (1994) Factors that contribute to genetic variation for nutrient efficiency of crop plants. Z. Pflanzenernähr. Bodenk. **157**, 215-224.

Schock, G., Ochs, G. and Wild, A. (1994) Glutamine synthetase from roots of Brassica napus. Plant Physiol. **105**, 757-758.

Schock, G., Ochs, G., Wild, A. (1994) Glutamine synthetase from roots of *Brassica napus*, Plant Physiol. **105**, 757 - 758.

Suzuki, A. and Rothstein, S. (1997) Structure and regulation of ferredoxin-dependent glutamase synthase from *Arabidosis thaliana.* Cloning of cDNA expression in different tissues of wild-type and glts mutant strains, and light induction, Eur. J. Biochem. **243**, 708 - 718.

Weißleder, K. (1996) Genetische Kartierung von Loci für züchterisch bedeutsame Merkmale beim Winterraps (*Brassica napus* L.). Dissertation Universität Göttingen. ISBN 3-89588-444-8.

Yau, S. K. and Thurling, N. (1987a) Variation in nitrogen response among spring rape (Brassica napus) cultivars and its relation to nitrogen uptake and utilization. Field Crop Res. **16**, 133.

Yau, S. K. und Thurling, N. (1987b) Genetic variation in nitrogen uptake and utilization in spring rape (Brassica napus L.) and its exploitation through selection. Plant Breeding **98**, 330-338.

Zhang, H. and Forde, B.G. (1998) Accession number Z 97058.

MOLECULAR MARKERS ASSOCIATED WITH PHOSPHORUS UPTAKE AND INTERNAL PHOSPHORUS-USE EFFICIENCY IN RICE

MATTHIAS WISSUWA AND NORIHARU AE

National Institute Of Agro-Environmental Sciences
3-1-1 Kannondai, Tsukuba, Ibaraki 305, Japan

Abstract

Phosphorus (P) deficiency is a major constraint to rice production. Increasing the P-deficiency tolerance of rice cultivars may represent a more cost-effective solution than relying on fertilizer application. The objective of this study was to identify putative QTLs for P-deficiency tolerance in rice, using 98 backcross inbred lines derived from a *japonica* * *indica* cross. Lines were genotyped at 245 RFLP marker loci and grown on P-deficient soil to evaluate their P uptake and internal P-use efficiency. Four QTLs were identified for P uptake, explaining 54.5% of the variation for this trait. One QTL linked to marker C443 on chromosome 12 had a major effect, accounting for about half of the explained variation. Two of the three QTLs detected for internal P-use efficiency, including the major one on chromosome 12, coincided with QTLs for P uptake as well, however, whereas *indica* alleles increased P uptake they reduced P-use efficiency. We concluded that this was not due to tight linkage of two genes in repulsion but rather due to an indirect effect of P uptake on P-use efficiency because most lines with high P-use efficiency were characterized by very low P uptake and low dry weight and apparently experienced extreme P-deficiency stress. In this study P-deficiency tolerance was thus mainly caused by genotypic differences in P uptake and genotypic differences in P-use efficiency and dry matter production were merely reflections of differences in P uptake. Lines with *indica* alleles at the major QTL on chromosome twelve had twice the P uptake compared to lines with *japonica* alleles. After further confirmation and fine mapping, this major QTL should be a prime target for isolating the gene or genes increasing P uptake in rice.

Key words: Rice, QTL, Phosphorus, Use efficiency, Deficiency tolerance

G. Gissel-Nielsen and A. Jensen (eds.), Plant Nutrition – Molecular Biology and Genetics, 433–439.

Introduction

Low levels of plant-available phosphorus (P) in soils are a major constraint for rice (*Oryza sativa* L.) production throughout the world (De Datta et al 1990). Developing cultivars with an improved ability to utilize this large but hardly plant-available P pool could offer a more sustainable solution than relying on fertilizer application alone. Majumder et al. (1989) and Chaubey et al. (1994) concluded that P deficiency tolerance in rice is a quantitatively inherited trait with outstanding parents being carriers of mostly additive genes. The availability of detailed molecular linkage maps in rice (Kurata et al. 1994) have allowed the dissection of such quantitatively expressed traits into Mendelian factors referred to as quantitative trait loci (QTLs), each linked to molecular markers of known map position. The detection of putative QTLs represents a crucial first step that could eventually lead to the identification of genes controlling P uptake and P use efficiency or to the identification of tightly linked markers to be used in marker-assisted selection. The objective of this study was to detect and map QTLs for P uptake and P-use efficiency. This was done in a rice population grown on a P-deficient soil with hardly soluble Fe- and Al-P as the predominant inorganic P forms.

Materials and Methods

PLANT MATERIAL

The Japanese variety 'Nipponbare' of the *japonica* subspecies was crossed with 'Kasalath', an *indica* type landrace from Assam, India. The resulting F_1 was backcrossed to 'Nipponbare' to increase fertility in the *japonica* * *indica* hybrid. Ninety-eight backcross inbred lines (BILs) were developed by advancing BC1F1 lines for five generations by the single-seed descent method (Lin et al. 1998). Leaf material of the BC_1F_5 was used for DNA extraction and RFLP analysis, whereas phenotypic data was collected on the BC_1F_6.

EVALUATION OF PLANT PERFORMANCE UNDER LOW-P STRESS

A fiberglass container of the dimensions 11.60*0.85*0.22m (length*width*depth) was filled with topsoil (Humic haplic andosol) from a highly P-deficient field situated at the NIAES campus in Tsukuba, Japan. The field was fertilized with the equivalent of 100 kg ha^{-1} of N and K_2O prior to topsoil removal for filling the container. Pre-germinated seeds were planted into the low-P soil in a randomized complete block design with five replications on 16 May 1997. Replications (rows) were spaced 15cm apart, spacing within rows was 10cm. Plants were grown under upland conditions but drought stress was avoided by watering plants during dry spells in summer.

On 18 September (125 days after planting) plants were cut and their dry weight determined after oven-drying samples at 65°C for 5 days. Samples were then ground in a Wiley mill to pass a 1mm mesh. The tissue-P concentration in 1 mg of plant sample was determined colorimetrically by the phosphovanadate method (Hanson 1950) after digestion in a mixture of HNO_3, $HClO_4$, H_2SO_4 (3:1:1). Total P uptake was calculated as the product of dry weight and tissue-P concentration, and P-use efficiency was expressed in g dry weight mg^{-1} P taken up.

RFLP MAPPING AND QTL DETECTION

A linkage map of the 245 RFLP markers used for QTL detection was obtained from the Rice Genome Project, Japan, as previously described by Lin et al. (1998). Data for plant P content were log transformed prior to analysis. QTL analysis was performed using the composite interval mapping (CIM) method proposed by Zeng (1994). Computations were done with the software package PLABQTL (Utz and Melchinger 1996). Because a specific model for BILs was not available, we analyzed our data as a F_2 backcross and treated heterozygous markers as missing data. In a first step, simple interval mapping was performed and cofactors selected. For cofactor selection F-to-enter and F-to-drop thresholds were set at 6.0 to avoid selecting multiple markers linked to one QTL as cofactors. Using these cofactors to reduce the residual variation, QTLs were detected with CIM. A LOD score >2.80 was considered significant for QTL detection. Support intervals were determined using a LOD fall-off of 1.0.

DEVELOPMENT AND EVALUATION OF NEAR ISOGENIC LINES

Near isogenic lines (NILs) were developed from selected BC_1F_2 lines, which were then backcrossed to Nipponbare an additional three times. Genotypic monitoring and selection against Kasalath alleles was done using 118 RFLP markers. A set of 10 NILs plus both parents were grown in 60-L buckets filled with the same P-deficient soil described above. Lines were grown in 5 replications for a 100-day period up to July 30, 1998, when shoots were cut and weighed after oven-drying at 65°C for 5 days. Roots of NILs were carefully separated by washing away the soil. Root length and surface area were determined on 1g of sample by using a digital scanner and the software package WinRhizo (Regent, Canada).

Results and Discussion

PHENOTYPIC RESPONSE

The P availability of the soil used in this study was low at 1.0 mg P kg^{-1} (Truog-P) or 4.5 mg P kg^{-1} (BrayII-P) (Otani and Ae 1996). Al and Fe-P were the predominant inorganic P-forms, the more plant-available Ca-P was only detected in traces. That low P availability was the growth limiting factor, reducing the dry weight of a set of standard lines by 50% on average, was previously shown (Wissuwa et al., in press).

Kasalath had a much higher P uptake than any of the BILs (Fig. 1) but the P-use efficiency of Kasalath was low. Nipponbare showed average P-use efficiency and was slightly below average for P uptake. Variability among the 98 BILs was much higher for P uptake (CV of 64%) than for P use efficiency (CV of 17%). Both traits were negatively correlated (r=-0.72).

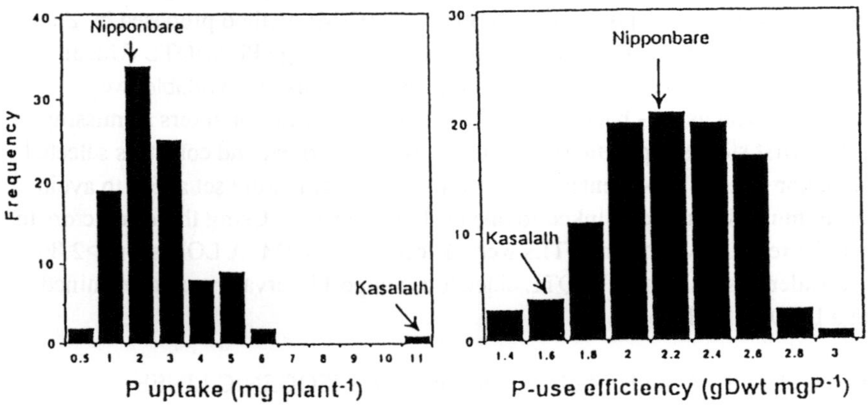

Figure 1 Frequency distribution of 98 backcross inbred lines for P uptake and P-use efficiency.

QTL DETECTION

Four putative QTLs were detected for P uptake on chromosomes 2, 6, 10, and 12 (Table 1). Together they explained 54.5% of the variation for P uptake observed among the BILs. One of these QTLs, linked to marker C443 on chromosome 12, had a major effect and accounted for half of the explained variation. For three of the four QTLs, including the major one, the positive allele came from Kasalath. Three putative QTLs were detected for P-use efficiency, together explaining 42.1% of the variation. The major QTL on chromosome 12 and the minor one on chromosome 2 coincided with QTLs for P uptake, however, Nipponbare alleles increased P-use efficiency whereas Kasalath alleles had increased P uptake. Significant two-way interactions involving markers flanking a putative QTL were not detected.

TABLE 1. Putative QTLs for P uptake and P-use efficiency under low-P stress

	Marker Interval[a]	Chr	Pos[b]	LOD	%[c] Variation	subst. effect[d]	positive allele[e]
P uptake	G227 - C365	2	106	2.82	5.8	+0.97	K
(mg plant^{-1})	C498 - R1954	6	13	3.52	9.8	+0.71	K
	R1629 -	10	32	4.70	7.7	-0.62	N
	G2140 - C443	12	30	10.74	27.9	+1.94	K
			total[f]	16.25	54.5		
P-use	G227 - C365	2	106	5.22	9.8	-0.35	N
efficiency	C946 - R1854	4	86	4.35	9.4	+0.30	K
(g Dwt mg^{-1}P)	G2140 - C443	12	28	6.57	19.1	-0.47	N
			total[f]	11.40	42.1		

[a] Marker nearest to QTL is underlined [b] Position on chromosome in cM
[c] Portion of phenotypic variation explained by QTL using a single-QTL model
[d] Effect of substituting both Nipponbare by Kasalath alleles, calculated for non-transformed data
[e] Allele increasing phenotype from Kasalath (K) or Nipponbare (N)
[f] Estimated using a multiple-QTL model including all putative QTLs

That two QTLs affected P uptake and P-use efficiency simultaneously but with opposite effects would pose problems in breeding for increased P-deficiency tolerance if both traits were of importance. This, however, was not the case in this study. Dry weight depended entirely on P uptake ($r = 0.96$), whereas P-use efficiency was negatively correlated to dry weight ($r = -0.60$), not positively as expected. The high use efficiency of lines with Nipponbare alleles at QTLs on chromosomes 6 and 12 was the consequence of their inferior alleles for P uptake. Insufficient P uptake then led to severe P deficiency and to dry weight production at highly sub-optimal tissue-P concentrations below 0.05%. These low P concentrations may represent the absolute minimum for survival rather than efficient P use in a sense that would be worth exploiting by plant breeders. We therefore consider the P-use efficiency QTLs on chromosomes 6 and 12 to be "Pseudo-QTLs", caused by the direct effect that genes for P uptake have on P-use efficiency.

In this study, dry weight 125 days after sowing was used as an indicator for P-deficiency tolerance instead of grain yield because the BILs differed in maturity, and grains could not be harvested at the same time for all lines. Dry weight production under P-deficient conditions largely depended on P uptake but reports on the dependence of grain yield on P uptake are less conclusive. Fageria et al. (1988) did not detect a significant correlation between these two parameters but the stress intensity in their study was low. Grain yield was only reduced by 14% relative to P fertilization and tissue-P concentrations remained well above 1mg P g^{-1}dry weight, which is considered the deficiency threshold for rice at the tillering stage. In a study conducted at IRRI (1994) a significant correlation between P uptake and grain yield of r=0.40 was detected and Koyama (1973) reported that grain yield was highly dependent on P uptake. This dependence increased with increasing stress intensity. Dry weight had a tighter correlation with P uptake than grain yield in all studies and the discrepancy could be explained by genotypic differences in the efficiency of retranslocation of P from vegetative to generative tissues. This efficiency largely depends on differences in harvest index (HI) and should therefore be independent of P uptake, unlike P-use efficiency in dry weight production. Based on our results it seems appropriate for breeders trying to develop genotypes with increased tolerance to P deficiency to combine high P uptake ability with high HI, rather than to concentrate on P-use efficiency.

To confirm the existence of QTLs for P uptake, near isogenic lines (NILs) for the QTLs linked to marker C443 on chromosome 12 (NIL-49 vs. NIL-82) and linked to marker C498 on chromosome 6 (NIL-14 vs. NIL-4) were evaluated on P-deficient soil. The NILs carrying Kasalath alleles at C443 (NIL-49) and at C498 (NIL-14) had twice the dry weight of their complementary lines with Nipponbare alleles (Table 2). Since

TABLE 2. Dry weight and root size parameters of NILs for the putative QTLs at markers C443 and C498

Genotype	QTL/ Chromosome	Genotypic Similarity[a] %	Nipponbare[b] %	Dry Weight (g)	Root Length (m)	Root Surface Area (cm^2)
Nipponbare				1.0 d	22.2 d	230 d
NIL-49	+C443[d] / 12		91.1	6.0 b	87.1 bc	1004 bc
NIL-82	-C443 / 12	92.0	88.8	3.3 cd	65.6 bcd	737 bc
NIL-14	+C498 / 6		93.7	5.1 bc	99.5 b	1106 b
NIL-4	-C498 / 6	96.2	96.9	2.6 d	53.8 cd	577 cd
Kasalath				19.3 a	209.5 a	2329 a

[a] Between NILs 49 / 82 and 14 / 4, based on 118 maker loci [b] Portion of the genome carrying Nipponbare alleles
[c] Values followed by different letters are significantly different at P< 0.05 (Tukey HSD)
[d] +C443/+C498 signifies NIL is carrier of allele increasing P uptake (from Kasalath)

dry weight was tightly correlated to P uptake (r = 0.96) on this highly P-deficient soil, these results represent a first confirmation for the presence of P-uptake QTLs on chromosomes 6 and 12. The superior dry weight produced by NIL-14 is potentially due to its two-fold bigger root system compared to NIL-4. Further indirect evidence that C498 is related to genes affecting root size is given by Ray et al. (1996), who mapped a root size QTL to the same interval on chromosome 6.

That QTL explained 10% of the variation for total root number in an *indica***japonica* cross, and as in our study, the *indica* allele increased root size. The root size of the NIL carrying Kasalath alleles at C443 (NIL-49) was only slightly smaller than that of NIL-14 but about 4 times as large as that of Nipponbare. However, other causes than root size must mainly be responsible for the superior dry weight accumulation of NIL-49 compared to NIL-82 (Nipponbare alleles at C443) because both NILs did not differ significantly in root size. To reduce the background effect of the few Kasalath alleles still remaining at loci unrelated to QTLs, the NILs used here are currently backcrossed once more to Nipponbare. In addition to extracting improved NILs, the F_2 will be used to fine-map the already confirmed QTLs on chromosomes 6 and 12.

References

Chaubey, C.N., Senadhira, D., and Gregorio G.B. (1994) Genetic analysis of tolerance for phosphorus deficiency in rice (*Oryza sativa* L.). Theor. Appl. Genet. 89:313-317.

De Datta, S.K., Biswas, T.K., and Charoenchamratcheep, C. (1990) Phosphorus requirements and management for lowland rice. In: Phosphorus Requirements for Sustainable Agriculture in Asia and Oceania. IRRI, the Philippines, pp 307-323.

Fageria, N.K., Morais, O.P., Baligar, V.C., and Wright, R.J. (1988) Response of rice cultivars to phosphorus supply on an oxisol. Fert. Res. 16:195-206.

Hanson, W.C. (1950) The photometric determination of phosphorus in fertilizers using the phosphovanado-molybdate complex. J. Sci. Food Agric. 1:172-173.

IRRI (1994) IRRI program report for 1993. International Rice Research Institute, Manila, The Philippines.

Koyama, T., Chammek, C., and Snitwonge, P. (1973) Varietal differences of Thai rice in the resistance to phosphorus deficiency. Trop. Agr. Res. Cent. Jpn. Tech. Bull. (4) 32p.

Kurata, N., Nagamura, Y., Yamamoto, K., Harushima, Y., Sue, N., Wu, J., Antonio, B.A., Shomura, A., Shimizu, T., Lin, S.-Y., Inoue, T., Fukuda, A., Shimano, T., Kuboki, Y., Toyama, T., Miyamoto, Y., Krihara, T., Hayasaka, K., Miyao, A., Monna, L., Zhong, H.S., Tamura, Y., Wang, Z.-X., Momma, T., Umehara, Y., Yano, M., Sasaki, T., and Minobe, Y. (1994) A 300 kilobase interval genetic map of rice including 883 expressed sequences. Nature Genetics 8:365-372.

Lin, S.Y., Sasaki, T., and Yano, M. (1998) Mapping quantitative trait loci controlling seed dormancy and heading date in rice, *Oryza sativa* L. using backcross inbred lines. Theor. Appl. Genet. 96:997-1003.

Majumder, N.D., Borthakur, D.N., and Rakshit, S.C. (1989) Heterosis in rice under phosphorus stress. Indian J Genet 49:231-235.

Otani, T., and Ae, N. (1996) Sensitivity of phosphorus uptake to changes in root length and soil volume. Agron. J. 88:371-375.

Ray, J.D., Yu, L., McCouch, S.R., and Champoux, M.C. (1996) Mapping quantitative trait loci associated with root penetration ability in rice (*Oryza sativa* L.). Theor. Appl. Genet. 92:627-636.

Utz, H.F., and Melchinger, A.E. (1996) PLABQTL: A program for composite interval mapping of QTL. J Quant Trait Loci http://probe.nalusda.gov:8000/otherdocs/jqtl.

Wissuwa, M., Yano, M., and Ae, N. (1998) Mapping of QTLs for phosphorus-deficiency tolerance in rice (*Oryza sativa* L.). Theor. Appl. Genet. 97:777-783.

Zeng, Z.B. (1994) Precision mapping of quantitative trait loci. Genetics 136: 1457-1468.

Subject index

MIX
Papier aus verantwortungsvollen Quellen
Paper from responsible sources
FSC® C105338

If you have any concerns about our products,
you can contact us on
ProductSafety@springernature.com

In case Publisher is established outside the EU,
the EU authorized representative is:
Springer Nature Customer Service Center GmbH
Europaplatz 3, 69115 Heidelberg, Germany

Printed by Libri Plureos GmbH
in Hamburg, Germany